D1706786

Urea-Formaldehyde Resins

Urea-Formaldehyde Resins

BEAT MEYER

Department of Chemistry
University of Washington
Seattle, Washington

1979

Addison-Wesley Publishing Company, Inc.
Advanced Book Program
Reading, Massachusetts

London · Amsterdam · Don Mills, Ontario · Sydney · Tokyo

Library of Congress Cataloging in Publication Data

Meyer, Beat.
Urea-formaldehyde resins.

Bibliography: p.
Includes indexes.
1. Urea-formaldehyde resins. I. Title.
TP1180.U7M48 668'.374 79-20727
ISBN 0-201-04558-3

Manufactured in the United States of America

Contents

Preface ix

1 **Introduction** 1

2 **History** 4

3 **Chemistry** 22

Formaldehyde 22

Properties and Equilibria 23

Reactions of Formaldehyde 45

Urea 72

Reactions of Urea 74

Thiourea 76

UF-Resins 79

Principles of Resin Formation 79

Reaction of Urea with Formaldehyde 88

Modified UF-Resins 95

Alcohols 95

Hemiacetals 96

Peroxides 96

Carbonic Acids 96

Sulfur Compounds 96

Ammonia and Amines 96

4 **Resin Manufacture** 97

Formaldehyde 97

Methanol Oxidation 97

Methane 99

Urea 99

Urea-Formaldehyde Resins 100

Manufacture 101

Reaction Control 103

Gellation Time or the Cure Test 104

Instrumental Chemical Analysis 105

Formulations 105

Adhesives 107

Resin Cure 112

Modified Resins 115

Late Addition of Reagent 115

Addition of Alcohols 115

Urea Substitution 117
Formaldehyde Substitutes. 118
Basic Co-condensation 118
Acidic Co-condensation 120
Co-condensation with Phenol, Resorcinol. 121

5 **Chemical Analysis and Sampling**. 123
Analysis of Raw Materials 124
Urea 124
Formaldehyde 124
Standard Quantitative Methods 125
Colorimetric Detection of Formaldehyde 127
Other Analytical Methods 130
Analysis of Resins 131
Free Formaldehyde 131
Total Formaldehyde 131
Total Nitrogen 132
Urea 132
Solids 132
Water and Volatiles 133
pH and Buffer. 133
Fillers 133
Individual Resin Species and Functional Groups 133
Older Methods 133
Nuclear Magnetic Resonance (NMR) Method 134
Infrared and Raman Spectroscopy 139
Formaldehyde Sampling 144
Air Sampling 145
Ambient Versus Indoor Air 145
Sampling Methods 145
Indoor Air 147
Aqueous Mixtures 149
Aqueous UF-Resins 149
UF-Bonded Materials 149

6 **Adhesives**. 156
Adhesive Properties and Requirements 157
Wood Bonding and Lamination 161
Plywood 162
Particleboard 164
Glue Properties 169
Additives 170
Particleboard Manufacturing 170
Medium Density Fiberboard 179
Wood Molding 180
Properties of UF-Wood Materials 181
Formaldehyde Standards 182
Formaldehyde Abatement. 183

7 **Foams** 185
Chemistry 185
Principles of Foam Preparation 191
Resins 191
Surfactants 191
Catalyst 191
Water 191
Foaming Gases 192
Additives 192
Foam Manufacture 192
Physical Properties 194
Foam Applications 196
Building and Home Insulation 196
Insulation of Ships 197
Electrical Insulators 197
Oil Absorbers 197
Medicinal Applications 197
Agricultural Uses, Plastoponics 198
Other Applications 198

8 **Resins Applications** 199
Textile Treatment 199
Coatings 203
Paper 203
Construction Materials 205
Fireproofing and Resistance 206
Binders for Metal Casting 207
Agriculture 207
Waste Treatment 210
Solid Resin Moldings 211
Other Applications 214

9 **Product Properties** 215
Performance and Standards 215
Molding 215
Foams 216
UF-Bonded Products 218
Formaldehyde Release 226
Economic Factors 227
Current Public Attitudes 228
Performance and Properties 229
Formaldehyde Release Mechanism 233
Formaldehyde Abatement 241
Coatings 241
Chemical After-Treatment 242
Resin Additives 245
New Resin Formulations 248

Foams . . . 248

10 Health and Environment . . . 250

Toxicity . . . 250

Urea . . . 250

Formaldehyde . . . 251

Acute Effects . . . 251

Chronic Exposure . . . 253

Formaldehyde Levels in Air . . . 255

Public Attitude . . . 255

Regulations and Recommended Limits . . . 257

Standard Ambient Indoor Air . . . 257

Nature of Source . . . 259

Total Vaporizable Formaldehyde . . . 260

Release Mechanism . . . 260

Abatement . . . 261

Health Effects and Dose Response . . . 262

Health Standards . . . 266

Formaldehyde in Water and Food . . . 267

11 Future Trends . . . 268

Production Trends . . . 268

General Trends . . . 268

Wood Adhesives . . . 277

Resin Chemistry . . . 283

Computer Aided Analysis . . . 283

Resin Manufacture . . . 285

Wood Adhesives . . . 285

UF Foams . . . 289

Other Applications . . . 289

Formaldehyde Release . . . 290

Health . . . 290

Regulations and Standards . . . 291

Conclusion . . . 292

Appendix . . . 293

Glossary . . . 295

Bibliography . . . 299

Author Index . . . 367

Patent Index . . . 385

Subject Index . . . 403

Preface

This is an interdisciplinary book for all those who come into contact with urea-formaldehyde resins or urea-formaldehyde resin bonded products. It is intended for industrial, university, and government scientists, administrators and regulators, builders, architects, manufacturers, designers, business people, consumers and their advocates. Accordingly, this book is written for readers with different interests and different background knowledge. It provides detailed references to guide technical and scientific readers to recent work covering the latest progress in chemistry and the related sciences. It contains tables and figures summarizing chemical data for hygienists, lawyers, and administrators who desire to quickly extract only the most important facts. It contains introductory and summarizing paragraphs to give the less directly affected reader a sufficient overview to determine whether he wants to learn more about a subject, how much is currently known about the subject, and where such information is available.

A great wealth of know-how and art exists on urca-formaldehyde (UF) resins, but it is not readily accessible. Unfortunately, the reader will have to collect much of the supplementary information from individual articles spread through the literature of several different fields. There have been no comprehensive books or sources on UF-resins available since the classic books by Blais, Snuparek, Tallet, Vale, and Virpsha. Much of the knowledge is proprietary and exists as trade secrets, or in sketchy information buried in the voluminous patent literature.

Obviously, a comprehensive coverage of this field would fill several volumes. There is a definite need for such an encyclopedia on urea-resin. However, that is not the goal of this book. Its purpose is to offer the reader an overview, to guide him to the literature, and to help him establish what sources are relevant for him. In order to maintain such an overview, much valuable information had to be omitted, and much of the information remains superficial. In fact, several important fields have been omitted or neglected. For example, the entire application of aminoplastics in the textile industry had to be limited to a few short pages. This seems justified, because today melamine resins are more prominent in this field than UF-resins, and the reader will have to consult reviews dealing with these resins, anyhow. Likewise, molding powders are not given adequate space, even though urea-resins have a great potential. In contrast, the problem of formaldehyde odor, a truly interdisciplinary problem, is covered at greater length, because it is believed that the solution to the odor problem will determine whether UF-resins will find a permanent market in the important construction market.

ISBN 0-201-04558-3

The introduction explains the current importance of urea-formaldehyde resins, and explains the reason for dealing with urea-resins as a field separate from amino-resins and plastics. Chapter 2 chronicles the history of urea and formaldehyde chemistry, the beginning of the resin application work, and the development of current uses. Chapter 3 reviews the basic chemistry of the reagents, of the resins and their derivation. Chapter 4 describes some of the many methods for preparing and modifying resins. Chapter 5 deals with the chemical analysis of resins and environmental air sampling. Chapter 6 is devoted to adhesives, especially in wood products. Chapter 7 deals with foams and their almost unlimited potential for application. Chapter 8 reviews some of the other uses, including molding powder, paper, and textile, as well as many other applications. Chapter 9 describes some of the properties and advantages of UF-resin products, and analyzes the odor problem. Chapter 10 summarizes the health and environmental effects of resin, and Chapter 11 reviews future trends. The Bibliography includes some 700 book and journal references and some 600 patents, all of which are thoroughly indexed in the author, patent, and subject index.

This book reflects personal views based on my training and experience. My interests in UF-resins were stimulated by many friends and colleagues in the chemical industry, the forest products industry, the textile industry and at universities in the U.S., Europe, and on other continents. My chemical interests were awakened by the late Professor P. Karrer, at whose institute in Zürich I earned my Ph.D. in chemistry. My understanding of the UF field is based on experience in other fields of chemistry which I gained by working with my Ph.D. supervisor, E. Schumacher, then, as postdoctoral fellow with H. Schiff, at McGill University, and especially by 18 years of cooperation at the Lawrence Berkeley Laboratory, University of California, Berkeley, with Leo Brewer, who always has untiringly encouraged me. Valuable information about resins came from friends in industry. Data on foams was obtained from H. Baumann, Frankenthal, and R. Blatter, Basel. Many data on resin stability and formaldehyde release were provided by B. Sundin, Stockholm. Numerous stimulating discussions with Prof. B. Bryant inaugurated me into the train of thought of forest product scientists. Joint work with Prof. W. Johns at the University of California Forest Products Laboratory and the College of Engineering at Washington State University always has been very stimulating and helped me better to understand experiments and theory. Dr. B. Gfeller of Novopan-Keller helped me gain insight into fundamental as well as production problems, and many companies, especially Aerolite, BASF, Borden, CIBA-GEIGY, Fahrni, and Pacific Resin Co. were most helpful. Doctors Deppe and Roffael made available their valuable data, as did Nobuya Minemura who also translated several Japanese papers. The list of those who stimulated and helped is too long to complete.

The book manuscript was prepared by Mrs. Marilee Kapsa, who coordinated the production of the manuscript, served as editor, collected the bibliography and prepared and proofed the Index section. Mr. Tim Burner helped prepare the indices and proof the entire manuscript. Miss Gina Papan helped prepare the bibliography.

ISBN 0-201-04558-3

This book is dedicated to all those in industry, government, and universities who work with urea-resins. It is hoped that their careers are as successful as that of Fritz Pollak, who promoted the technical and commercial development of UF-resins for more than half a century, and lived 98 years.

BEAT MEYER

ISBN 0-201-04558-3

Urea-Formaldehyde Resins

1. Introduction

This book deals with the chemical properties, and the commercial application and uses of urea-formaldehyde resins. Since 1870 when these resins were accidentally discovered, interest in these materials has steadily increased. Parallel with it, the production and consumption of resin grew from one thousand tons in 1930 to over one million tons today. However, during that time the end uses of the resins have drastically changed. These changes reflect progress in resin chemistry, progress in chemistry at large, as well as changes in the status of the economy, lifestyles, and in society. When the first commercial products were manufactured, UF-resins were a luxury and twice as valuable as phenol resins. Their performance was measured by their mechanical performance. Today, most of their uses are based on their unique availability, their low price, and their environmental compatability; and their use is more heavily influenced by such factors as toxicity or odor than by physical performance. Originally, the major uses proposed were as molding resin, and for some time substantial efforts were made to produce a glass window. UF moldings are still used, for example, in ceiling fixtures, but the main use has shifted to adhesives, especially for wood products. Among the latter, the most important application is in particleboard, in which UF-resin makes about 7% of weight and often constitutes 30% of the final product value.

Particleboard was commercialized in the 1950s, twenty-five years after the methods for basic chemical formulation and manufacturing of UF-resin were perfected. Thus, the main application of resin is in areas for which its use was never contemplated. Today, in Europe, particleboard has replaced plywood in most uses for which the original Kaurit adhesives were designed. Likewise, many of the markets for particleboard were not yet in existence when UF-resin were designed. For example, mobile homes did not exist in 1920. In the U.S. today over 300,000 of the 3.5 million cubic meters of particleboard produced are for mobile home decking. On the average, each of these homes contains 800 square meters of particleboard.

Another important application of particleboard is as floor underlayment and for home or commercial furniture. In both fields, demands and requirements for UF

ISBN 0-201-04558-3

Beat Meyer, Urea-Formaldehyde Resins

reflect deep changes in lifestyle. In 1900, homes were built to last for several generations, and furniture was purchased for a lifetime. Today, U.S. government economists classify both homes and furniture as consumer products. According to Webster, this means that such materials are meant to be quickly reduced in value and rendered useless. These changes have affected the quality and design of UF-bonded products, and they show the depth of changes both in product demand and in the societal goals during the current adjustment to the recent industrial, electronic, and communication revolution.

The primary purpose of this book is to take a fresh look at UF-resins from the viewpoint of current and possible future uses, and to identify and define their potential. Strangely, the rapid development of new and larger markets has only slowly stimulated new research in resin chemistry. Except for gradual improvements in manufacturing methods, the resins are by and large the same as twenty years ago. There are several reasons for this. For one, UF-resins were brought to an advanced stage and near perfection in the 1930s, and patents of that time have now expired and are available for free use. Furthermore, the field of polymer chemistry has greatly expanded during the last fifty years. A very large number of new competitors for UF have been invented, each of which has some stronger points. For example, melamine gives better tableware and molded wood products and fixtures, isocyanates give superior bonds, and epoxies are stronger. Customers can choose from a large variety of material. The main advantage of UF-resins is their low price, but this is not necessarily an advantage for the resin industry, because this keeps the potential for profit small and forces it to sell large volumes to compensate for the limited mark-up. Therefore, the driving force to improve UF comes only from the users. Currently the forest products industry is the largest user.

This book is written in the belief that UF-resins offer unique potential technical advantages in a variety of applications, in an abundance unmatched in the competing products. Among those advantages are the above mentioned low price, the nontoxicity of resin and resin products, and the environmental compatibility of resin bonded products, such as particleboard, which can serve as soil conditioners, and can biologically degrade into plant food. Furthermore, UF-resin can be readily modified or copolymerized with many materials, yielding resins fitting almost any purpose. A final advantage of UF-resins is that they are not as directly dependent on petroleum product user trends as are other resins, such as the phenol resins. Thus, in a time of increased concern for energy conservation, UF-resins are likely to remain widely available at a reasonable price, and accordingly, the current production capacity is not limited by raw materials, but rather by profits.

However, in order to draw maximum benefit of UF-resin, their application must be optimized. This calls for proper design and reliable quality control. The recent revolution in analytical chemistry is now providing an entirely new arsenal of tools which are suitable to aid the research chemist as well as the practitioner. These instruments are still complex and expensive, but they are reliable and widely available. Many of these tools make it possible to observe chemical reactions *in situ,* i.e.,

ISBN 0-201-04558-3

in the reaction, during manufacture. This allows quick and reliable quality control, and it helps chemists understand resin reactions. Thus, it becomes possible to better formulate the art of resin manufacture. Soon it will be possible to start a better and more direct correlation between resin chemistry and the performance properties of the finished product. These two factors will surely help bring the age-old dream of converting the chemical art into a science closer to fulfillment, and this, in turn, will yield better resins.

Thus, the current situation poses new challenges both to academe and to practitioners. The odor problem should be of acute interest to all. Bulk UF moldings, the original main use, did not exude much formaldehyde. The odor problem is specific to particleboard and foams. In these products the resin forms thin UF films, with a large surface area that remains periodically exposed to moisture from the weather. Particleboard is especially vulnerable, because the comminuted wood can act almost like a sponge. It is obvious that the solution to the odor problem will decide whether this material can find permanent use in the construction industry. If the odor can be economically suppressed, the otherwise desirable properties will make UF-bonded particleboard and UF foam mainstays of the construction industry. If the odor problem is not solved, other more expensive material will be used and construction costs will increase. This could adversely influence housing starts and, thus, the national economy.

It is unreasonable to expect that this problem could be efficiently solved by any one of the affected parties alone. The research effort does not hold enough profit for the resin industry; the particleboard industry is not in a position to conduct the necessary chemical research; and consumer pressure produces mainly legislation and regulation.

UF-resins deserve truly interdisciplinary attention. For example, modern science and technology, and especially the above mentioned modern analytical tools, make it now possible for research chemists to study the synthesis, thermodynamics, and kinetics of the formation of ether, hemiacetal, and methylene groups in conjunction with polyhydroxy groups and carbonyl-amines; analytical chemists can now learn how to simultaneously follow several different chemical reactions *in situ* and in mixtures. Resin chemists can use microelectronics to aid resin synthesis. For this, UF-resins are especially interesting, because they constitute a simple two-system model. Forest products scientists can study *in situ* the correlation between chemical structure and mechanical performance of end products. Resin manufacturers can use microelectronics to achieve a new level of reaction and quality control. Particleboard manufacturers will be able to take advantage of the new nondestructive analytical methods for routine product quality analysis, making it possible to identify and reserve top quality boards for special premium use, and to eliminate the few substandard boards that account for almost all the problems. The purpose of this book is to assist basic and applied workers in all these and similar ventures which recent progress has now made feasible.

ISBN 0-201-04558-3

2. History

The history of UF-resins can be divided into six partly overlapping periods. During the first period, chemists learned how to synthesize raw materials. This period started with the synthesis of urea by Wöhler (1824), an event which shook the intellectual world, because it broke the barrier between the chemistry of living and inanimate matter. This period ended with the observation of resinous products in UF reactions, around 1880. During the second period the reaction products were identified and the conditions for making resins were explored. This exploratory work took place between 1884 when B. Tollens studied the condensation and the 1955 decisive paper of Jong and Jonge which brought an understanding of basic resin chemistry. The third period was devoted to reducing to practice the commercial potential of resins. It started probably in 1887 with C. Goldschmidt's patent, peaked with John's application of 1918 and extended through Curs's foam patent to the formulation of particleboard by Fahrni and others during the Second World War. The fourth period was the commercialization of UF-resins. It began in the early 1920s by Pollak in Austria, S. Goldschmidt in Germany, and Ellis in the U.S. The development was stepped up by E. C. Rossiter's "Beetle," CIBA and I.G. Farbenindustrie who quickly secured and maintained early dominance, peaking with the international agreements for licensing many inventions through Pollak's patent pool. The fifth period was one of maturing and adjustment; it extended from about 1930 to about 1975. During this period UF formulations remained unchanged, but UF slowly penetrated into newly developing markets. The sixth, and current, period is characterized by renewed interest in basic resin chemistry, to adapt resins to modern use and control odor. At the same time, this period is bringing deeper insights into resin reaction, because modern analytical instruments allow *in situ* study of reactions.

Urea was first identified by G. F. Rouelle in 1773 in urine. In 1824 Wöhler obtained it, together with oxalic acid and other products, during the reaction of cyanogen with liquid ammonia. He immediately recognized the importance of his discovery and, during the same year, published a second paper in which he reported a comparison of the full elemental analysis of natural urea, as reported by Proust,

---

Beat Meyer, Urea-Formaldehyde Resins

ISBN 0-201-04558-3

and his own for the synthetic material, Table 2.1. He produced larger quantities of urea by reacting silver cyanide with aqueous ammonia, or lead cyanide with liquid ammonia, and he correctly recognized that this synthesis established the "curious fact" that animal, organic matter could be synthesized from inorganic materials. In fact, his synthesis opened the way for organic chemistry, signaled the start of a new chapter in chemistry, and was probably the most important milestone in the chemists' yet unfulfilled dream to synthesize life. The synthesis of commercial quantities of urea is based on A. Basaroff's invention (1870) which followed H. Kolbe's method of reacting liquid ammonia with carbon dioxide. The reaction involves 30 kcal/mole. The resulting ammonium carbamide can be dissociated into urea and water. The reaction is conducted at 100 atm and 150°C. More details are provided in Chapter 4. Originally, the manufacture was hampered by corrosion problems. The reaction is catalyzed by acids, bases, sulfate, charcoal, and ammonium sulfide. Until World War II the world supply of urea was controlled by one company, I. G. Farbenindustrie (Ellis 1935), which maintained stockpiles of 20,000 tons and more. Today, U.S. production is about 5 million tons per year.

Table 2.1

Elemental Analysis of Urea (weight percent)

Element	Proust[a]	Wöhler[b]	Theoretical	Mol Ratio
Nitrogen	46.650	46.78	46.66	4
Carbon	19.975	20.19	20.00	2
Hydrogen	6.670	6.59	6.66	8
Oxygen	26.650	26.24	26.00	2
Total	99.875	99.80	100.00	16

[a] Ann. Phil. **11**, 354 (1827)

[b] Pogg. Ann. **12**, 253, (1828); Wöhler's atomic weights are based on Berzelius: O = 100.000; H = 6.2398; C = 76.437 and N = 88.518

Doebereiner discovered aldehydes in 1826 and called them oxyethers, because of their characteristic properties and odors. He produced them by distillation of alcohols in the presence of sulfuric acid and manganese peroxide. He reported that their most characteristic property was their reaction with acids, forming yellow or brown resins. Liebig repeated these experiments in 1835 and dexcribed the properties and reactions of aldehyde resins. In 1848 Weidenbusch studied both basic and acid systems and described the formation of thioaldehydes.

Formaldehyde was found by Butlerov in 1859 when he set out to synthesize methylene glycol. Butlerov immediately recognized his discovery and much of the complexity of its aqueous solutions. He then further pioneered the study of its

ISBN 0-201-04558-3

reactions. For example, he is the discoverer of hexamethylene tetramine. The history of the formaldehyde discovery has been well described by Walker (1964). The potential for formaldehyde as a polymer precursor was readily recognized.

The history of synthetic resins started shortly after Wöhler's above mentioned synthesis of urea in 1824. In 1827 Bonastie prepared monostyrene, in 1832 Lüdersdorf reacted sulfur with natural rubber, and in 1837 Liebig discovered acetaldehyde resins. In 1872 Baeyer observed that formaldehyde reacted with phenol yielding resinous solids, and in 1877 C. Goldschmidt commenced work on urea-formaldehyde. It took another ten years before UF chemistry was seriously studied. In fact, urea-formaldehyde reactions remained still poorly understood well after 1901, when Blumer began commercial production of phenol-formaldehyde varnishes. In fact, the first aniline-formaldehyde patent of Allers (1904), Baekland's phenoplast patent of 1907, Raschig's fluid PF-resins of 1909, as well as the commercial production of cellophane by Brandenburger (1911), vinylchloride by Klatte (1912), and butadiene by BASF (1915) were all history before the first important patent application for UF-resins was filed by John in 1918. The main reason for this was probably economical: until 1925 UF-resins were more than twice as expensive as phenol-formaldehyde resins.

In 1885 Tollens published three papers detailing derivatives of formaldehyde. He restudied Butlerov's hexamethylene tetramine, the formaldehyde-aniline reaction, the reaction with phenylhydrazine, *p*-toluidine, *o*-toluidine, naphthylamine, xylidine, and other aromatics. His coworker, Holzer, proposed (1884) methyleneurea with the fine structure $NH_2 - CO - N = CH_2$. In 1885 Lüdy studied Schiff's analytical reaction, published in 1877, according to which urea and aldehydes yield purple dyes. Lüdy reported that a Mr. Polikier, a fellow student of his teacher Nencki, had found crystalline solid white $C_2H_4N_2O$ as a reaction product of urea and formaldehyde, and that he determined the structure to be cyclic methyleneurea. In 1891 Hemmelmayr reacted a new product, chloromethanol, with urea and thiourea. He described reaction products which were similar to Lüdy's, i.e., cyclic methylenes, which he hydrolyzed with acids yielding urea and formaldehyde. He reported that, when heated, the reaction products behaved strangely and yielded resinous materials. Furthermore, he noted the unusually low solubility of the crystals. In 1896 C. Goldschmidt reported similar insoluble products which he obtained by reacting urea with excess formaldehyde in the presence of dilute hydrochloric acid for an hour. He proposed hexaisourea, or a similar chain structure. In 1897 he reported dimethylolurea which he obtained at a U:F ratio of 3:7 by careful pH control. He also applied for and was issued a patent for his invention, GP 97,164. Tollens promptly published a note indicating his preference of Hölzer and Lüdy's interpretation.

A. Einhorn and A. Hamburger published a series of papers in which they scrutinized Goldschmidt's work and reported their own observations. They were the first to unambiguously describe the synthesis of mono- and dimethylol urea. They worked with U:F = 2 first in BaOH and then in neutral solution. The reaction

ISBN 0-201-04558-3

of ethylurea and formaldehyde in potassium hydroxide gave a soluble solid melting at 69°C, while a U:F = 1 mixture yielded only oils. Dimethylolurea melted at 126°C. Girsewald (1914) reacted urea formaldehyde and hydrogen peroxide and found condensation products in which he identified methylene bridges. He considered several cyclic products. Dixon and Taylor published some thirty papers on urea and thiourea and described a substantial number of new compounds. They confirmed Hemmelmayr's methylene urea and gave detailed synthetic instructions for mono- and dimethylolurea (MMU, DMU).

On May 16, 1918 Hanns John, Magister of Pharmacy of Prague, filed his application for "Manufacture of aldehyde condensation products capable of technical utilization." He was granted Austrian Patent 78,251; German Patents (GP) 392,183 and 394,488, and eventually U.S. Patent (USP) 1,355,834. In this application, Figure 2.1, he referred to Goldschmidt, Einhorn, and other earlier authors and claimed a novel product which was liquid in hot forms, but gelled upon cooling. He worked at U:F = 4. In his ten U.S. claims he described a transparent, colorless, and elastic material which was chemically resistant and insoluble. As possible uses he contemplated glues, lacquers, impregnating materials, and substitutes for rubber, ebonite, and celluloid. Later suggestions included impregnation of fabrics, as well as filling root canals. John's patent is considered the beginning of UF-resin manufacture. His work initiated a flurry of activity which is reflected both in the patent and the scientific literature.

From this point on, the history cannot be easily chronicled, because the great commercial value of UF-resins was recognized and researchers as well as companies became secretive. Much of the data was kept as trade secrets, and has never been released. Some of the findings are summarized in patents, but the patent literature makes it hard to correlate the chemical and commercial development, because patent issue dates have little relation to the date of the inception of the invention. Thus, many of the important facts remain obscure or appear in review articles and book chapters written by the key inventors several decades after their work was completed, and after patents have expired. In the following the author heavily relies on the patent literature, and on summaries released by leading companies.

M. van Laer (1919) recognized that the UF condensation gave complex products and was more complicated than that of aromatics, and that Lüdy and Hölzer's products were tautomers. In GP 563,037 Rothera, Blythen, and Gillespie 1921 describe the acidic condensation of U:F = 3. In GP 568,629 the same group (1921) includes acidic salts. In USP 1,899,109 K. Ripper describes the condensation of hot thiourea in aqueous formaldehyde. The product is separated, cooled, and becomes a fine powder. In the same year, Pollak in GP 499,792 controls condensation by addition of a strong base and followed by a weak acid. In British Patent (BP) 171,096 Pollak used ammonia, pyridine, urea, or hexamethyl-tetramine as a base to induce polymerization, slowly heat the intermediates, and reduce excess formaldehyde. He proposed fillers, such as wood-pulp, asbestos chalk, sand, and carbonundum. He proposes further the impregnation of paper. In BP 181,014 Pollak proposes the addition of salts to delay curing for use as lacquers, artificial threads, molded

ISBN 0-201-04558-3

UNITED STATES PATENT OFFICE.

HANNS JOHN, OF PRAGUE, CZECHOSLOVAKIA.

MANUFACTURE OF ALDEHYDE CONDENSATION PRODUCT CAPABLE OF TECHNICAL UTILIZATION.

1,355,834. Specification of Letters Patent. **Patented Oct. 19, 1920.**

No Drawing. Application filed October 25, 1919. Serial No. 333,377.

To all whom it may concern:

Be it known that I, HANNS JOHN, magister of pharmacy, a citizen of the Czechoslovakian Republic, residing at Prague, Manesgasse, Czechoslovakia, have invented certain new and useful Improvements in Manufacture of Aldehyde Condensatiton Products Capable of Technical Utilization, of which the following is a specification.

My invention relates to the manufacture of products obtained by condensation of an aliphatic aldehyde, preferably of formic aldehyde, with carbamid (urea) or thiocarbamid or other carbamid derivatives, they being substantially equivalent.

The action of formic aldehyde on urea has been studied by various chemists. C. Goldschmidt, for instance, has caused formic aldehyde to act upon urea in the presence of potassium hydroxid. By this reaction he obtained an amorphous substance which readily disengaged formic aldehyde, and which he assumed to be dimethylolcarbamid.

All the products obtained up to the present by condensation of formic aldehyde with carbamid, were of no technical value or could not be used for industrial purposes, owing to their physical and chemical properties.

Now I have found that by carrying out the reaction between formic aldehyde and carbamid or its derivatives in a special manner, it is possible to produce different substances fit for a variety of technical applications.

By causing the condensation to take place either in the presence of alkali or other condensing mediums or making the starting substances to react upon each other at a low temperature during a more or less long time, the prior authors only obtained products such as the dimethylolcarbamid mentioned above, whereas by the process according to my invention substances are produced which were up to now unknown, and which are quite distinct from dimethylolcarbamid.

I have found that, in order to obtain condensation products fit for industrial applications, the raction is to be carried out without addition of condensing agents and at a higher temperature. According to the quantities of formic aldehyde and of carbamid present, as well as according to the duration and height of heating applied, products of very different nature are obtained.

If the ingredients are caused to react only during a comparatively short period, the product of condensation is still soluble in water. This solution is highly adhesive and, when dried, leaves the condensation product in the shape of a perfectly colorless product which is transparent like glass and insoluble in all inert solvents.

In this stage therefore a glue is produced which sticks in a cold state to all objects, of preference with smooth surfaces such as glass, metals, etc. The substance may also, besides other applications, be used for filling root-canals in dental surgery, or as a lacquer which can be directly laid on without addition of spirituous or such like solvents. This colorless and brilliant lacquer may be mixed with any organic or mineral pigments. Furthermore, it may advantageously be utilized for impregnating materials or fabrics of any kind, especially for covering the supporting planes of flying-machines, and such like.

If, from the beginning, the ingredients are brought together on such conditions that the process will be achieved only in a longer time than stated above, the resultant product of condensation will be still fluid in a heated state, but it will gelatinize when being cooled. In this way prepared, the product forms a colorless, transparent, tensile and elastic mass, insoluble in water as well as in alcoholic solvents, and which is acted upon only by acids or alkali liquors. This mass can be used as a substitute for india-rubber or such like.

The substance produced in the manner described, is moreover capable of being hardened. This hardening process is carried out most advantageously by heating the product up to say 80° C. According to the duration of the hardening process, the final products show different properties. The main advantage is that also the hardened product is colorless and transparent, as well as rather resistant to chemical agents, that it is readily to be worked and possesses a sufficient degree of elasticity.

Figure 2.1. Title Page of USP 1.355,834, by Hanns John, Magister of Pharmacy, a Citizen of the Czechoslovakian Republic.

ISBN 0-201-04558-3

When strongly heated, the substance is carbonized, but does not flare up.

The reagent substances may be used in various proportions as shown the following examples of carrying out the process under notice:

1. 5 parts of a commercial solution of (say 40 per cent.) formic aldehyde, and 1 part of carbamid are heated in a distilling vessel or still. When an adequate proportion of the liquid is distilled off, the glue-like mass as described above, is obtained.

2. 6 parts of a commercial solution of formic aldehyde and 1 part of carbamid are heated in a distilling vessel until nearly half the liquid has been distilled off. After cooling the gelatinous product described above, is obtained.

3. The substance obtained in accordance with example 2, is heated up to about 80° C. during such a period until it has attained the desired degree of hardness.

The carbamid mentioned in Examples 1 and 2, can be replaced by thiocarbamid.

Furthermore, by heating other derivatives of carbamid with formic aldehyde in the manner described, products of like or similar nature and properties are obtained.

4. 5 parts of a commercial (say 40 per cent.) solution of formic aldehyde and 1 part of acetylcarbamid are heated in the manner stated above. The resultant substance is colorless and liquid in a heated state, but sets when cooling, to form a white corny mass.

The acetylcarbamid cited in Example 4, can be replaced by benzoylcarbamid.

The different products as described may be used, according to their special properties, as a glue, lacquer, impregnating material or as a substitute for rubber, ebonite, celluloid, corn or for other industrial purposes.

What I claim and desire to secure by Letters Patent of the United States is:

1. The process of manufacturing condensation products capable of technical utilization which comprises reacting with an aliphatic aldehyde upon substances containing carbonic acid amids, in a highly heated state, without adding any condensing medium.

2. The process of manufacturing condensation products capable of technical utilization which comprises reacting with formic aldehyde upon carbamid at the temperature of distillation without adding any condensing medium.

3. The process of manufacturing condensation prod ucts capable of technical util tion which comprises reacting with fon aldehyde upon thiocarbamid in a heated state, without adding any condensing medium.

4. The process of manufacturing condensation products capable of technical utilization which comprises reacting with formic aldehyde upon acidulated carbamid in a heated state, without adding any condensing medium.

5. The process of manufacturing condensation products capable of technical utilization which comprises reacting with formic aldehyde upon acidulated thiocarbamic in a heated state, without adding any condensing medium.

6. The process of manufacturing condensation products capable of technical utilization which comprises reacting with an aqueous solution of formic aldehyde upon substances containing carbonic acid amid, the components being present in a ratio of about 5 parts to 1 part, and heating the fluid during a comparatively short period.

7. The new product obtained by reacting with an aqueous solution of formic aldehyde upon substances containing carbonic acid amid, the components being present in a ratio of about 5 parts to 1 part, and heating the fluid only a comparatively short time, which product is a glue-like mass, leaving, when dried, a colorless sheet transparent like glass and insoluble in all inert solvents, and which is applicable for sticking objects of any material, preferably with smooth surfaces, as well as for lacquering, varnishing or impregnating purposes.

8. The process of manufacturing condensation products capable of technical utilization which comprises reacting with an aqueous solution of formic aldehyde upon substances containing carbonic acid amid, the components being present in a ratio of not to exceed 6 parts to 1 part, and heating the mass until about half the reacting liquid is removed.

9. The process of manufacturing condensation products capable of technical utilization which comprises reacting with an aqueous solution of formic aldehyde upon substances containing carbonic acid amid, the components being present in a ratio of not to exceed 6 parts to 1 part, heating the mass until about half the reacting liquid is removed, and submitting the resultant product to a hardening process, preferably by heating it for some time at a temperature of about 80 degrees centigrade.

10. The new product obtained by reacting with an aqueous solution of formic aldehyde upon substances containing carbonic acid amid, the components being present in a ratio of not to exceed 6 parts to 1 part, and heating until about half the reacting liquid is removed, which product, still in a heated state, is fluid, but, when cooled, forms a tensile and elastic material, insoluble in all inert solvents, and applicable as a substitute for india-rubber or, after being subjected to a hardening process, for ebonite, corn, celluloid and the like.

In testimony whereof I affix my signature in presence of two witnesses.

HANNS JOHN.

Witnesses:
STEPAU HAAS,
CORÉNE KOLER.

Figure 2.1. (Continued)

ISBN 0-201-04558-3

articles, water proofing, and adhesives. Retarding salts, such as sulfates, nitrates, chlorates, rhodamide, and iodide, enhance glossy appearance. In 1922 Pollak, as founder of his own company, and Nerrn, in GP 484,972, describe molding powders. At the same time, I. G. Farbenindustrie described mineral acid curing in GP 409,847 and organic acid curing in GP 535,851. H. Goldschmidt, to whom John had sold his patent, described turbid materials, resembling meerschaum and porcelain in BP 187,605. In BP 208,761 the same author describes lower U:F ratios and acid concentrations of 0.5%. He also shows that urea can be replaced with cyanamide or $Ca(CN)_2$. In USP 1,460,606 Ripper proposed sodium acetate to delay gelling.

In 1923 Pollak, in BP 193,420, described the addition of urea, thiourea, phenols, or hydrogen peroxide to UF to react with excess formaldehyde. DMU can be used to replace UF. In BP 213,567 he described two reaction stages: (1) a first reaction with neutral or alkaline solution, and (2) a second reaction in the range $2 < pH < 5$. A lower pH yields brittle products. Boric acid, acetic acid, benzoic acid or formic acid are recommended as buffer systems. In BP 201,906 he described salts capable of absorbing free formaldehyde. In BP 206,512 Pollak described ten cations suitable for preparing transparent, rather than cloudy solid. O. Neuss described hot condensation with and without solvent in BP 202,651. In BP 238,904 Pollak described additives suitable to absorb water.

In 1924 and 1925 E. C. Rossiter described a condensation mixture comprising urea and thiourea in a ratio of 1:1 and cellulose fiber. The latter serves to absorb formaldehyde and water liberated in the condensation reaction. His material, BP 248,477, 258,950, and 266,028, is called Beetle, because "it beats all others." These patents were assigned to British Cyanides Co. Ltd. which later merged with American Cyanamide and which would greatly contribute to further developments. Rossiter showed that UF-resins are suitable as a basis for laminate and molding powders. I. G. Farbenindustrie conducted parallel studies, and in GP 493,988 described hot curing with hydrogen peroxide.

At the same time, Pollak and Ripper (1924) published a review of their own and earlier work, which influenced many other workers. Their summary of earlier chemistry is probably the best available. However, their description of other authors' patent work is colored by their own interest, namely their development of Pollopas, a glass substitute. They questioned the usefulness of John's patent because they recognized that it was important to reduce the U:F ratio. They explained the need for a two-stage condensation process, and explained the "second addition" of urea, which has recently come into vogue again. Furthermore, they recognized the danger of excess acid catalyst which can lead to decomposition of the cured resin, and release of formaldehyde. They also explained the effect of pH control in the initial condensation, and detailed the effect of iron, aluminum, potassium, ammonium, copper, magnesium, zinc, calcium, and sodium ions, and the corresponding amide series comprising sulfate, citrate, oxalate, acetate, chlorate, carbonate, nitrate, cyanide, rhodamide, iodide, and salicylate. Their main objective

ISBN 0-201-04558-3

was to produce large transparent panels, free of cracks and flaws. The colorless panels are transparent to 300 nm, i.e., they are better than normal glass. The refractive index is D = 1.5, very close to quartz, with D = 1.55. The density is 1.44, half that of glass. Compressibility is 2230 kg/cm^2, and elasticity is likewise comparable to glass. Hardness of UF is 2.95, comparable to mother of pearl. UF panels could be readily dyed in any desired color. Thus, their products were suitable for manufacturing safety glass, jewelry, buttons, piano keys, pool balls, and the like. The addition of UF glass to regular glass to prevent splintering was also proposed. The material was used to build greenhouses, since it transmits more ultraviolet light than regular glass. Molded products included among other items table legs, ashtrays, molded artifacts, such as busts, and the like. Since Pollak succeeded in stabilizing the liquid resin, he also proposed its use as a varnish called Schellan for coating furniture for use in the textile industry; as a wound dressing material; as well as a carrier for printing dyes and photographs, and as an impregnating material for paper and for felt and straw hats.

In the U.S. Carleton Ellis independently worked on UF-resin as a casting material. While the first edition of his book, *Synthetic Resins and their Application,* published in 1923, contained only 6 pages of a general chemical discussion of UF and thiourea resins, and he did not mention his own interest in this field, even though, during the previous year, he had already been granted USP 1,482,358, 1,536,881, and 1,536,882, all dealing with important developments of John's invention. Several others followed in 1925, among them USP 1,846,853 in which he showed that organic acids complexed with the resin in a variety of ways; benzoic acid, yielding an easily curing fairly white product; citric acid, producing a slightly yellow cast; acetic anhydride, causing snow-white glossy materials; propionic, a hard white mass; gallic acid, a glossy yellow resin; lactic, malic, and tannic, all yellow solids; and salicylic, tartaric, mucic, and trichloroacetic acid, all white products. Stearic acid, finally, yielded opaque and brittle materials. Ellis preferred a pH between 4 and 8. These and other UF materials received more justice in his second book, published in 1935, in which he devoted 126 pages, twenty times more space, to the subject of UF-resins. Such expanded coverage was well justified in view of the furious worldwide activity which took place during the intermediate years!

In 1925 A. Gams and G. Widmer filed a new procedure for condensing urea and formaldehyde, Swiss P 114,289, USP 1,674,199, BP 246,126, and GP 487,871. Their hot autoclave was widely adapted, and their employer, The Society of Chemical Industry in Basel, CIBA, entered the field and quickly became one of the leaders in UF-resins until 1935 when it discovered an economical method for the synthesis of melamine from calcium cyanamide and switched its production to that field.

In 1925 G. Walter, BP 284, 272, described the manufacture of methylol ureas. W. S. Rothera, S. Blythen, and H. R. Gillespie, GP 567,271, described UF condensation products, F. Pollak, BP 261,409, described porous, cloudy materials, and

ISBN 0-201-04558-3

I. G. Farbenindustrie, BP 260,253, described condensation in ethylene chlorohydrin, ethylene glycol, and other organic solvents. M. Luther and C. Henck, GP 535,852 and USP 1,791,062, described condensation methods for I. G. Farbenindustrie. CIBA, GP 499,589, reported that insoluble products can be rendered soluble by autoclaving with excess formaldehyde at 100°C. The same company claimed in BP 253,094 that dimethylurea can be synthesized by shaking urea and formaldehyde in the presence of charcoal. Refluxing of the latter *in vacuo* yields varnishes which can be stabilized with alcohols, ketones, or ethers. BP 249,101 of CIBA. Finally, G. Walter, BP 291,712, described methylol acetamides, also to be used as varnishes.

In 1926, A. Gams and G. Widmer of CIBA reported in GP 523,181 the synthesis of porous meerschaum-type products. This patent constitutes the first foam formulation, and yields a density of 0.4 g/cm^3. BP 281,717 reported the synthesis of the same by extended boiling with excess formaldehyde beyond the methyl stage by addition of salts. The latter are later washed out with water. Pollak, Austrian P 109,532, described the manufacture of shaped articles. I. G. Farbenindustrie, GP 537,611, described that resins can be separated without evaporation of water. Its FP 641,420 deals with condensation in organic solvents; BP 288,346 signaled the beginning of a long series of descriptions of wood adhesives, which eventually lead to the Kaurit glues. GP 510,513 described sulfur dioxide gas as curing agent, and GP 511,979 listed as other volatile acids formic acid, phosphoric acid, hydrogen bromide, hydrogen chloride, and hydrofluoric acid. Finally, BP 288,346 described continuous pH control, with a pH of 4 to 6 during the first stage and a pH of 6 to 7 being the preferred value for the second stage. Cellulose esters and ethers are used as plasticizers, and sodium phosphate as a buffer.

1926 signals the entry of H. Kadowaki in this field with a report on optimum conditions for producing glass. He recognized that water is formed during condensation, and that it is the origin of cracks produced during aging and flatly states that it is impossible to overcome this intrinsic defect. At the same time, H. Staudinger tackled the problem of polyoxymethylenes and studied their degradation with acetic anhydride. He described the properties of the first 22 members of the polyoxymethylene diacetate series and of 6 methylene derivatives.

In 1927 smaller companies entered the field, for example, L. Smidth with BP 294,253 for E. G. Budd Manufacturing Co., and Silur Products Co., with GP 569,342. But, the basic improvements were made by the original leaders: Pollak, FP 637,318, which described albumin as an additive; Pollak, BP 301,798, which described U:F ratios below 2; Pollak, BP 291,366, which revealed the art of reducing formaldehyde by ammonia or dry ammonia salts; Pollak, BP 248,729, which gave new condensation procedures; Pollak, E. C. C. Baly, and E. J. Baly, BP 299,487, which described slow condensation at 15–20°C leading to a product melting at 134°C; and BP 301,626 by the same authors which described various products obtained from dimethylolurea. K. Ripper, in BP 323,047 described molding materials made from dicyanodiamide, and in BP 287,568 he described

ISBN 0-201-04558-3

thiourea products. E. C. Rossiter, Canadian P 274,738 described products made by precondensation of thiourea and urea first separately and then co-condensing the intermediates. CIBA, FP 644,648, described more porous materials, A. Gams and G. Widmer, Can. P 274,266 described room temperature condensation in the presence of "active" carbon. S. Goldschmidt, GP 551,422, described vacuum concentration of resins, and I. G. Farbenindustrie in FP 641,770 described oily condensation products obtained by addition of alcohols and ketones.

In 1928 H. Scheibler and coworkers described several organic derivatives of UF-resins and their conversion. CIBA continued its string of patents, while I. G. Farbenindustrie obviously preferred working with trade secrets. A. Gams and G. Widmer reported for CIBA in USP 1,674,199 that a pressure of 2–3 atm at 115°C accelerates condensation to such a degree that the resin is ready within an hour. The same authors, in USP 1,676,543, revealed that they used charcoal as a catalyst, and in Can. P 281,327 that a second addition of urea to a U:F = 2 resin is possible. The preceding patent, Can. P 281,326 deals with the recovery of cured resin wastes by boiling in 35% formaldehyde. Their Can. P 281,325 compared the curing time with and without pressure. CIBA's BP 342,767 described co-condensation with Schiff's base. BP 342,730 described phenylamine modifications and BP 342,723 described other aromatic amines. In the meantime CIBA entered the tradename CIBANOID for its entire resin class for uses ranging from drugs, toiletries, resins, adhesives, mirrors, other manufactured goods, teaching aids, optical tools, and art objects to wood conservation aids.

BP 316,194, issued to Allgemeine Elektrizitäts Gesellschaft, described the use of UF and other resins to prepare bonded layers of material, with the hardening effected by heating for 30 seconds to 150°C. In the same year, the French chemical company of Laire, Malet, and Armenault, BP 340,114, reported the catalytic effect of zinc carbonate and in FP 679,321 that of the corresponding oxides. S. Goldschmidt and R. Mayrhofer, BP 316,144 added phthalic acid products with glycols and other polyalcohols to UF-resins. The same authors, in GP 561,050 and GP 553,161, found a step-wise co-condensation method for UF yielding a material that could "substitute for plate glass." Pollak marketed this product as luxite in the U.S. and as Pollopas in Europe. K. Ripper, in Austrian P 111,534, described more porous products; in USP 1,687,312 he outlined desirable pH values; in USP 1,625,283, he revealed that the addition of the ammonium salts of rhodamide, chloride, and sulfate accelerate curing, yielding a glass-like product; and finally, in Can. P 284,790, he described a hydrophobe resin made in neutral solution by mixing U:F in a ratio of 1:1. An anonymous author reported for Pollak in FP 657,794 step-wise procedures for another type of resin. Still in 1928, F. E. K. Steppes described the synthesis of UF-resin from paraformaldehyde. He proceeded by dissolving the latter with the help of ammonia in water and adding urea. F. Lauter, of Rohm and Haas, who purchased U.S. rights to John's original patent from I. G. Farbenindustrie, in USP 1,671,596 and USP 1,672,848 co-condensated with toluene-sulfonamide, and condensated to avoid methylene bridges. I. G.

ISBN 0-201-04558-3

Farbenindustrie, in BP 314,908, described hard condensation products obtained in the presence of formamide, and in GP 536,733, W. Pungs, K. Eisenmann, and J. Kuchenbuck reported anhydrous condensation procedures. M. Luther, W. Pungs, R. Griessbach, and C. Henck, again for I. G. Farbenindustrie, reported the use of sulfur dioxide as gas or in pressurized solutions as a curing agent. Sodium phosphate was used as a buffer. The method can be used to harden surfaces of impregnated articles, to spin aritificial silk by expanding a thread of UF-resin through a sulfur dioxide atmosphere, for curing moldings, and similar purposes. At the same time, H. Scheibler (1928) reported a series of publications on the synthesis, identification, and properties of UF-resin components.

In 1929 Vierling, Schmiking, and Klingenberg, GP 550, 647, described the use of aqueous UF-resins as thermosetting resin for plywood. The product was called Kaurit. Production started immediately and it quickly and permanently replaced phenol resins as the leading wood adhesive. A. V. Keller, BP 333,763, proposed the impregnation of paper with the same material. I. G. Farbenindustrie, in FP 675,398, proposed the manufacture of an artificial material by mixing UF-resin with cork before curing; in GP 528,582, it described a method for making glass, in GP 519,892 and GP 540,071, W. Pungs and coworkers described modified UF-resins. CIBA, in GP 562,943, described a hot-molded mass containing fibrous and other fillers. A. Gams and G. Widmer, in USP 1,718,901, improved the U:F ratio. Pollak, in BP 339,601, described new condensation methods, and in FP 680,110 described a truly modern resin with U:F between 1:1,4 and 1:1,8. In FP 674,999, Pollak minimized the acid catalyst. In the U.S., L. Smidth in USP 1,704,347 described the addition of thiourea at the second stage. A. V. Keller and J. Taylor in BP 331,428 also described a mixed thiourea resin; Toledo Manufacturing, FP 674,589, described a high U:F resin; L. Auer, FP 674,703, added fatty acids to UF producing cleansing agents or vulcanized rubber substitutes; and H. Barthelemy, in USP 1,691,427 added acetic acid anhydride to facilitate mold release. In the open literature, S. Basterfield and E. C. Powell (1929) described new isoureas and their condensation, and L. E. Hinkel and D. H. Hey explored benzaldehyde and ethyl-acetoacetate and condensation partners for urea and thiourea.

In 1930 CIBA in Swiss P 154,520 described sulfite pulp, cellulose, leather strips, and other fillers to make materials which imitate natural materials. The Établissements Lambiottes Frères, FP 722,289, described the use of hydrogen sulfide as curing agent, which they believed to act most effectively on methylol urea. In FP 697,874 I. G. Farbenindustrie described improvements in cold setting aqueous UF-adhesives, and in GP 552,264 described the synthesis of lacquers and molding compounds in hydroxylated organic solvents. Water is extracted with calcium chloride or phosphorous pentoxide. In FP 38,157 I. G. Farbenindustrie disclosed nonflammable varnishes. A. V. Keller and J. Taylor, FP 699,968, described synthesis of an aqueous paste for preparing transparent gels and syrups. K. Ripper, USP 1,762,456 described the use of solvents which enchance the vaporization of

ISBN 0-201-04558-3

condensation water. G. Walter, Austrian P 130,030, described the manufacture of mono- and dimethylolurea, and their conversion products from paraformaldehyde. In Austrian P 121,999 he described the preparation of resins from dehydrated MMU and DMU. A. Einhorn, A. Hamburger, and coworkers (1930) described N-methylol compounds of acid amides. L. Monti (1930) described the condensation of aromatic methylols.

In 1931 O. A. Cherry, USP 1,790,461 described pH and other condensation conditions corresponding to the state of the art. British Cyanides in FP 714,101 recognized the importance of built-in bases to neutralize latent acid release in the final product. I. G. Farbenindustrie described in FP 721,828 the manufacture of solid dimethylol urea by vacuum evaporation. CIBA in Swiss P 154,831 described mixed resins prepared by second addition of formaldehyde. A. Gams and G. Widmer in USP 1,831,706 described porous materials obtained by condensation of UF-resins with sulfuric acid using chemical rather than external heat. At the same time, in USP 1,844,570, H. Goldschmidt and O. Neuss described a similar procedure, using phenol and sulfuric acid. Toledo Synthetic Products, in FP 711,333 and FP 711,395, described grinding of molding resin with phenol or resorcinol to produce copolymers. Société Nobel Française, FP 736,744, described solvent extraction of condensation water to prevent cracking of final products. G. Walter of Vienna published three papers in 1931 describing a yellow oil obtained from copper-thiourea resins, and analyzing the resin structure, as well as the theoretical basis of the UF-resins. He recognized the existence of methylene bridges and identified methylol-methylene-urea and proposed ring structures and crosslinking structures.

In the meantime, commercial production of UF-resins began in earnest. The U.S. production was estimated to have reached 2500 tons. Ellis's patents were combined with British know-how when British and American Cyanamide merged. Since Pollopas held worldwide patent control, the German I. G. Farbenindustrie, the French Ugine-Kuhlmann, and others entered license agreements. This lead to a consolidated, worldwide industry which in the remaining years before World War II would increase UF production to about 2000 tons in Germany, 1000 tons in Great Britain, 4000 tons in the U.S., and maybe 500 tons both in France and Russia (K. Thinius 1977). This compares with 400,000 tons in 1978 in the U.S. However, UF-resins were still twice as expensive as phenols and their usefulness was to be challenged before they were fully developed. The main goal was still to produce better casting materials and glass, as well as varnishes. Since the potential importance of the UF-resins was fully recognized in 1932, broader interest awoke in the scientific community. In 1932 E. B. de Chesne, a student of Fierz at the Swiss Federal Institute of Technology published an excellent review in which he analyzed the synthesis of the methylol reagents, and described the best procedures for the condensation stage and the polymerization stage. He related concentration and viscosity, as a function of temperature and used Staudinger's classification of colloids to describe the products. He also applied Staudinger's relation between

ISBN 0-201-04558-3

viscosity and chain length and discussed Scheibler's analysis, but found the results inconclusive. He followed the pH during condensation, discussed redox reactions, and finally described synthetic procedures for methylol-urea, methylene urea, and resins. He provided tables correlating the gellation time as a function of temperature and pH and even attempted to isolate intermediates. Finally, he provided microphotographs of cured resin samples, and a list of some 150 patents. The latter gives an interesting insight into the level of information exchange available at a major university of that time, and it reflects the difficulties of obtaining patent information. During the same year, G. Walter and K. Oesterreich (1932) published another review in the same journal. They mainly summarized their own work on preparing higher molecular weight polymers. They analyzed the crosslinking during condensation at high temperature, using primarily thiourea. They studied conductivity measurements as a function of concentration to derive information on the condensation process, and thoroughly explored the effect of copper salts and other electrolytes. In a second paper. G. Walter and M. Gerwing (1932) explored the reaction mechanism and analyzed the difference between the various published procedures. They reaffirmed their belief, based on Scheibler's paper of 1928, that crosslinkage via methylene bridges yields the best cured products. They designated the procedures of John, Pollak, and Walter as mild reactions, which would yield methylene bridges if conducted more acidic. They correctly excluded a variety of functional groups from bonding in varnishes, but did not yet recognize the predominance of ethers. Their experiments are also described in detail. In the general area of basic research, F. D. Chattaway reported the reactions of urea with bromal. In the industrial field, I. G. Farbenindustrie in Ludwigshafen, today BASF, officially introduced their Kaurit wood adhesive lines. Their Kaurit-W showed sufficient penetration to secure good adhesion and yet remain viscous enough not to soak away from the interface. In 1936 already 2118 tons were made and sold, and UF-resins now challenged the predominance of phenol adhesive. In due time the acceptance of this material was to change the industrial application of UF-resins from molding materials to adhesives, but this change was not reflected in patent literature for almost two more decades. In 1932 O. Neuss, in BP 379,154, described a chalky solid which could be filled with phenol, to be made by reaction of liquid urea with formaldehyde solution, or by cold condensation as described in BP 372,847. In USP 1,833,868, K. Ripper described the manufacture of intermediates. In BP 325,843, assigned to Ignaz Kreidl, who operated a factory that also produced modified UF-resins, G. Walter described an artificial resin; and in USP 1,863,426, assigned to I. G. Farbenindustrie, he described a transparent mass prepared from DMU and thiourea.

In 1933 A. Curs and H. Wolf of I. G. Farbenindustrie obtained GP 636,658 describing the first commercially viable foam. This patent goes far beyond the work of Gams and Ripper and gives a chemically accomplished description of the art of producing small-cell low density foam. Naphthaline sulfonate is used as a foaming agent. Sulfur dioxide, sulfuric acid, or chloralhydrate is used as catalyst. The latter

ISBN 0-201-04558-3

acts as a built-in delayed action acid. The foam can be dyed with oxide pigments. The foam is cast in slabs which are used to enforce metal foil, carpets, wallpaper, wood veneer, and the like. The main use is as a thermal or sound insulator. In the same year the development of traditional resins continued. A. Benteli, in Swiss P 161,052, described nitrocellulose as filler. W. v. Knilling in USP 1,920,451, assigned another UF formulation to I. G. Farbenindustrie, and L. Smidth, USP 1,893,911 described a second addition of urea to bring U:F to 1:1. C. A. Redfarn (1933) explained the effect of a formic acid buffer at pH 4.5 to ensure good resin properties.

In 1934 G. Walter continued his review of UF-resins in a 47-page paper which covers all aspects of the topic. M. Tokuoka (1934) reported a new synthesis of urea in Japan, signaling increased interest in Japan. B. I. Levi did the same in Russia. I. G. Farbenindustrie, in FP 769,588, described the laminating process with UF adhesives. Minnesota Mining and Manufacturing, in BP 419,812, described the use of UF as flexible binder for sandpaper. O. Neuss, in GP 604,194, described another improvement of resins; F. Pollak, in USP 1,950,746 assigned to Synthetic Plastics Co., proposed the removal of excess formaldehyde during the second stage by air blowing. K. Ripper, USP 1,967,261, described a molding powder prepared by floculation of resin in aqueous solution. He also renewed his USP 1,762,456, as did L. Smidth with USP 1,704,347. Thus, UF-resin technology and chemistry started its second generation! M. Marks, in USP 1,991,765, described mixed UF-H_2S resins made at a pH between 2 and 12. C. Ellis, in USP 1,952,060, assigned to his Ellis-Foster Co., describes reinforced UF-resins containing quanidine and similar chemicals. The Société Nobel Française, in FP 804,714, described suitable solvents such as cyclic acetals, and W. Pungs and K. Eisenmann assigned USP 1,967,685 to Unyte Co. for preparing alkaline UF under pressure.

At this stage the UF-resin art had clearly reached maturity. The main product lines were varnishes and coating compositions, such as described by I. G. Farbenindustrie in their patent series extending from BP 260,235 (1925), BP 261,029 (1925), BP 262,818 (1925), BP 301,696 (1926), and BP 319,251 (1928), to BP 459,788 (1935) and by Ellis (1935). The impregnation of textiles is demonstrated by the patent series issued to Tootal Broadhurst Lee Co. Ltd. with BP 291,473 (1926), BP 499,818 (1927), and BP 449,243 (1934). The development of molding and casting materials, as explained in detail in Ellis's excellent book, *The Chemistry of Synthetic Resins* published in 1935, had reached its peak. Many companies had been set up with hydraulic hot presses for hot molding UF-resins. The Unyte Co. made bric-a-brac and a variety of home decorations, and entire bathroom and washroom combinations were molded. Nonshattering cups, glasses, and other UF kitchenware were available. U.S. railway coaches were equipped with translucent lamp shades, 1 mm thick, which transmitted 50% of the light. Telephones and boxes were molded from UF, and even soap dispensers. The impregnation of paper was firmly established by a long series of patents extending from F. C. Rossiter and C. H. Cooper's BP 246,886 (1924) to the work of K. Atsuki and

ISBN 0-201-04558-3

K. Matsuoka (1930), and American Reinforced Paper, FP 821,840 (1937), which described wet-strength paper. The failure of UF-resins as glass substitutes was evident, and many of their successful miscellaneous uses had become evident, including their use as preservative for biological specimens, leather, skins, and other animal products (I. G. Farbenindustrie, BP 507,175 (1939)). I. G. Farbenindustrie in BP 359,163 (1930) had already proposed UF for coating urea fertilizer for preparation of "slow release" fertilizers, which reached the commercial market only twenty-five years later, in J. Jung, GP 1,081,482 (1959), assigned to BASF, which introduced the tradenames Floramin and Ureaform.

Only the wide potential of UF as an adhesive was greatly underestimated. For example, Ellis (1935) used only 10 lines of some 150 pages to describe their use, lumping them together with nonshattering glass. The Kaurit adhesives were expanded by I. G. Farbenindustrie, BP 459,788 (1937). In 1949 BP 460,275 (1937), FP 847,953 (1939), J. V. Nevin, USP 2,247,764, introduced UF for patching plywood veneer, using phosphoric acid as a hardener. In England, N. A. de Bruyne from 1935 to 1937 developed an adhesive which he called aerolite (de Bruyne 1945). The patents, BP 536,493, BP 549,496, and BP 545,409, were issued in 1941 and 1942 during the war when these adhesives were furiously used to build boats and airplane frames. He used formic acid to cure aqueous resins, especially for "gap joints," because older resins had a tendency to develop cracks and fissures and lose strength after curing. He preferred U:F = 2 or 2.5 and added thiourea to his resin. Paper pulp, gypsum, or tripoli powder and starch were used as filler. H. Scheuermann (who later improved UF foams) and J. Lenz assigned FP 729,029 to I. G. Farbenindustrie in 1942, describing the addition of ammonia to obtain clear resins. O. R. Ludwig assigned GP 733,710 to Rohm and Haas in 1943; CIBA obtained GP 732,498 in 1943 describing a very acidic condensation; E. C. Rossiter in 1944 assigned USP 2,097,895 to American Cyanamid Co., describing co-condensation. Such "modified UF" compositions were the main thrust of most UF work between 1935 and today. A typical example is given by M. Segond (1936) who used saccharides and zinc salts, and by H. Scheibler (1940) who modified UF with acrylic.

The use of UF adhesives was quickly threatened by the discovery of a cheap melamine synthesis by workers at CIBA in 1936. Melamine, first studied by Liebig in 1834 is a triazine and offers six active hydrogen atoms for reaction with formaldehyde. This yields better crosslinking and thus better water-resistant adhesives. Thus, CIBA, Heubel (BP 455, 008 by W. Hentisch and R. Koehler 1936), and others shifted production to melamine-formaldehyde (MF) resins (G. Widmer 1965). However, melamine resins remained comparatively expensive, and left the bulk plywood and particleboard market to UF. In fact, MF established itself mainly in Europe, and even as of today, has not yet reached a comparatively great importance in the U.S.

ISBN 0-201-04558-3

The development of UF foams, outlined by Gams and Widmer (1926) and Ripper (1927), was chemically perfected in 1933 by Curs, GP 636,658. According to H. Baumann, Curs was also responsible for the use of UF foam as insulating material in the smoking saloon of the dirigible Zeppelin. The material was manufactured and sold in slabs under the trade name Iporka, as it was used as a substitute for cork. H. Scheuermann in 1943 in GP 734, 094 described the manufacturing method: a 2% alkyl-naphthaline sulfonic acid was used as foaming agent. The acid catalyst and foaming agent were discharged and mixed from a pressure vessel into an equally foamed resin solution. A long tube was used to ensure full mixture of the two. During the war, some 8000 tons were sold each year as insulating material. However, in 1949, when Scheuermann modified the equipment for truly continuous production of foams and assigned GP 800,704 to BASF, which was then formed as a result of the liquidation of I. G. Farbenindustrie. BASF de-emphasized the use of UF foam in favor of polystyrene. Fortunately, W. Bauer invented a portable foaming machine, primarily for insulating walls in coal mines, and laid the groundwork for on-site insulation of residential and commercial housing (H. Baumann 1959-1979).

A truly outstanding summary of the basic chemistry before World War II is given by H. Kadowaki (1936) who single handedly made and isolated some thirty UF intermediates in his Osaka laboratory. Table 2.2 summarizes the knowledge at this time. The best summary of technology is Ellis's book of 1935, and especially the excellent book *Chemie und Technologie der künstlichen Harze,* by J. Scheiber, published in 1943.

After the Second World War, the housing industry boomed again, and the demand for plywood adhesives increased. At the same time, particleboard was perfected, Chapter 6, and was widely and quickly accepted. This development is discussed in Chapter 6, page 164. Thus started a new period in the history of UF-resins. While the wood adhesive market picked up, the use of molding resin decreased. The period from 1935 to 1975 was characterized by a maturing of the resin market, with resin use shifting first to wood adhesives, and then increasingly from plywood to particleboard.

The current period is characterized by renewed interest in resin chemistry. This has become possible because of the revolution in analytical chemistry which has yielded many new tools which allow simultaneous *in situ* observation of several species in mixtures. The commercial interest in such work is based on the recognition that resins should be adjusted for modern applications which were not known during the 1930–1950 period, the golden years of resin chemistry. The current period has started to yield better resins with lower odor, higher weather resistance, and shorter curing period, combined with more reliable reaction control and quality control.

ISBN 0-201-04558-3

Table 2.2

Eleven Urea-Formaldehyde Condensation Products known 1936
(after Kadowaki, 1936)

No.	Name	Constitutional formula	Properties
I	Methylene-diurea	$H_2C<$ $NH\text{-}CO\text{-}NH_2$ $NH\text{-}CO\text{-}NH_2$	Needle-crystals m.p. 218°C. (effervesces)
II	Trimethylene-tetraurea	$H_2C<$ $NH\text{-}CO\text{-}NH_2$ $NH\text{-}CO\text{-}NH$ $>CH_2$ $H_2C<$ $NH\text{-}CO\text{-}NH$ $NH\text{-}CO\text{-}NH_2$	White powder, at 230°C. turns yellow.
III	Penta-methylene-hexaurea	$H_2C<$ $NH\text{-}CO\text{-}NH_2$ $NH\text{-}CO\text{-}NH$ $>CH_2$ $H_2C<$ $NH\text{-}CO\text{-}NH$ $NH\text{-}CO\text{-}NH$ $>CH_2$ $H_2C<$ $NH\text{-}CO\text{-}NH$ $NH\text{-}CO\text{-}NH_2$	White powder, at 236°C. turns yellow.
IV	Methylol-methylene-diurea	$H_2C<$ $NH\text{-}CO\text{-}NH_2$ $NH\text{-}CO\text{-}NH\text{-}CH_2OH$	White powder.
V	Methylenebis-methylolurea	$H_2C<$ $NH\text{-}CO\text{-}NH\text{-}CH_2OH$ $NH\text{-}CO\text{-}NH\text{-}CH_2OH$	White powder m.p. 228°C. (effervesces)
VI	Methylenebis-acetylurea	$H_2C<$ $NH\text{-}CO\text{-}NH\text{-}COCH_3$ $NH\text{-}CO\text{-}NH\text{-}COCH_3$	Needle-crystals m.p. 156°C.

ISBN 0-201-04558-3

Table 2.2 (Continued)

VII	Methylenebis-methylurea	$H_2C<$ $NH\text{-}CO\text{-}NH\text{-}CH_3$ / $NH\text{-}CO\text{-}NH\text{-}CH_3$	Prism-crystals m.p. 184°C.
VIII	Mono-methylolurea-methylether	$OC<$ $NH\text{-}CH_2\text{-}OCH_3$ / NH_2	Plate-crystals m.p. 91°C.
IX	Mono-methylolurea-ethylether	$OC<$ $NH\text{-}CH_2\text{-}OC_2H_5$ / NH_2	Needle-crystals m.p. 111°C.
X	Dimethylolurea-dimethylether	$OC<$ $NH\text{-}CH_2\text{-}OCH_3$ / $NH\text{-}CH_2\text{-}OCH_3$	Needle-crystals m.p. 101°C.
XI	Dimethylolurea-diethylether	$OC<$ $NH\text{-}CH_2\text{-}OC_2H_5$ / $NH\text{-}CH_2\text{-}OC_2H_5$	Plate-crystals m.p. 124°C.

ISBN 0-201-04558-3

3. Chemistry

This chapter deals with the basic chemistry of reagents and resins. The first part deals with formaldehyde; the second with urea and thiourea; and the third with the reactions between the two. In this part, a short section on the principle of resin modification is included. The topic of this chapter is so extensive that only some of the most important reactions could be covered. For more details, the reader is referred to basic chemistry texts, to handbooks, or to the original literature which is quoted in this chapter.

Formaldehyde:

Work before 1961 is described in the comprehensive book of Walker which has become an unrivaled classic in this field. Formaldehyde occurs in the atmosphere of many stars and in interstellar space. The troposphere contains 0.12 to 0.39 parts per billion (ppb) (Warneck 1978). In ambient air, the concentration of formaldehyde strongly depends on that of other species, because formaldehyde is highly reactive, and, thus, it is in equilibrium with a large number of precursers and derivatives. Ambient urban air can contain up to 0.12 parts per million (ppm) (Andersen 1979). Pure formaldehyde can exist only as a dilute vapor. At higher concentrations it polymerizes. With water it forms glycols; with methanol it forms hemiacetals. These reactions are reversible. It is common to use the term *formaldehyde* as a synonym for *latent* formaldehyde, regardless of whether the chemical species is an aldehyde, a glycol, a monomer, or a polymer. This custom is convenient for those who synthesize and use resins and for anyone interested in the total formaldehyde equivalent available for chemical bulk reactions. However, the global use of the word formaldehyde is not adequate, and it may be misleading for those who are dealing with reactivity, kinetics, and reaction mechanisms, for example toxicologists, meteorologists, and hygienists. The following sections first describe our current knowledge of the composition of these systems, and then their chemical reactivity.

Beat Meyer, Urea-Formaldehyde Resins

ISBN 0-201-04558-3

PROPERTIES AND EQUILIBRIA

Pure monomeric formaldehyde, CH_2O, is a colorless, pungent gas with a boiling point of −19°C, and a molecular weight of 30.03. It readily and reversibly converts into a mixture of any of several different derivatives. In pure form it readily polymerizes; in aqueous or alcoholic solutions it undergoes polymerization as well as solvolysis. Aqueous solutions contain less than 0.1% of formaldehyde in monomeric form. Table 3.1 lists the five most common forms.

Table 3.1

Species in Commercial Formaldehyde

Name	Formula
Formaldehyde	CH_2O
Methylene glycol	$CH_2(OH)_2$
Polyoxymethylene glycol	$H(CH_2O)_nOH \quad (2 < n < 100)$
Formaldehyde hemiformal	$H{-}O{-}CH_2{-}O{-}CH_3$
Polyoxymethylene hemiformal	$H{-}O{-}(CH_2{-}O)_m{-}OCH_3 \quad (2 < m < 80)$
Trioxane	$(CH_2O)_3$

Of these species, monomeric formaldehyde is best known, even though it is everywhere, except in the vapor, only a minor species in the equilibrium mixture. In contrast, methylene glycol is probably the most ubiquitous species, but it is least known. In the following, a short summary of some select pertinent data on the individual species is presented. It is followed by a discussion of the aqueous system, the most important one for the context of this book. For a fuller and more detailed review, the reader is referred to the comprehensive and excellent, though already quite dated, reviews of Walker (1966), Imoto (1965), and others.

Monomeric Formaldehyde

Monomeric formaldehyde melts at −118°C and boils at −19°C, but in the temperature range between −90°C and 100°C it spontaneously polymerizes, and thus, below 100°C it does not obey the ideal gas law. The monomer can be stabilized by quenching to −100°C. The gas readily dissolves in polar solvents, but immediately solvolyzes. Bercovici (1972) studied formaldehyde in several different solvents, using NMR, absorption, and emission spectroscopy and found that the products remain monomeric in nonpolar solvents below −90°C. Above −90°C it polymerizes in dimethyl sulfoxide, DMF, and THF. It can act as an hydrogen donor as well as acceptor. Above 400°C it decomposes rapidly into CO and H_2. The best vapor pressure data is still that published by R. Spence and W. Wild in 1935. They also

ISBN 0-201-04558-3

gave the best description on how to produce the monomer. The partial pressures listed in Table 3.2 are calculated from their data.

Table 3.2

Partial Pressure above Monomeric Formaldehyde (after Walker, 1964)

T (°C)	p (Torr)
−100	3.1
− 90	8.1
− 80	19.7
− 70	41.6
− 60	86.7
− 50	157.4
− 40	276.2
− 30	474.5
− 20	731.0

The vapor pressure is represented by the equation:

$$\log p = -1429/T + 1.75 \log T - 0.0063T + 3.0177. \qquad \text{Eq. 3.1}$$

The heat of vaporization is 5.570 kcal/mole. The Trouton constant is 21.9 entropy units.

Formaldehyde gas is flammable and forms explosive mixtures with air and oxygen. At room temperature the explosive range is 0.39 to 13 volumes of air per volume of formaldehyde, i.e., 7–70% by volume for formaldehyde.

The heat of formation,

$$\Delta H_1 = 27.7 \pm 1.5 \text{ kcal/mole at } 25°C, \qquad \text{Eq. 3.2}$$

$$C\text{ (graphite)} + H_2 + \tfrac{1}{2} O_2 \rightarrow CH_2O + 27.7 \text{ kcal} \qquad \text{Eq. 3.3}$$

The formaldehyde molecule has a C–O bond distance of 1.21 Å; the C–H distance is 1.09. The H–C–H bond angle is about 120°C. Several theoretical calculations have been conducted. Recent references can be found in the papers of Gordon (1978), Lehn (1974), and Tapia (1974). Spectral and molecular data has been summarized by Herzberg (1966). Formaldehyde vapor has a characteristic ultraviolet absorption which is well known, but complex and by no means well understood (Meyer 1969). The ultraviolet spectrum consists of some 40 bands between 250 and 550 nm which are not suitable for qualitative analysis. The dissociation energy of CH_2O is D_0 = 120 kcal. In contrast, the infrared and Raman spectra are

ISBN 0-201-04558-3

now well known and the assignment of all frequencies have been confirmed by isotope substitution, Table 3.3. The gas phase values reported by Chapput (1973) have been confirmed by Nixon (1973, 1974) in quickly quenched monomeric solid and in rare gas matrices, and by Meyer (1979). The spectrum is shown in Figure 3.1.

Table 3.3

Vibrational Modes of Formaldehyde (in cm^{-1})
(after Chapput 1973)

Frequency (cm^{-1})	Assignment	Symmetry	Species
1180	υ_4	b_1	γ CH_2
1257	υ_6	b_2	$\delta_\alpha CH_2$
1507	υ_3	a_1	δ_s CH_2
1748	υ_2	a_2	γ C–O
2780	υ_1	a_1	υ_s CH
2847	υ_5	b_2	υ_α CH

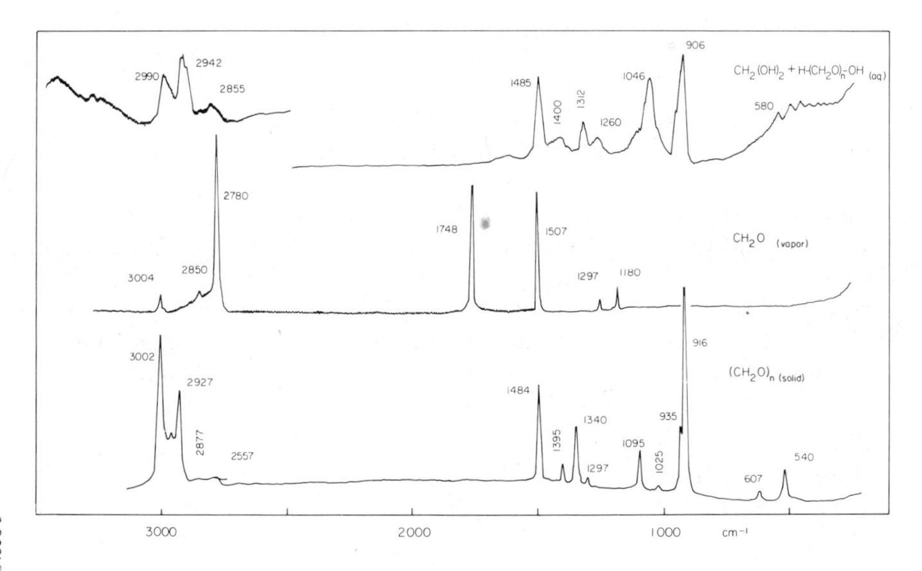

Figure 3.1. Raman Spectra of Formaldehyde Species. Top: Aqueous Formaldehyde; Middle: Formaldehyde Vapor; and Bottom: Para-formaldehyde (Meyer, 1978)

ISBN 0-201-04558-3

The chemical properties of monomeric formaldehyde are described below.

Methylene Glycol

Butlerov discovered formaldehyde in 1859 when he attempted to prepare methylene glycol, $CH_2(OH)_2$. The latter is formed whenever formaldehyde dissolves in water. Methylene glycol is the primary member of the homologous series of regular glycols, but it is also the first geminal glycol. Its identity was established by Schorr in 1929, who found that formaldehyde loses the characteristic gas phase ultraviolet absorption band of the carbonyl group in aqueous solution. Nielsen confirmed this interpretation in 1937 by Raman spectroscopy, but we know today that their spectrum is largely due to polyoxymethylene, and not due to monomers. The structure of methylene glycol has not been experimentally established, but *ab initio* calculations by Lehn (1974) indicate that the C–O bond distance is 1.41 Å, i.e., 0.03 Å shorter than in methanol, but 0.20 Å larger than in formaldehyde. The force constants for the CO motion were predicted to be 6.0, 5.8, and 0.3 mdyne/Å, respectively. Calculations based on the CNDO theory indicate that intramolecular hydrogen bonding is expected to be stronger than intermolecular forces (Tapia, 1974). The first acidity constant of methylene glycol is estimated to be pK 1 = 14 at 25°C and 13.5 at 50°C. The equilibrium between formaldehyde and methylene glycol,

$$CH_2O + H_2O \rightarrow CH_2(OH)_2 \qquad \text{Eq. 3.4}$$

has been studied by Bieber and Truempler (1942) who determined the equilibrium constant:

$$(CH_2O)\,(H_2O)/(CH_2(OH)_2) = K \qquad \text{Eq. 3.5}$$

At 0°C it is 10^{-4}, at 30°C it is 5.7×10^{-4} and at 60°C it is 2×10^{-3}. Iliceto obtained similar values. The conversion is quite quick. Rudnev (1977) established the constants for pure water and methanol listed in the next section.

The mole fraction of total formaldehyde present as methylene glycol and the solubility of methylene glycol strongly depend on the total concentration of formaldehyde and temperature. A 50 wt % formaldehyde solution contains 15.6% methylene glycol. At room temperature the solubility is 17%. In solutions containing methanol, the concentration of methylene glycol is not strongly altered.

Early studies have been reviewed by Walker (1964). In 1949 Iliceto discovered that the kinetics of acidic bisulfite reactions, described in the section on analysis,

ISBN 0-201-04558-3

can be used for the analysis of formaldehyde solutions, because the kinetics differ for each solute species. At pH 4.75 and 0°C, the reaction of methylene glycol is so much faster than the depolymerization of polymeric species that the monomer concentration can be determined at 5–10-minute intervals, and the values can be extrapolated to the "instant-of-dilution" time. In 5 wt % formaldehyde solution, 82% methylene glycol is formed. In a 30 wt % solution, 35% is methylene glycol, and in a 50 wt % solution, about 15–25% is glycol. The equilibrium is not sensitive to temperature. Recently, *in situ* NMR spectroscopy has made more accurate analysis possible. The 220 MH_2 NMR signal of methylene glycol is at 4.65 ppm (Dankelman 1976), the ^{13}C-NMR signal at 83.8 ppm (de Breet 1977). Table 3.4 shows the composition of aqueous solution as a function of total formaldehyde. Figure 3.2 shows the ratio of monomer to polymer.

Table 3.4

Vapor Pressure of Aqueous Formaldehyde Solutions (in Torr)
(after Hall and Piret 1949)

	25°C				110°C			
Total F wt %	$CH_2(OH)_2$ as % of F[a]	P_F	P_{H_2O}	P_{total}	MG%	P_F	P_{H_2O}	P_{total}
0			24	24			1075	1075
5	82	0.4	23.6	24	(99)			
10	66	0.7	23.3	24	95	77	928	1006
20	46	0.9	23.1	24	80	139	866	1004
30	36	1	23	24	69	183	819	1002
40	b				56	209	791	1000
50					43	240	730	970
60						250	680	930
70						255	585	840
80						260	435	695

[a] Methylene glycol as wt % of total dissolved formaldehyde.
[b] Solubility at 25°C is 37%.

ISBN 0-201-04558-3

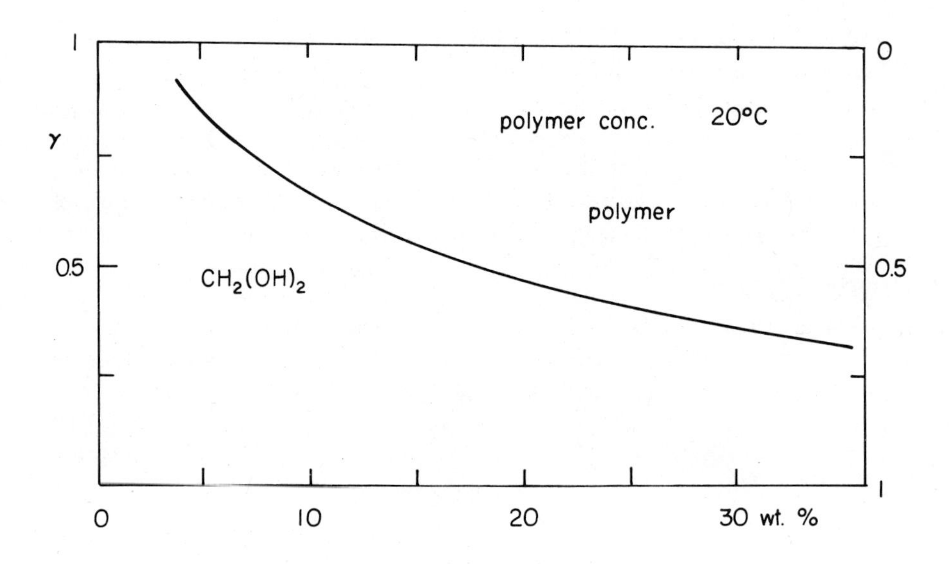

Figure 3.2. Molfraction of Methylene Glycol *vs.* Polyoxymethylene Glycol in Aqueous Solution at 20°C

Dimers, Trimers, and Tetramer Formaldehyde

Dimeric formaldehyde is probably present in equilibrium vapor between −19°C and 100°C, but its concentration is small and its properties have not yet been described. It is unstable and will polymerize or dissociate depending on conditions. Dioxane is not dimeric formaldehyde but has the formula, $C_4H_8O_2$, and is not a relative of formaldehyde, despite the similarity of its name with trioxane, described below.

Cyclic Compounds

Trioxane was discovered in 1885 by Pratesi when he heated formaldehyde in the presence of sulfuric acid. It has a pleasant, chloroform-like odor. Commercially, it is prepared by distillation of 60% formaldehyde solution containing 2% sulfuric acid. The yield is 60%. The product can be extracted with methylene chloride, or similar water-immiscible solvents and separated by fractional distillation.

Trioxane is a cyclic trimer with the formula $(CH_2O)_3$. It has a chair structure with oxygen and carbon atoms in different planes. The C–O distance is 1.41 to 1.43 Å. The valence angle is 110°. The dipole moment is 2.18. Trioxane is used as an intermediate in the manufacutre of commercial acetal resins, as fuel tablets, and in many other reactions where it substitutes for formaldehyde.

ISBN 0-201-04558-3

Pure trioxane melts at 64°C, boils at 115°C and forms an azeotrope with water containing 70 wt % trioxane which boils at 91.3°C. Its solubility in water is 17 wt % at 18°C and 21 wt % at 25°C. The vapor pressure of trioxane is listed in Table 3.5. The heat of vaporization is 9.8 kcal/mole.

Table 3.5

Vapor Pressure of Trioxane
(after Walker, 1964)

T (°C)	p (Torr)
25	12
35	27.5
50	100
75	225
90	330
100	505
115	760

Liquid trioxane is an excellent solvent for urea, and many aromatics. The reaction with the latter is so exothermic that a mixture of solid trioxane and solid phenol spontaneously melts, yielding a clear solution.

The heat of combustion of trioxane is 4 kcal/g at 23°C. This corresponds to 156.7 kcal/mole. The heat of formation is about 123 kcal/mole. The heat of conversion of monomeric formaldehyde to trioxane is about 45 kcal/mole, i.e., 15 kcal per formaldehyde unit.

Trioxane is stable up to 245°C. Above 200°C it slowly dissociates, especially in the presence of charcoal, silicon carbide, or potassium sulfate catalysts. In neutral or alkaline solutions it is stable and thus does not respond to the sulfite or peroxide analysis. In concentrated acids it depolymerizes. The hydrolysis constants are 1.5×10^{-7}/min at 15°C, 1.17×10^{-5}/min at 35°C, and 2.4×10^{-4}/min at 55°C. They are comparable to methylol, but about 10^6 slower than acetal and about 10^5 times faster than diethyl ether.

Tetraoxane was obtained by Staudinger (1931) by sublimation of high molecular weight diacetate. It melts at 112°C. This chemical has no practical importance in the context of this book.

Polyoxymethylene Glycols, Paraformaldehyde

Solid and liquid monomeric formaldehyde is unstable. At −90°C both spontane-

ISBN 0-201-04558-3

ously polymerize, forming solid polymeric oxymethylene chains $-(CH_2O)_n-$. These are hygroscopic and form glycol $H(CH_2O)_nOH$. The reaction is reversible. In contrast, the formation of polyhydroxy aldehydes yielding sugars is not spontaneous and not easily reversible.

The heat and entropy of polymerization of solid paraformaldehyde from gaseous monomer was determined from vapor pressure measurement. The best values appear to be $\Delta H = -17.2$ kcal/monomer unit, and $\Delta S = -43.8$ cal/deg·monomer unit (Melia 1967). The NMR spectra of small, soluble polymers with $2 < n < 8$ have been described in the section on aqueous systems. These species can be separated by quenching. Their melting point lies between 80 and 120°C. Staudinger prepared and studied these species, including octo-oxymethylene glycol and was the first to interpret their structure correctly (Staudinger 1936).

Paraformaldehyde comprises species with a polymerization range of 6–100 melting at 120–170°C. Higher molecular weight species with a polymerization range of 100–500 are obtained by reaction of paraformaldehyde. They are called alpha-polyoxymethylene. Beta-polyoxymethylene is obtained by polymerization in the presence of catalytic amounts of sulfuric acid.

The structure of paraformaldehyde, investigated by Staudinger (1944), consists of unbranched chains. According to Pierce (Walker 1964), the unit cell contains 30 formaldehyde units. The chain spiral makes 21 turns per unit cell. The C–O–C distance is 1.92 Å; and the chains are separated by 4.46 Å. Schneider and Bernstein (1956) and Philpotts (1955) and coworkers studied paraformaldehyde almost simultaneously and independently. Both groups used infrared spectroscopy and confirmed a spiral structure with a rotation angle of 120° rather than a zig-zag configuration. This group correctly identified a band at 920 cm^{-1} as C–O–C motion, and by comparison with propane acid CH_2F_2 assigned most frequencies. The O–H stretching bands at 3170 and 3370, and the bending modes at 1282 and 1040 cm^{-1} were assigned to terminal hydroxyl groups. A more recent Raman spectrum of paraformaldehyde with higher spectral resolution is shown in Figure 3.1, bottom. The C–O stretching at 1748, corresponding to the double bonded carbonyl in free formaldehyde, Figure 3.1, middle, is absent, and the C–H stretching and bending motions appear markedly shifted (Meyer 1979). Nixon (Khoshkhoo 1973, 1974) studied the spectra of solid and matrix isolated formaldehyde at −200°C and −250°C, using isotopic substitution to confirm assignments of all bands. They concluded that the unit cell contains at least four molecules.

The vapor pressure of solid formaldehyde is shown in Figure 3.3 together with that of solutions. The curve reflects the fact that paraformaldehyde must depolymerize before vaporization, i.e., that the vaporization is a complex process. This fact is possibly responsible for some of the spurious behavior of formaldehyde in air.

If paraformaldehyde is produced from methanol solutions, hemiacetals remain in the polymer and increase its solubility. Commercial processes for preparing paraformaldehyde almost always involve the evaporation of solutions (Walker 1964).

ISBN 0-201-04558-3

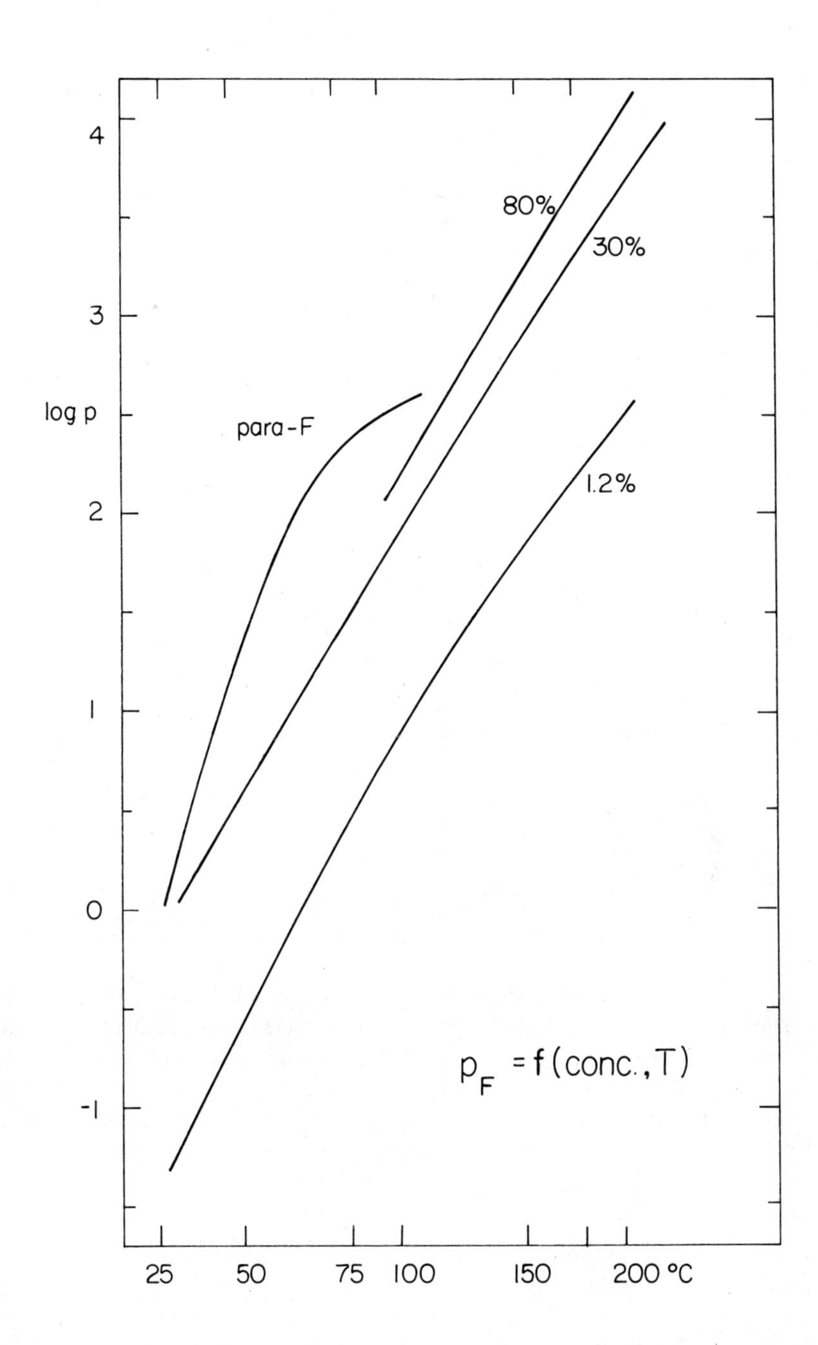

ISBN 0-201-04558-3

Figure 3.3. Vapor Pressure of a) Para-formaldehyde, b) 80% Aqueous Solution, c) 30% Aqueous Solution, and d) 1.2% Aqueous Solution of Formaldehyde between 25°C and 250°C (unpublished data, Meyer, 1979)

Aqueous Formaldehyde Solutions

This section consists of three parts. The first deals with general properties of solution; the second with the composition of dilute solutions and the third with the composition of concentrated and saturated solutions. Reagent grade and commercial formaldehyde solutions contain 5–15% methanol to increase solubility. This section deals first with pure, aqueous solutions which are obtained by dissolving solid paraformaldehyde or gaseous formaldehyde in water.

General Properties. Pure formaldehyde solutions are clear and colorless and exude a pungent odor. The heat of formation is ΔH_f = 43 kcal/mole; the value is practically independent of concentration over the measured range. The density and refraction index of formaldehyde solution is reviewed in Table 3.6.

Table 3.6

Density and Refractive Index of Aqueous Formaldehyde (after Walker, 1964)

Concentration wt %	Density at 18°C	Refractive Index η^{18}
water	0.9986	1.3330
5	1.0141	1.3388
10	1.0299	1.3445
20	1.0600	1.3559
30	1.0910	1.3676
40	1.1220	1.3795

The dielectric constant is a function of concentration and is influenced by methanol. The magnetic properties are not in agreement with calculated formaldehyde content; this facet has been used to confirm the existence of methylene glycol. The best value for the magnetic susceptibility appears to be -0.5×10^{-6} cgsm for concentrations between 3 and 30 wt %. The surface tension does not differ much from that of water. The viscosity of a 35 wt % solution is 1.87 centipoise at 25°C. It increases with increasing methanol concentration. It is 2.69 centipoises at 25°C and 12% methanol.

For all practical purposes, the boiling point of aqueous formaldehyde is independent of the formaldehyde concentration. The boiling point has a slight minimum at 99.10°C at a concentration of 25% . A 55% solution boils at 99.90°C.

In contrast, Table 3.7 shows that the freezing point is quite sensitive to concentration. Further more, the freezing point is dependent on the history of the solution, because the equilibria are only slowly established at low temperature.

ISBN 0-201-04558-3

Table 3.7

Freezing Point of Pure Formaldehyde Solutions (after Walker, 1964)

Concentration (wt %)	mp (°C)
10	- 6.30
15	- 9.50
20	-12.95
25	-16.45

Thus, it takes up to 50 hours to equilibrate the solution at room temperature. If a solid phase is involved, equilibrium takes several weeks, unless catalysts are used.

The solubility of formaldehyde depends on temperature and purity. Thus, in a pure aqueous solution formaldehyde is less soluble than in the presence of methanol. Some approximate solubilities are listed in Table 3.8. The solubility is also dependent on pH.

Table 3.8

Solubility of Formaldehyde in Aqueous Solution at 8 Temperatures (after Walker 1964)

T(°C)	Concentration	
	p (gF/100 g)	(gF/100 ml)
0	28.6	31.25
20	36.2	40.2
35	35.9	39.5
45	35.5	39.2
60	58	62
70	82	86
80	~100	~104
90	~115	~120

Since formaldehyde readily equilibrates between vapor and aqueous solution, its behavior during phase changes is important. Figure 3.3 shows the vapor pressure above solid paraformaldehyde. The distribution between a boiling aqueous solution and the vapor phase is shown in Table 3.9.

Table 3.9, also reviews total vapor pressure and approximate partial pressures of formaldehyde and steam as a function of temperature and concentration. The same

ISBN 0-201-04558-3

values are plotted Figure 3.3. The vapor pressure is approximately

$$\log p_{F(\mathrm{Torr})} = \mathrm{alpha} - 2905\mathrm{T} \qquad \text{Eq. 3.6}$$

where alpha is a function of concentration and is 9.48 for a 10 wt % solution; 9.71 at 20%, 9.83 at 30%, and 9.89 at 40%. The concentration dependence of alpha is due to the complex equilibrium in solution as discussed below.

Table 3.9

Vapor Pressure of Saturated Vapor in Equilibrium with boiling Solutions at 760 Torr (after Walker, 1964)

		FORMALDEHYDE CONCENTRATION IN PERCENT BY WEIGHT								
°C Temp.	P_{H_2O}	10%	20%	30%	40%	50%	60%	70%	80%	
20	18	0.4 17.6 18.0	0.7 17.3 18.0	0.9 17.1 18.0	1 17 18	1 17 18				
25	24	1 23 24	1 23 24	1 23 24	1 23 24	1 23 24				
30	32	1 31 32	1 31 32	2 30 32	2 30 32	2 29 31				
35	42	1 41 42	2 40 42	3 39 42	3 37 40	3 36 39	4 34 38	4 33 37		
40	55	2 53 55	3 52 55	4 50 54	4 49 53	5 47 52	5 45 50	6 42 48		
45	72	2 70 72	4 67 71	5 65 70	6 62 68	7 59 66	8 56 64	8 54 62		
50	93	3 90 93	6 86 92	7 83 90	8 80 88	9 76 85	11 71 82	11 67 78		
55	118	4 112 116	7 108 115	10 103 113	11 100 111	12 96 108	14 90 104	15 84 99		
60	149	6 141 147	10 135 145	13 130 143	15 125 140	17 119 136	19 112 131	21 105 126		
65	188	8 175 183	14 167 181	17 161 178	20 154 174	23 147 170	26 137 163	27 128 155		

HEAVY LINE MARKS REGION IN WHICH SOLUTIONS PRECIPITATE POLYMER.

ISBN 0-201-04558-3

Table 3.9 (Continued)

FORMALDEHYDE CONCENTRATION IN PERCENT BY WEIGHT										
°C Temp.	P_{H_2O}	10%	20%	30%	40%	50%	60%	70%	80%	
70	234	10 222 232	18 212 230	24 201 225	27 193 220	32 183 215	34 166 200	37 153 190		
75	289	13 272 285	24 259 283	33 247 280	36 239 275	40 230 270	46 214 260	48 202 250		
80	355	18 337 355	33 307 340	43 292 335	51 279 330	58 262 320	62 253 315	63 237 300		
85	434	24 406 430	43 377 420	57 353 410	61 339 400	70 320 390	78 302 380	82 288 370		
90	526	30 480 510	52 453 505	68 432 500	80 410 490	92 388 480	95 365 460	103 337 440		
95	634	38 577 615	66 544 610	90 510 600	102 483 585	120 445 565	130 410 540	133 367 500	133 277 410	
100	760	49 700 749	87 653 740	113 607 720	130 585 715	150 540 690	165 505 670	168 432 600	168 337 505	
105	906	62 838 900	108 767 875	150 700 850	163 667 830	195 610 805	210 560 770	210 510 720	210 390 600	
110	1075	77 928 1005	138 866 1004	183 819 1002	209 791 1000	240 730 970	250 680 930	255 585 840	260 435 695	

All figures in Torr; Top Figure = P_{CH_2O}; Second Figure = P_{H_2O}; Bottom Figure = Total Vapor Pressure.

The behavior during steam distillation is similar to that under normal distillation. However, in the pressure distillation of dilute solutions, formaldehyde is increasingly enriched with increasing pressure. Thus, a 2% solution yields at 1 atm at 100°C a 3.0% solution, at 20 atm at 125°C a 4.6% solution, at 60 atm at 153°C a 6.2% solution and at 100 atm at 170°C a 7.6% solution. The vapor pressure of methanol containing solutions is of practical importance. Typical values are summarized in Table 3.10. The distribution curves, Figure 3.4 and Table 3.9 show that at high concentration, formaldehyde enriches in solution during boiling; in dilute solutions, the reverse is true. This is due to the composition of the solutions, described below and is significant in connection with boiling of UF-resins in hot presses. The vaporization process will be discussed in Chapter 5.

ISBN 0-201-04558-3

Table 3.10

Liquid and Vapor Composition of Formaldehyde-Methanol-Water System at 760 Torr for 6 Formaldehyde Concentrations

Liquid Composition (Wt. %)			Vapor Composition (Wt. %)			Boiling Point (°C)
CH_2O	CH_3OH	H_2O	CH_2O	CH_3OH	H_2O	
5.0	10.0	85.0	4.0	38.5	57.5	92.6
	20.0	75.0	3.0	57.6	39.4	87.4
	40.0	55.0	1.9	75.5	22.6	79.1
	60.0	35.0	1.6	85.4	13.0	74.4
	90.0	5.0	1.3	96.3	2.4	67.1
10.0	10.0	80.0	7.5	34.5	58.0	93.8
	20.0	70.0	5.9	53.5	40.6	88.8
	40.0	50.0	3.9	72.8	23.3	80.8
	60.0	30.0	3.3	84.7	12.0	74.8
	80.0	10.0	2.9	92.6	4.5	69.2
20.0	10.0	70.0	15.8	32.2	52.0	95.2
	20.0	60.0	13.2	45.2	41.6	91.5
	40.0	40.0	9.0	66.9	24.1	83.5
	60.0	20.0	7.6	83.0	9.4	76.4
30.0	10.0	60.0	22.1	23.6	54.3	96.0
	20.0	50.0	19.9	38.5	41.6	93.2
	40.0	30.0	15.7	59.7	24.6	86.1
40.0	10.0	50.0	27.8	22.4	49.8	97.1
	20.0	40.0	26.0	32.9	41.1	94.2
50.0	5.0	45.0	35.6	13.3	51.1	97.6

Acidity. The pH of aqueous formaldehyde solutions lies in the range 2.5 to 4.4. The acidity is due to formic acid which is formed in pure solution as a consequence of the Cannizzaro reaction (Geissman, 1944) which consists of the spontaneous auto-redox decomposition of two formaldehyde molecules in the presence of water:

$$2CH_2O + H_2O \rightarrow HCOOH + CH_3OH \qquad \text{Eq. 3.7}$$

The first dissociation constant of formaldehyde is $k_1 = 10^{-14}$ at 0°C and 3.3×10^{-13} at 50°C.

ISBN 0-201-04558-3

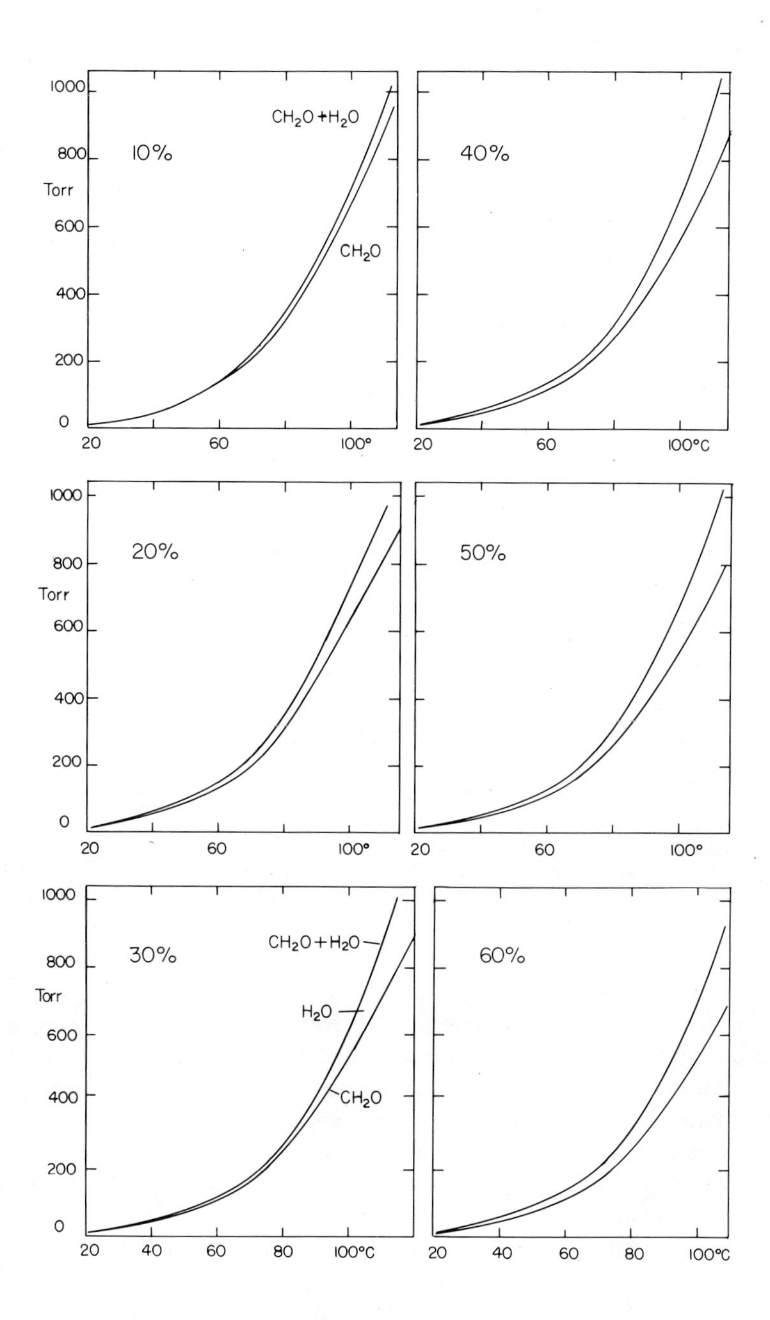

Figure 3.4. Partial Pressure of Steam and Formaldehyde as a Function of Concentration and Temperature (after Walker, 1964)

ISBN 0-201-04558-3

Composition of Dilute Solutions. The predominant species in solutions containing up to 10 wt % of formaldehyde is methylene glycol, $CH_2(OH)_2$. Figure 3.2 shows the ratio of monomer to the sum of all polymers as a function of concentration at 20°C. At the boiling point the concentration of monomer is only about 10% higher. The concentration of free formaldehyde in solution varies from $k = 10^{-4}$ at 0°C to 3×10^{-3} to 65°C (Truempler, 1948).

$$k = \frac{CH_2O_{aq} \cdot H_2O_{liq}}{CH_2(OH)_{2\,aq}} \quad \text{Eq. 3.8}$$

With increasing concentration methylene glycol dimerizes forming dioxymethylene glycol:

$$CH_2(OH)_2 \rightarrow HO(CH_2O)_2H + H_2O \quad \text{Eq. 3.9}$$

the latter can add further monomers and yield:

$$HO(CH_2O)_2H + nCH_2(OH)_2 \rightarrow HO(CH_2O)_{n+2}H + nH_2O \quad \text{Eq. 3.10}$$

Compounds of the type $HO(CH_2O)_nH$ are linear chains and are called polyoxymethylene glycol or paraformaldehyde. The equilibrium 10 shifts to the right if the concentration is increased. Thus concentrated solutions constitute complex multicomponent systems.

Concentrated Formaldehyde Solutions. Concentrated formaldehyde solutions contain a mixture of methylene glycol and polyoxymethylene. The concentration of the various oligomers is not yet accurately known. It changes with total formaldehyde concentration, temperature, and as a function of additives. As Figure 3.2 shows, the concentration of methylene glycol decreases with increasing concentration. In early studies investigators tried to analyze the composition of solutions from molecular weight measurements. At 20°C the average molecular weight increases from 31.9 for a 5 wt % solution to 54.2 for a 35% solution. Later studies using ultrasonic sound gave similar data. Dankelman (1976) used a combination of gas chromatography and nuclear magnetic resonance to measure the concentration of species up to heptaoxymethylene glycol. For this he reacted the solutions with N, O-bis-(trimethylsilyl) trifluoroacetamide (BSTFA). This reaction does not disturb the equilibrium significantly. A typical equilibrium composition for 50% solution at 65°C is listed in Table 3.11. Old values are available for a large range of concentrations from 5–50 wt % (Walker 1964). Brandeni (1976) computed the thermodynamics of these equilibria by regressive nonlinear analysis. Dankleman's NMR values yield somewhat higher concentration for the medium-sized polymers than earlier studies. Figure 3.5 shows the resolution obtained by gas liquid chromatography using a mass spectrometer as the detector (Dankelman 1976). The proton

ISBN 0-201-04558-3

Table 3.11

Distribution of Polyoxymethylene Glycols in 50% Formaldehyde Solution at 65°C (after Dankelman, 1976)

	$HO(CH_2O)_\eta H$							
η =	1	2	3	4	5	6	7	8
Relative % (m/m)	14.1	30.1	26.0	15.4	9.0	4.1	1.2	0.1

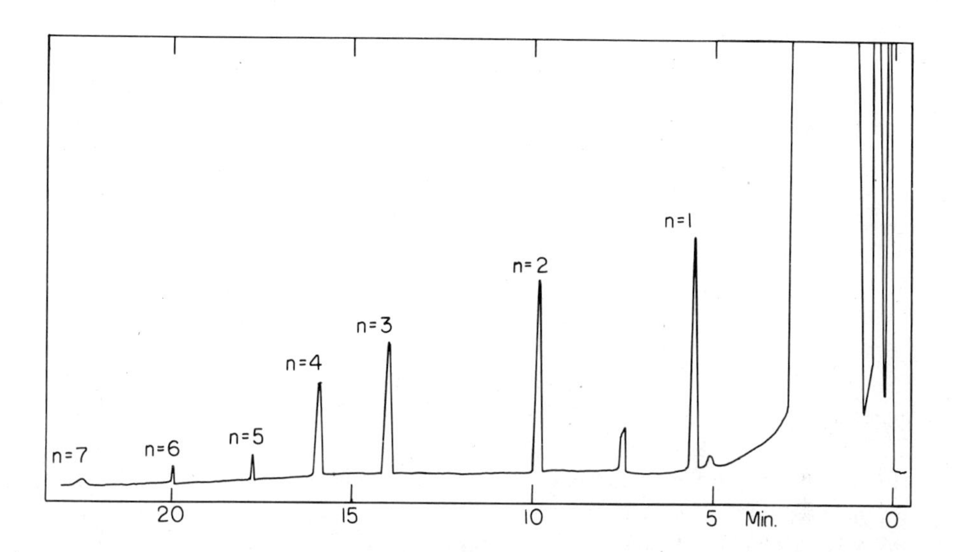

Figure 3.5. Gas Chromatogram of Silylated Polyoxymethylene Glycols in 50% Formaldehyde Solution (after Dankelman, 1977)

NMR chemical shifts for the methylene group are 4.65 ppm for the monomer, 4.72 ppm for the dimer, 4.71 ppm and 4.78 ppm for the trimer. Higher oligomers are not well resolved (Dankelman 1976). The $-CH_2-OH$ hydroxyl group lies at 5.2–5.3 ppm (Chiavarini 1975). The shifts are temperature dependent (Tomita 1976). The problems of overlap with water and oligomers is greatly reduced by ^{13}C-NMR analysis. The corresponding shifts are listed in Table 3.12. More accurate equilibrium data can be expected to result from this method. Another powerful *in situ* method is Raman laser spectroscopy. Figure 3.1 shows the Raman spectrum of formaldehyde vapor, paraformaldehyde and aqueous formaldehyde (Meyer 1979). The assignment of formaldehyde and paraformaldehyde peaks has been discussed in the corresponding sections earlier in this chapter. Figure 3.6 shows the

ISBN 0-201-04558-3

Table 3.12

^{13}C-NMR Shifts of Glycols and Hemiacetals in 37% Formaldehyde Solutions (after Dankelman 1976)

Degree of Polymerization *n*	NMR Shifts (ppm) Terminal Carbon	Second Carbon	Central Carbon
1	83.8		
2	86.9		
3	87.5	90.0	
4	87.7	90.4	90.8
5		90.5	91.0
6,7	90.8	91.0	91.1

Number of Monomer Units	Shift (ppm) Glycol	Hemiacetal
1	83.8	91.6
2	87.0	94.5
3	87.5	95.2

Raman spectrum of formalic solutions as a function of concentration. The frequencies in the 2990, 2940, and 2800 cm^{-1} range are due to C–H stretches. The C–H bending motion falls into the 1400 cm^{-1} range. The O–H stretch of $H(CH_2O)_nOH$ is in the 1000 cm^{-1} range. The strong feature at 900 cm^{-1} is due to the $-CH_2-O-CH_2$ frequency. Figure 3.6 shows that the relative intensity of the 920 to 1050 cm^{-1} region changes drastically upon dilution. In a concentrated room temperature solution containing 37 wt % total formaldehyde the 920 cm^{-1} feature is by far the strongest. In a 1 wt % solution the peak is a mere fraction of the C–OH band. Direct comparison of the intensity of these two peaks yields equilibrium compositions very similar to those obtained with other techniques (Meyer 1979).

Equilibrium Kinetics. While the kinetics of the gas solution equilibrium is reasonably fast (Rudnev 1977), as shown in the following section, the depolymerization of polyoxymethylene glycol and the dissolution of solid glycols can be very slow. If a 36% formaldehyde solution at 0°C is diluted to 3%, it takes more than three days to reach equilibrium. The reaction rate decreases by a factor of three for each temperature drop of 10°C. The equilibrium is strongly catalyzed by acids and bases. Figure 3.7 shows that the rate is lowest at pH 4 and increases by a factor of more than a hundred at pH 2 and pH 7. However, as described above, the pH also influences the total solubility. Figure 3.8 shows that it takes five hours at room temperature and pH 4.6 to prepare a 2 wt % solution. At pH one or 7 it takes only

ISBN 0-201-04558-3

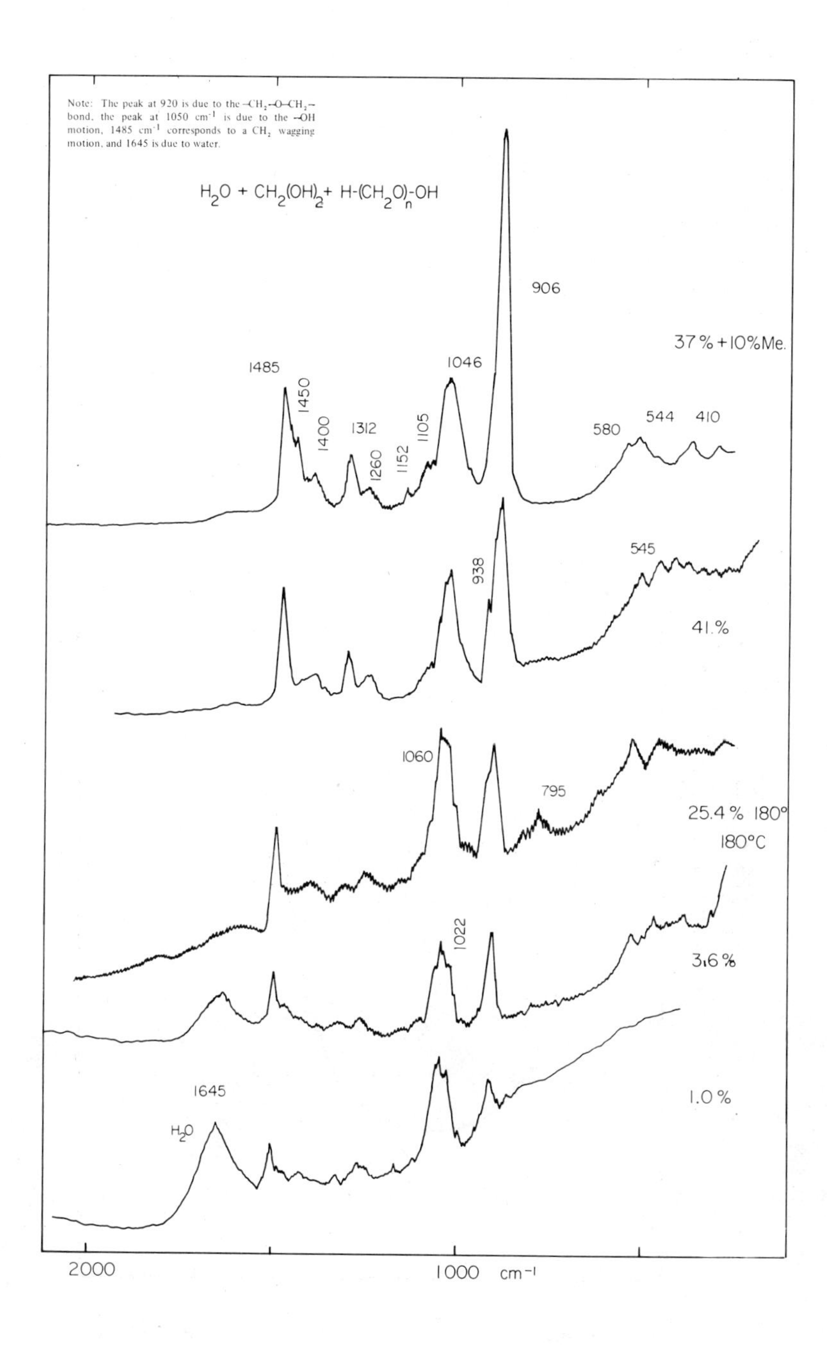

Figure 3.6. Raman Spectrum of Aqueous Formaldehyde at Five Concentrations (Meyer, unpublished work, 1979)

ISBN 0-201-04558-3

a few minutes, and after five hours, the concentration can reach 60% (Loebering 1936). The influence of the pH is due to ionic species. In acid medium the carbenium ion, $CH_2{}^+OH$ is conceivably involved as an intermediate, but not in significant concentrations. In neutral and alkaline solution the oxyion $CH_2(OH)O^-$ is formed. Since the acid dissociation constant of methylene glycol is $k = 10^{-14}$, the ratio of diprotic to monoprotic glycol is 10^{-7} at pH 7, and 10^{-4} at pH 10 and increases to 1 at pH 14. Thus, at pH 14, half of all glycol is dissociated, and the total solubility is roughly doubled. However, at high pH auto-redox decomposition by the Cannizzaro reaction becomes prominent.

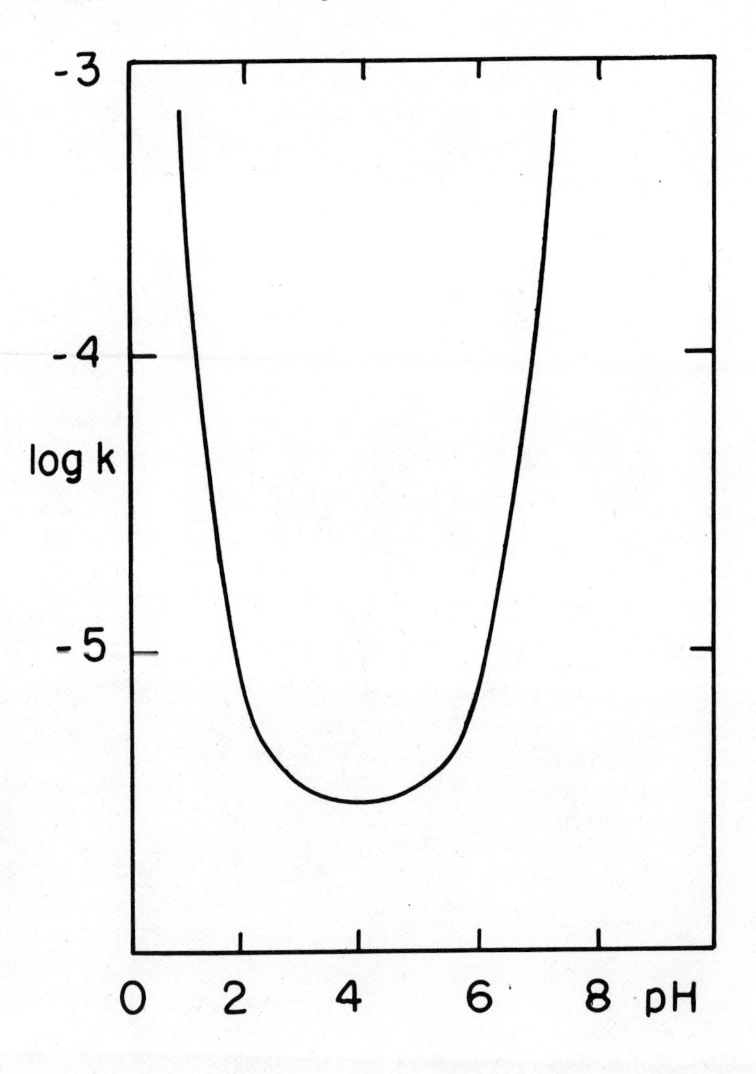

Figure 3.7. Solubility of Formaldehyde as a Function of pH (after Lobering, 1936)

ISBN 0-201-04558-3

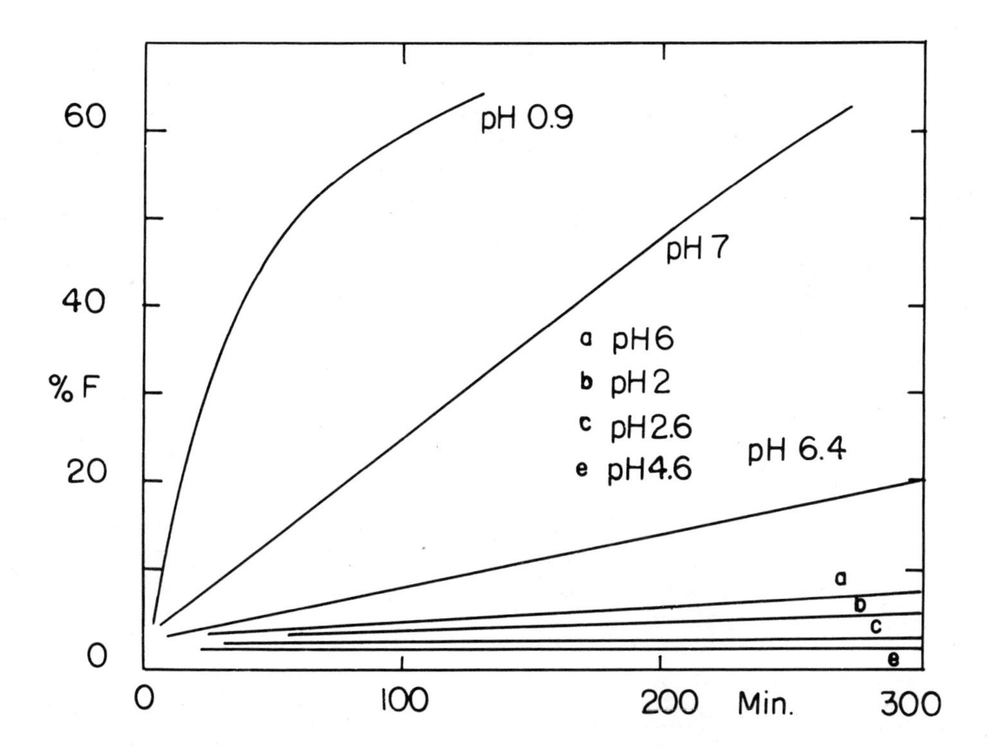

Figure 3.8. Kinetics of Formaldehyde Solvation as a Function of pH (after Lobering, 1936)

The kinetic mechanism can be complex and can lead to a metastable system. For example, in a quickly quenched solution the higher polymers precipitate first, and the short and intermediate chains equilibrate very slowly, yielding liquids with lower than equilibrium average molecular weight. In contrast, upon heating, solid polymers polymerize despite contact with solution. This decreases their solubility. Thus, heating does not clarify solutions as quickly as expected. The poor solubility and the poor kinetics are two reasons why reagent and commercial formaldehyde solutions contain stabilizers.

Commercial Formaldehyde Solution

USP formaldehyde, reagent grade formaldehyde, or commercial formaldehyde all contain about 30–40 wt % total formaldehyde and enough methanol, usually 5–15%, to prevent precipitation. The action of methanol is chemical, not physical. Dankelman (1976) showed with the help of proton NMR spectroscopy that methanol reacts quantitatively, forming the hemiacetal:

$$HO(CH_2)_nH + CH_3OH \rightarrow HO(CH_2O)_{n+1}H + CH_3OCH_2OH \qquad \text{Eq. 3.11}$$

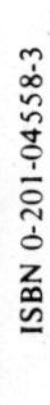
ISBN 0-201-04558-3

These hemiacetals are soluble in water, and thus methanol increases the solubility of glycols. The formation of hemiacetals depletes low molecular weight polyoxymethylene glycol. This causes deploymerization and thus shifts the equilibria towards lower molecular weights. Typically, in a 37 wt % solutions at room temperature, about 40% of the methanol is converted to monomeric hemiacetal, 30% to dimeric, and the rest to polymeric hemiacetals. Less than 1% of the methanol remains unreacted. Thus, the solution contains only some 0.1% free methanol (Dankelman 1976). If 6% more methanol is added to the stabilized solution, most of it reacts to yield to former monomeric hemiacetal, some yields dimer and trimer, and only a trace remains unreacted. The equilibrated solution contains less than 0.2% free methanol. The composition of these solutions can now be studied with the help of ^{13}C-NMR. The glycols and hemiacetals exhibit distinctly different shifts, Table 3.12. The observed shifts correspond closely to those predicted with the help of additivity rules and comparison with other functional groups (Dankelman 1976).

Due to the formation of hemiacetals, the kinetics of solvation of formaldehyde in solution are improved in the presence of methanol. The equilibrium constants for both pure aqueous and methanolic solutions between 70° and 130°C are shown in Figure 3.9 (Rudnev 1977).

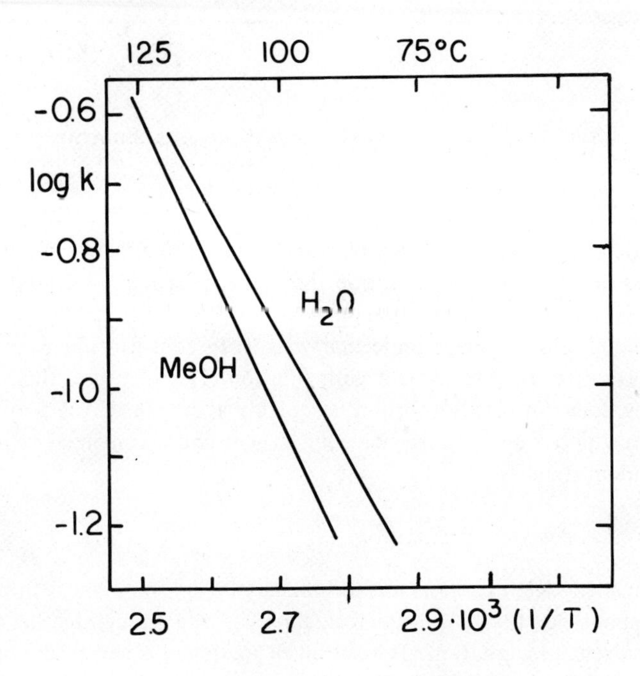

Figure 3.9. Vaporization Rates of Formaldehyde from Water and Methanolic Solutions between 50°C and 150°C (after Rudnev, 1977)

ISBN 0-201-04558-3

Figure 3.10 shows the Raman spectrum of an 80% methanolic formaldehyde solution. The methanol frequencies are C–O = 1033 cm^{-1}; C–H_{sym} = 2837 cm^{-1}; C–H_{asym} = 2942 cm^{-1}; and O–H = 3683 cm^{-1}. The band at 920 cm^{-1} is substantially changed. It appears shifted to 910 cm^{-1}. This frequency belongs to the methylol C–O–C motion (Meyer 1979). The band at 1030 cm^{-1} is due to the O–H of both methanol and the various glycols and their derivatives.

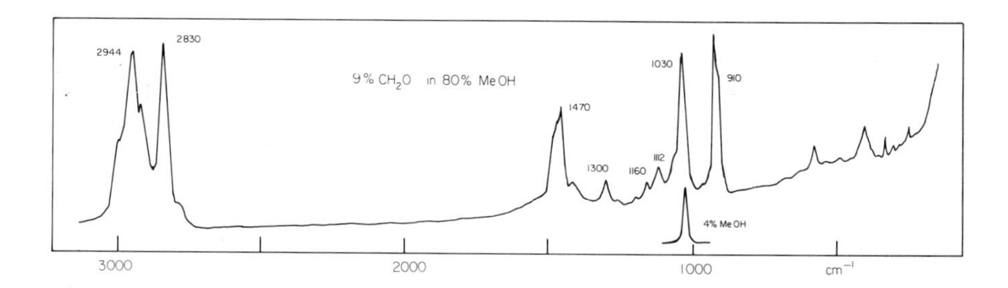

Figure 3.10. Raman Spectrum of Para-formaldehyde in 80% Methanol at 25°C (Meyer, 1979)

The physical properties of methanolic formaldehyde solution have been described by Walker (1964). USP formaldehyde contains 37.0–37.5 wt % formaldehyde, 6.0–15.0 wt % methanol, 0.01–0.03 wt % formic acid, less than 0.8 wt % iron, and it is colorless.

Commercial solutions have similar specifications. Their boiling point is 210°F; their density is 9.1–9.2 lbs/gal (1.09–1.15 g/ml); the flash point is 150–180°F; and their specific heat is 0.6–0.8. Accurate specific gravity tables for commercial solutions are presented by Walker (1964) as a function of formaldehyde and methanol concentration. The formaldehyde vapor pressure is increased by methanol. This effect is greatest at high temperature.

During storage, formaldehyde solutions age due to polymerization, the Cannizzaro reaction (see below) and condensation. To prevent precipitation, formaldehyde solutions at all times must be kept above the recommended storage temperature, usually 7–21°C (40–70°F). If solid precipitates, a density gradient establishes in the solution and the upper solution layers will be, and will remain, irreversibly under-strength.

REACTIONS OF FORMALDEHYDE

As this book is not intended as a review of formaldehyde reactions, this chapter deals only with some of the most important reactions necessary for understanding UF chemistry, and with the problems of formaldehyde abatement processes. The

ISBN 0-201-04558-3

chemistry of UF with urea is the topic of the last section of this chapter.

As stated above, monomeric formaldehyde is only the predominant species in air. In aqueous or methanolic solutions and in solid paraformaldehyde, the predominant species is a derivative and monomeric formaldehyde is only a precursor. Thus, chemical reactions nominally attributed to it often reflect the chemical properties of other reagents, such as glycols or polymeric glycols.

Atmospheric Reactions

Pure air contains between 0.12 and 0.39 ppb formaldehyde (Warneck 1978). It is believed that the main source is the oxidation of methane. Formaldehyde photolyzes at an average tropospheric rate of 1.8×10^{-5}/sec into carbon monoxide and hydrogen. Some reacts with sulfur dioxide and transfers via rain into oceans and onto the earth. Formaldehyde from mobile and stationary combustion sources in urban or industrial areas can increase local concentrations to 0.12 ppm, but such anthropogenic formaldehyde can decay rapidly, because it can undergo photochemical reactions with other pollutants or with oxygen, especially singlet oxygen.

Formaldehyde decomposes above 400°C into carbon monoxide and hydrogen. The reaction is catalyzed by many common catalysts (Walker 1964). Decomposition in the presence of ultraviolet light yields formyl radicals, HCCO. If water is also present, sugars can be synthesized.

If oxygen or ozone is present, the performyl radical can be formed. Several reactions:

$$CH_2O + O_3 \rightarrow CH_2O_2 + O_2 \qquad \text{Eq. 3.12}$$

$$CHO + O_3 \rightarrow CHO_2 + O_2 \qquad \text{Eq. 3.13}$$

$$CH_2O + O_2 \rightarrow CO + H_2O \qquad \text{Eq. 3.14}$$

$$CH_2O + O_2 \rightarrow CO_2 + H_2O \qquad \text{Eq. 3.15}$$

can proceed parallel, all competing at different reaction rates. In solution formaldehyde can be destroyed with hydrogen peroxide:

$$CH_2O + H_2O_2 \rightarrow CH_2OOH + H_2 \qquad \text{Eq. 3.16}$$

An intermediate in this reaction, dimethylol peroxide, $HOCH_2OOCH_2OH$, has been isolated. It melts at 62°C and explodes at 70°C.

$$CH_2O + H_2O_2 \rightarrow CO_2 + H_2O \qquad \text{Eq. 3.17}$$

ISBN 0-201-04558-3

The reduction of formaldehyde to methanol is not efficient unless excess hydrogen and special catalysts are used.

The Cannizzaro Reaction

In alkaline solutions, and in a highly acidic medium, formaldehyde can undergo spontaneous autoredox reactions:

$$2CH_2O + H_2O \rightarrow CH_3OH + CH_2OOH \qquad \text{Eq. 3.18}$$

The reaction in alkali involves very likely the nucleophilic attack of the methylene glycol ion, $HO{-}CH_2{-}O^-$ upon the carbon of another formaldehyde molecule, followed by intramolecular dismutation of the intermediate. The reaction of formaldehyde with other aldehydes can lead to crossed Cannizzaro reaction (Geissman 1944). At temperatures below 60°C the Cannizzaro is the predominant decomposition reaction of alkaline solutions. Above 70°C aldol-condensation, leading to sugars, become important.

The methanol reacts further with residual formaldehyde, yielding hemiacetal or methylal:

$$CH_2O + CH_3OH \rightarrow HOCH_2{-}O{-}CH_3 + H_2O \qquad \text{Eq. 3.19}$$

$$HO{-}CH_2{-}OCH_3 + CH_3OH \rightarrow CH_3{-}O{-}CH_2{-}O{-}CH_3 \qquad \text{Eq. 3.20}$$

The Tischenko Reaction

Paraformaldehyde, alone, and in the presence of sulfuric acid can lead to methyl formate:

$$2CH_2O \rightarrow HCOOCH_3 \qquad \text{Eq. 3.21}$$

Aldol Condensations

In hot, alkaline solution formaldehyde can form sugars:

$$2CH_2O \rightarrow CH_2(OH) \cdot CHO \qquad \text{Eq. 3.22}$$

Among the reaction products are glyceraldehyde, $CH_2(OH)CH(OH) \cdot CHO$; dihydroxyacetone, $CH_2(OH)COCH_2OH$; and erythrose, $CH_2(OH)CH(OH)CH(OH)CHO$. Butlerov had already obtained brown tarry products in 1861. In 1886 Loew produced formose (hexose) by the reaction of a hot 4% formaldehyde solution with excess calcium hydroxyde. The reaction has a substantial induction period. The

ISBN 0-201-04558-3

reaction can be controlled at will to yield either 2, 3, 4, 5, or 6 carbon hydroxy-aldehydes. On ion-exchange resins, such as phenacyl polyaminostyrene, an approximately equimolar yield of glycol, glycerol, and erythritol can be obtained.

Reduction Reactions

By oxidation to formic acid, formaldehyde can reduce all reagents with half potential below 1.14 V:

$$CH_2O + 3OH^- \rightarrow HCO_2^- + H_2O + 2e + 1.14\ V \qquad \text{Eq. 3.23}$$

$$CH_2O + H_2O \rightarrow HCOOH + 2H_3O + 2e + 0.01\ V \qquad \text{Eq. 3.24}$$

With ammonium chloride, one can obtain trimethyl amine hydrochloride:

$$6CH_2O + NH_4Cl \rightarrow (CH_3)_3N + 3CHOOH \qquad \text{Eq. 3.25}$$

Alkaline silver reduction is an example in point. In organic cross-Cannizzaro reactions formaldehyde is almost always oxidized.

Condensations and Addition Reactions:

In aqueous systems formaldehyde is quantitatively converted to the methylol, i.e., methylene glycol. This reaction is reversible. Urea and phenol undergo analogeous reactions.

$$CH_2O + H_2O \rightarrow CH_2(OH)_2 \qquad \text{Eq. 3.26}$$

$$CH_2O + NH_2{-}CO{-}NH_2 \rightarrow NH_2CONH \cdot CH_2OH \qquad \text{Eq. 3.27}$$

$$CH_2O + C_6H_5OH \rightarrow HO{-}C_6H_4 \cdot CH_2OH \qquad \text{Eq. 3.28}$$

These reactions are the first step, and thus of basic importance in the preparation of resins. The reactions are very sensitive to pH. Normally, the reaction is conducted in neutral or slightly alkaline medium. In acidic solution methylene derivatives are produced, either directly, or via the methylol intermediate. These reactions will be discussed below.

Reaction with Ammonia

Butlerov observed in 1860 that formaldehyde reacts quantitatively to form hexamethylene tretramine, $(CH_2)_6N_4$ or 1,3,5,7 tetra-aza-tricyclo-(3.311)-decane. Other names are urotropine, aminoform, formin, methenamine, hexamine, etc. Delepine studied this reaction extensively in 1898. He also studied the reaction

ISBN 0-201-04558-3

involving ammonium sulfide described in the section below dealing with sulfides. This chemical is used as a special anhydrous form of formaldehyde. Thus, in the final hardening of resins, it can be used to prevent cracking caused by water release during condensation. The structure is shown in Figure 3.11. The structural data is as follows: The C–N distance is 1.48 Å, the C–N–C angle is 109.5°, and the N–C–N angle is also 109.5°. The C–H distance is 1.09 Å.

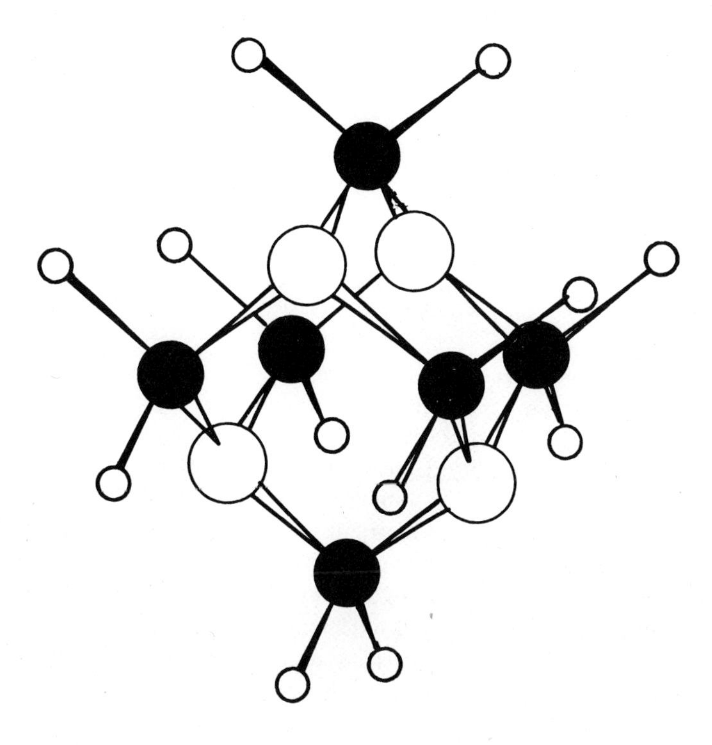

Figure 3.11. The Structure of Hexamethylene Tetramine (after Dickinson and Redmond, 1943)

The heat of formation is $\Delta H = -81$ kcal/mole.

$$6CH_2(OH)_{2\,(aq)} + 4NH_{3\,(g)} \rightarrow C_6H_{12}N_{4\,(aq)} + 6H_2O + 81 \text{ kcal/m} \qquad \text{Eq. 3.29}$$

Cyclic trimethylene-triamine might occur as an intermediate in this reaction, but the end product is always hexamethylene tetramine. With ammonium carbonate,

ISBN 0-201-04558-3

carbon dioxide can be recovered.

$$6CH_2(OH)_2 + 4NH_4Cl \rightarrow (CH_2)_6N_4 + 4H_3O^+ + 2H_2O + 4Cl^- \qquad \text{Eq. 3.30}$$

The acid produced in the reaction of formaldehyde with excess ammonium chloride can be titrated to determine the exact amount of formaldehyde present. With excess formaldehyde a by-product is obtained which does not convert to hexamethylenetetramine, possibly methylamine, or symmetrical trimethylcyclotrimethylene triamine. These reactions will be discussed in the next chapter.

Hexamethylene tetramine forms colorless and odorless orthorhombic crystals which have a sweet taste. It can be sublimed *in vacuo* at 230°C without decomposition. Below 13.2°C a hexahydrate can exist.

The heat of formation is $\Delta H_{(25)}$ = 28.8 kcal/m; ΔF = 102.7 kcal, ΔS = 32 cal/deg·mol and the specific heat, C_p is 36.4 cal/deg·mol. The vapor pressure follows the formula:

$$\log p = -3937/T + 10.01 \qquad \text{Eq. 3.31}$$

The heat of combustion is ΔH_{comb} = 1000 kcal/mol.

Above 330°C it decomposes, yielding methane and hydrogen cyanide. In a double oven kept at 260°C and 1000°C, 90% hydrogen cyanide can be produced. In aqueous solution the solubility decreases with increasing temperature, as it does for some other tertiary amines. The solubility is 47.3 wt % at 0°C and 45 wt % at 50°C. Aqueous solutions have a pH of about 8.5; the acid constant is 1.4×10^{-9}. In acidic solution hexamethylene tetra amine hydrolyzes slowly. Steam treatment, such as used in sterilization can lead to 0.1% hydrolysis. Hexamethylene tetramine was used for many years as urinary disinfectant with the trade names Urotropin, Helmitol, and Formanol. Its effectiveness results from acid hydrolysis in the stomach. However, side reactions, possibly caused by impurities, have been reported.

Hexamethylene tetramine reacts as a monobasic compound and can be titrated as such against a methyl orange indicator. It forms acid salts with hydrochloric acid, bisulfate, perchlorate, and phosphates. The chromates react explosively. Heming reacted the dinitrate with nitric acid and obtained cyclo-trimethylene trinitramine, also called RDX, or cyclorite, an explosive, used extensively during the Second World War. With hydrogen peroxide an additional product is formed which can also be converted to an explosive.

With elemental sulfur, reaction sets in at 160°C yielding hydrogen sulfide. In hot solutions, hydrogen sulfide produces amorphous products. Sulfur dioxide produces an addition compound.

Reactions with Inorganic Sulfur Compounds

The reaction of formaldehyde with sulfides and sulfites were studied by Walker

ISBN 0-201-04558-3

(1964) and workers in the 1920s, but relatively little is known about these reactions.

Formaldehyde does not react readily with elemental sulfur. Methylene glycol reacts readily in acid solutions with hydrogen sulfide forming trithiane:

$$3CH_2(OH)_2 + 3H_2S \rightarrow (CH_2 \cdot S)_3 + 6H_2O \qquad \text{Eq. 3.32}$$

Trithiane, first obtained by Homann in 1886, is a stable, odorless compound melting at 208°C (Schmidt 1976). NMR work indicates that it has a rapidly oscillating chain structure. The vapor density indicates that it remains unchanged in the gas phase. Trithiane can be polymerized under conditions similar to those effective with trioxane (Gipstein 1963). With sulfur dichloride it reacts quantitatively to symmetrical dichloromethylsulfide:

$$(CH_2S)_3 + 2SCl_2 \rightarrow (CH_2Cl)_2S + CS_2 + 2HCl + 1/4S_8 \qquad \text{Eq. 3.33}$$

Tetrathiane, $(CH_2S)_4$ is among the products in the direct reaction of hydrogen sulfide and formaldehyde. It is best prepared by the reaction of methylene chloride with methylmercaptan (Schmidt 1959).

Hydrogen sulfide can enter polyoxymethylene glycol chains as substitutes for oxygen:

$$H_2S + CH_2(OH)_2 \rightarrow CH_2\,(SH)(OH) + H_2O \qquad \text{Eq. 3.34}$$

$$HSCH_2OH + CH_2(OH)_2 \rightarrow HO{-}CH_2{-}S{-}CH_2{-}OH + H_2O \qquad \text{Eq. 3.35}$$

In neutral solution, the reaction is slow; at 10°C it takes several weeks. If 37% formaldehyde is saturated with H_2S at 40–50°C four thionals are formed in an exothermic reaction, yielding solids melting at 80°C.

Methanedithiol, $CH_2(SH)_2$ is obtained in 33% yield if 37% formaldehyde is reacted with hydrogen sulfide at 42°C at 30 atm for 16 hours (Cairns 1952). It is a colorless liquid which boils at 80 Torr and 58°C.

Under alkaline conditions, polymethylene sulfides are obtained when formaldehyde is reacted with metal sulfides or polysulfides. The reaction was first studied by Wohl (1886) who prepared several derivatives. These polymeric compounds have chain structure and the formula $HS(CH_2S)H$, in analogy to the polyoxymethyl glycols. Depending on the reaction conditions, these polymers have properties ranging from oils to rubber as well as brittle tars. Pure polymethylene sulfide was prepared by Lal in 1961. It converts to trithiane by sublimation *in vacuo* at 190°C. Kolino (1941) reacted sodium sulfide with formaldehyde and obtained methylene polysulfides, colorless amorphous powder. Rubber-like polymers were obtained by Walker (1964) who used alkali polysulfides with a rank of about 4. These polymethylene polysulfides contain polysulfide chain linkages (Meyer 1977). Contrary

ISBN 0-201-04558-3

to some earlier suggestions, branched chains are unstable and are not formed. These resins are yellow to dark olive, depending on the degree of polymerization . They melt at about 140°C. They ae insoluble in all common solvents except carbon disulfide. Similar products were described by Baer (1936), Ellis (1943), and Patrick (1940). Hills (1939) used a mixture of elemental sulfur and sodium sulfide to prepare an odorless, light colored powder melting at 140°C. Harmon (1962) obtained rubber-like compounds by polymerization of methanedithiol. Hydrogen sulfide is released in the latter reaction. This makes the reaction undesirable. Ammonium sulfide and polysulfide yield mixed products containing both C–S and C–N–S bonds (Wohl 1886). Delepine (1898) made an elemental analysis and recognized that products contained 36% C, 6% H, 17% N, and 39% S. These systems were carefully studied by LeTerre (1931, 1932) who noticed that the induction period preceding the reaction was not even approximately inversely proportional to the concentration of reagents. He concluded that at least two different reactions had to be involved. He tentatively identified some exotic intermediates, such as methylthioformaldin, $(CH_2)_3S_2NCH_3$, first proposed by Wohl. He finally proposed a compound as a major product. He also studied the reaction with methylamine and other intermediates with formaldehyde and hydrogen sulfide.

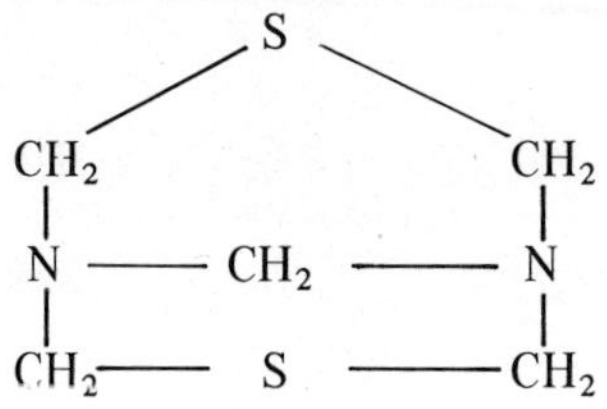

```
            S
        /       \
   CH2             CH2
    |               |
    N  —— CH2 ——    N
    |               |
   CH2 ——  S  ——   CH2
```

Reaction with Thiosulfate

Raschig (1929) studied the reaction of thiosulfate with formaldehyde and reported the formation of $HO{-}CH_2{-}S_2O_3^-$. This reaction has been repeatedly reported. It does not proceed in neutral or alkaline solutions. It is most vigorous under acidic conditions, especially those which lead to decomposition of pure thiosulfate solution (Meyer 1977). However, the author found that the bulk of the solid precipitated from thiosulfate solutions consists of paraformaldehyde. The structure of the species is not yet fully understood. It might be a mixture of redox products.

Reaction with Sulfur Dioxide and Sulfite

Sulfur dioxide and sulfite react vigorously with aqueous formaldehyde in an exothermic reaction, first described in 1853 by Bertagnini. Ripper recognized in 1900 that this reaction was suitable for quantitatively determining aldehydes. This procedure is still the most widely used and accepted analytical method. It is

ISBN 0-201-04558-3

described in Chapter 5. The reaction proceeds over the entire pH range, and the products are quite stable.

$$CH_2(OH)_2 + SO_2 \rightarrow HOCH_2{-}SO_3H \qquad \text{Eq. 3.36}$$

Originally, many difficulties were encountered in attempting to decide between the alpha-hydroxyester structure and the alpha-hydroxysulfonic acid formula. The constitution of the sulfite was extensively studied by Raschig and Prahl (1926) in a brilliant paper. They used elaborate chemical reasoning and prepared several organic derivatives to argue that the sulfur atom was bonded to the methylene glycol carbon, rather than via the oxygen atom, as would be expected on the basis of other convincing arguments, including modern electronegativity. They also studied the reaction with acetone, and drew on the analogy of other functional groups and adducts, especially the analogy between the formaldehyde addition products with water, hydrogen chloride, and ammonia.

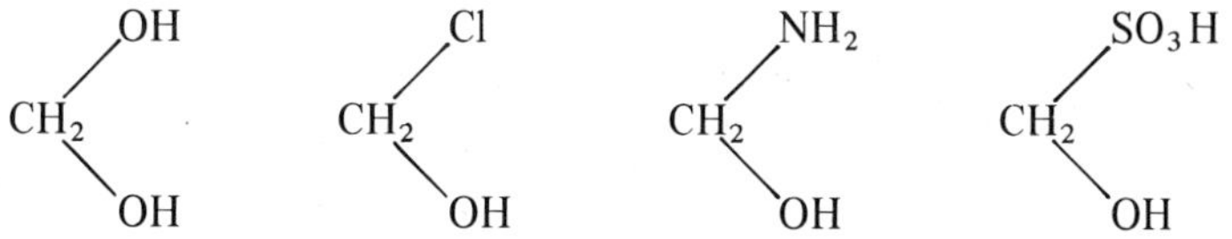

$CH_2(OH)(OH)$ $\quad$ $CH_2(Cl)(OH)$ $\quad$ $CH_2(NH_2)(OH)$ $\quad$ $CH_2(SO_3H)(OH)$

Furthermore, they used their structure to explain why the bisulfite compound is stable against oxidation, but vulnerable towards reduction to the sulfuric compound, Rongalite, which is of commercial importance. Finally, their structure explained why oxymethane sulfonic acid easily reacts with dilute ammonia, yielding amidomethyl sulfonic acid, while other hydroxy sulfonic acids stubbornly resist reaction with ammonia.

$$CH_2(OH)SO_3H + NH_3 \rightarrow CH_2(NH_2)SO_3H \qquad \text{Eq. 3.37}$$

Their work was later checked by Lauter (1935) who used four partly interlocking reaction chains to verify their conclusions. Shriner (1941) demonstrated the C–S linkage in acetone addition compounds in a similar manner. Today, modern instrumental techniques such as Raman spectroscopy make it comparatively easy to verify their work by *in situ* observation of the reaction products in aqueous solutions.

Figure 3.12 shows the Raman spectra of a titration of aqueous formaldehyde with sulfite. The frequency at 777 cm^{-1} is characteristic for the C–S bond. Thus, the sulfite is linked via the sulfur atom as argued by Raschig, and not via an oxygen bridge as had been postulated intermittently. The figure shows intermediate points of the titration of solutions containing 37%, 32%, 27%, 18.5%, and 9.25% formaldehyde. The absolute intensity of the peaks at 777, 915, and in the 1040 cm^{-1} region can be used to determine the concentration of all species (Meyer 1979).

ISBN 0-201-04558-3

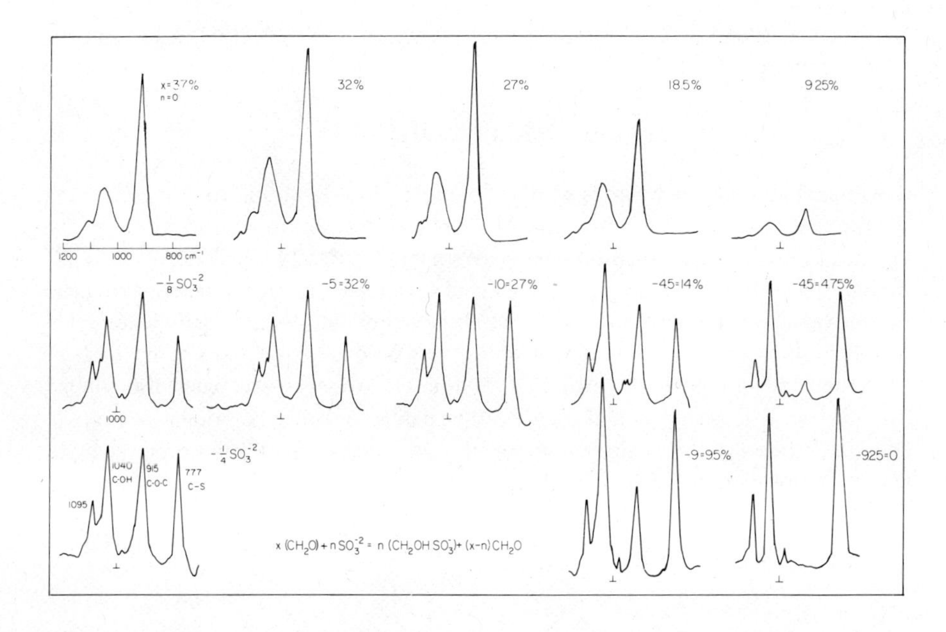

Figure 3.12. Titration Points of Formaldehyde with Sulfite. Raman Spectra of Five Different Solutions (Meyer, 1979)

Blackadder and Hinshelwood (1958) studied the reversible decomposition kinetics as a function of pH by adding step-wise iodine solution and observing the time lapse before the blue color disappeared. They found the reaction to be first order. The rate increases with increasing pH. At pH 5 they found an activation energy of 22.35 kcal/m and the log of the rate constant, log $k = -0.26$/sec.

Alpha-hydrogen methylene sulfonic acid can be reduced with zinc dust and acetic anhydride or by hydrogenation in the presence of nickel, yielding formaldehyde sulfoxylate:

$$OH{-}CH_2{-}SO_3^- + H_2 \rightarrow HO{-}CH_2{-}SO_2H \qquad \text{Eq. 3.38}$$

The properties of this compound have been described in other references (Meyer 1977). It has been used in large quantities for stripping and discharging dyed textiles. Our own work, Figure 3.13, indicates that it normally forms in less than 50% yield, and that the properties attributed to it reflect the properties of mixtures containing both sulfonic and sulfoxalate salts.

ISBN 0-201-04558-3

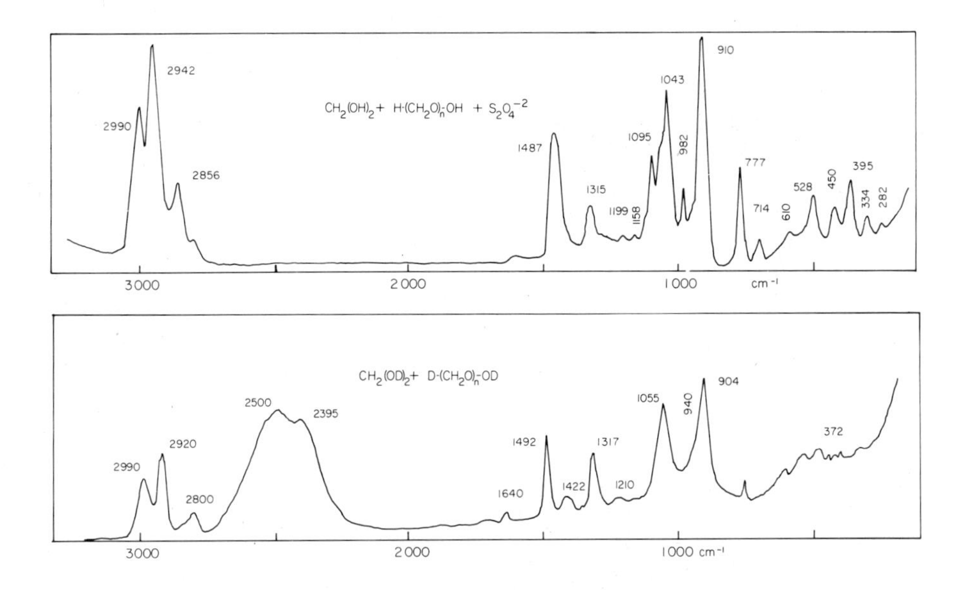

Figure 3.13. Raman Spectrum of Aqueous Formaldehyde. Top: with Dithionite; bottom: in D_2O (Meyer, unpublished, 1978)

Reactions with Acids

Sulfuric acid reacts with formaldehyde only because it is a strong acid. Sulfate, with sulfur in the oxidation state 6, does not react with formaldehyde. Sulfuric acid catalyzes the polymerization of formaldehyde to beta-polyoxymethylene; it also catalyzes the formation of trioxane. Fuming sulfuric acid reacts at 60°C yielding methylene sulfate, $(CH_2SO_4)_2$, an eight-membered cyclic compound. It forms a colorless powder which decomposes at 155°C. It is insoluble in water. Anhydrous phosphoric acid reacts at about 140°C, and various reaction products and polymers are formed.

Nitric acid reacts violently with formaldehyde forming carbon dioxide, water, nitric oxide, and some nitrogen. Thus, formaldehyde can be used to decompose nitric acid. Nitrite reacts similarly. This reaction is used to determine nitrite analytically:

$$4HNO_2 + 3CH_2O \rightarrow 3CO_2 + 5H_2O + 2N_2 \qquad \text{Eq. 3.39}$$

ISBN 0-201-04558-3

Reaction with Halogens

Chlorine can react with formaldehyde in two ways. In sunlight it forms phosgene, in artificial light or in the dark it produces carbon monoxide:

$$CH_2O + 2Cl_2 \rightarrow COCl_2 + 2HCl \qquad \text{Eq. 3.40}$$

$$CH_2O + Cl_2 \rightarrow CO + 2HCl \qquad \text{Eq. 3.41}$$

Bromine yields carbon dioxide and ormic acid. Iodine reacts only upon extended heating in a sealed tube at 125°C.

Reactions with Aliphatic Hydroxy Compounds

The first reaction step between aldehyde and all hydroxy compounds is the formation of hemiacetals:

$$HOCH_2{-}OH + CH_3OH \rightarrow HO{-}CH_2{-}O{-}CH_3 + 15\ \text{kcal/m} \qquad \text{Eq. 3.42}$$

$$HO(CH_2O)_nH + ROH \rightarrow HO{-}(CH_2O)_n{-}OR \qquad \text{Eq. 3.43}$$

These reactions are reversible. The heat of reaction with methanol $\Delta H = 15$ kcal/m, with *n*-propanol $\Delta H = 14.2$, with *n*-butanol $\Delta H = 14.9$ kcal/m. As explained above, the equilibria are in favor of hemiacetal; thus, in methanolic solutions more than 99% of methanol is reacted with methylene glycol, and thus the solubility of total formaldehyde is increased.

In neutral and alkaline solution the reaction produces merely hemiacetals. Only in acidic solutions are formals produced; the reactions are reversible and resemble esterifications.

$$pH \geqslant 7{:}\ CH_2(OH)_2 + ROH \rightarrow HOCH_2OR \qquad \text{Eq. 3.44}$$

$$pH \leqslant 7{:}\ CH_2(OH)_2 + ROH \rightarrow R{-}O{-}CH_2{-}OR \qquad \text{Eq. 3.45}$$

Formals of aliphatic alcohols can also be obtained by gas phase reactions at 350°C.

Methylal, $CH_2(OCH_3)_2$, also known as formaldehyde dimethyl acetal, formal or dimethoxymethane, is the dimethyl ether of methylene glycol. It melts at −105°C, boils at 42.3°C and has a density of 0.86 g/ml. It forms a 30 wt % solution in water and it is fully miscible with alcohols and ethers. Methylal is obtained best by heating paraformaldehyde in methanol in the presence of ferric chloride or hydrogen chloride.

ISBN 0-201-04558-3

Diethyl formal boils at 89°C, diisopropyl formal at 119°C, dipropyl formal at 141°C, diisobutyl formal at 164°C, ditertiary butyl formal at 184°C, dibenzyl formal at 280°C, and dioctyl formal at 360°C.

Mixed formals are obtained from a mixture of alcohols. Methyl-ethyl formal boils at 67°C.

Polyoxymethylene formals are important components in the commercial formaldehyde solutions just described. Modern ^{13}C-NMR spectroscopy makes it possible to better analyze the behavior of the compounds. They consist of unbranched chains. Presently, not much more is known about them than what Staudinger described in 1929. Short chains form colorless liquids. Their boiling and melting points are given in Table 3.13. Their vapor pressure is given by:

$$\log p = A - B/T - C \log T \qquad \text{Eq. 3.46}$$

Table 3.13

Melting Point, Boiling Point, and Vapor Pressure Constants of the Dimethyl Ethers of Polyoxymethylene (after Staudinger 1928)

			Vapor Pressure Constants		
n	mp (°C)	bp (°C)	A	B	C
1	-105	42.3			
2	- 70	105	74.7	7.2	8.3
3	- 43	156	70.3	8.0	7.4
4	- 9	202	87.8	10.0	9.8
5	18	243	93.6	11.3	10.4

Hexoxymethylene dimethyl ether apparently melts at 31°C and boils at about 150°C (Walker 1964). Longer oligomers are no longer soluble in water.

Polyoxymethylene dimethyl ethers are stable up to 160°C. At that temperature oxidation by air become significant. However, acids induce hydrolysis at room temperature. Long polyoxymethylene diethers with 100–500 formaldehyde units are often called gamma-polyoxymethylene. Such polymers are odorless, insoluble in water but soluble in dimethylformamide. They are prepared by addition of concentrated sulfuric acid to commercial formaldehyde solution at 20°C. Upon boiling in water, intrachain rearrangement occurs, yielding some carbohydrate functional groups. This polymer is called delta-polyoxymethylene.

ISBN 0-201-04558-3

Glycols and Glycerols. Formaldehyde reacts with glycols yielding both cyclic and polymeric formals. If ethylene glycol is heated with paraformaldehyde in the presence of an acid catalyst dioxolane is produced:

$$(CH_2\text{–}OH)_2 + CH_2O \rightarrow \begin{matrix} CH_2\text{—}O \\ | \\ CH_2\text{—}O \end{matrix} \!\!> CH_2 + H_2O \qquad \text{Eq. 3.47}$$

Dioxolane, also called glycol formal, boils at 76°C, has a density of 1.06 g/ml and is fully miscible with water. The formals of trimethylene glycol and tetramethylene glycol boil at 112°C and 117°C, respectively. Polyformals with a molecular weight of up to 10,000 can be synthesized.

Polyhydroxy formals are produced in analagous manners. Thus, pentaerythritol reacts with methyleneglycol in the presence of zinc chloride yielding dimethylene pentaerythritol, with a melting point of 50°C and a boiling point of 230°C. Likewise, marmitol and sorbitol react readily with formaldehyde yielding 1,3; 2,4; and 5,6 acetal bridges. Finally, polyvinyl alcohols also react with formaldehyde in the presence of acids to give polycyclic formals which are used in industry:

$$\begin{matrix} \text{—} CH\text{—}CH_2\text{—}CH\text{—}CH_2\text{—}CH\text{—}CH_2\text{—}CH\text{—}CH_2\text{—} \\ O\text{—}CH_2\text{—}O \qquad\qquad O\text{—}CH_2\text{—}O \end{matrix}$$

Sugars. Starch and sugars react very much like smaller hydroxy molecules and form both hemiacetals and the more stable methylene ethers. A total of up to 5 moles of formaldehyde can be absorbed by each mole of sugar. It seems that these formaldehyde-rich systems contain loosely adsorbed formaldehyde, as well as acetals.

Adducts are obtained by evaporating concentrated formaldehyde-sugar solutions *in vacuo* at 70°C. The products are stated to be "without sharp odor" (Lauch 1913). These substances dissolve readily in water, releasing both formaldehyde and sugar. This type of adduct might well be formed by a mechanism similar to one responsible for formaldehyde retention in cellulose in particleboard after thermal polymerization of UF-resin. Similar compounds were obtained by Quade (1913) and Rosenberg (1907) and by heating of sugars with solid paraformaldehyde by Jones-Sefton (1911). Heiduschka (1916) claimed these were unreacted solutions, Contardi (1936) proposed they were hemiacetals. It would be well worth to study the reaction conditions and products in these systems using modern *in situ* methods, as these reactions likely hold clues to formaldehyde retention on the surface of industrially important materials. Pure acetals and hemiacetals of these compounds are best obtained by acid catalysis. This reaction was studied by Schwenkler (1947) using birch wood pentosan yielding methylene ethers. Tollens (1899), finally, produced methylene glucose with the formula $C_6H_{10}O_6 \cdot CH_2 \cdot H_2O$.

Starch reacts with formaldehyde in neutral or alkaline solution yielding hemi-

ISBN 0-201-04558-3

acetals. Upon heating resins are produced. Formaldehyde reduces the gellation temperature of starch, and prevents the formation of the blue iodine complex. Ammonia can be used to recover formaldehyde from the systems, and the starch can thus be regenerated. Acid catalysis yields water-resistant compounds. Resins are obtained if the reagents are heated under pressure to 80°C.

Cellulose reacts with formaldehyde very much like smaller hydroxy molecules. In a neutral and mildly alkaline medium hemiacetals are formed, in an acid medium diethers are formed, and under most conditions excess formaldehyde can be absorbed. The treatment changes the physical and chemical properties of cellulose. This effect is used for the modification of cotton, wood, paper, rayon, and similar materials. These reactions have been described in a large body of patent literature. The subject will be touched in chapter 6. The hemiacetal formation in neutral and alkaline solutions seems reversible. In the acidic system crosslinking between molecules occurs. This reduces swelling in water, increased elasticity, and resistance to cotton dyes. A 2% total formaldehyde suffices to render rayon viscose insoluble; 7.5% formaldehyde causes methylene glycose to become so brittle that it can be powdered. The first commercial use was proposed by Eschalier in 1906 in a series of patents dealing with the strengthening of artificial silk. He washed fibers in aldehyde in the presence of lactic acid at 55°C. His *sthenosage* (Greek for strength) process is the basis on which all further work has been built. The literature has been carefully reviewed by Pacsu (Wagner 1952) who from his own work proposed the use of 1% boric acid to bind 6% formaldehyde to cellulose at 110–150°C. It was observed that each formaldehyde reacted with two hydroxy groups, the latter being secondary hydroxy. The interpretation pointed towards crosslinking between neighboring molecules.

Gruntfest and Gagliardi (1948) found similar reactions. They studied the temperature dependence between 30 and 150°C and its dependence on pH. At low acidity the products remained little changed, and formaldehyde absorption was reversible. Table 3.14 shows the influence of the acid cónstant of the catalyst. Even though these studies were made on rayon, it should be noted here that the above reaction conditions are almost identical to those during the press cycle of particleboard, during which UF-resin with excess formaldehyde is acidified in order to induce UF-resin polymerization and then heated for several minutes at 100–150°C until about half of the 5–20% water in the wood is driven out as steam. From the results of Gruntfest (1948) it seems likely that the mild resin catalyst preferentially leads to formation of hemiacetals which can be slowly hydrolyzed. This process might contribute to the formaldehyde odor of particleboard. An excellent and comprehensive review on textile cellulose was published by Roff in 1958. He recognized that free formaldehyde in UF-resins, first applied to textile in 1926, might greatly influence the properties of products. Woo, Dillan, and Dusenbury (1956) showed that the relation between initial formaldehyde and formaldehyde uptake for each baking time gave a straight-line plot. An activation

ISBN 0-201-04558-3

energy of 16 kcal/mole was deduced for producing methylene linkage at pH 2; the corresponding value for cotton was 6 kcal/mole. Steele(1955) and Lineken, Davis, and Jorgensen (1956) studied the infrared spectra of cellulose exposed to formaldehyde and registered a distinct reduction in the intensity of the O–H bond at 3300 cm^{-1}. Depending on the condition of the formaldehyde treatment, the bond recovered after washing in water.

Table 3.14

Weight Gain of Rayon Treated with Formaldehyde as a Function of Acid Strength (Gruntfest, 1948)

Catalyst	Acid Strength at 25°C *K*	Weight-gain after Cure (%)	Weight-gain after Wash (%)
None	–	6.25	-0.53
$NaHCO_3$	6.0×10^{-11}	11.20	-0.41
Lactic acid	1.4×10^{-1}	8.80	-0.24
Formic acid	2.1×10^{-4}	5.70	-0.64
Citric acid	8.0×10^{-4}	7.90	+0.31
Fumaric acid	1.0×10^{-3}	8.45	-0.11
Phosphoric acid	1.1×10^{-2}	9.40	+4.10
$NaHSO_4$	2.0×10^{-2}	7.90	+3.20
Oxalic acid	3.8×10^{-2}	9.20	+3.90
NH_4Cl	–	6.80	+5.00
$AlCl_3$ (0.5%)	–	7.40	+2.70

Stable methylene bridges are formed by catalysis with acids with acid constants of $\geqslant k = 10^{-4}$. Boric acid, with $k = 10^{-10}$, is an exception as it is extremely effective. Methylene bridge formation is enhanced by dehydration, as the reaction products include water. Thus, the concentration of the solution is a factor. It is not yet possible to influence the ratio of intramolecular cyclic linkage to crosslinking among neighboring molecules.

In 1959 Roff examined the absorption of formaldehyde from aqueous solution by cellulose in textile yarn. He found that the equilibrium was reached only slowly at room temperature. At 5°C it took over 40 hours before equilibrium was approached. At 25°C it took eight hours and at 45°C, two. The activation energy was found to be 13 kcal/mole. Temperature and pH did not influence the equilibrium. However, the absorption increased with the square root of concentration and depended on the nature of the cellulosic material. Roff also studied the influence of pretreatment with methanol, sodium chloride, urea, boric acid, copper sulfate, lissapol, sappamine, and ethylamine. The latter is known to influence the cellulose

ISBN 0-201-04558-3

structure by opening crystallites of cotton and rayon. This increased formaldehyde absorption by 40%. If wood cellulose reacts equally, the current practice of binding the excess formaldehyde in UF-resin by the addition of inorganic or organic ammonium might increase, rather than decrease, the formation of labile methylol cellulose. This would increase the release of formaldehyde, because of methylol cellulose hydrolysis during weathering and aging. In cotton, Roff found that formaldehyde absorption was diffusion limited, and directly proportional to the moisture regain. He found the process formed primarily methylol cellulose. The best catalysts for the reaction are ammonium chloride, aluminum chloride, and boric acid. In the presence of alcohols, urea, and phenols, formaldehyde can lead to substituted oxymethyl groups $-O-CH_2-R$.

The reaction of wood cellulose with formaldehyde reduces swelling considerably. For example, 4% formaldehyde can reduce swelling by 70%. Treatment can be achieved by leaching at room temperature, or by heating of wood in contact with formaldehyde vapor. Acid catalysts are necessary for obtaining maximum effectiveness. Stamm (1959) studied the behavior of Sitka spruce and redwood and recorded the correlation between treatment parameters and abrasion resistance, toughness, and swelling.

With excess hydrogen chloride alpha-chloro-methyl ethers can be formed:

$$CH_2O + CH_3OH + HCl \rightarrow CH_3-O-CH_2Cl + H_2O \qquad \text{Eq. 3.48}$$

The same reactions can occur with glycols.

Mercaptans

Formaldehyde reacts with mercaptans in the same way as with alcohols.

$$RSH + CH_2O \rightarrow RS-CH_2OH \qquad \text{Eq. 3.49}$$

The product can be oxidized to sulfones.

Reaction with Amines

Primary and secondary amines react with formaldehyde yielding methylolamines, also called alkylaminomethanols. They are the nitrogen analogue of the hemiacetals discussed above. Henry discovered in 1893–1894 that they form readily when mono- or dialkylamines are gradually added to a commercial formaldehyde solution. Upon addition of potassium carbonate, they form a separate phase which can be isolated. The overall reaction is:

$$R-NH_2 + CH_2(OH)_2 \rightarrow RNHCH_2OH + H_2O \qquad \text{Eq. 3.50}$$

The methylamines are labile and can undergo further reactions. The reaction

ISBN 0-201-04558-3

path and the product depend on the reagent, pH, temperature, and many other factors.

The Mannick Reaction: In this type of reaction first described by Mannick (1910) formaldehyde links two other reagents, and amine which acts as a base and a proton donor. For example:

$$CH_3O + HN(CH_3)_2 + (CH_3)_2CO \rightarrow CH_3{-}CO{-}CH_2{-}CH_2{-}N(CH_3)_2 + H_2O \quad \text{Eq. 3.51}$$

In the above case acetone acts as acid. Other suitable bases are all primary or secondary amines and ammonia. The acids can include phenols, alcohols, esters, etc. The Mannick reaction can take place both in acidic and alkaline medium. In bases the carbonium ion of the proton donor is an intermediate. In acidic medium the carbenium ion of methylolamine has been invoked. In some Mannick reactions trioxane can be used instead of formaldehyde. If the amine concentration is insufficient, tertiary amines may be formed. For more details, the reader is referred to the article of Geisse (1959).

With hydrogen sulfide and methylamine formaldehyde forms methylthioformaldine, $(CH_2)_3S_2NCH_3$. Hydrogen selenide reacts similarly.

Aniline reacts with formaldehyde yielding mixtures in which anhydroformaldehyde aniline can be made to be the main product. It has the structure:

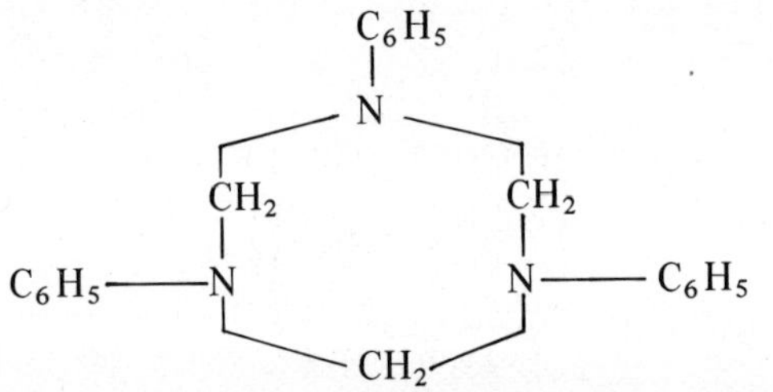

```
                    C6H5
                     |
              ┌───── N ─────┐
             CH2           CH2
              |             |
   C6H5 ───── N             N ───── C6H5
              └──── CH2 ────┘
```

Pure liquid formaldehyde reacts with aniline to a colorless material melting at 180°C. This is apparently an isomer of the above anhydroformaldehyde aniline. The reaction rate for aniline substituents is m $CH_3 > o$ $CH_3 >$ unsubstituted $> m$ $Cl > p$ $CH_3 > p$ Cl. In these reactions monomethylol arylamines are intermediates. Dimethylol aniline forms a colorless liquid which decomposes above 40°C. In strongly alkaline solutions bis(phenylamino)-methane is obtained. In acidic solution resins are formed. These resins contain nuclear methylene linkages, similar to those found in phenol resins. The reactions are primarily determined by the aromatic skeleton, rather than by the amine function which serves merely as a facilitator. Thus, even tertiary amines form resins. In hot acid rearrangement readily takes place. Often complex heterocycles are produced. Thus, phenylhydrazine and

ISBN 0-201-04558-3

formaldehyde react to form a series of heterocyclic compounds. Secondary amines are not as reactive as primary amines, but with acid they still form brittle resins.

Amides

The reactions of formaldehyde with amides touch the main subject of this book, and are covered in the next section. The prime products of these reactions are methylol and methylene derivatives. The methylol amides are more stable than the methylolamines described above. Some can be isolated as pure substances. The best reaction conditions are pH 7 and warm solutions. Often methyl or ethylether is employed as a solvent, especially when paraformaldehyde is used as a solvent. Methylene diamides are generally stable, high melting solids formed when an acidic hot formaldehyde solution is dehydrated.

Einhorn observed the formation of monomethylol in 1900 and believed that further reaction was not possible. Today we know the dimethylol derivatives of formamide, acetamide, and other amides. They act as formaldehyde donors.

Monomethylol formamide is prepared by heating equimolar quantities of formamide with paraformaldehyde quickly to 120°C. The product forms a colorless oil which is fully miscible with water and alcohol. Methylolacetamide, $CH_3CONHCH_2OH$, a solid melting at 52°C, is obtained by the reaction of acetamide with formaldehyde and potassium carbonate. The dimethylol compounds are prepared by heating four-fold formaldehyde with formaldehyde at pH 8–9 to 60°C. The yield is over 80% after two hours. Methylolamides can be reacted with primary alcohols to yield ethers. With thiols sulfides are obtained, and with carboxylic acids esters. If excess formamide is used, methylene-diformamide is obtained. It is a crystalline substance with a melting point of 143°C. In potassium carbonate solutions, benzamide yields methylol benzamide. In acidic solution it converts to methylolmethylene-dibenzamide:

$$2C_6H_5CONHCH_2OH \rightarrow C_6H_5CON(CH_2OH)\cdot CH_2NHCOC_6H_5 + H_2O \quad \text{Eq. 3.52}$$

Methylene-diamine, $CH_2(NH_2)_2$, is the ammino-analogue of methylene glycol. It is obtained by the reaction of methylene diformamide with excess acid in cold solution. The reaction yields hydrochloride salts. The free amine is only stable in solution.

$$CH_2(NHCOH)_2 + 2HCl + 2H_2O \rightarrow CH_2(NH_2)_2 2HCl + 2CHOOH \quad \text{Eq. 3.53}$$

Under the same reaction conditions methylene diacetamide hydrolyzes completely to ammonium chloride. Mixed methylene derivatives can be obtained by similar reactions. Even sulfonates can be obtained that way:

$$C_6H_5CONH_2 + CH_2O + NaHSO_3 \rightarrow C_6H_5CONHCH_2SO_3Na + H_2O \quad \text{Eq. 3.54}$$

ISBN 0-201-04558-3

The reactions with urea make the content of a later section. The primary reaction products are mono- and dimethylol derivatives, and if eight-fold formaldehyde excess is used also trimethylol urea.

Monomethylol urea, $NH_2-CO-NHCH_2OH$, first prepared by Einhorn and Hamburger in 1908, is a colorless liquid melting at 111°C. Dimethylol urea, first obtained by the same authors melts at 126°C. The latter is made and sold as a commercial chemical which is used in the textile, adhesive, and resin industries. It is often traded in form of a solution containing excess formaldehyde at pH 7–9. In acidic solution resins are obtained.

Methylene-diurea was first prepared by Kadowaki in 1936 in mildly acid medium. It crystallizes in needles which melt at 218°C. It readily reacts further forming linear condensates. At 30°C methylenediurea can be methylolated yielding mono, di, tri, and tetra derivates, at 60°C penta- and hexamethylol can be produced. While the methylolated ureas easily liberate formaldehyde, their ethers are quite stable. The latter are prepared by acid treatment with corresponding alcohols. The ethers can be reduced with hydrogen over nickel at 100°C and yield 1,3 dimethyl urea.

In 1936 Kadowaki discovered that evaporation of a 4:1 alkaline urea-formaldehyde solution yields N′, N-dimethyloluron which distills at 0.1 Torr at 83°C. It has the structure:

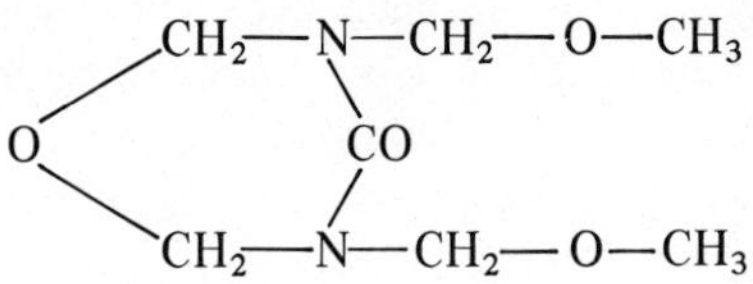

```
   CH2—N—CH2—O—CH3
  /     \
 O      CO
  \     /
   CH2—N—CH2—O—CH3
```

In strong acids urea-formaldehyde forms highly crosslinked resins, i.e., cyclic compounds. Dixon and Goldschmidt described species such as:

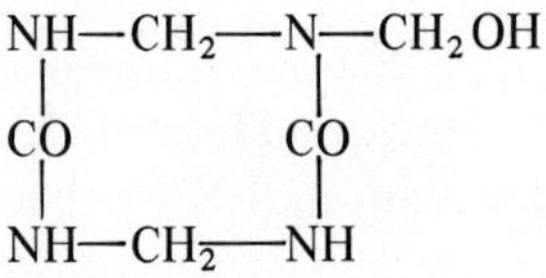

```
NH—CH2—N—CH2OH
|      |
CO     CO
|      |
NH—CH2—NH
```

Substituted ureas also react with formaldehyde, but their tendency to undergo condensation decreases with substitution. Thus, Einhorn prepared monomethylols of both symmetrical and asymmetrical dimethylurea, but dimethylol derivatives could not be prepared. Triethylurea did not react at all. This reluctance has been

ISBN 0-201-04558-3

attributed to tautomerism of the urea:

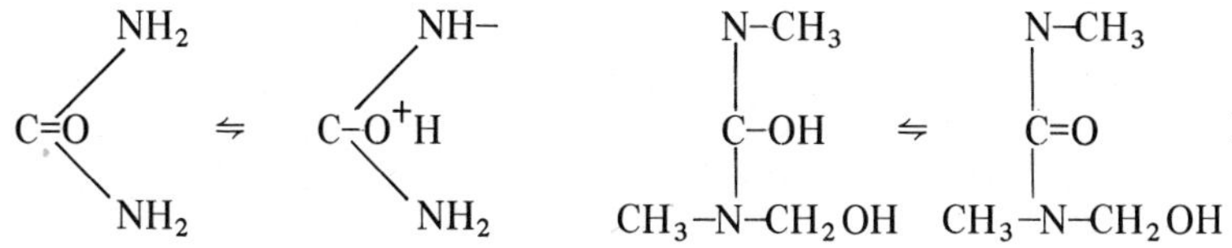

$$
\begin{array}{c} NH_2 \\ / \\ C{=}O \\ \backslash \\ NH_2 \end{array}
\;\rightleftharpoons\;
\begin{array}{c} NH- \\ / \\ C{-}O^{+}H \\ \backslash \\ NH_2 \end{array}
\qquad
\begin{array}{c} N{-}CH_3 \\ | \\ C{-}OH \\ | \\ CH_3{-}N{-}CH_2OH \end{array}
\;\rightleftharpoons\;
\begin{array}{c} N{-}CH_3 \\ | \\ C{=}O \\ | \\ CH_3{-}N{-}CH_2OH \end{array}
\qquad \text{Eq. 3.55}
$$

In sulfuric acid, acetylurea reacts with formaldehyde yielding bis-acetylurea. The latter converts by hydrolysis to methylene bis-urea which loses ammonia and forms a cyclic product.

Cyclic ureides, such as hydantoin, likewise form mono-methylol derivatives. Walker (1964) described low-molecular weight glassy resinous products obtained from 5-alkyl and 5,5-dialyl hydantoins which can be modeled to be soluble or insoluble in water or alkalis.

Diamides of dibasic acids such as oxamide, adipamide, and succinamide also form methylols, but they do not readily polymerize. Malonamide, having a methylene group adjacent to two carbonyls forms a trimethylol. Polyamides react with formaldehyde forming first methylols, and then methylene or polyoxymethylene bridged compounds. Further reaction with alcohols yields N-alkoxymethyl compounds. Mercaptans form N-alkylthiomethyl derivatives.

The N-methylol of polyhexamethylene adipamide, Nylon 66, is obtained in aqueous formaldehyde at 60°C. The reaction must be carefully controlled to prevent crosslinking. This can be achieved by using pyridine or other inert solvents. Alkoxymethyl derivatives, prepared in formic acid at 60°C are softer and more elastic than the original polymers. Upon heating the alkoxymethylated polymers in acid, methylene crosslinkage results and the products become insoluble and lose their thermoplasticity.

Carbamates, or methanes, react with formaldehyde very much like the corresponding amides. Thus, ethyl methane forms methylol urethane which melts at 52°C. Upon heating these compounds give off formaldehyde. In acid, they convert to methylenediurethane.

Thiourea

A colorless crystalline compound melting at 173°C was made by Claus in 1927 by reaction of cyanamide with hydrogen sulfide. Walter recognized in 1927 that thiourea reacts analogous to urea. The simple addition products are more soluble and less stable than their urea analogues. Pollak (1939) prepared mono- and dimethylolthiourea and noted differences between acidic and alkaline forms which

ISBN 0-201-04558-3

he attributed to isomer formation. He distinguished them by their reactivity towards silver nitrates. The N-methylol compound is not very reactive, but the S-methylol reacts speedily. The N-methylol is formed in acidic reactions.

The addition of thiourea to acidic formaldehyde yields a polymer with the formula $(NH_2 \cdot CS \cdot NCH_2)_n$. Thiourea polymers are hydrolyzed by strong hot acids.

Cyanamides react with formaldehyde yielding white solids which can be polymerized. Dicyanamide forms a monomethylol melting at 118°C.

$$NH_2C(NH)NH \cdot CN + CH_2O \rightarrow HO{-}CH_2{-}NHC(NH)NH{-}CN \qquad \text{Eq. 3.56}$$

At pH 8–10 it polymerizes.

Melamine

Melamine, cyclic trimer of cyanamide, first studied by Liebig in 1834, melts at 350°C. It reacts with one to six moles of formaldehyde yielding mono- to hexamethylols. The latter is obtained by short boiling of the proper M:F ratio:

```
          NH2                                              N<
          |                                               /
          C                                              C
        //  \                                          //  \
       N     N                                        N     N
       |     ||       + 6CH2(OH)2   →                 |     |         Eq. 3.57
NH2—C       C—NH2                              >N—C       C—N<
       \\   /                                          \   //
          N                                              N
```

These methylol compounds can be used for preparing melamine resins, first described in 1935 and commercialized in 1936. The chemistry of melamine is similar to that of urea. The condensation reactions have been studied by Gams (1941). The methylol groups form easier than in urea, and are less soluble. De Breet (1977) studied ^{13}C-NMR spectra of melamine reactions. The methylols readily convert to methoxymethyl groups. Gas chromatography, infrared and H-NMR all confirm that contrary to the older literature methylene bridges are absent in basic condensation products. The ^{13}C-NMR shifts of the various functional groups were measured by de Breet (1977) and are summarized in Table 3.15. The chemistry of melamine resin has been reviewed in an excellent chapter by Widmer (1965).

The reaction of other commercially important amines is discussed below, together with heterocycles.

ISBN 0-201-04558-3

Table 3.15

Chemical NMR Shifts of Carbon Atoms in Melamine
(after de Breet, 1978)

Observed (ppm)	Assignment
167.4[a]	$\gtrdot C{-}N(CH_2OCH_3)_2$
166.7	$\gtrdot C{-}NHCH_2OCH_3$
166.3	$\geqslant C{-}NHCH_2OH$
165.7	$\geqslant C{-}NH_2$
90.1	?
77.3[a]	$\geqslant C{-}N(CH_2OCH_3)_2$
72.9	$\geqslant C{-}NHCH_2OCH_3$
69.9	$\geqslant C{-}NHCH_2OCH_2NH{-}C<$
64.7	$\geqslant C{-}NHCH_2OH$
55.4[a]	$\geqslant C{-}N(CH_2OCH_3)_2$
55.0	$\geqslant C{-}NHCH_2OCH_3$
49.3	CH_3OH

[a] Values similar to hexamethoxymethyl melamine (167.2, 77.1 and 55.5, respectively).

Amino Acids

Amino acids and esters react similarly to amines. Schiff discovered in 1891 that glycine, alanine, and asparagine react as strong acids after treatment with formaldehyde. Sorensen used this fact to develop in 1908 a method for titrating amino acids. The reaction is as follows:

$$NH_2CH_2COOH + CH_2(OH)_2 \rightarrow CH_2NCH_2COOH + 2H_2O \qquad \text{Eq. 3.58}$$

Glycine is neutral; methylene aminoacetic acid is a strong acid. In strongly acidic solution methylene diglycine can be formed. It was suggested by Galeotti (1913) that formaldehyde is used in the synthesis of peptides. In 1960 Feldman reacted adenine and adenosine with formaldehyde. Ninhydrine and idantrione convert

ISBN 0-201-04558-3

glycine to formaldehyde. Enzymes apparently can produce serine from formaldehyde and glycine.

Aminouracil, adenine, and adenosine have been used as model compounds to explain why formaldehyde efficiently deactivates viruses in vaccines. Apparently, both methylol and methylene linkages form. Obviously, formaldehyde would be very toxic for amino acids. Fortunately, the liver produces enzymes which consume it, and consequently the danger is small for humans (Andersen 1979).

Proteins

It has been suggested for some time that the origin of life proceeded via reaction of formaldehyde with ammonia and hydrogen cyanide yielding eventually polyglycines.

Formaldehyde hardens proteins and makes them water resistant. This tanning effect is due to crosslinkage of protein chains. Formaldehyde is by far the best tanning chemical of all aldehydes. Primary and secondary amines react over a wide pH range. Epsilon nitrogen of lysine reacts at pH 6–8. Calcium chloride and other salts increase the formaldehyde retention by collagens. Formaldehyde containing proteins give characteristic stains when treated with sulfuric acid. This fact is used in the analytical test for formaldehyde in milk. Formaldehyde treated proteins are very stable against alkalis. Thus, wool becomes resistant to caustic when treated with formaldehyde. In acids, formaldehyde is liberated.

Nitriles

Formaldehyde and trioxane react with nitriles in sulfuric acid yielding amides of cyclo-trimethylene tetramine. Acrylonitrile reacts with formaldehyde forming acrylamides. Hepp and Spiers observed the formation of bis-methylene made in 1876. In the last thirty years, polyamides have become important commercial materials, but their discussion is beyond the goal of this short review.

Heterocyclic Compounds

Furfuryl alcohol condenses with acidic formaldehyde yielding insoluble, infusible resins. If carefully reacted at pH 1.5 to 3.5, the resins remain fusible. Furfural likewise yields resins. Tetrahydrofuran reacts already at room temperature. The ring opens and chain polymers are formed. Thiophene condenses with formaldehyde yielding methylene-crosslinked resins. This reaction has been used to purify benzene, as thiophene reacts selectively. In acidic media resins are formed. In contrast, thiophene thiol reacts in alkali, exactly like phenol. Thiophene and phenol resins can be mixed. Thiophene can be reacted with ammonium chloride according to Mannick's reaction yielding either polymers or 2-phenylamine, according to reaction conditions:

ISBN 0-201-04558-3

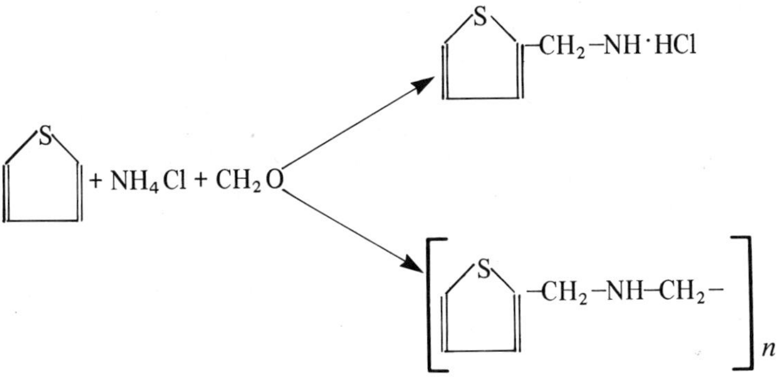

$$\text{thiophene} + NH_4Cl + CH_2O \rightarrow \text{thienyl}-CH_2-NH\cdot HCl \quad ; \quad \rightarrow [\text{thienyl}-CH_2-NH-CH_2-]_n \qquad \text{Eq. 3.59}$$

Selenophenes are too toxic for having any practical importance.

Pyrroles react readily with alkaline formaldehyde yielding alpha-methylol pyrroles.

$$\text{pyrrole (NH)} + CH_2O \rightarrow \text{pyrrole}\ N-CH_2OH \qquad \text{Eq. 3.60}$$

Pyrrolidine reacts only upon prolonged heating in a sealed tube at 150°C. Pyridine forms an adduct with formaldehyde. Beta-hydroxy pyridine forms an alpha-methylol derivative. The monomethylol derivative of alpha-picoline has been known for ninety years. Quinoline only reacts in strongly acidic medium.

The reactions of imidazolidone, i.e., methylene urea, $(-CH_2-NH)_2CO$, have been carefully studied by Slonim (1977) and coworkers who reported the ^{13}C-NMR spectra of some 23 formaldehyde derivatives which might be important analogues to compounds occurring in UF-resins used in the textile and paper industry, where cyclic compounds are desirable. They also studied triazine rings with:

Imidazolidine: CH_2-CH_2, NH, NH, $C=O$

triazinone: R–N, NH, NH, $C=O$

urons: O, NH, NH, $C=O$

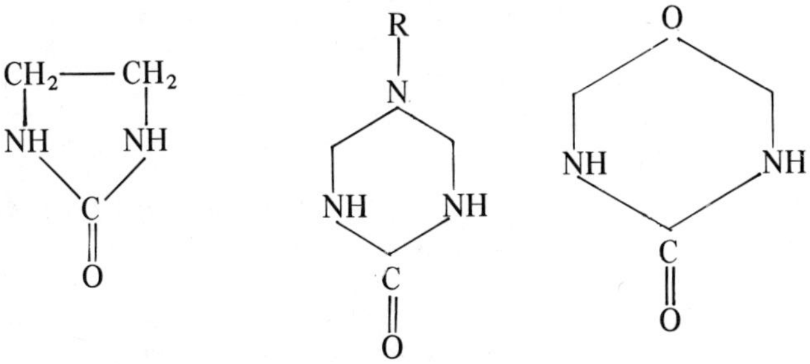

Eq. 3.61

ISBN 0-201-04558-3

$R = -CH_2-CH_2-OH$, and $R = CH_3$. Staudinger (1954, 1956) had suggested that such rings could form directly from urea and formaldehyde. ^{13}C-NMR (Slonim 1977) indicates that this occurs only in the presence of excess ammonia. The uron rings, finally, form whenever urea and formaldehyde are reacted with a large formaldehyde excess. They will be discussed in the chapter on UF-resins.

Reactions with Phenol

The literature on phenol-formaldehyde reactions and resins is far too large to be reviewed even superficially. The reader will find at least half a dozen truly outstanding reviews in this field. The phenol-aldehyde reactions were discovered and exploited some thirty years before the urea-formaldehyde system. Early work has been thoroughly reviewed by Ellis (1935) and Scheiber (1941). Every handbook on polymer chemistry devotes at least one major chapter to the phenol systems.

Phenol reacts in acidic as well as in basic medium both with formaldehyde and paraformaldehyde. With trioxane it forms a liquid adduct which reacts only in acidic medium. The reactions fall into two groups: (1) the formation of methylol phenols, usually called phenol alcohols, and (2) the formation of polynuclear methylene derivatives. Hemiformals are unstable. The reaction involves possibly a hemiformal intermediate:

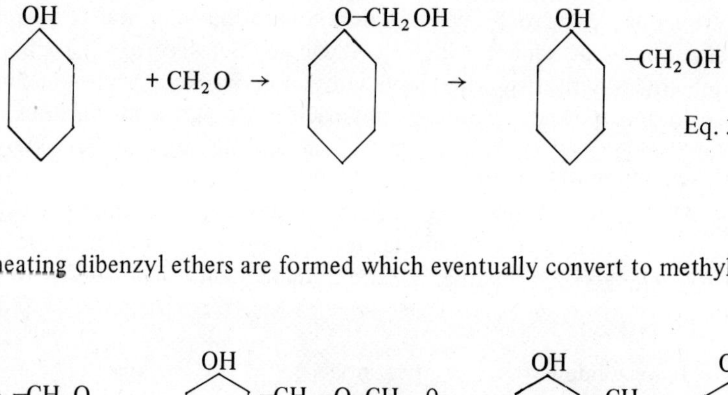

$$C_6H_5OH + CH_2O \rightarrow C_6H_5-O-CH_2OH \rightarrow HO-C_6H_4-CH_2OH$$

Eq. 3.62

Upon heating dibenzyl ethers are formed which eventually convert to methylene phenols:

$$HO-C_6H_4-CH_2O \rightarrow HO-C_6H_4-CH_2-O-CH_2-O \rightarrow HO-C_6H_4-CH_2-C_6H_4-OH$$

Eq. 3.63

The reactions occur preferably in ortho or para position. The influence of substituents on the reaction velocity, in decreasing order of reactivity with formaldehyde at 98°C in the presence of triethanolamine is:

ISBN 0-201-04558-3

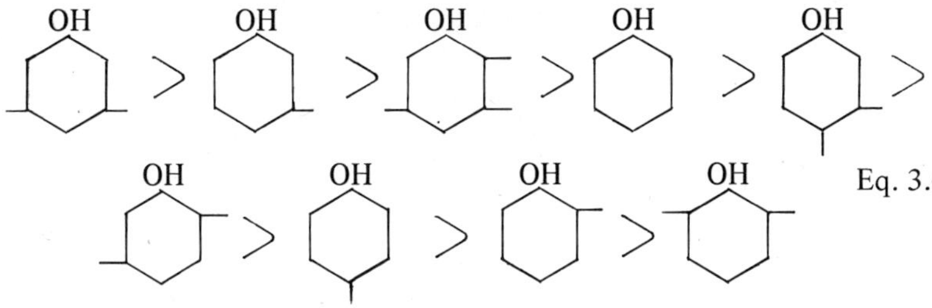

Eq. 3.64

All reactions are catalyzed by hydroxyl and hydronium ions. Obviously, a wealth of products is possible. Manasse (1884) was the first to use equimolar base to prepare methylolphenols; during the same year Lederer made the same products by heating with less alkali. Since then phenol alcohols have been generally prepared by the first method. The products are called resole. The simplest products are ortho-methylolphenol, called saligenin, and paramethylolphenol, called homosaligenin.

Saligenin melts at 84°C. It was first described by Piria (1843). He isolated it from natural salicin. The three most common dimethylol phenols are 2,4-dimethylphenol which melts at 92°C; 2,6-dimethylolphenol which melts at 96°C; and 2,4,6-trimethylolphenol which melts at 80°C. Substituted phenols form similar products. Cresol and meta-cresol is highly reactive. Ortho-nitrophenol produces methylols even in acidic medium.

As mentioned above, methylolphenols form hydroxybenzyl ethers as primary products. The latter convert eventually to diphenylmethylene. In the presence of a third reagent, formaldehyde can link phenols with RH, where RH = HCl, H_2SO_3, $(CH_3)_2NH$, etc.

$$C_6H_5OH + CH_2O + H_2SO_3 \rightarrow C_6H_4(OH)CH_2HSO_3 + H_2O \qquad \text{Eq. 3.65}$$

The chloro-derivatives form best if cool, concentrated hydrochloric acid is used to prevent resin formation. Upon boiling in water, the chlorides convert to methylols. The first sulfomethylation was patented by Bayer in 1896 using beta-naphthol.

Dimethylamine, formaldehyde and phenol react via the Mannick mechanism to *p*-hydroxydimethylbenzylamine. Bis- and tris(dimethylaminomethyl)phenol can be made by similar reactions.

Methylene derivatives are prepared by using large excess of phenol. The first monomer clearly identified in the phenol-formaldehyde system were 2,4′, 4,4 and

ISBN 0-201-04558-3

2,2′ dihydroxydiphenyl methane. They melt around 119°C:

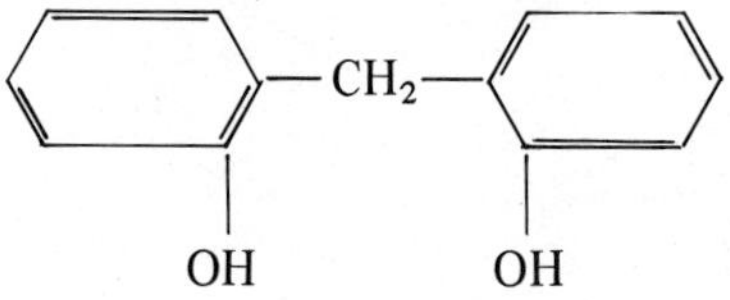

$$\text{C}_6\text{H}_4(\text{OH})\text{—CH}_2\text{—C}_6\text{H}_4(\text{OH})$$

OH OH

The 2,2′ isomer is the most reactive, the 2,4′ is least reactive. Tri- and tetranuclear methylenephenols are also known. Polymers are called Novolaks. Analysis of these systems is difficult, as hexanuclear Novolaks can form 439 different ortho and para isomers.

All the above reactions also are known to occur with resorcinol, cresols, and other similar aromatics. Polyhydroxyphenols reacts quickly and quantitatively with formaldehyde. Most of these products can be converted to resins. Two classes are distinguished: linear resins which are fusible, and infusible resins, characterized by crosslinking.

Novolaks, linear fusible resins, derived from monomethyl phenols were made by Backeland by acid catalysis, using a P:F ratio greater than 1. The presence of hydroxy groups in these resins causes them to be soluble in hot alkali. In commercial use, Novolaks are cured by the addition of formaldehyde.

Infusible resins form step-wise: the resole stage, also called A-stage, the resitol or B-stage, and the resite or C-stage. The A resins are soluble in alcohol and contain alcohols; B resins are insoluble but soften upon heating.

Resorcinol yields "cold-set" adhesives which become infusible at room temperature when mixed with excess formaldehyde. Such resins are used as dip adhesives for tire-cord. Dankelman and de Wit 1977 reported the ^{13}C-NMR spectrum of such a resin and the functional groups observed. From these spectra average chain length and the molar ratio R:F can be computed. Gas liquid chromatography, high pressure liquid chromatography and gel-permeation chromatography by the same authors indicate that the average chain length is only 2,5 molecular units and that the R:F ratio can be measured to within ± 5%.

Urea

Carbamide, or urea, was first identified by Rouelle in 1773. The adult human body discharges almost 50 grams of urea daily. In 1828 Wöhler synthesized urea from ammonium cyanate. In doing so he broke the barrier between inorganic chemistry and the chemistry of living matter. Within thirty years of Wöhler's discovery, urea became a popular fertilizer. Today, six million tons of urea are manu-

ISBN 0-201-04558-3

factured each year in the U.S. alone. Three quarters of this is used as fertilizer. Despite the great commercial importance of urea, very little basic chemical research has been conducted during the last hundred years, and there is not even one recent review chapter on urea chemistry available in this author's library.

Urea has a molecular weight of 60.06. It is colorless, odorless, and exudes a soothing taste. Its density is 1.335 g/cm^3; its melting point is 132.7°C; its solubility in water increases from 78 wt % at 5°C to 100 wt % at 17°C, 110 wt % at 21°C, and 733 wt % at 100°C. The apparent dissociation constant is 1.15×10^{-11} at 25°C and 1.74×10^{-14} at 30°C. The solubility in aqueous ethanol is 20 wt % at room temperature; pure ethanol dissolves 5 wt % at 20°C.

A boiling urea solution slowly decomposes and ammonium carbonate is formed. Acid and base catalyze decomposition in solution. Pure liquid urea slowly decomposes above 135°C, yielding ammonia, carbon dioxide, and cyanuric acid. Between 150 and 170°C biuret, NH_2–CO–NH–CO–NH_2, is formed. Biuret also forms as a side product during vacuum evaporation. Above 200°C, cyanuric acid triureide is formed.

The structure of urea has been determined by neutron diffraction, x-ray, infrared and other methods. The molecule has C_{2v} symmetry. The C–O bond length is 1.243 Å, C–N is 1.351 Å, N–H–(O) = 0.988 Å, and N–H–(N) = 0.995 Å. The angles are N–C–O = 121.5°. N–C–N = 117°, C–N–H(O) = 119.8°, C–N–H(N) = 118.1°, and H–N–H = 122.1°. The crystals are tetragonal with the space group $P_{42,m} = D^3{}_{2d}$. The unit cell measures a = 5,661 Å, c = 4,712 Å. Each unit contains two molecules. The vapor pressure has been measured by Suzuki (1956) between 73°C, where it is 1.6 Torr, to 95°C, where it is 10.2 Torr.

The infrared and Raman spectra have been measured by Spinner (1959), Mecke (1959), Schradere (1959), and Yamaguchi (1957). Table 3.16 lists the frequencies and the assignment of the nine normal modes. The H-NMR spectrum has been published by Andrew (1955), Dankelman (1976), Tomita (1978), Chiavarini (1975), and Duclairoir (1976); the ^{14}N-NMR by Schmidt (1959); and the ^{13}C-NMR by Tomita (1978), Dankelman (1978), and Ebdon (1977).

Urea is an amazing chemical. It can react in three fundamentally different ways.

1. With acids, such as sulfuric acid, phosphoric acid, hydrochloric, and nitric acid, urea forms salts of the type: $NH_2 \cdot CO \cdot NH_3 \cdot NO_3$. The vibration spectra show that these ions have the structure $NH_2CON^+H_3$ (Spinner 1959). Sodium and potassium salts are formed from the amides in liquid ammonia: Na $NH \cdot CO \cdot NH_2$.

2. With many substances, urea forms clathrates and adducts. Examples are phenol, hydrochinone, formic acid, salicylic acid, trichloroacetic acid, ammonia, hydrogen peroxide, phenol, and antipyridine.

3. Urea undergoes chemical reactions typical for the carbonyl-amido functional groups.

ISBN 0-201-04558-3

Table 3.16

Calculated and observed vibrational frequencies of Urea and perdeutero-Urea (after Yamaguchi, 1957)

Symmetry Type	$OC(NH_2)_2$			$OC(ND_2)_2$		
	Calc.		Obs.	Calc.		Obs.
A_1	ν_1	1648	1686	ν_1'	1597	1610
	ν_2	1615	1603	ν_2'	1265	1245
	ν_3	1143	1150	ν_3'	986	1001
	ν_4	1012	1000	ν_4'	874	887
	ν_5	555	556	ν_5'	476	475
B_2	ν_6	1639	1629	ν_6'	1477	1490
	ν_7	1461	1464	ν_7'	1174	1154
	ν_8	1152	1150	ν_8'	893	887
	ν_9	558	570	ν_9'	523	512

REACTIONS OF UREA

The reactions of formaldehyde with urea are discussed in the chapters on formaldehyde chemistry and UF-resin chemistry. With chlorine, $Cl-NH-CO-NH_2$ is formed, and NHCl–CO–NHCl. Nitrous acid decomposes urea to N_2, CO_2, and water. Sulfurylamide yields $NH_2-SO_2-NH-CO-NH_2$. Tertiary alcohols give $R_3-C-NH-CO-NH_2$. Oxonium compounds, such as $R_3O \cdot BF_4$, produce $NH_2-C(OR) = NH$. Chloral, CCl_3-CHO, yields $CH(OH)-NH-CO-NH_2$ and $CCl_3-CH(OH)-NH-CO-NH-CH(OH)-CCl_3$. Dioxo compounds, such as $(CHO)_2$, yield cyclic urones $CO(NH-CH-NH)_2CO$. Acid chlorides or anhydrides yield $R-CO-NH-CO-NH_2$. Phosgene, $COCl_2$, forms cyclic $(CO-NH)_2CO$. Malonic ester yields $CH_2(CO-NH)_2CO$.

Amines produce $R-NH-CO-NH_2$ and R–NH–CO–NH–R. 1,2-diamine yields cyclic $(CH_2-NH)_2CO$. Aminocarbonic acid leads to $HOOC-CH_2-NH-CO-NH_2$. Hydrazine leads to $H_2N-NH-CO-NH_2$ and $NH_2-CO-NH-NH-CO-NH_2$. Phenylhydrazine produces $O-NH-NH-CO-NH_2$. Urea can form resonance mesomers; The isourea:

$$NH_2-\overset{\overset{\displaystyle O}{\|}}{C}-NH_2 \leftrightarrow NH_2-\overset{\overset{\displaystyle OH}{|}}{C}=NH \qquad \text{Eq. 3.66}$$

form might be a very important intermediary, but it is not a prominent species in aqueous UF systems. A summary of urea reactions is shown in Figure 3.14.

ISBN 0-201-04558-3

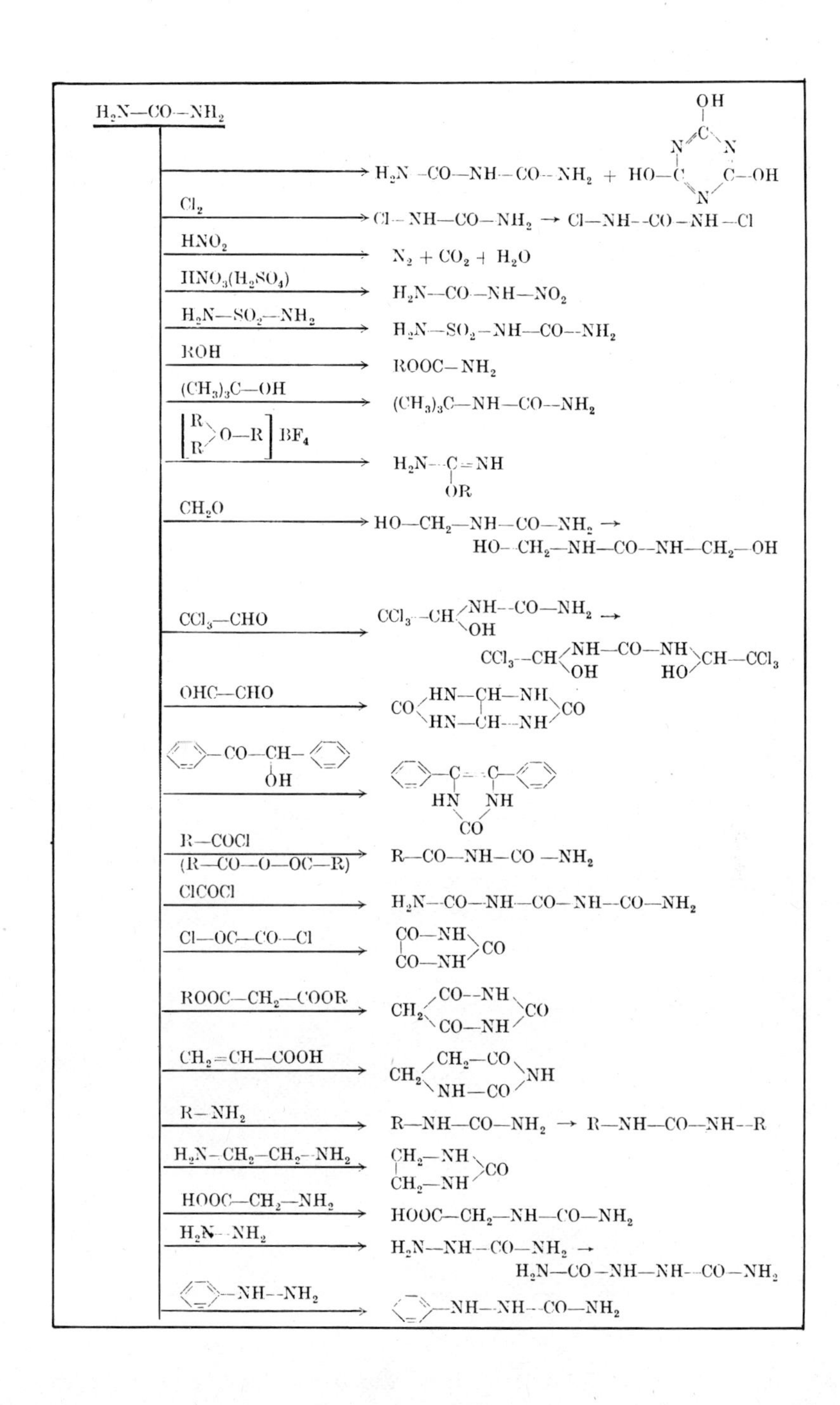

Figure 3.14. Summary of Characteristic Reactions of Urea (after Petersen, 1952)

ISBN 0-201-04558-3

THIOUREA

Thiourea has many similarities to urea. Thus, thiourea can be incorporated or substituted for urea in resins. The resins are usually upgraded by such treatment. The difference between the two becomes evident from the fact that the isomer is more readily formed:

$$\begin{array}{c} NH_2 \\ | \\ C{=}S \\ | \\ NH_2 \end{array} \rightleftharpoons \begin{array}{c} NH \\ | \\ C{-}H \\ | \\ NH_2 \end{array} \rightleftharpoons \begin{array}{c} NH_3^{\,+} \\ | \\ C{-}S^- \\ | \\ NH \end{array} \qquad \text{Eq. 3.67}$$

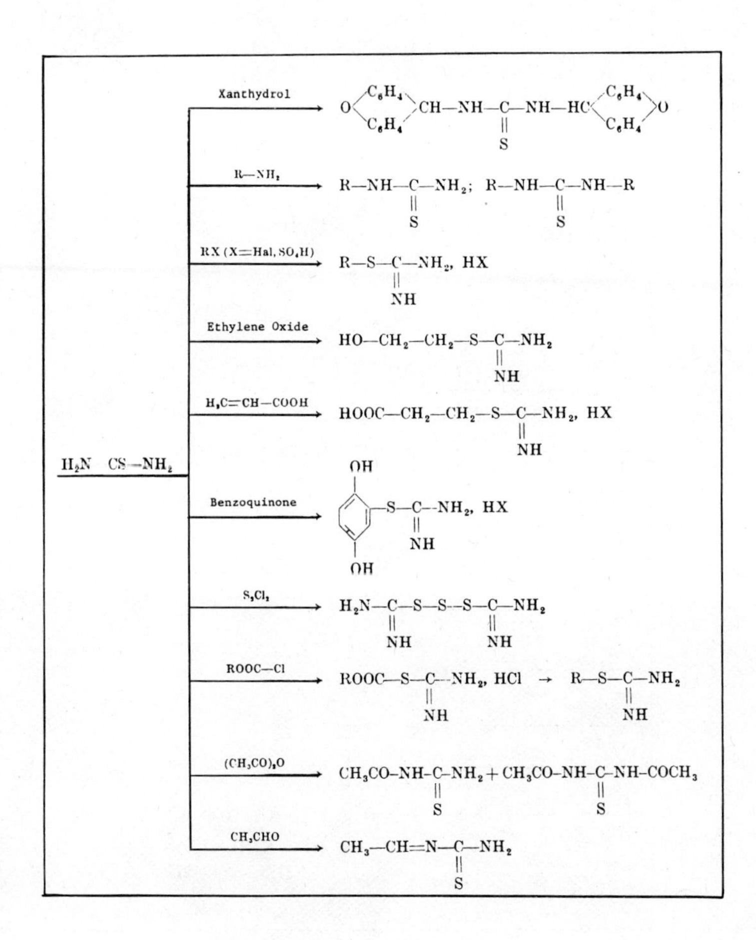

Figure 3.15. Summary of Characteristic Reactions Thiourea (after Bogemann, 1955)

ISBN 0-201-04558-3

The solubility of thiourea is 13 wt % at 20°C and 62 wt % at 60°C. Thiourea forms clathrates, like urea. In thiourea the sulfur atom can be methylolated. A short summary of other chemical reactions shows that: silver oxide converts thiourea to NH_2–CN; with amines the product is NH_2–C–(NH)–NH_2; 6% hydrogen peroxide oxidizes sulfur and yields NH–C(NH_2)–SO_2H and NH–C(NH_3)–S–S–C (NH)NH_2; permanganate yields urea; and C_6H_5–NCO forms C_6H_5–NH–CO–NH–CS–NH_2. R–NH_2 reacts to R–NH–CS–NH_2 or R–NH–CS–NH–R; alkylchlorides form R–S–CNH–NH_2; ethyleneoxide forms HO–CH_2–CH_2–S–C(NH)–NH_2; quinone produces ϕ $(OH)_2$–S–C(NH)NH_2, disulfur dichloride yields NH_2–C(NH)–S–S–S–C(NH)–NH_2; acetic anhydride reacts to CH_3CO–NH–CS–NH_2 and CH_3 CONH–CS–NH–$COCH_3$; and acetaldehyde forms CH_3–CH = N–CS–NH_2. As stated in the formaldehyde chapter, aldehydes probably attack the sulfur atom first and then rearrange to the nitrogen-substituted compound. Many reagents form cyclic compounds. Some of these are outlined in the schematics of Figures 3.15 to 3.17. These reactions are not currently important, but they might be useful for modifying resins with sulfur compounds.

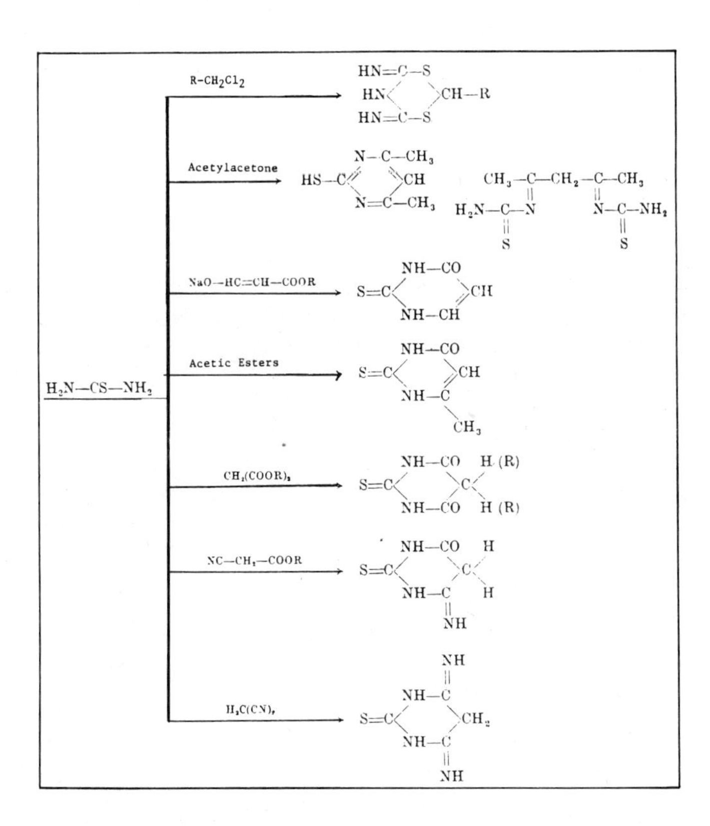

Figure 3.16. Characteristic Reactions of Thiourea (after Bogemann, 1955)

ISBN 0-201-04558-3

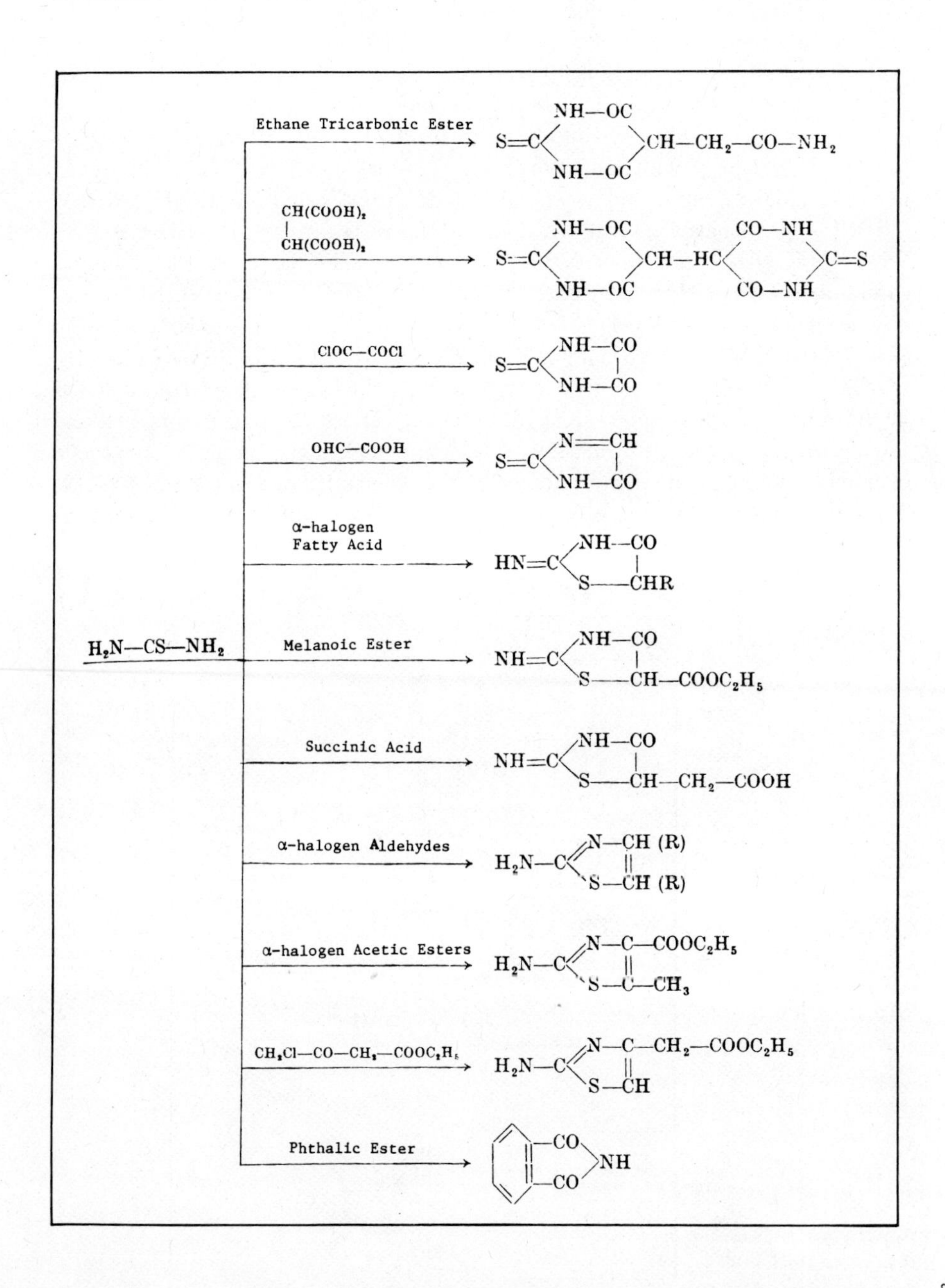

Figure 3.17. Characteristic Reactions of Thiourea (after Bogemann, 1955)

ISBN 0-201-04558-3

UF-Resins

Basic UF-resin chemistry consists of a step-wise addition of urea and formaldehyde to a resin of modest molecular weight. The manner of interaction is determined by the functional groups. Urea reacts as an amine, formaldehyde as a glycol. During the formation of the initial reaction products, water is released. The reaction path depends on pH, concentration, molar ratio, temperature, reaction time, and on the sequence of reactions. Obviously, the seven parameters permit a large number of permutations, and make for a complex system, belying the fact that only two reagents with simple functional groups are present.

For a more detailed description of the UF-system, the reader should consult specialized reviews of the field, for example the chapters by Wegler (1963), Widmer (1965), Baumann (1967), Petersen (1968), Schoenthaler (1972), Sandler (1977), or Updegraff (1978). Older work has been reviewed by Ellis (1935) and Scheiber (1943).

Currently, a new wave of interest has drawn attention to UF reactions. This is due to progress in the nuclear magnetic resonance method, an analytical tool that makes it possible to determine the composition of resins during manufacture. Key papers in this field are quoted in the text, and include work by Chiavarini, Dankelman, de Breet, Duclairoir, Kumlin, Slonim, and Tomita, all published since 1977.

PRINCIPLES OF RESIN FORMATION

Petersen, 1964, described the condensation of urea-formaldehyde as alpha-ureido alkylation involving three reagents:

$$\underset{\text{nucleophile}}{H{-}X} + \underset{\text{carbonyl}}{R_2CO} + \underset{\text{amino}}{R'NH_2} \rightleftharpoons \underset{\text{condensation product}}{R'{-}NH{-}C{-}X \cdot R_2 + H_2O} \qquad \text{Eq. 3.68}$$

This reaction principle is the common chemical basis of all aminoplasts. Some of the most common reagents are summarized in Table 3.17. The first reaction step almost always involves the alpha-methylolation of an amine via nucleophilic reaction. The reactivity of the system depends on how strongly the reagents are able to enter nucleophilic reactions. Figure 3.18, a typical example of the kinetics of such a reaction, represents the basic principle of all these reactions which states that the formation of monomethylol-ethyl urea from ethyl-urea and formaldehyde is a second order reaction, both in an acidic as well as in a basic medium. The C–N bond in the resulting hemiacetal is far less stable than normal C–N amine bonds. Thus, the reaction is reversible. The dissociation reaction of a similar compound,

ISBN 0-201-04558-3

Table 3.17

NH- Compounds Suitable for Preparing Aminoplasts

$H_2N\text{-}CO\text{-}R$	Acetamides	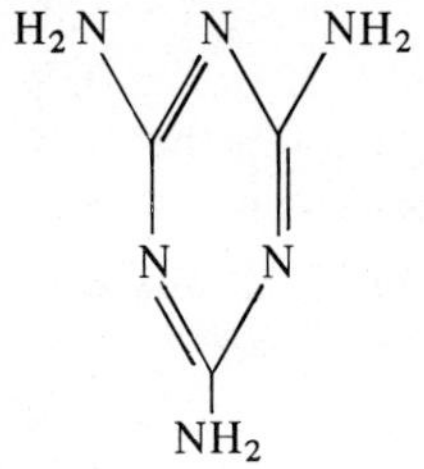 H_2N, N, NH_2 / N, N / NH_2 (triazine ring)	*Melamine*
$H_2N\text{-}CO\text{-}OR$	Urethane		
$H_2N\text{-}SO_2\text{-}R$	Sulfonic acid amides		
$H_2N\text{-}CO\text{-}NH_2$	Urea		
$H_2N\text{-}CS\text{-}NH_2$	Thiourea		
$H_2N\text{-}C(=NH)\text{-}NH_2$	Guanidine	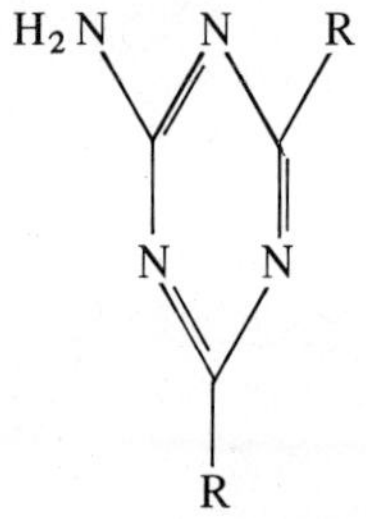 H_2N, N, R / N, N / R (triazine ring)	Guanamine and other Aminotriazines
$H_2N\text{-}SO_2\text{-}NH_2$	Sulfurylamide		
$H_2N\text{-}CN$	Cyanamid and Dicyandiamide		

dimethylol-propyl-urea is shown in Figure 3.19. The reaction is first order. The equilibrium between the reagent and the methylol is not sensitive to pH, but it depends on the molar ratio of the reagents, total concentration, temperature, and, of course, on the chemical reactivity of the reagents and products. The reaction mechanism does not affect the equilibrium. In acids, the reaction probably proceeds via an intermediate carbenium ion, $CH_2{}^+OH$. The actual reaction intermediates are not known. Nominally, most chemists write equations of the type shown in Equation 3.68, but in reality:

$$CH_2O + H^+ \rightleftharpoons CH_2{}^+\text{–}OH + H_2O \qquad \text{Eq. 3.69}$$

$$CH_2(OH)_2 + H_3O^+ \rightarrow CH_2(OH)(OH_2{}^+) + H_2O \rightarrow CH_2{}^+\text{–}OH\text{–}2H_2O \qquad \text{Eq. 3.70}$$

formaldehyde exists as methylene glycol, and H^+ as H_3O^+, and thus reactions of the type in Equations 3.69 and 3.70 seem more realistic, even though the free carbenium ion has never been identified in aqueous solution, and cannot possibly

ISBN 0-201-04558-3

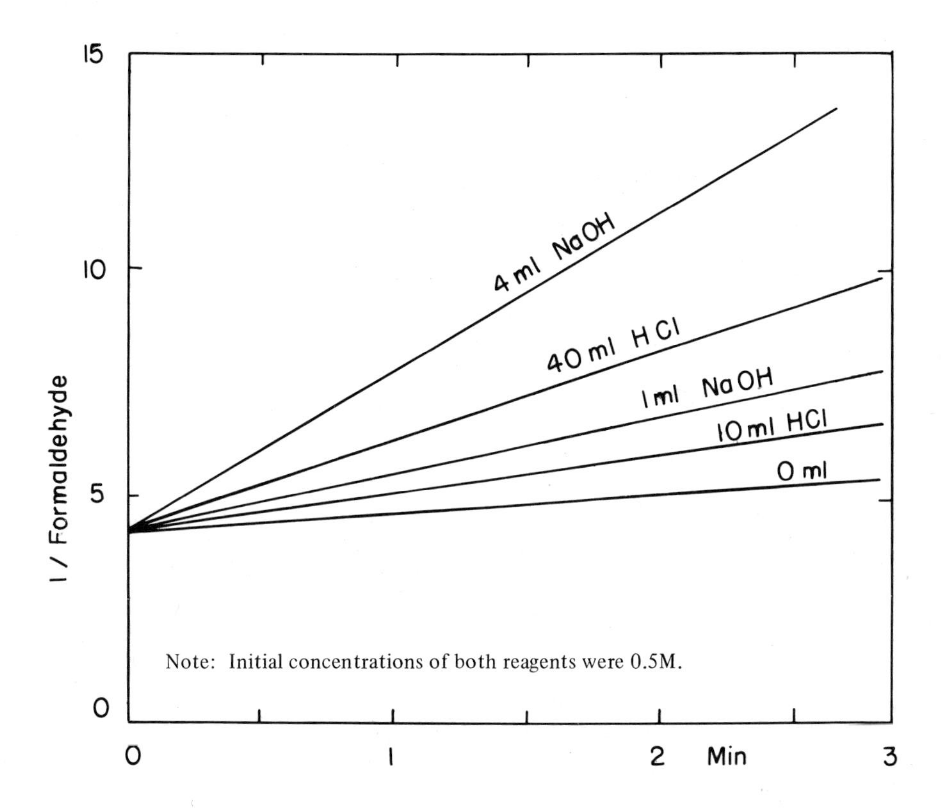

Figure 3.18. Kinetics of the Formation of Ethyleneurea Monomethylol as a Function of Acid or Base Catalyst at 50°C (after Petersen, 1965)

exist in any appreciable concentration. It is then assumed that the carbenium ion attaches to the nitrogen electron and expels an amine proton:

$$R{-}NH_2 + CH_2{}^{+}OH \rightarrow R{-}NH_2{}^{+}{-}CH_2OH \qquad \text{Eq. 3.71}$$

In basic solution the commonly invoked reaction mechanism involves base attack on the amine group, which, in turn, attacks methyl glycol and:

$$R{-}CO{-}NH_2 + B \rightarrow R{-}CO{-}NH^{-} + BH \qquad \text{Eq. 3.72}$$

ISBN 0-201-04558-3

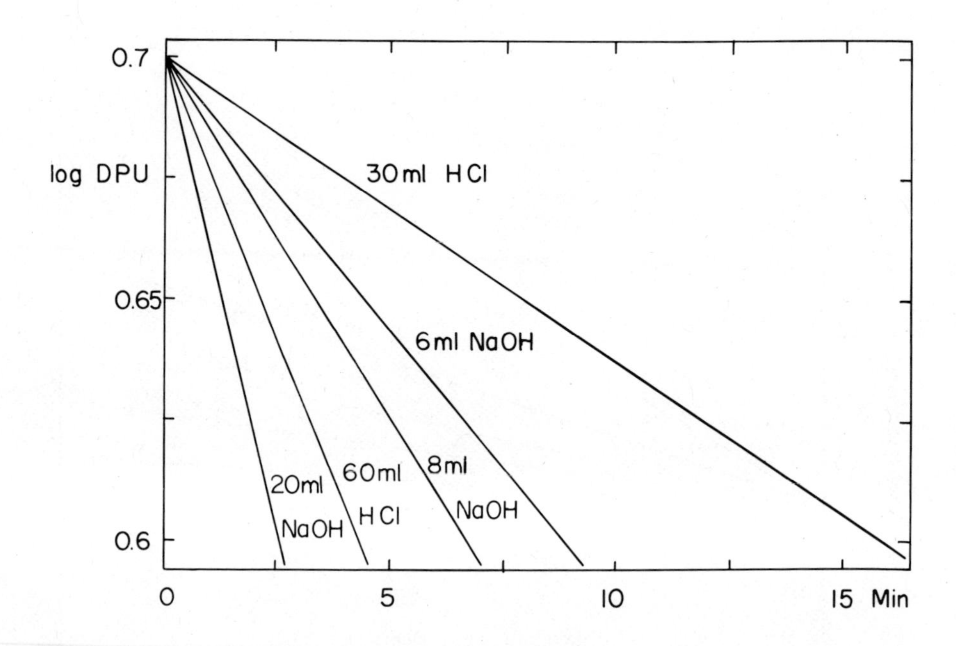

Figure 3.19. Dissociation of Propylene-urea Dimethylol at 70°C in 0.5 M Solution as a Function of Acid and Base Catalyst (after Petersen, 1965)

forms the methylol ion. These processes are aided by other supporting mechanisms. For example, in the acidic system, ureas can form mesomers in which the carbonyl group is protonated, and thus decreases the reactivity of the nitrogen atom via resonance. This effect is far more prominent in urea than, for example, in the planar ethylurea molecule in which the nitrogen electrons are locked out of the coplanar position necessary for resonance. Thus, the reactivity of the cyclic compound is more than ten times higher (Petersen 1964). An example of the reverse is provided by 4,5-hydroxy ethylurea which has basically the same ring structure as ethylurea, but in which the hydroxy groups enable the nitrogen electrons to participate in mesomeric ammonium ions. This reduces the reactivity of the molecule towards formaldehyde by a factor of a hundred, i.e., 4,5-hydroxy ethylurea is less reactive than urea. This book does not discuss in detail all the effects that might influence the reaction mechanism. Most of these mechanisms are hypothetical, anyhow. However, the relative reactivity reveals itself in the rate constants shown in Table 3.18 and in the C–O bond strength, as reflected in infrared or Raman spectra. In the seven compounds studied by Petersen (1964) the correlation is striking. High C–O frequency, i.e., high bond strength of C–O corresponds to low reactivity of the amino group. The correlation between kinetics and C–O bond is shown in Figure 3.20.

ISBN 0-201-04558-3

Table 3.18

Rate Constants for the Formation of the Monomethylol of 7 Aminocompounds at pH7 and 50°C, and the Corresponding Vibrational Frequencies of the Reagent Carbonyl (after Petersen, 1965)

Compound	υ_{co} cm^{-1} in $CHCl_3$	k_1 [Lit/Mol·sec]
4,5-Dimethoxy-ethylene urea	1722	10^{-1}
Ethylene urea	1704	$95 \cdot 10^{-4}$
Urea	1680	$4 \cdot 10^{-4}$
5-Methyl-1,3,5-hexahydro-triazinon	1669	$1.4 \cdot 10^{-4}$
5-Methyl-propylene urea	1667	$0.88 \cdot 10^{-4}$
Propylene urea	1665	$0.82 \cdot 10^{-4}$
Cyclobutylene urea	1663	$0.5 \cdot 10^{-4}$

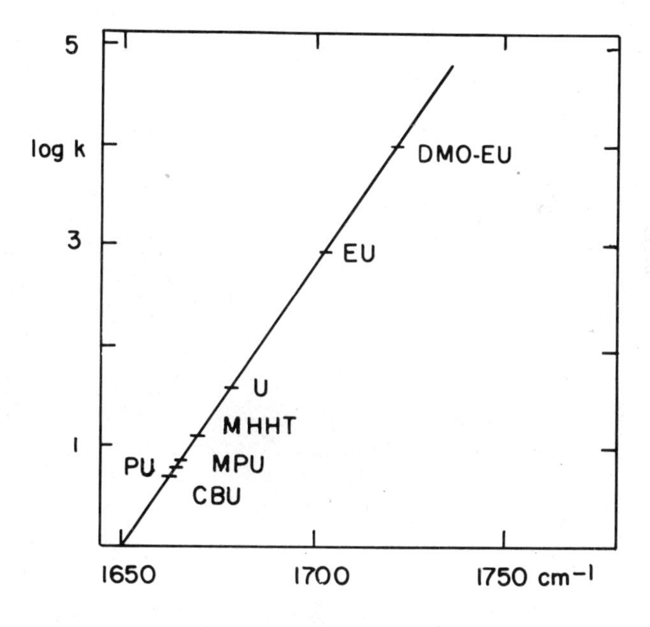

Figure 3.20. Correlation between Methylol Formation Rate Constant and the Vibrational Frequency of the Carbonyl Function, at 50°C and pH 7 (after Petersen, 1965) DMO-EU = 4,5-Dimethoxy-ethylene-urea; EU = Ethylene Urea; U = Urea; MHHT = 5-Methyl-1,3,5-Hexahydrotriazinone; MPU = 5-Methyl-propylene-urea; PU = Propylene-urea; CBU = Cyclobutylene-urea

ISBN 0-201-04558-3

It is well established that all reactions are ionic, and thus catalyzed by hydroxyl and hydronium ions. Figure 3.21 shows the rate constants of four aminocarbonyls as a function of pH. The reagents were imidazolidone urea, 4,5-dihydroxy-imidazolidone and hexahydro-pyrimidone (Petersen 1964). Typical aldehyde groups are listed in Table 3.19. Formaldehyde is still the predominant commercial reagent. However, in the last fifty years many other carbonyl compounds have been investigated. If some of the more reactive higher aldehydes could be cheaply produced in quantity, they would now undoubtedly be considered as substitutes to avoid the unpleasant odor which formaldehyde imparts on its products under some conditions.

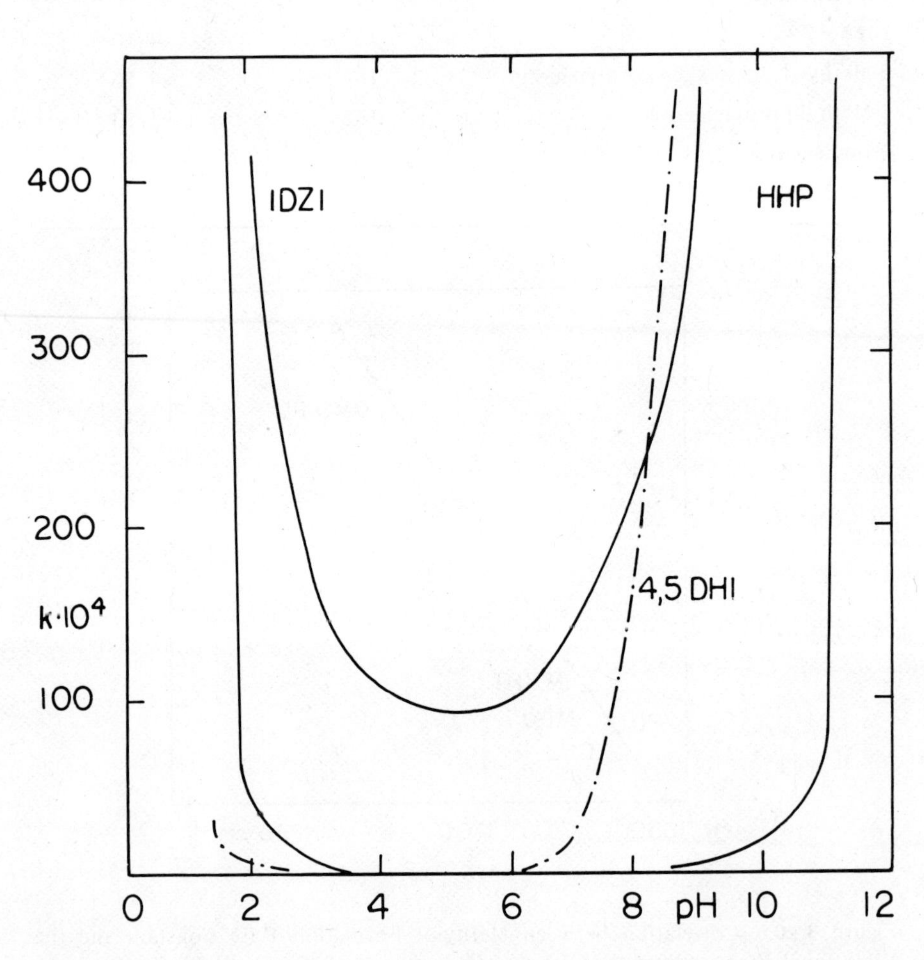

Figure 3.21. Rate Constants of the First Methylolation of Urea as a Function of pH at 50°C in 0.5 M Solution (after Petersen, 1965) IDZI = Imidazolidone; 4,5 DHI = 4,5-Dihydroxy-Imidazolidone; HHP = Hexahydro-pyrimidone

ISBN 0-201-04558-3

Table 3.19

12 Aldehydes and Ketones Suitable for Aminoplast Formation

Compounds	Examples
CH_2O	Formaldehyde
$OHC{-}CH_3$	Acetaldehyde Propionaldehyde-n- and iso-Butyraldehyde
$OHC{-}CHO$	Glyoxal
$OHC{-}CH_2{-}CHO$	Malonic-dialdehyde
$OHC{-}COOH$	Glyoxylic acid
$OHC{-}CCl_3$	Chloral
$OHC{-}C(CH_3)_2{-}CH_2OH$	Hydroxypivalaldehyde
$OHC{-}CH_2{-}CHOH{-}CH_3$	Acetaldol and Aldoles
$OHC{-}CH{=}CH_2$	Acrolein, Crotonaldehyde
$CH_3{-}CO{-}CH_3$	Acetone and Ketones, such as Cyclopentanone or Cyclohexanone
$CH_3{-}CO{-}CH_2{-}C(OH)(CH_3)_2$	Diaceton alcohols
$CH_3{-}CO{-}CH{=}CH_2$	Vinylmethylketon and other unsaturated Ketones
$(CH_3)_2C{=}CH{-}CO{-}CH{=}C(CH_3)_2$	Phoron

Glyoxale has two functional groups and thus tends to form cyclic compounds which yield resins with different properties.

$$NH_2{-}CO{-}NH_2 + 2CH_2O \rightleftharpoons \text{cyclic: } C(=O) \text{ bonded to two NH; NH}{-}CH(OH){-}CH(OH){-}NH \qquad \text{Eq. 3.73}$$

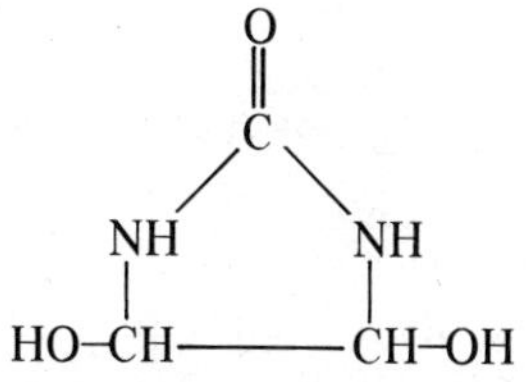

Aldehydes and ketones with alpha-CH react not only as carbonyls, but also undergo nucleophilic reactions at the acidic C–H yielding aldols.

ISBN 0-201-04558-3

The nucleophilic component necessary for amino-carbonyl condensations can be provided by any of a variety of proton donor reagents. The most common classes are mineral acids, OH-acidic compounds, acidic S–H, NH and CH groups, and some olefins. The classic compounds are hydrogen chloride or hydrogen bromide which act as a source of hydronium ion; acidic alcohols, hemiacetals, peroxide, or carbonic acids. Among the sulfur compounds, hydrogen sulfide, sulfites, mercaptans, thiophenols, and thiourea are very effective. Their use has been limited because handling toxic sulfides requires skills, and the sulfites influence the chemical and mechanical properties of both the resin and products. Nitrogen-acid compounds are the most widely used. Among these are ammonia, primary and secondary amines, urethane, sulfonamides, urea, thiourea, guanidine, melamine, and cyanamide. Effective C–H groups are provided by aromatic phenols or amines, heterocycles, aldehydes, and ketones. Among the olefins, styrole, isobutylene, and isoprene are the best known examples.

Condensation Reaction

In basic chemical terms, the key for the condensation of methylol compounds is their ability to accept a proton and to form a alpha-ureidol-alkyl-carbenium-ammonium ions. They are formed with a low activation energy due to their mesomeric structure:

$$R_2N\text{–}CO\text{–}NH\text{–}CH_2OH_2{}^+ \rightarrow R_2N\text{–}CO\text{–}NH\text{–}CH_2{}^+ \rightarrow R_2N\text{–}CO\text{–}NH^+ = CH_2 \quad \text{Eq. 3.74}$$

The actual alpha-ureido alkylation consists then of a complicated sequence of step-wise reactions, many of which are reversible, and many of which are paralleled by side reactions. Petersen (1964) used mercaptans and N-methoxy-methyl urea as model compounds to study the kinetics. In this reaction mercaptan is used by addition and methanol is formed by replacement. Thus, he could distinguish between the addition and dissociation reactions, and he could compute the concentration of the intermediate carbenium ion, curve B in Figure 3.22. The figure shows that the product appears only after a significant induction period. Such induction periods are characteristic for all condensations.

A further basic characteristic of all aminoplast condensation reactions is that N-alpha-alkylol compounds readily exchange their nucleophile partners in favor of stronger nucleophiles; normally, X = OH, but X can be any other group:

$$\text{–}NH\text{–}C\text{–}H_2\text{–}X + HY \rightarrow \text{–}NH\text{–}CH_2\text{–}Y + HX \quad \text{Eq. 3.75}$$

This process has been called trans-alpha-ureido alkylation. Such rearrangement reactions occur readily upon heating or in the presence of strong acids. Zigeuner (1951, 1952, 1955) studied this reaction for the example of diureamethylene and phenol. Two modifications of resin properties are directly due to such trans-

ISBN 0-201-04558-3

ureido alkylation. Phenosulfonic acids react in this manner and thus render polymethylene ureas soluble; and methylene urea reacts with Novolaks by the same mechanism, and thus causes their hardening. However, such reactions occur in all systems, and even if they do not immediately result in dramatic changes of properties, they deeply affect all applications, regardless of whether it is in textiles, varnishes, or adhesives.

In normal UF-resins, the nucleophile in Equation 3.68 is again urea or an urea derivative. These reactions will be discussed in the following section. In modified UF-resins, either the nucleophile or the carbonyl is exchanged for one, or a mixture of the reagents listed in Tables 3.17 and 3.19. Such modified systems are discussed in a separate section.

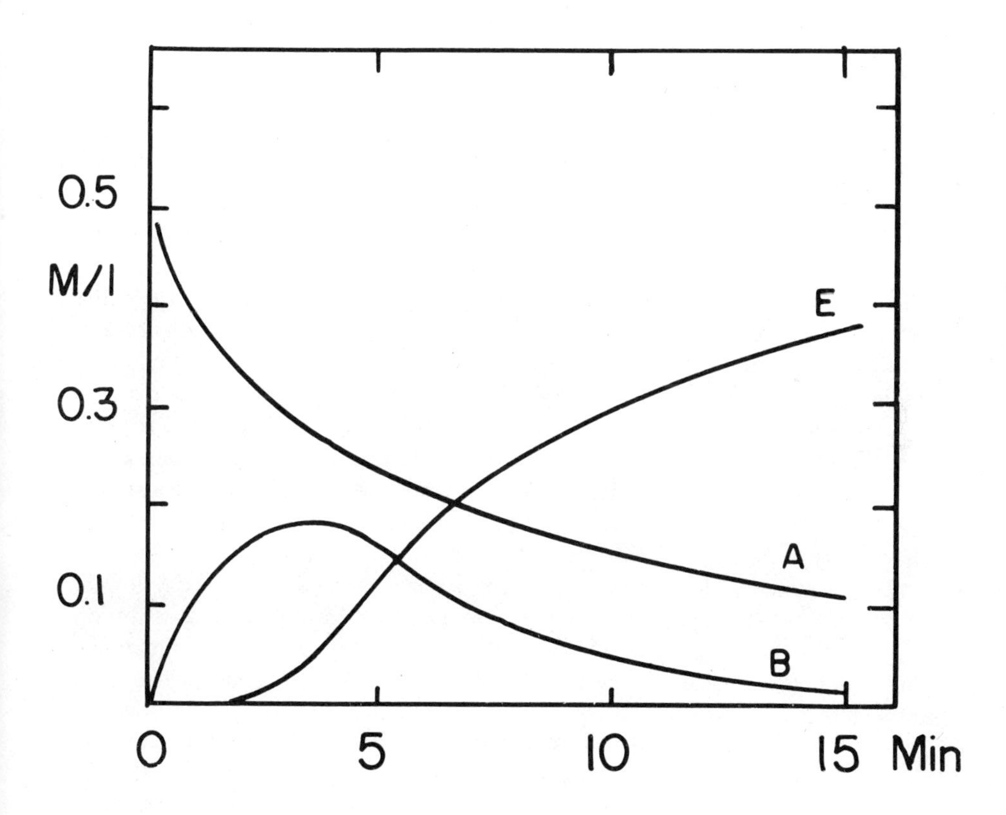

Figure 3.22. Kinetics of the Reaction of *N*– Methoxymethyl-*N'* Methyl Ethylene-Urea (A) with n-Dodecylomercaptan (after Petersen, 1965) (A) $R{-}C_3H_2N_2O{-}CH_2OCH_3$; (B) $R{-}C_3H_2N_2O{-}CH_2^+$; (E) $R{-}C_3H_2N_2O{-}CH_2{-}S{-}C_{12}H_{25}$

ISBN 0-201-04558-3

REACTION OF UREA WITH FORMALDEHYDE

Tollens discovered in 1884 that formaldehyde forms resins with urea. It took another thirty years before better reaction conditions were formed. Only in the last twenty years have the reaction mechanisms been established experimentally. The reason for the slow progress is the intrinsic complexity of the products, caused by the sensitivity of the reaction to temperature, concentration, and pH. Recently a major breakthrough in the understanding of the U:F system has occurred due to modern analytical tools which make *in situ* analysis of the various mixtures possible. Among the methods, ^{13}C-NMR has probably offered the most dramatic progress. Until some thirty years ago, analysis had to rely mainly on elemental analysis, which yielded the ratio of the various reagent atoms in the product. Such analysis necessitated complex deductive analysis. Work before 1943 has been reviewed by Scheiber (1943). Later, it became possible to identify reactive groups in resins, such as primary amine $-NH_2-$, secondary amine $-NH-$, and carbonyls CO. From these, some educated guesses of the average degree of crosslinkage were possible.

In 1951 at the Pharmaceutical Chemistry Institute in Graz, Zigeuner started a very productive study leading to a series of important papers with a short study on co-condensation of 2,4-dimethylphenol with urea-formaldehyde. Almost simultaneously, at the Phillips Research Laboratory in Eindhoven, J. de Jong and J. de Jonge began a parallel series of experiments on equilibria and kinetics. Their first study, in 1952, dealt with the formation and decomposition of dimethylolurea. They found that the equilibrium between urea, formaldehyde, and dimethylolurea was independent of the pH and could be established from either side; that the synthesis was bimolecular and the dissociation monomolecular, as expected; that both the hydroxyl and hydronium ions catalyzed the reaction; that buffers influence kinetics; and that the activation energy was $\Delta H = 19$ kcal/mole. In a second paper they showed in 1952 that concentrated systems react in the same manner as dilute systems. A study on monomethylurea followed, showing similarities between mono- and dimethylolurea kinetics. In 1953 a paper on trimethylolurea followed, and another paper on the formation of methylene bridges. Again, the activation energy was found to be about 15 kcal/mole. Finally, they analyzed the functional groups in urea-formaldehyde systems at pH 2–7 and 20–76°C, and deduced an average molecular weight of 300–500 for methylene urea.

In 1955, Zigeuner identified and confirmed biscarbamidomethyl ether in alkaline solutions, and, in another paper, reported the reactions of methylene diurea. In 1957 Lundqvist at Rydboholms Laboratory, Sweden, quantitatively studied the equilibria of urea and formaldehyde by ultraviolet spectroscopy, using absorption at 4300 Å. He studied the influence of methanol, pH, and considered the possible structure of formaldehyde and methyl glycol zwitterions and tautomers of urea. In 1974 Ding showed that UF and MF can be distinguished in the ultraviolet. In 1961 Zigeuner explored the correlation between chemical structure of reagents and their relative tendencies to form linear versus cyclic products. In his

ISBN 0-201-04558-3

series of modified urea-formaldehyde systems a study on acetaldehyde and methylurea in acidic medium revealed aldoecondensation leading to 2-oxo-1,4-dimethyl-6-(3-methylureido)-hexahydro-pyrimidone and similar heterocycles. In the same year, Zigeuner (1961) prepared beta-ureidoaldehyde acetals such as 2-oxo-6-ureido-hexahydro-pyrimidine and 2-oxo-6-hydroxy-phenyl-hexhydro-pyrimidine, and studied an entire series of homologues. In 1965 this series continued, including alpha, beta unsaturated aldehydes.

In the 1960s infrared spectroscopy became easier and more popular. Potutkin (1972) studied the lignin UF-resin system; Baranowski (1972) studied coatings and enamels; and Svrdiik (1973) conducted both qualitative and quantitative analysis of UF- and MF-resins. In 1974 Baranowski studied lacquers. Ivanov (1975) used infrared to correlate the properties of modified UF-resins to chemical structure. Suetaka (1976) studied thin films with an infrared spectrometer; Fourier-transform infrared was used by Liebman (1976), and Pshenitsyna (1978) used the secondary amide bands to deduce crosslinking and other properties.

Lee (1972) and many others employed paper chromatography; Schliefer used thin-layer chromatography, as did Ludlam (1973) who successfully separated urea, monomethylol urea, and dimethylol urea. Hope (1973) used gel-permeation chromatography to study molecular weight distributions. Tsuge (1974) used the same method and found average molecular weights of 200–500 in UF-resins, and found the molecular weight distribution to be much larger than in PF and MF. El'bert (1977) used gel chromatography to determine thermal hardening and crosslinking.

In recent years, H-NMR made it possible to recognize reactive groups *in situ* without the necessity for disturbing the subtle equilibria. Richard (1975) distinguished methylol and ether bridges. Chiavarini and coworkers were probably the first to fully demonstrate the use of this method which has stimulated many publications by other authors worldwide. The drawback of H-NMR is that water interferes and that the method is only partly selective. Thus, it was necessary to resort again to isolation of species, but now more sophisticated tools had become available. Thus, equilibria were frozen by the reaction of resins with BSTFA (*N,O*-bistrimethyl-silyl-trifluoro-acetamide) which blocks amino groups and alcohol, and makes gel-permeation chromatography, gas-liquid chromatography, and similar techniques possible. In very recent years, the situation has been drastically simplified, because ^{13}C-NMR gives excellent resolution of functional groups, and makes it even possible to distinguish between neighboring functions of a given functions group. Since 1977 this work has already lead to important results by several authors. Among these workers are Dankelman's group in Holland, Slonim's group in Russia, Tomita's group in Japan, Ebdon in England, and Duclairoir's in France, to name only a few. Thus, in the coming years, it will be possible to unravel the composition of UF-resins at various stages of reaction. However, this is not all. The NMR method opens two more possibilities. It makes quantitative work, and thus the observation of equilibria *in situ,* possible. And finally, time resolved NMR,

ISBN 0-201-04558-3

Raman, and similar work make it possible to determine absolute and relative rates of reaction. This chapter can give only a very superficial summary of what has been observed so far.

Acid Catalyzed Reactions

It is likely that the acidic reaction involves the carbenium ion $CH_2{}^+{-}OH$ formed by proton action of methylene glycol:

$$CH_2(OH)_2 + H_3O^+ \rightarrow CH_2{}^+{-}(OH){-}OH_2{}^+ \rightarrow CH_2{}^+{-}OH + H_2O \qquad \text{Eq. 3.76}$$

the intermediate ions have not been identified. The first reaction step is probably the addition of the carbenium ion to urea:

$$NH_2{-}CO{-}NH_2{-}CH_2{}^+{-}OH \rightarrow NH_2{-}CO{-}NH{-}CH_2OH \qquad \text{Eq. 3.77}$$

followed by deprotonation. The resulting product is the methylol compound (Bois de Chesne 1932). Acidic reactions are characterized by the fact that methylolurea is unstable and vulnerable to protonation. The resulting oxonium ion dehydrates promptly and forms again a carbenium ion:

$$\begin{array}{ccc} & A & \\ U{-}CH_2OH & \rightarrow & NH_2{-}CO{-}NH{-}CH_2OH_2{}^+ \\ B \updownarrow H_2O & & \updownarrow \\ H_2N{-}CO{-}N{=}CH_2 & \rightarrow & NH_2{-}CO{-}NH{-}CH_2{}^+ \end{array} \qquad \text{Eq. 3.78}$$

Further reaction depends on the concentration, pH, and temperature of the methylenium urea solution. With excess urea the products include methylenediurea, $U{-}CH_2{-}U$. If this reaction proceeds via path A, above, it is nominally an addition. If path B, prevails, it is nominally a condensation. In the presence of concentrated acid, the equilibria shifts backwards in favor of the reagents. Kadowaki (1936) claimed that he could continue the reaction via methylol methyl-diurea up to products containing five urea units. However, these products are insoluble, and thus Staudinger (1955) found that methylene-diurea is not normally capable of further addition or condensation. Higher molecular weight products are formed by cross-linking and ring closure. Intermediate in such reactions is, for example, cyclic tetra-methyldimethyl-diurea, first proposed, again, by Kadowaki (1936). In very strong acid, U:F = 1 reacts exothermically. The products are not known, but dissociate upon dilution. In the presence of excess formaldehyde, methylurea can dimerize and form diureamethyl ether.

$$NH_2{-}CO{-}NH{-}CH_2{-}O{-}CH_2{-}NH{-}CO{-}NH_2 \qquad \text{Eq. 3.79}$$

ISBN 0-201-04558-3

With high urea-formaldehyde, dimethylol urea is formed. The latter dissociates at very low pH. Excess formaldehyde also yields acetals, among them cyclic products which are called mones (Kadowaki 1936). They can form from both N,N and N,N′-dimethylol:

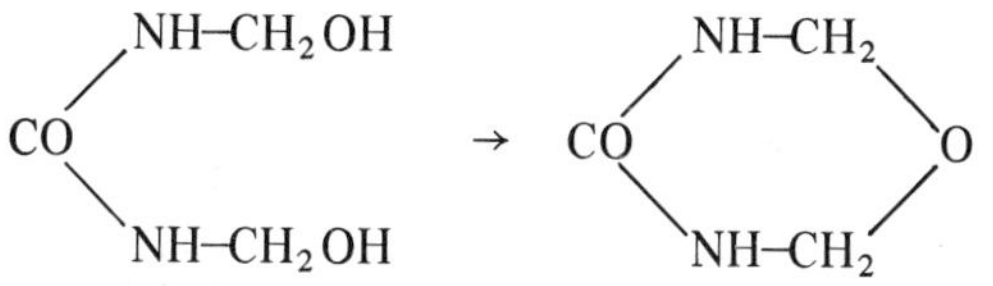

$$CO(NH{-}CH_2OH)_2 \rightarrow CO\begin{matrix}NH{-}CH_2\\ \\NH{-}CH_2\end{matrix}\!\!>O$$

Eq. 3.80

$$NH_2{-}CO{-}N\langle\text{ring: }{-}O{-}\rangle N{-}CO{-}NH_2 \leftarrow N(CH_2OH)_2 \rightarrow NH_2CO{-}N\begin{matrix}CH_2O\\CH_2O\end{matrix}\!\!>CH_2$$

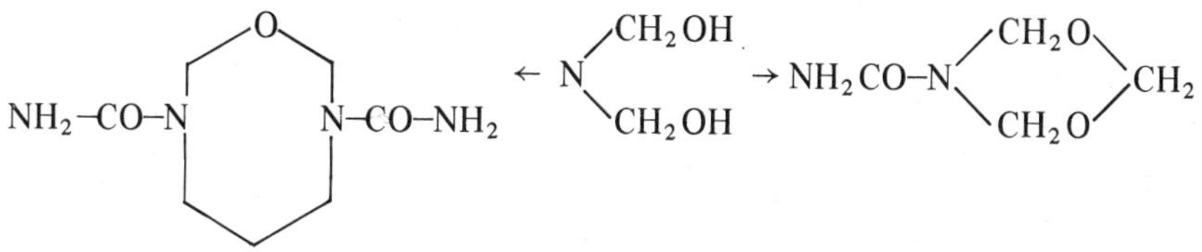

Condensation is enhanced if U:F = 1 and the solution is mildly acidic. Products obtained by this method have low to medium molecular weights, because the condensation products are insoluble. The same effects hamper polymerization of thiourea. Linear polymethylene ureas with a rank of 5 are obtained by acidulation of diureamethylene. Methylenediurea is also reactive at high pH, but far less than the hemiacetal. Zigcuner (1951–55) observed that dimethylurea can react with phenol, yielding:

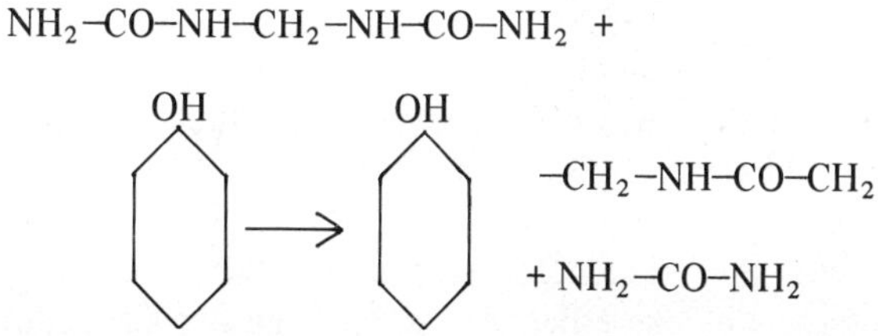

$$NH_2{-}CO{-}NH{-}CH_2{-}NH{-}CO{-}NH_2 \; +$$

$$C_6H_5OH \rightarrow HO{-}C_6H_4{-}CH_2{-}NH{-}CO{-}CH_2 \; + NH_2{-}CO{-}NH_2$$

Eq. 3.81

In the overall reaction one urea molecule appears merely as a catalyst. That is why early workers thought the species $NH_2{-}CO{-}NH{-}CH_2$, methylurea, was stable.

Upon heating to 80°C, excess formaldehyde reacts promptly at pH less than 3 to a water soluble product. During sulfite titration, the products decompose slowly over a period of several hours.

At a pH greater than 5, the products contain measurable quantities of methylolurea. Tomita (1976) used H-NMR to identify transient hemiformals in the reaction of urea with formaldehyde at pH 7 and 70°C. He measured the ratio of the equilibrium constants K_1/K_a for the addition of methylene glycol to methylolurea (K_1) and for the dimerization of methylene glycol (K_a) and obtained a value between two and three. This ratio is comparable to that for the benzyl alcohol

ISBN 0-201-04558-3

reaction, and some 25 times larger than for the corresponding phenol reaction. Thus, hemiformals of methylolureas are important intermediates in the formation of the dimethylene ethers under these conditions.

Alkaline Solutions

In alkaline solutions the primary products are methylols and ethers. Methylene bridges form only very rarely. Kadowaki succeeded in preparing mono-, di-, tri-, and allegedly tetramethylolurea, by using slightly more than the stoichiometric quantity of formaldehyde. The most reactive U:F ratio is about 1.5, because at this ratio the concentration of the resulting methylol groups equals the concentration of reactive nitrogen functions. In contrast, a ratio of 2.5 yields predominantly dimethylolurea and the corresponding ether derivatives which have insufficient free nitrogen functions for polymerization. Ito (1961) used paper chromatography and Tomita (1976) used H-NMR to establish the existence of trimethylolurea. While Ho (1961), Zigeuner (1952), de Jonge (1953), and Kadowaki (1936) proposed tetramethylolurea, Tomita (1976) found under the same conditions no indication of tetramethylolurea, but observed urons to be the main products. Thus, a U:F = 1.9 ratio yielded dimethyloluron, rather than tetramethylolurea. Likewise, neither Chiavarini (1977) nor Kumlin (1978) found evidence for tetramethylolurea.

The effect of the U:F ratio on the concentration of amine groups has been measured by Chiavarini and coworkers (1978). Figure 5.3a, page 137, shows the concentration of NH, NH_2, and NH_3 as functions of U:F. Figure 5.3b, page 137, shows the concentrations of $(CH_2{-}O)_n$, $(N{-}CH_2{-}N)_n$, and $(N{-}CH_2OH)$ as the same function. These figures suggest that the addition of urea to UF in acidic medium causes polyoxymethylene to decrease in favor of methylene bridges and methylol. Methylene bridges increase more rapidly than methylol. At the same time the free primary amine concentration increases, but not as rapidly as secondary amine. In Chiavarini's work (1978), the first condensation at 95°C and pH about 6 proceeded quickly to methylene bridges $-N{-}CH_2{-}N-$, but the second stage, at pH 8 and room temperature, was very slow. During this stage all oxymethylene formaldehyde converted to oxymethylene ether. At high pH and low temperature methylene ethers are formed. At high temperature, they convert to methyl bridges.

The kinetics of the primary reaction steps have been studied by Kadowaki (1932), Imoto (1948), Smythe (1953), Crowe (1948, 1949) de Jonge (1952), Sofue (1952), and Tomita (1976). The reactions involve complex equilibria which are correlated in sequences, for example, where MMU = monomethylolurea:

$$\begin{array}{ccccc} U + F & \leftrightharpoons & MMU & \leftrightharpoons & DMU \\ & & \updownarrow & & \updownarrow \\ & & D'MU & \leftrightharpoons & TMU \end{array} \qquad \text{Eq. 3.82}$$

DMU = N,N′-dimethylolurea, D′MU = N′,N-dimethylolurea, and TMU = trimethylolurea. The variation of molar concentration at 60°C and pH 9 are shown in

ISBN 0-201-04558-3

ISBN 0-201-04558-3

Table 3.20

23 Cyclic Reaction Products of Ethylene-urea with Formaldehyde and Their ^{13}C-NMR Frequencies (after Slonim, 1977)

δ, p.p.m.		Fragment
Experimental	Calculated	Structure
38·5	–	$HOCH_2N$ NH, O (ring)
38·6	–	$>NCH_2N$ NH, O (ring)
41·4	–	$-OCH_2N$ NCH_2O, O (ring)
41·5	–	HN NH, O (ring)
41·9	–	$HOCH_2N$ $NCH_2N<$, O (ring)
42·1	–	$-CH_2N$ $NCH_2N<$, O (ring)

δ, p.p.m.		Fragment
Experimental	Calculated	Structure
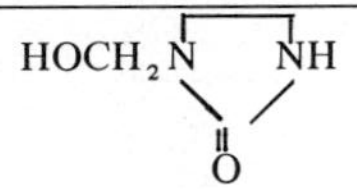 67·6	67·99	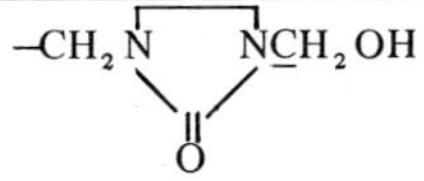 $-CH_2N$ $N\underline{C}H_2OH$, O (ring)
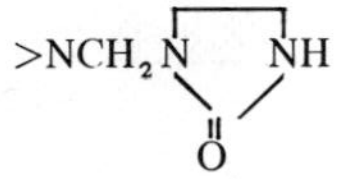 72·5	72·55	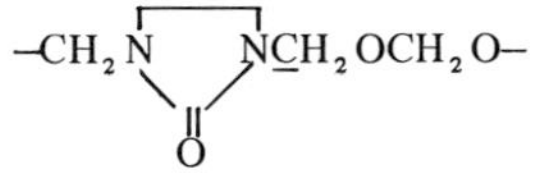 $-CH_2N$ $N\underline{C}H_2OCH_2O-$, O (ring)
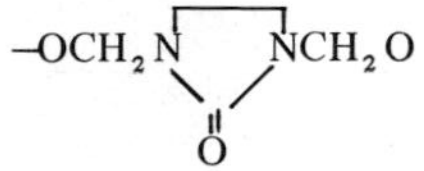 73·3	73·31	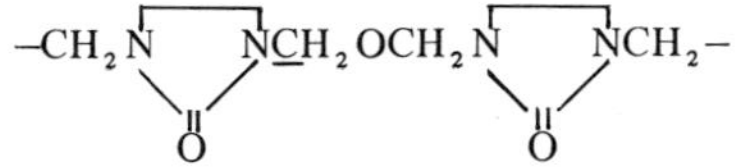 $-CH_2N$ $N\underline{C}H_2OCH_2N$ NCH_2-, O O (rings)
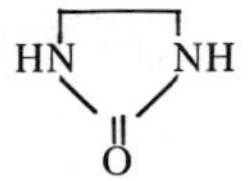 161·2	161·16	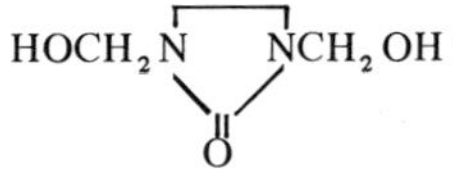 $HOCH_2N$ NCH_2OH, O (ring)
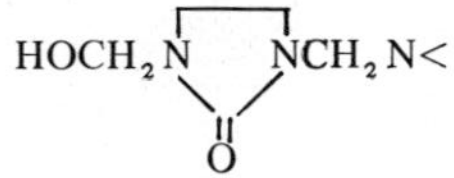 161·3	161·33	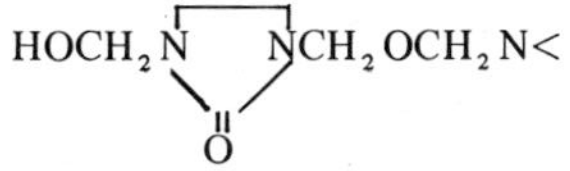 $HOCH_2N$ $NCH_2OCH_2N<$, O (ring)

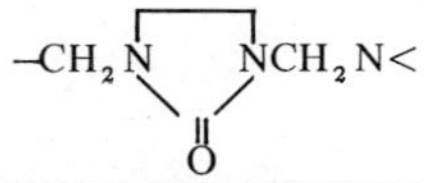

Table 3.20 (Continued)

δ, p.p.m.		Fragment	δ, p.p.m.		Fragment
Experimental	Calculated	Structure	Experimental	Calculated	Structure
43·9	–	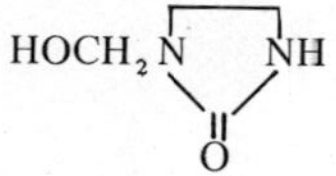 $HOCH_2N$ NH O	161·5	161·50	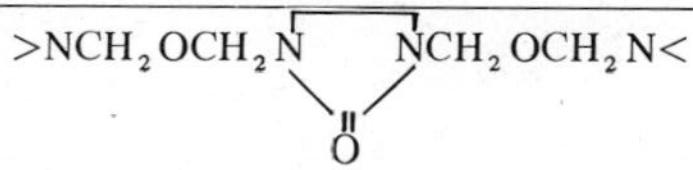 $>NCH_2OCH_2N$ $NCH_2OCH_2N<$ O
44·6	–	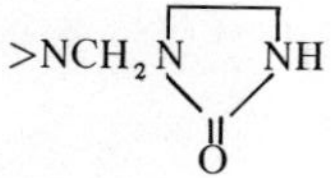 $>NCH_2N$ NH O	161·8	161·71	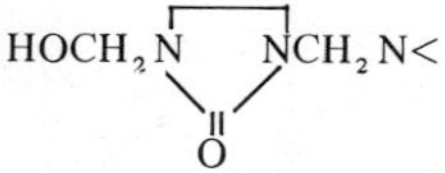 $HOCH_2N$ $NCH_2N<$ O
52·4	52·33	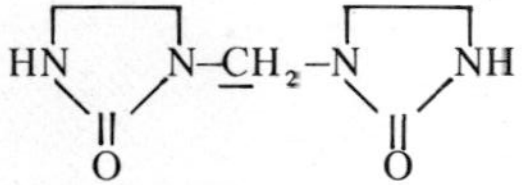 HN $N–\underline{C}H_2–N$ NH O O	162·3	162·26	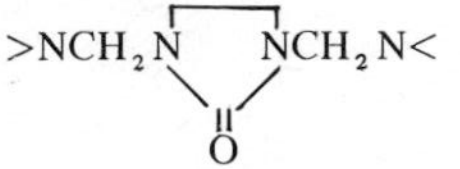 $>NCH_2N$ $NCH_2N<$ O
52·8	52·74	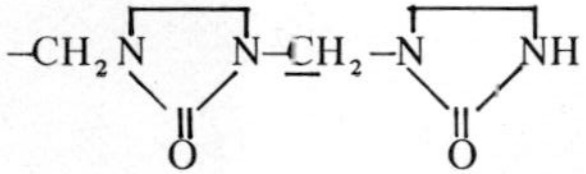 $–CH_2N$ $N–\underline{C}H_2–N$ NH O O	164·1	164·23	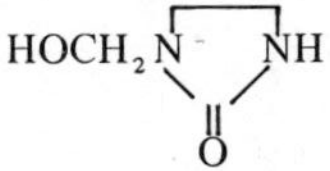 $HOCH_2N$ NH O
53·2	53·14	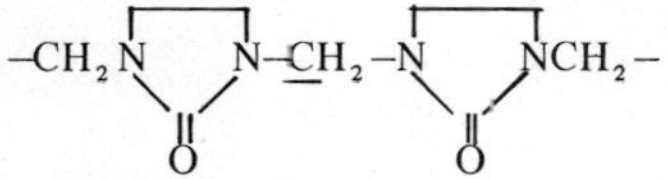 $–CH_2N$ $N–\underline{C}H_2–N$ $NCH_2–$ O O	164·7	164·78	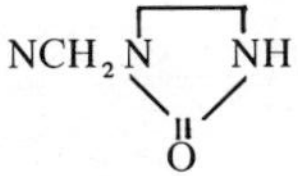 NCH_2N NH O
67·3	67·51	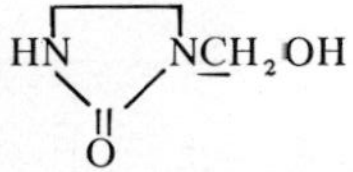 HN $N\underline{C}H_2OH$ O	167·3	167·30	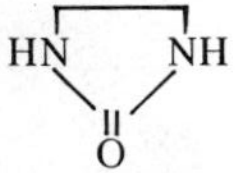 HN NH O

ISBN 0-201-04558-3

Figure 5.7 under these conditions, the Cannizzaro reaction is still negligible. The equilibrium constants under these conditions were determined by studying both formation and decay of each product starting with its pure precursor which had been synthesized and isolated separately for this purpose. The rate reactions were in the order of:

$$k_{1,2} > k_{3,4} \sim k_{9,10} > k_{7,8} = k_{5,6} \qquad \text{Eq. 3.83}$$

Typical values at 60°C and pH = 9.8 were $k_{1,2}$ = 27, $k_{3,4}$ = 5.4, $k_{5,6}$ = 0.5, $k_{7,8}$ = 0.6, $k_{9,10}$ = 5 1/mole.

Cyclic derivatives, which are specially important in the paper and textile industry, have been studied by ^{13}C-NMR by Slonim (1977) and his coworkers, who reacted urons, imidazolidone, and two triazinone ring derivatives with formaldehyde and studied some 23 products. Their ^{13}C-NMR spectra are summarized in Table 3.20. The relative concentrations of intermediates and products at two different U:F ratios were followed by Kumlin and Simonson (1978) at pH = 8.5, who started the reaction at 40°C and then raised the temperature to 70°C over a period of 30 minutes. Figure 5.2, page 137, shows typical compounds.

Modified UF-Resins

In modified resins either urea or formaldehyde is partly replaced by a reagent with similar functions which can substitute for the original reagent. Resins are formed by combination of three functions: a carbonyl, an NH-component, and a nucleophile. Modifiers are normally introduced via trans-alpha-ureido alkylations (Petersen 1964). In fact, normal urea condensation can be regarded as proceeding via trans-alpha-ureido alkylation of the methylol.

ALCOHOLS

Replacement of the methylol hydroxy group by an alcohol is one of the most common substitutions. Formation of ethers, especially of cyclic ureas, melamine, and mono- and diurethane is enhanced by acid catalysis in resins for producing crease-resistant textiles. The most important compounds are ethylene-urea, propylene-urea, glyoxale-mono-ureine, 5-alkyl-hydro-triazone, and urethanes. In all, the goal is to establish ether links with cellulose hydroxide. This leads to dimethyl-ether bridges between neighboring cellulose molecules. With some especially reactive reagents such as methylolated melamine, the reaction can be initiated in alklaine medium.

Ethylene glycol forms highly soluble products. The reaction products of higher alcohols are insoluble and are used in varnishes. Unsaturated alcohols yield ethers which can be mixed with unsaturated oils and can be used as air-curing varnishes. Polyepoxides yield soft and pliable resins.

ISBN 0-201-04558-3

HEMIACETALS

These have been described in the section on formaldehyde chemistry.

PEROXIDES

Hydrogen perioxide reacts readily with all N-methylol compounds. Kadowaki (1936), already produced and identified monomethylurea peroxide, a solid which melts at 153°C.

CARBONIC ACIDS

The compounds usually have high boiling points and can be made by reaction with ethers, from which the dissociate alcohol can be separated by distillation.

SULFUR COMPOUNDS

A very large group of acidic sulfur compounds have been successfully introduced into resins. Hydrogen sulfide reacts in a variety of ways. Zigeuner (1956) produced bis-ureido-alkyl-sulfides in which the sulfur atom links two rings. The reactions of mercaptans are quite well known (Petersen 1964), because they are more stable than their oxygen analogues and thus often form quantitatively, and yield products which can be isolated. Already Kadowaki (1936) prepared *N,N'*-bis-ethyl-mercapto-methylurea from dimethylolurea with ethyl mercaptan. Methyl mercaptan, butyl-mercaptan, thiophenol, or *p*-nitro thiophenol can be used to react with melamine, making insecticides, antioxidants, inhibitors, and vulcanization catalyzers.

Sulfite

Sulfite, sulfur dioxide, bisulfite, and disulfite form alkylsulfonic acids which yield adhesives (Scheuermann 1954), ion exchangers, and paper reinforcers. Sulfite can be used to solubilize UF-resins. With UF-resins and fatty alcohols one can obtain ion exchange resins and emulsifiers.

Thiourea

Thiourea reacts very much like urea, except that the sulfur atom also reacts, yielding sulfur bridges.

AMMONIA AND AMINES

Co-condensation of ammonia yields basic resins which are used to increase the wet-strength of paper. Ammonia and primary amines enhance crosslinking. Basic aminoplasts also serve as ion exchange resins. The UF-resin formulation and modified resins are discussed in Chapter 4.

ISBN 0-201-04558-3

4. Resin Manufacture

This section deals with the basic principles of the production of formaldehyde, urea, and their resins. The basic chemistry was discussed in Chapter 3. For a more detailed description the reader should consult more specialized literature, for example, the *Encyclopedia of Polymerchemistry,* Kirk Othmer's *Encyclopedia of Chemical Technology,* Ullmann's *Handbuch der Chemie,* Volume 7, or Houwink's "Synthetic Organic Adhesives" (Rayner 1965).

Formaldehyde

Formaldehyde production has increased from 50,000 tons of 37% solution in 1938 to over one million tons today. The properties of commercial and reagent grade formaldehyde are described in Chapter 3. Formaldehyde can be derived from five precursors: methanol by oxidation with excess air or from oxygen deficient mixtures; methane; carbon dioxide; formic acid; or methylene chloride. Methanol by oxidation (the common method) and methane are discussed in this section. The carbon dioxide reaction is not suitable because formaldehyde converts directly to methanol. The formic acid and methylene chloride processes have not yet been commercially successful.

METHANOL OXIDATION

This process, introduced in 1888, was the first and is still the best. Methanol and air form an explosive mixture in the range from about 60 to 90%, Figure 4.1. The explosion limits shift in the presence of other chemicals. With steam, the air component can be increased to the stoichiometrically desirable ratio.

Oxidative Dehydrogenation

This process involves two steps:

$$\text{oxidation: } CH_3OH + \tfrac{1}{2} O_2 \rightarrow CH_2O + H_2 + 38 \text{ kcal/m} \qquad \text{Eq. 4.1}$$

Beat Meyer, Urea-Formaldehyde Resins

ISBN 0-201-04558-3

$$\text{dehydration: } CH_3OH \rightarrow CH_2O + H_2 - 20 \text{ kcal/m} \qquad \text{Eq. 4.2}$$

Silver, copper, or plastic catalysts are used in the processes developed by BASF, Degussa, DuPont, ICI, and Monsanto. An oxygen deficient mixture (1 mol methanol to 2.2 mol air) is passed through the reactor at 600–720°C. At the lower temperature limit, the yield is poor. In the presence of steam, at the higher end, it is far better. About 45% of the reaction follows Eq. 4.1. The theoretical yield is 93.75%, but in practice a yield of 88–90% is obtained. The steam prolongs the catalyst life substantially. The reaction products are quenched to 150°C. One ton of formaldehyde requires 1.22 tons of methanol. The product can be enriched to yield 37% stabilizing solutions, or dried to form paraformaldehyde.

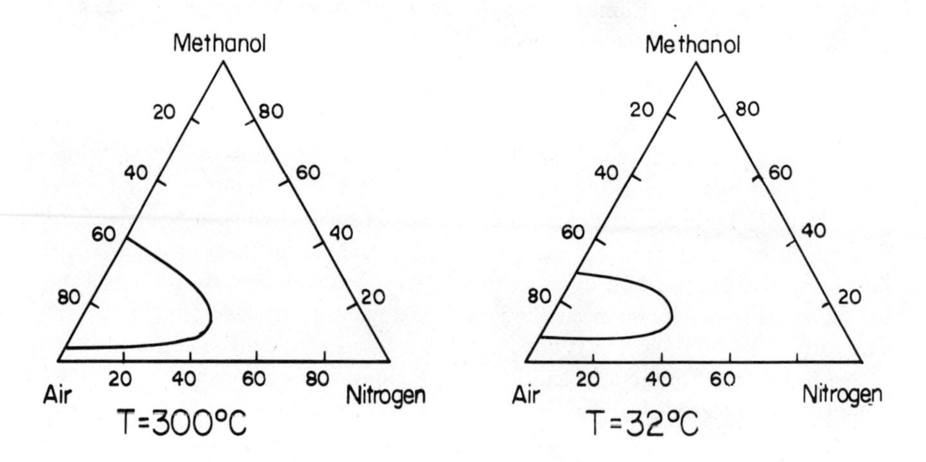

Figure 4.1. Explosive Limit of Methanol-Air-Nitrogen Mixtures a) at 32°C and b) at 300°C (after Baccaredda, 1950)

Oxidation

In this process excess oxygen is used. A 5–8 wt % methanol vapor is oxidized between 250–400°C on a mixture of iron oxide and molybdenum oxide. Chromium and cobalt oxide are used as promoters. The conversion is 95–97%. The product is condensed at 100°C and scrubbed with water yielding a 37–60 wt % solution. Hiag-Lurgi, Reichold, Montecatini, Lummus, and the Adkin-Peterson methods use this technique. Bakelite Corporation (1931) was one of the first to use this method.

ISBN 0-201-04558-3

METHANE

The methane oxidation follows these equations:

$$CH_4 + \tfrac{1}{2}\, O_2 \rightarrow CH_3OH + 29.73 \text{ kcal/m} \qquad \text{Eq. 4.3}$$

$$CH_3OH + \tfrac{1}{2}\, O_2 \rightarrow CH_2O + H_2O + 20.64 \text{ kcal/m} \qquad \text{Eq. 4.4}$$

In the U.S. about 8% of the formaldehyde is produced from methane. The problem is temperature control. One must work above 600°C, where dry formaldehyde decomposes. Huebs proposed reaction at 60 atm, with a reaction period of mere milliseconds. The Celanese process uses propane, or propane-butane mixtures, and yields a mixture of aldehydes. Hybernia Mining Company proposed reaction with a 1% ozone mixture.

Urea

The chemistry is described in Chapter 3. The technical production is based on the reaction of ammonia with carbon dioxide:

$$2NH_3 + CO_2 \rightarrow NH_2\text{–}COONH_4 + 30 \text{ kcal/m} \qquad \text{Eq. 4.5}$$

$$NH_2\text{–}COONH_4 \rightarrow (NH_2)_2CO + H_2O - 7.7 \text{ kcal/m} \qquad \text{Eq. 4.6}$$

The intermediate, ammonium carbinate, is directly formed from the gases. The method was developed by A. Basaroff (1870). The reaction conditions are usually 100 atm at 150–200°C. The reaction Eq. 4.5 proceeds backwards above 60°C at atmospheric pressure. At elevated pressure, urea is formed and water condensed. The main problem in this reaction is to increase the yield, by recirculation of the unreacted reagents (Lützow 1974). This can be achieved by many different methods. BASF (1914) originally isolated the intermediate carbonate, but changed the procedure during the same year. The current methods were patented in 1948.

Until 1955 the reactors had to be lined with silver, lead, or monel to reduce corrosion by sulfur impurities in carbon dioxide. Today austenitic Cr-Ni steels have virtually solved the problem. These steels can tolerate sulfur in the presence of oxygen. The latter forms a corrosion-resistant surface oxide. Today, main process improvements are aimed at heat recovery and yield. The equilibrium between ammonia, carbon dioxide, and water was carefully studied by Otsuka (1960) at 40°C and 100°C. Yudina (1975) explored the influence of the purity of urea on the UF-resin.

ISBN 0-201-04558-3

Urea-Formaldehyde Resins

The basic chemistry of resin formation has been discussed in Chapter 3. For a more thorough review of this subject, the reader should consult Wegler's chapter in Houben-Weyl (1963) Volume XIV/2 or any of the references given in Chapter 3. The reaction proceeds in two steps:

$$\text{Addition: } U + F \rightarrow U\text{–}CH_2OH \qquad \text{Eq. 4.7}$$

$$\text{Condensation: } U + U\text{–}CH_2OH \rightarrow U\text{–}CH_2\text{–}U + H_2O \qquad \text{Eq. 4.8}$$

The first reaction, often called methylolation or hydroxymethylation is catalyzed by both acids and bases. The condenstion, Eq. 4.8, is only acid catalyzed. The quality of aqueous resin and of the finished product depends primarily on the control of the two reactions, Eqs. 4.7 and 4.8. The reaction rates are shown in Figure 4.2, as a function of pH.

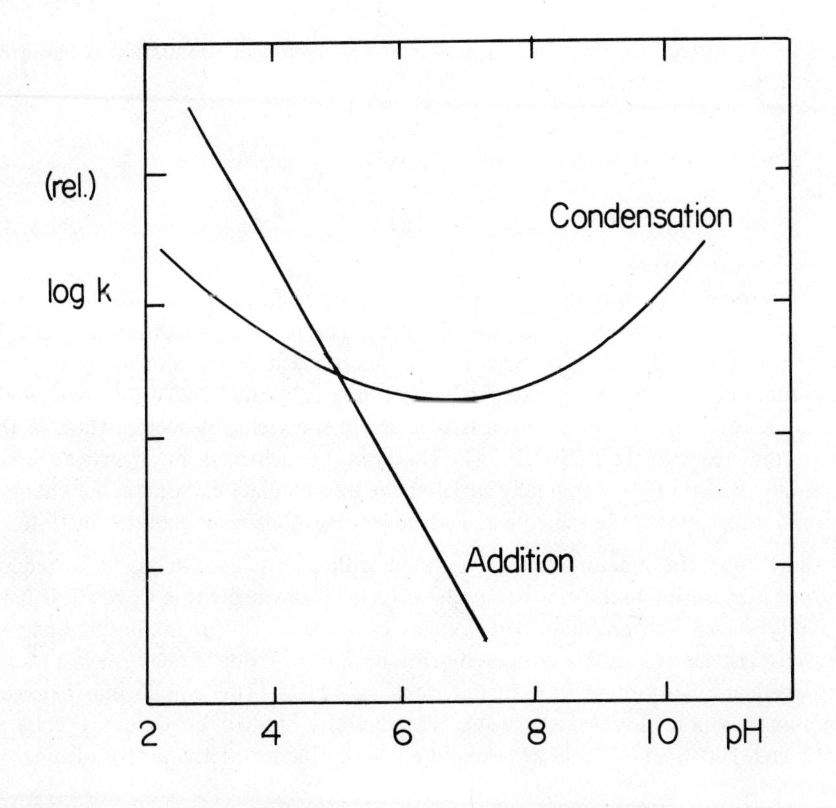

Figure 4.2. pH-Dependence of the Reaction Rates for the Initial Urea-Formaldehyde Steps (after Jong and Jonge, 1953, and Updegraff, 1978)

ISBN 0-201-04558-3

The methylol compound can be prepared in a neutral or alkaline medium or in pure form and remain stable. These compounds are extensively used in the textile industry. They can be directly condensed. Monomethylolurea, MMU, is produced at 20–50°C at pH 7–8 from a 1:1 molar reagent mixture. It is not very soluble at room temperature. It is not commercially used. The melting point is 110°C.

Dimethylolurea, DMU, is formed from a U:F = 1:2 mixture (i.e., a 1:3 weight ratio) at pH 7–8 at 35°C. The reaction is exothermic and the temperature must be kept below 45°C for 50 hours. The melting point is 123°C. DMU is sold to the textile industry under various trade names. With prolonged reaction, higher methylol compounds can be formed. The reaction strongly depends on pH, molar ratio, temperature, and time. The analysis of reaction mixture presented great difficulty, until the ^{13}C-NMR method and other modern analytical tools were developed.

Manufacture

UF-resins are made in a batch process, or by continuous manufacture (Brunnmüller 1972). Continuous processes have been described by Jaccard and Eklund, Swiss P 542,893, FP 2,068,286, and Ger. Offen. 2,046,496 (1972), for CIBA-Geigy; by Brunnmüller, GP 2,109,754 (1972), GP 2,241,995 (1974), USP 3,816,376; by Boehm in East Ger. P 88,196 (1972), Ger. Offen. 2,202,202; Vargin, USP 3,842,039 (1973); and Allyn, USP 4,065,421.

The conversion of resin to solids makes it possible to store resin and ship it over long distances, without the excess water burden. Such dry resins are manufactured in many countries, especially in England, for export. The use of dry resins in North America has been smaller than anticipated, and is still essentially restricted to insulating foam, probably because most resin is manufactured for in-house use. Examples of methods were published by Jaccard, Hostettler, and Siegrist, GP 2,131,717 (1972), who described an apparatus to convert liquid phases into solids; CIBA-Geigy (1974) which disclosed a method for solidifying resin at 65°C; and Jaccard, USP 4,051,208 (1977), who again described dry resin. Renner, S. Africa P 70:5,676, described manufacture of a colloid.

A typical reactor for batch manufacture of the type still widely used is shown in Figure 4.3. The set up is basically the same as that used on a laboratory scale, Figure 4.4. The latter arrangement makes it possible to distill or reflux condensate. A separation funnel makes it possible to divide the condensate, which is desirable if alcoholic phases are to be separated. Graduated burettes make it possible to add more reagent during the reaction, or adjust the pH. On a commercial scale, the reactor can be several stories tall.

ISBN 0-201-04558-3

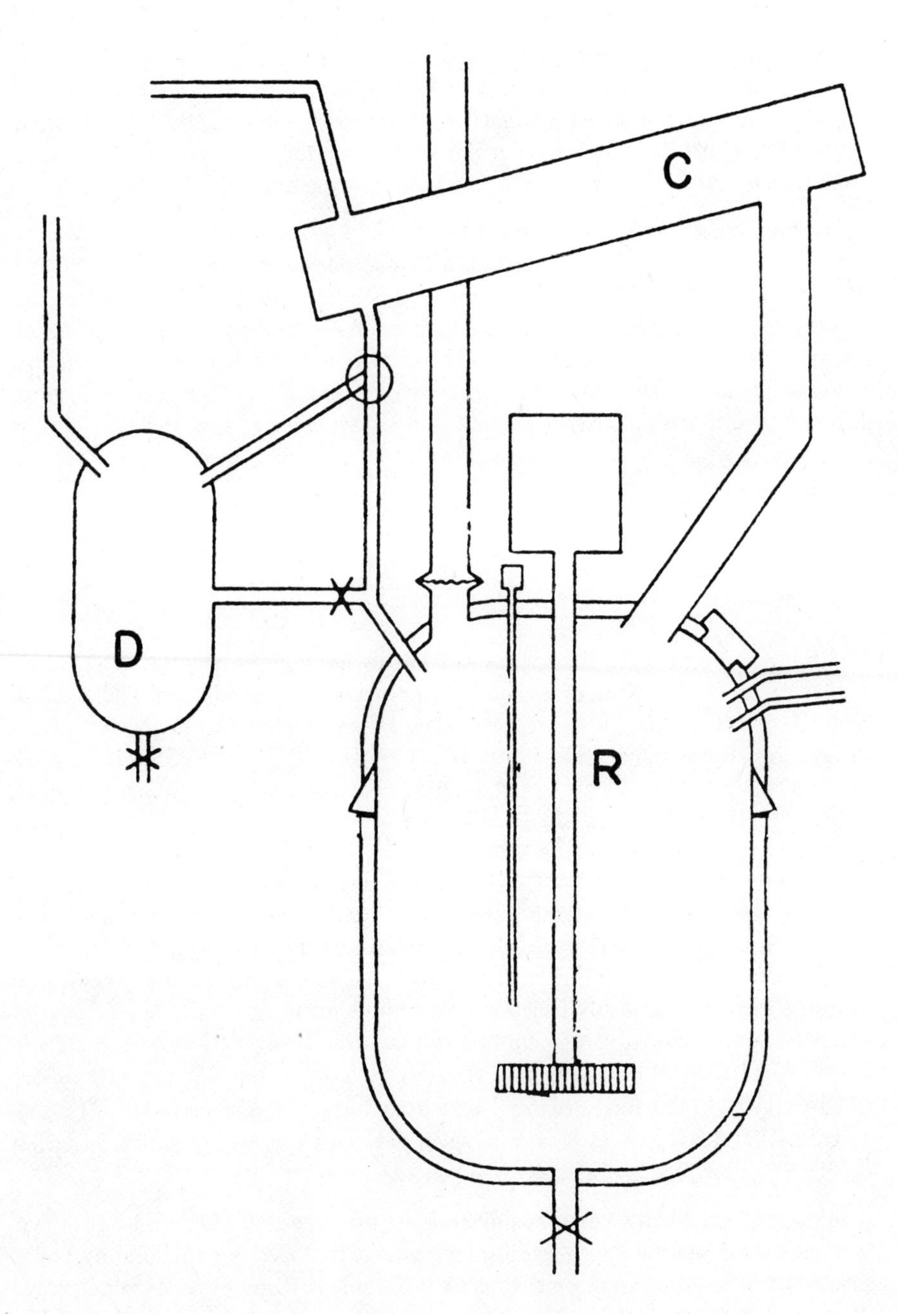

Figure 4.3. Schematic of Industrial Batch Reactor used for UF-manufacture
R = Reactor; C = Condensor; D = Decantor

ISBN 0-201-04558-3

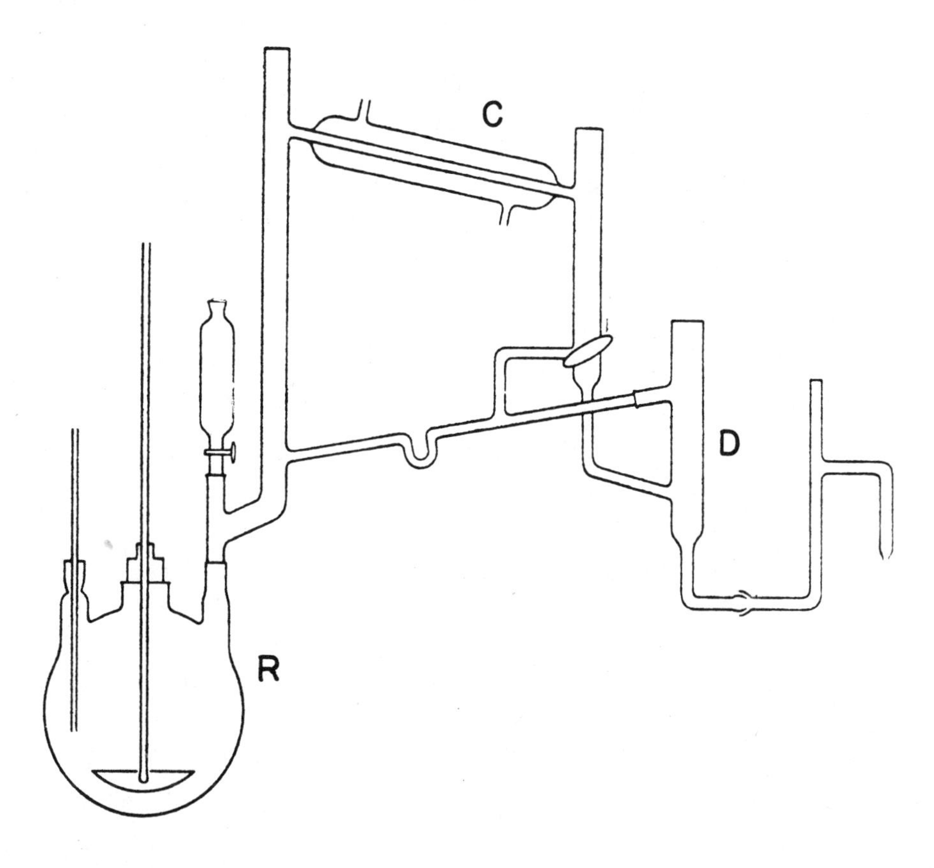

Figure 4.4. Laboratory Glass Reactor for UF-Resin Synthesis R = Reactor; C = Condensor; D = Decantor

REACTION CONTROL

On a commercial scale, the chemical reaction is still controlled by measurement of bulk behavior, rather than by testing for functional groups. This will likely change, now that NMR techniques allow quick and reliable qualitative and quantitative analysis. During the initial stage, the free formaldehyde content is estimated, usually with the sulfite method; Kaesbauer *et al.* (1976) have developed a method for determining free formaldehyde and *N*-methylol at the same time, by modifying the de Jong method, as described in Chapter 5.

At an intermediate stage, viscosity becomes the decisive property. Both the Gardner-Holdt bubble tube method and the Brookfield Viscosimeter are widely

ISBN 0-201-04558-3

used. The temperature must be carefully controlled, as the viscosity changes reversibly, as well as irreversibly, with temperature.

As the molecular weight continues to increase and the "turning point" or hydrophobe point is reached, the mixture becomes cloudy. Again, the temperature is a crucial factor. At this point, the mixture is still homogenous, but the high molecular fraction is no longer soluble in dilute solution and thus precipitates upon dilution with water. In practice, often a few drops of resin are dumped into water, and a white streak is observed. The temperature at which water clouds is called the hydrophobe temperature.

GELLATION TIME OR THE CURE TEST

The time required to reach the gellation point depends on temperature, concentration, pH, the resin volume tested, and the rate of heating. A variety of methods are used to test gellation times. The latter are only a relative measure of actual gellation time under working conditions.

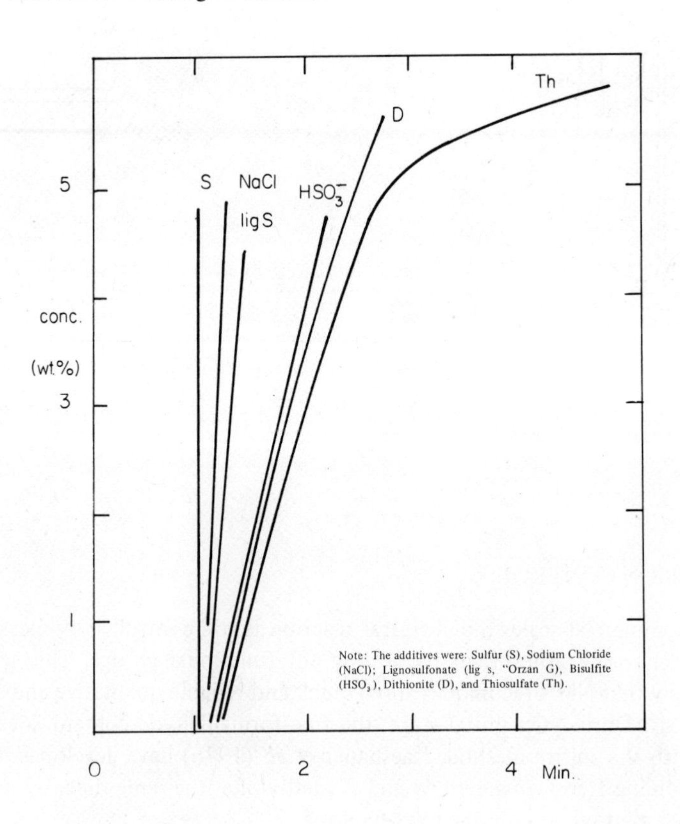

Figure 4.5. Gellation Times of UF-Particleboard Resin as a Function of Additive and Additive Concentration (Meyer, unpublished work)

ISBN 0-201-04558-3

The gel test is simple, but the underlying principle is not. The test works well, because the reaction is of a higher order. The gellation is preceded by an incubation period during which the resin viscosity usually changes little, or even drops, and the solidification process takes place within a few seconds. Figure 4.5 shows some gellation times for resins which were modified by step-wise increases in additive. The curves show how sensitive the gellation test responds to the chemical nature of the additive, even upon relatively small additions of materials.

INSTRUMENTAL CHEMICAL ANALYSIS

Today, chromatography (Kumlin 1978). and NMR (Tomita 1978, Chiavarini 1978) are quickly becoming recognized methods, Chapter 5.

FORMULATIONS

Older work is reviewed in the many excellent review works cited throughout this book, especially by Ellis, 1934 and Scheiber, 1943. In this chapter only a few advances are selected for discussion.

Ham (1963) in USP 3,080,343 described diamine-urea-thiourea copolymers for producing fibers; Gavrilov (1969) introduced etheylenediamine. In 1971 Kozlov, USSR 317,804 described a chalking composition; Lander, Ger. Offen. 2,105,253, introduced zirconium and hafnium chloride, melamine, and acrylamide into UF-resins, imparting heat resistance to the products; Tajima, Japan 71:28,783, described cyclic primary compounds; Sharphouse, Indian P 120,113, described a resin; Nakanishi, Japan 71:31,015, introduced glycidyl methacrylate, Newmann, Ger. Offen. 2,130,577, used ammonium polysulfide to obtain tough products.

In 1972 Hamada, Japan 72:47,293, obtained a resin with exceptional adhesive power, Nagao, Kokai 72:22,441 added guanamine as copolymer; Anokhin, USSR 364,426, employed melamine; Schmidt-Hellerau, Ger. Offen. 2,043,440, added coconut shell powder; Wirpsza, Pol. P 66,154, used zinc phthalate, zinc butyrate, and zinc fluoride. Szemiotowicz, Pol. P 78,613, added melamine and urea; Nagao, Kokai 72:43,128, used cyclic nitrogen compounds for resins; Kugriklov, USP 3,689,463, added aryl or alkyl sulfonic acids; Bonner, USP 3,645,981, made fibers; Bayersdorf (1972) described the silylation of UF-resins which increases their solubility in benzene, tetrahydrofuran (THF), and allows chromatographic analysis. Hensley (1972) described fast curing surface coatings; Kamenski (1972) made fusible furfural copolymers; Petrusenko (1972) mixed higher alcohols with resin; Tsvetskov (1972) added 2–4% epichlorohydrin; Wirpsza (1972) reviewed uses.

In 1973 Kotani, Kokai 73:60,140, added methylene acetoacetate to reduce odor; Putlyaev (1973) modified a Russian resin and introduced polyethylene polyamines to hold sand and similar fillers; Purcell, USP 3,761,339, laminated paper to board; Akutin, USSR 384,843, added acrylic acid; Tanaka, Kokai 73:58,032, disclosed phenol copolymers; Higuchi, Kokai 73:79,241, prepared odorless resins

ISBN 0-201-04558-3

with phosphate; Tiedeman, Ger. Offen. 2,225,478, included acylhydrazide; Clarke, Australian P 441,798, made spheroidal aminoplasts; Takahashi, Japan 73:17,877, added toluene sulfonamide; Sundie, Ger. Offen. 2,261,800, disclosed a melamine copolymer; Takeshita, Japan 73:19,865, reinforced resin with polyamide fibers; Penczek, Pol. P 67,486, made a U–F–P–M resin; Tsukamoto, Kokai 73:18,338, used zinc stearate to make an impact-resistant resin; Ueda, Kokai 73:58,048, made the same with hexamethylene tetramine; Sato, FP 2,161,936, made a granular composition; Kitta, Kokai 73:31,290, made an amorphous precondensate; Kotani, Kokai 73:60,140, added calcium malonate and monobutyl malonate; Erhardt, Ger. Offen. 2,207,921, described a resin for general purpose; Lenz, Ger. Offen. 1,570,240, added thiourea to make an impregnating resin; Vargin, USP 3,830,783, made a plywood resin; Biondi, Ger. Offen. 1,645,016, obtained stable aqueous concentrated solutions; Inverarity, Ger. Offen. 2,318,802, used dipropylene glycol to crosslink resin; Junkermann, Ger. Offen. 2,218,582, made solid concentrates; BASF, FP 2,163,578, disclosed that diethanolamine and triethanol amine improve the stability of resin.

In 1974 Behar, Ger. Offen. 2,347,401; Kitsuda, Japan 74:40,617; Suzuki, Kokai 74:30,504; Shofuda, Kokai 74:128,031; Ohhara, Kokai 74:92,154; Kitta, Kokai 74:63,789; and Freeman, USP 3,786,025, all made novel resins; Aries, FP 2,230,348, added sulfamoylanthranilic acid; Nowak, Pol. P 83,871, obtained a water insoluble resin; Christensen, Can. P 1,029,915, mixed methylolureas with resin; Olszowski, Ger. Offen. 2,400,908, made copolymers; Anderson, USP 3,862,914, made coatings; Kodama, studied heat degradation.

In 1975 Lindberg, Ger. Offen. 2,507,586, used lignosulfonates. Yamamoto made a wood resin in Kokai 75:76,145, a wood laminate binder in Kokai 75:4,133, and soluble resins in Kokai 75:76,146. Central Laboratory in Sofia, BP 1,499,285, disclosed an improvement for curing; Sausman, USP 3,893,963, prepared a resin with built-in hardener, consisting of 2-amino-2-ethyl 1,3-propanediol hydrochloride; Lapitskii, USSR 476,296, reported an epoxy copolymer.

El'bert (1976) described a dry molding composition for fiberboard; Prilukov (1976) proposed on-site mixing formulations; Aignesberger, Ger. Offen. 2,505,578, added disulfite; Stenzel, Rom. P 60,253, used furfuryl; Yui, Ger. Offen. 2,546,668, made UF self-extinguishing by adding polypropylene; Lenke, USP 3,997,612, obtained polythiaformal by use of mercaptans; Kinastowski, Pol. P 91,790, added carboxylic acid; Blommers, USP 4,032,515, made resorcinol-terminated resins; Moore, USP 4,097,419, crosslinked resins; Demko, USP 4,018,959, incorporated vinyl acetate; Lajoie, USP 4,064,307, made molding resins; Monsanto, BP 1,486,342, made a storable syrup. Laqua, Ger. Offen. 2,620,478; Brial, FP 2,338,956; British Industrial Plastics, Neth. P 76:373; Gavirati (1977), Pizzi (1977), Whiteside, S. African P 77:345, Osipova, USSR 557,859, and Pritchard, Australian P 481,197, all describe novel resins.

In 1978 Hayashi, Kokai 78:41,354, Kokai 78:41,393, Kokai 78:41,392, and Kokai 78:33,253, disclosed various useful resins; Izumi, Kokai 78:17,690; Herbert,

ISBN 0-201-04558-3

Ger. Offen. 2,754,525, disclosed a filler, and Hubbard, Ger. Offen. 2,726,617, made a concentrate.

Adhesives

Chapter 6 describes the requirements for plywood and particleboard resins. A basic resin is individually modified with filler, extenders, and the like to impart the desired properties. Often, phosphates or triethanolamine are added to neutralize the catalyst after curing.

Eisele, Petersen, Mayer, and Wittmann (1977) in USP 4,021,413, described a most simple and elegant basic glue composition. It is prepared from a condensate containing U:F ratios between 2.5 and 4.5. These are converted into wood adhesives by adjusting the pH to a value between 1 and 3, just above room temperature, until the desired viscosity is reached. Urea is then added at a pH between 7 and 9. Figure 4.6 shows the viscosity change in this resin at seven temperatures as a function of time. Below 50°C, the viscosity gradually increases; above 60°C, the viscosity rapidly reaches a maximum and then drops. Obviously, the resins degrade at high temperature. The final resin can be adjusted to an U:F ratio between 1.2 and 2.

Early work has been reviewed by Updegraff (1978,1979), Sandler (1977), Widmer (1965), C. P. Vale (1964), Blais (1959), and many others. More specialized articles were published by Baumann (1967). Muravev (1971) and Perevertov (1971) measured the storage life of adhesives; Sparkes (1972) compared UF with other wood adhesives; Preobrashenskaya (1972) modified UF with polyvinyl acetate emulsions; Alekseev studied the impact of UF on tool wear; Neusser modified it with melamine; Plath investigated optimum starch concentrations in plywood by microscopic investigation; Yoshihiro (1972) measured the heat of reaction of ammonium-chloride-cured UF-resin containing beechwood flour with 5–6 kcal/g. Mostovoy related the toughness to the curing temperature. Kawahara compared internal bond strength as a function of press temperature. Sergeeva measured the gellation times of free adhesive films *in vacuo*. Potutkin (1972) reacted dioxane lignin; Black, USP 3,697,355, added alkali and alkaline earth halides to improve the surface activity and thus the efficiency of resin; in FP 2,098,692, he reported that the bond strength increased from 4.5 to 8.5 kg/cm^2. Mayer, Ger. Offen. 2,020,481, made waterproof resin from M–U–P. Davidson, USP 3,642,042, treated wood with sulfur dioxide and then with ammonia, and claimed the wood did not stain; Brunnmüller, Ger. Offen. 2,215,947, described a phenolic adhesive; Boehm, Ger. (East) P 88,196, disclosed a continuous process; Erhardt described an adhesive in Ger. Offen. 2,243,857 and Ger. Offen. 2,207,921; Koch, Ger. Offen. 2,213,269 sprayed resin and hardener separately; Kubitzky, Ger. Offen. 2,206,696, disclosed a low odor resin; Parnell, USP 3,666,597, used encapsulated sodium hydroxide and phosphate buffer. The encapsulation crushed in the press. Raizman, USSR 327,224, made a foaming agent from malt root, bicarbonate, and alum. Riesmeier, Ger.

ISBN 0-201-04558-3

Offen. 2,111,651, mixed a hot melt with comparatively slow curing UF.

In 1973, Lehmann evaluated catalysts versus press time effects. Wirpsza evaluated an entire series of organic acids, derived from poly(vinylpyridine). Krach made a furfural resin; Gaponenko analyzed resin; Temkina described the Russian resin KS-68M; Rikunov applied vacuum curing; El'bert added polyvinyl acetate; Temkina tested the effect of U:F ratio on odor; Mukudai measured internal stress in adhesive films as a function of curing time and found it to be double that found in laminates; Adamca imparted electric conductivity to films; Chow measured the softening temperature of various glues; Schneider compared absorption of UF with PFF; Weeraratne used natural rubber latex as a blend for wood; Vargin, Ger. Offen. 2,262,197, described a methanolic UF-resin with good storage properties; Iwata, Kokai 73:56,731, added guanine-phenol precondensate; Iwaki, Kokai 73:12,386, copolymerized benzoguanime; Lambru, Rom. P 56,115, made a hot curing resin with a setting time of 10 seconds. Okamura, Japan 73:14,793, cured UF with radioactivity; Takahashi, Kokai 73:55,223, added dicyandiamide and melamine; Tanaka, Kokai 73:58,032, mixed phenol; Vargin, USP 3,842,039, made continuously storage stable adhesives.

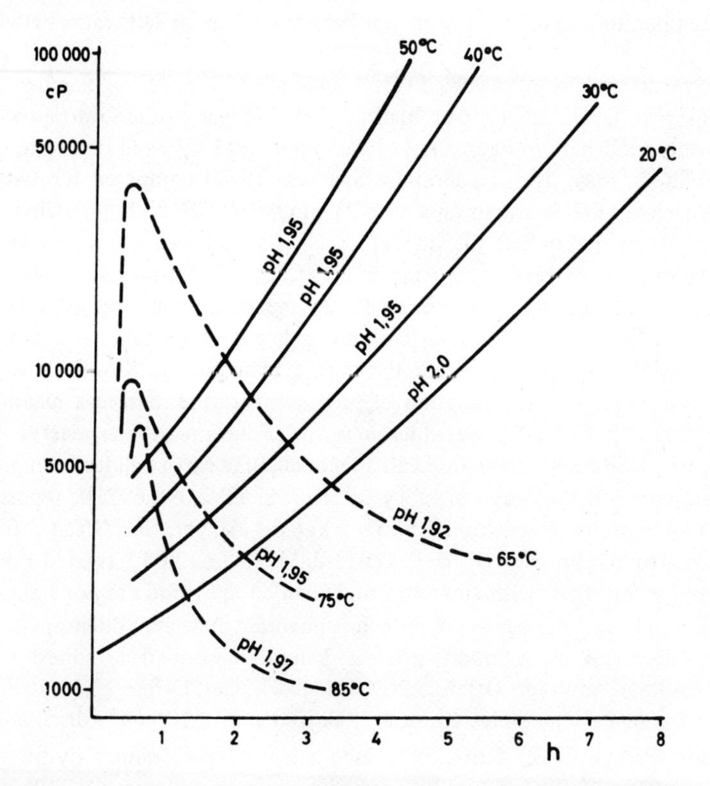

Figure 4.6. Viscosity of UF-Resin Intermediate as a Function of Time at Seven Temperatures and pH 1.9 (Eisele, USP 4,021,413, May 3, 1977)

ISBN 0-201-04558-3

In 1974 Krach disclosed an MUF-furfural resin. Hirata used starch as filler; Chow systematically investigated the morphology of a series of RF-, PF-, and UF-resins; Kreibich described resin analysis by using chromatography paper instead of wood to measure the durability. Starzynska made furfural resins; Ivancheva measured aging; Hayashi used ionizing radiation to induce polymerization; Grabowsky, Ger. Offen. 2,435,793, described a resin; Denisova studied the kinetics of hardening; Rikunov hardened resin *in vacuo;* Roffael used lignin wastes; Kolaric used radiation; Krach used furfural; Potekhina described a quick curing resin; Afanasev measured the viscous properties; Ando, Japan 74:44,581, used glycerol monolaureate to make a coating; Billingsfors, Neth. Appl. 74:6,753, described coatings made from 60 g/m^2 paper and UF; Brunnmüller, Ger. Offen. 2,241,995, described a continuous manufacturing process; Buschfeld, Ger. Offen. 2,337,233, made a storage stable phenol; Erhardt, Ger. Offen. 2,243,857, disclosed a UF-resin for BASF; Eisele, USP 4,021,413, described a wood cement for BASF; Grabowsky, Ger. Offen. 2,454,769, reported a stable, soluble, and reactive adhesive for BASF; Hata, Japan 74:4,542, improved shelf life with etherized starch; Lenz, Ger. Offen. 2,324,440, described a bisulfite resin for BASF; Maeda, Kokai 74:128,943, described a phenolic resin for impregnating textiles, fishing nets, and making adhesives. Masuda, Kokai 74:130,934, mixed saporified acrylamide-vinyl acetate; Nakajima, Kokai 74:97,744, used polyvinyl-acetate to add clay as filler and make a stronger bond; Nomura, Japan 74:461, used butyl-phenol and toluene sulfonamide to increase adhesion on wood; Tanioka, Kokai 74:110,585, made a waterproof vinyl-acetate coating; Tsvetkov, USSR 411,108, mixed epoxy resin; Willert, Pol. P 73,289, mixed U, M, and dicyandiamide under pressure; Yamashita, Japan 74:44,936, improved flowability.

In 1975, Brancato and Herman described dispensed microparticles with a surface of 10–90 m^2/g and an average diameter of 0.2 to 4 micron; Klebeko described a triethanolamine resin; Boehm described a continuous process. Steiner and Chow described melamine, furfuryl, resorcinol, and 2-imidazolidinone graft copolymers which increased the softening points; they established that the durability was related to the softening point. In another paper, Steiner and Chow compared UF with PRF, UMF, PF, polyvinyl-acetate, and casein as plywood adhesives under moderate (–20°C) and extreme frigid (–65°C) conditions and found that the first two were the poorest performers in shear strength. In a third paper they studied the use of western hemlock bark extractives, together with formaldehyde, as adhesives. Gvozdev, *et al.* used tri(hydroxymethyl) phosphine. This led to reduced flammability, strength, and elasticity of adhesives, and DTA showed that it acted as a plasticizer and not as a crosslinking agent. Kodama added resorcinol to improve the thermal durability. Anokhin investigated the effect of hardener and hardening time, especially on toxicity; Zirnin described the rheology of resins using Kohlrausch's equation. Gillespie studied the effect of dry heat on products. Pashkov used kaolinite, with oxalic acid hardener; Bergsund, Norw. P 131, 891, used alum-citric acid granules as solid hardener; Black, USP 3,905,847, added 1–30% alkaline salts

ISBN 0-201-04558-3

and increased the internal bond strength by a factor of two; Blank, USP 3,907,740, used dipropylene glycol and methylate of acrylate, and the like; Gordon, USP 3,962,166, prepared a precondensation methylol paste adhesive; Higashimura, Kokai 75:9,693, used *p*-toluene sulfonic acid as hardener; Higuchi, Kokai 75:34,330, made a high tack adhesive; Iwata, Kokai 75:109,933, used a boiled rice with 3% UF adhesive as mixture. Kawamura, Kokai 75:35,230, made an odor free adhesive; Krueger, GP 1,719,151, used hexamethoxymethyl melamine, ricinemic acid, and propylene glycol derivatives as solvents. Kusushita, Kokai 75:148, used dispensable polybutadiene; Lambuth, USP 3,909,470, used cellulose paper mill sludges as extenders; Merkel, GP 2,550,739, formulated a resin for BASF; Nogi, Kokai 75:126,792, used dimethyl-sulfoxide as solvent; Oda, USP 3,892,709, used aromatic compounds; Ohhara, Kokai 75:104,293, added isobutylene-maleic anhydride; Pace, Ger. Offen. 2,516,327, used alum-phosphate mixtures to control hardening; Renner, USP 4,010,132, and USP 4,064,088, made a naphthaline sulfonic acid resin, also suitable as filler for elastomers for CIBA; Reuss, Ger. Offen. 2,403,443, made a cationic resin; Shiota, Kokai 75:112,430, added yeast, chlorella, beer sludge, canned fish, and dairy product microorganisms yielding higher strength and lower odor than flour as a filler. Siegler, Ger. Offen. 2,401,554, added propanediamine, ethanolamine, or diethylenetriamine to give better cold adhesive strength; in Ger. Offen. 2,351,981, he added polybasic carboxylic acids for the same purpose. Takahashi, Ger. Offen. 2,421,445, improved aging with citric acid; Terporten, Ger. Offen. 2,401,440, made laminating paper resins, Yamamoto, Kokai 75:76,147, disclosed copolymers; Yamashita, Ger. Offen. 2,413,362, added humic acid to produce a soluble adhesive; Zabrodkin, USSR 496,291, made low-odor resins with UF ratios between 0.9 to 1.35; and Hubbard and Atkinson, USP 4,035,456 used calcium formate as hardener.

In 1976, Richard used NMR and GPC to identify the properties of resins. Minemura used starch wastes as extenders; Catoni analyzed application factors; Adamca tested dielectric heating; Mattsson reviewed preparation and uses of wood adhesives; Eades, BP 1,421,119, made a foam; Foris, USP 4,001,140, made microencapsulated inks; Holtschmidt, Ger. Offen. 2,448,472, used taurine, and NH_2SO_3; Hubbard, Ger. Offen. 2,532,753 added sodium silicate; Koyama, Kokai 76:88,595, added ammonium phosphate. Renner, USP 3,931,063, made a free-flowing resin with sulfamic acid; Sano, Kokai 76:89,537, added pulping waste liquors; and Smirnova, USSR 509,615, added organic amines and carboxylic acids.

In 1977, Matsumoto discussed the stability of resin. The interaction of resin and wood is more thoroughly discussed in Chapter 6. In the following some select papers are cited which explain the type of work currently under way. Since 1954 when Fahrni in USP 2,686,143 described his process for making boards with high face density and strength, a continuation of his Swiss P 232,065 of 1944, a large amount of literature has accumulated. Earlier work was reviewed by Deppe (1977) and Maloney (1977) in their authoritative books on particleboard. The following examples of the early 1970s represent typical current interest.

ISBN 0-201-04558-3

In 1971, Wittman described the use of paraffins; Bergin described the result of a 15-year weathering test; Boklen compared shear strengths; Clad compared the weather resistance of phenol, melamine, and urea and found the first superior. Erins described the action of ammonia on wood; Gerlachov added 15% of creosol to resin; Ginzel observed that swelling and shrinkage, rather than UF degradation, damaged board. Hirotsugu evaluated the influence of wood extractive on gellation time. Koromyslova described structural panels; Lielpeteris evaluated weight loss of wood under ammonia treatment; Plath made a microscopic study of the depth of penetration; Raknes studied accelerated aging for six years. Sakaguchi evaluated the influence of water extractives from *fillia* species wood on bonding ability; Sakuno studied the effect of surface aging. Shishkov determined optimum press conditions; Splawa-Neyman examined resistance of particleboard to biological corrosion.

In 1972, Glukhikh reviewed internal stress in veneer bond lines, and, in a second paper, related bond strength to stress. Halligan and Schneiwind reviewed the effect of moisture on particleboard swelling, springback, and creep. Jain reviewed problems in gluing *Prinus roxburghii;* Johns compared the absorption of binary liquids on board; Katuscak compared swelling in organic liquids; Khrulev measured the resistance of wood joints to water; Mitrofanov compared hardness and wear; Narasaki analyzed polymers by ashing. Neusser described the effect of small quantities of asbestos, gluten, blood albumin, and other extenders. Schorning described polyvinyl alcohol bonds, and, in another paper, fire tests; Seifert studied swelling; in a series of articles, Tsyvin modified UF with vinyl alcohols; Wnuk proposed gas-shock treatment to accelerate curing; Yamagishi studied the bonding ability of tropical woods.

In 1973, Chang related manufacturing techniques to performance; Havinga, Ger. Offen. 2,317,846, reported photocuring; Ivanov described quick hardening resins; Koval'chuk proposed a method for standardizing measurements; Lehmann reviewed resin effects on dimensional stability; Lutoshkina evaluated factors influencing the brightness of UF coating on plywood; Stashevski examined tannin resins; Wittmann studied the effect of alkali in phenols; Wnuk used ammonia-ammonium chloride hardener to accelerate resin cure.

In 1974, Bartashevich coated furniture panels, Chow and Steiner evaluated cyclic freezing, Gamova described wood-polymers; Hse evaluated the catalyst effect on final product properties; Imura added bark powder; Oblivin determined the volatile products emanating from the press. Paszner used radiation to make veneer-polymer composites in one step. Soto Urbina Co. evaluated fireproofing; Tsutsumoto compared nine tropical species; Wu evaluated thickness swelling; Tsai evaluated hardwood extractives; Swarfsman applied hot resin to wood chips to speed penetration and curing; Teveb did a similar study; Sholokhova studied wood joints; El'bert studied wood-glue interaction in the press; Azarov evaluated hydrolytic and thermal stability; Anderson studied indoor air pollution due to formaldehyde; Strelkov determined optimum conditions; and Wittmann discussed bonding with isocyanate.

ISBN 0-201-04558-3

RESIN CURE

The condensation, Eq. 4.8, is usually initiated by catalysts or hardeners. The term catalyst is not always justified, as ammonia and organic amines normally react with the resin during the condensation. The gellation of resins occurs at low pH. The process can be induced by any method that lowers the pH. The best pH value is about 2–4. Below pH 2 the resin degrades rapidly. In cold setting adhesives, free acid or a buffer must be added, as described in Chapters 3 and 7. The curing reaction is complex. Until recently, very little was known about the processes during condensation. During the last decade, instrumental analysis has made dramatic progress. Steiner (1974) used x-ray to follow curing. Differential thermal analysis (DTA) made it possible to identify processes, and their duration. The use of this method is demonstrated by the work of Troughton and Chow (1975) and Chow and Steiner (1975). They demonstrated that the cure of resin is accompanied by an exothermic peak which occurs at a characteristic temperature, and reflects upon the resin cure, and thus the resin performance. Figure 4.7 shows the DTA chart of a UF-resin with a ratio of 1.3 with and without the addition of 5% ammonia chloride hardener. The peak at 90°C appears only in the presence of ammonia salts. The second addition of urea also is visible in the DTA; Figure 4.8 shows a comparison of differential thermal analysis (DTA), thermal gravimetric analysis (TGA), and infrared spectroscopy (IR). The latter reflects the ratio of the peaks at 1020 and 1640 cm^{-1}. The first is due to the methylol group, the second due to the carbonyl plus the water band. In the last five years, nuclear magnetic resonance has made it possible to observe the resin cure *in situ,* and follow it quantitatively. This powerful method is described in Chapter 5. The power of the method is obvious from Tables 5.3 and 5.4, pages 138 and 140, from work by Tomita (1978) who identified some 16 products simultaneously, as a function of a variety of parameters and time!

If heat is released, the reaction proceeds at an accelerating rate. In a cold-setting resin, the incubation period should be short, but the cure should be thorough. This can be achieved by applying the resin and the acid separately; in laminates each component can be applied to a separate surface. When they are brought into contact, the reaction will then proceed promptly.

In thermosetting resin, the hardener is usually premixed. Upon heating the cure proceeds quickly, since the reaction speed doubles whenever the temperature is increased by approximately 10°C. Very often, latent hardeners are built into resins. As explained in Chapter 3, they act by chemical addition of the cation with formaldehyde, for example:

$$6CH_2O + 4NH_4Cl \rightarrow (CH)_6N_4 + 6H_2O + {}_4HCl \qquad \text{Eq. 4.9}$$

Quite frequently *p*-toluene sulfonic acid is used for hardening resins (Higashimura 1975). Sulfite, sulfur dioxide, all weak organic acids (Ellis 1935), and chlorides or sulfates of the organic amides are suitable. Phthalic acid, beta-bromohydro-

ISBN 0-201-04558-3

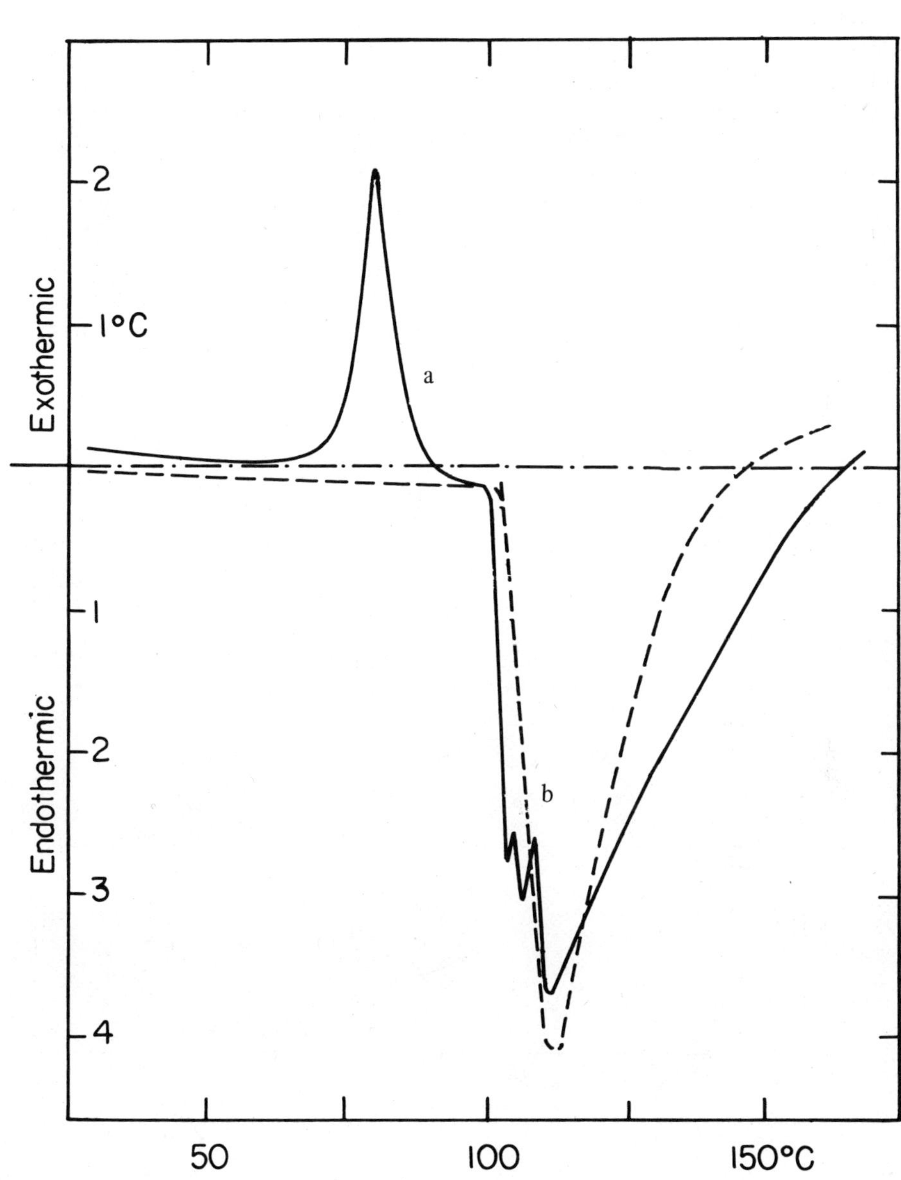

Figure 4.7. Differential Thermal Analysis Thermogram of UF-Resins (F/U = 1.3) a) With and b) Without 5% Ammonium Chloride (after Chow and Steiner, 1975)

ISBN 0-201-04558-3

cinnamic acid and succinic anhydride become soluble only at elevated temperature, and thus can be employed. More details are contained in references cited above.

Wellons (1971) used radiation, Brose, Ger. Offen. 1,966,796, employed ultraviolet light by adding a light sensitive pigment; Oetgen (1972) used ammonium fluoride; Beck (1973) used N-cyclo-hexylsulfamate; Sarrut, FP 2,203,847, employed aluminum chloride as a rapid hardener. Kinastowski (1976) used pine shavings; Simeonov (1977) employed phosphogypsum; and Teukros Company, Ger. Offen. 2,745,951, applied alkali chlorides.

Typical hardener action is described by Dubrovin (1970), Reuther (1971) in GP 2,110,264; by Sakovich (1971), El'bert (1971), Pollak (1971), Yoshihiro (1972), Wirpsza (1972), Suwaska (1972), Bogosavlejevic (1972), El'bert (1973), Mochalova (1973), Panov (1973), Ishida (1973), Erykhov (1973), Temkina (1974), Bagley, BP 1,380,919, Denisova (1974), Inoue (1974), Higashimura (1975), Andrezen (1976,1977), and Moore USP 3,970,625.

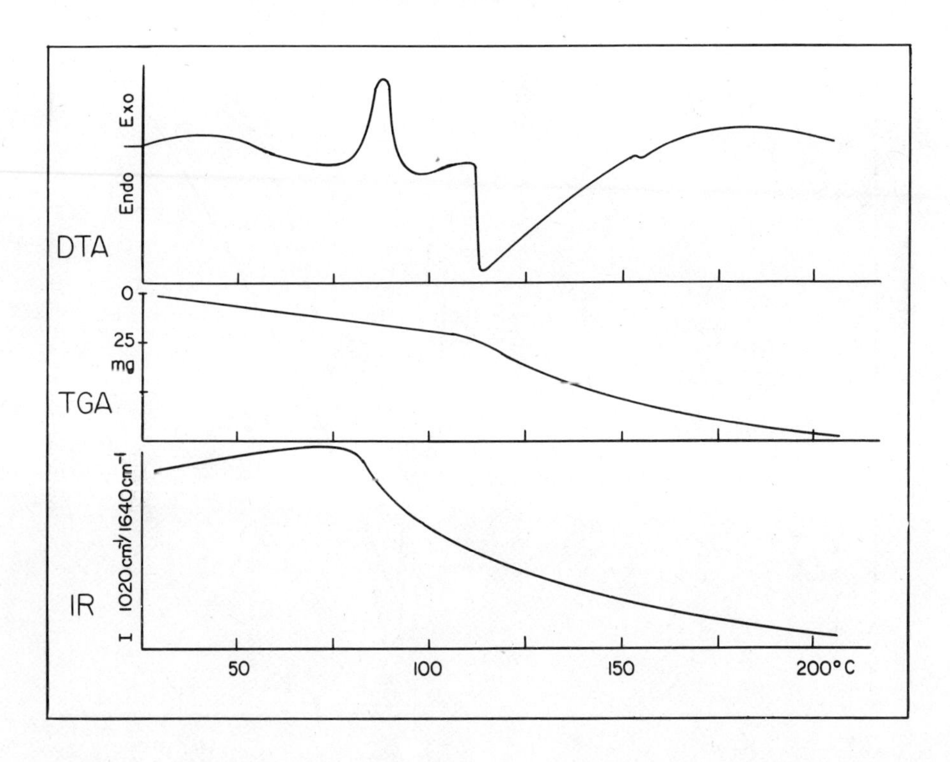

Figure 4.8. Changes During the Curing of UF-Resin Containing 5% Ammonium Chloride: a) DTA Thermogram, b) TGA Weight Loss, and c) Change in Relative Intensity of Infrared Bands at 1020 cm^{-1} and 1640 cm^{-1} (after Chow and Steiner, 1975)

ISBN 0-201-04558-3

Modified Resins

UF-resins are almost always modified before their final use. The purpose of such alterations is to either improve the performance or to save expenses. Unmodified resins are often too brittle and crack because of trapped condensation water. Among the many modifications which can be made, the most frequent are reactions with alcohols, yielding ethers; high molecular weight modifications; substitution of urea or formaldehyde by homologous reagents with equivalent reactive groups; basic addition, usually of amines; acidic addition, often of sulfite; and finally copolymerization with phenol. Scheiber (1943) gave an excellent summary of the results obtained by modifications.

LATE ADDITION OF REAGENT

In many cases, the ready-to-use resin is reinforced with a reagent to modify the behavior of the resin. Such a modification can affect the cured resin greatly. Almost all particleboard resins are now reinforced with either urea or hardener to reduce free formaldehyde and odor. Chow and Steiner (1975) have shown that urea is not simply a formaldehyde scavenger, but it reacts with the resin with release of heat, Figure 4.9. Such urea addition also changes the viscosity, Figure 4.10. In some cases, formaldehyde, or even solid formaldehyde, is added to secure full curing at the later resin stage. Petersen (1974) has extensively studied the effect of late urea addition on the performance of board and recommends the use of some additional 2% total resin solid to offset the reduced mechanical performance which is the trade-off for reduced odor.

ADDITION OF ALCOHOLS

Commercial formaldehyde contains some methanol. Alcohols cause the formation of ethers or esters. During synthesis these groups block reactive functions of the reagent, and reduce crosslinking. Thus, the cured resin remains plastic, and contains mainly long chains. The best reaction conditions are pH 4 to 6.5 and a hot medium, but the reaction also proceeds at other pH values and at room temperature. In the commercial manufacture of resin, the ethers often precipitate. Their melting points range from 90°C for the methylether of MMU and 111°C for the ethylether of MMUT to 240°C for the dimethyl-ether of dimethylol-methylene-diurea. A popular reagent is furfuryl alcohol, a cyclic compound with a five membered ring and the formula $C_4H_3O{-}CH_2OH$. Low molecular weight ethers are water soluble and are used for leather treatment. Ethers of higher alcohols are insoluble and are used in varnishes. Such alcohols are best added by exchange reaction of methyl ethers. This is done by refluxing at pH 3–4. Monoethers of ethylene glycol are well soluble, and are used as resins. Higher alcohols produce oil soluble resins suitable for varnishes and coatings.

ISBN 0-201-04558-3

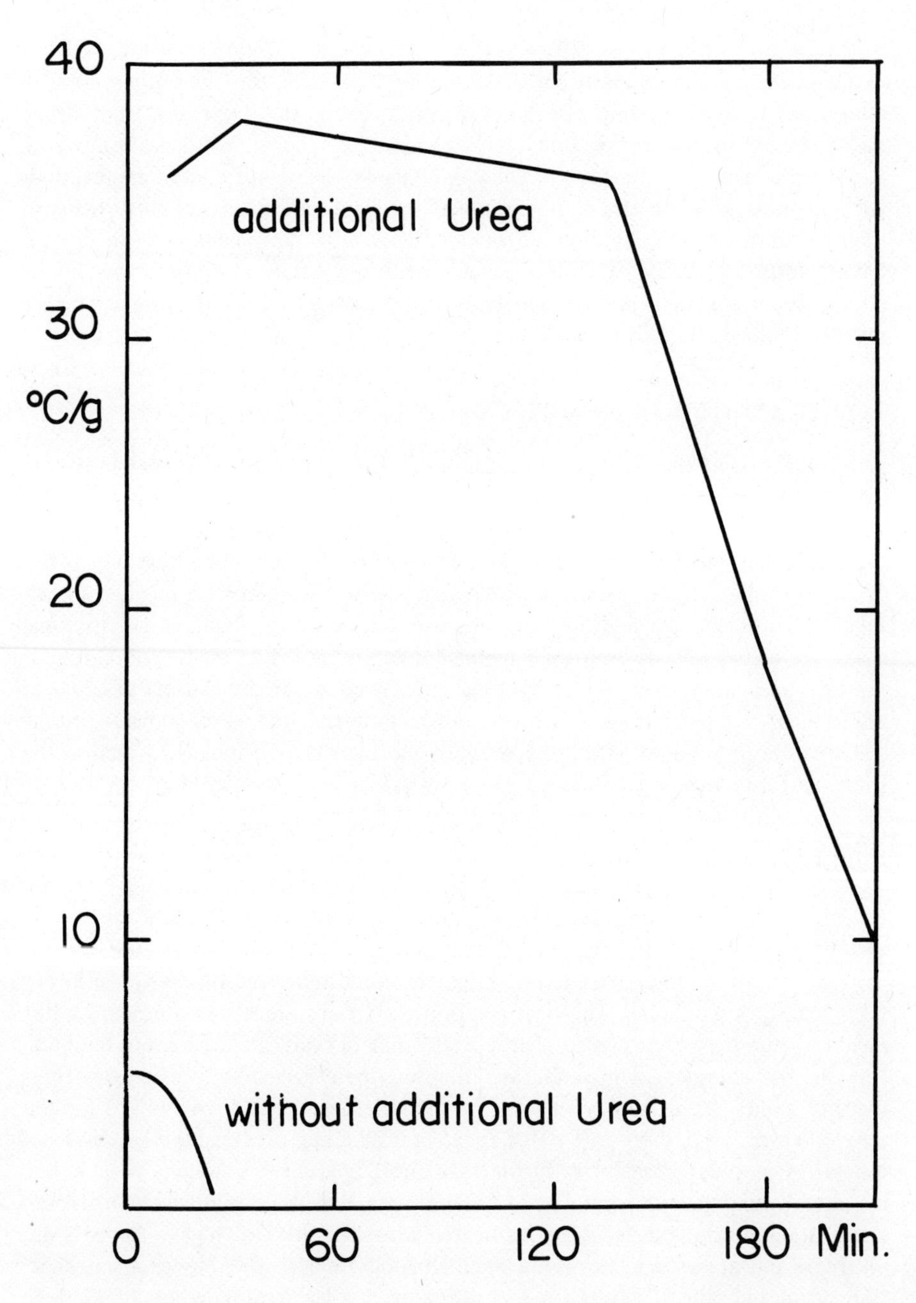

Figure 4.9. Heat Release from UF-Resin Containing 5% Ammonium Chloride at 23°C, a) With Addition of 25% Urea, b) Without Second Urea

ISBN 0-201-04558-3

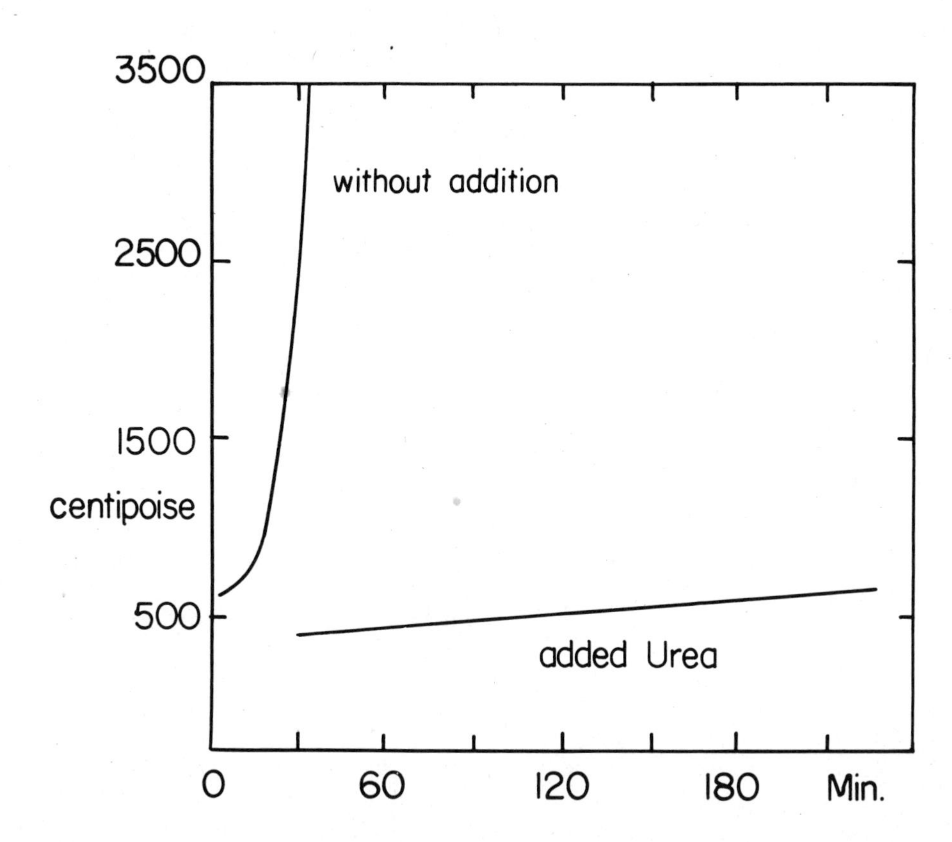

Figure 4.10. Viscosity Change of Plywood UF-Resin Containing 5% Ammonium Chloride as a Function of Time, a) With Second Addition of 25% Urea, and b) Without Second Addition (after Steiner and Chow, 1975)

Plasticized condensates of high molecular weight are obtained by reaction with adipic acid, trimethylol propane, and glycerin. Monoglyceride of linseed oil or other unsaturated compounds, acrylic acid can be copolymerized with UF. Reactions in these mixed systems are limited by solubility.

UREA SUBSTITUTION

The reactivity of substituted urea is low. Thus, the condensation proceeds only to lower molecular weight species. For example, cyclohexylurea yields merely esters of the substituted methylolurea, regardless of the excess formaldehyde used. Likewise, thiourea is less reactive, as discussed in Chapter 3. However, thiourea yields highly desirable reduced sensitivity to hydrolysis of the cured resin. Thiourea readily forms rings. With benzaldehyde it can form triazine.

ISBN 0-201-04558-3

In acidic solution thiourea can form mono-methylthiourea and N,N′-dimethylthiourea. In alkaline solution, the latter rearranges to N,S′-methylolthiourea. In the presence of amines, S-dimenthylol products of isothiourea are formed.

Diurea can yield highly elastic polymers. Hydrazodiurea and acetylene diurea can react with formaldehyde at high pH and yield tetramethylol compounds which yield resins with excellent weather resistance. Acetylene diurea can be transformed into methylol ethers which are useful in the textile industry. Polyvinyl alcohols yield products suitable for impregnation. Quaternary ammonia compounds can also be used to link urea. Thus, dimethyl-diethyl, diurea ammonium can be used on textiles.

The best substitute is melamine, a six-membered ring with six active hydrogen functions and the formula $(C_3N_3{-}NH_2)_3$.

FORMALDEHYDE SUBSTITUTES

Glyoxal reacts with urea forming cyclic glycols. These can be converted to methylols. Acrolein and furfural yield useable resins, but the latter give colored products. Acetaldehyde autopolymerizes so readily that it does not react well with urea. Acetone does not react with alkaline urea. In an acidic medium it reacts only above 150°C.

Thioformaldehyde

Since thioformaldehyde is not stable, it must be formed *in situ.* It is obtained as an intermediate whenever sulfur in the oxidation state-2 is introduced in an aqueous system in the presence of formaldehyde. Suitable sulfur precursors are hydrogen sulfide, polysulfides, or oxyacids with mixed oxidation states, such as thiosulfate, polythionate, or, under some conditions, elemental sulfur. Reactions of formaldehyde with these precursors have been studied and reviewed by Walker (1964) and Meyer (1977). These reactions were first proposed by Kreidl, USP 2,113,485, and by Pfenning-Schumacher Industries, BP 313,455, BP 366,009, BP 366,065, GP 587,643, and GP 588,879. The reaction forms thioether and similar compounds. Pollak, FP 689,410, used thiourea and hydrogen sulfide. In acidic solution trithiane $(CH_2{-}S)_3$ is also formed. It can polymerize with itself. With elemental sulfur some type of vulcanization takes place, and UF-resins can be improved by elemental sulfur, in regard to mechanical properties (Figure 4.11), odor and fire resistance (Meyer and Johns 1978). Mavoks, Can. P 351,666, proposed incorporation of urea into a mixture of formaldehyde and hydrogen sulfide.

BASIC CO-CONDENSATION

Ammonia, primary and secondary amines, dicyanodiamide and guanidine, and their derivatives can co-condense with UF. Such resins have their main application

ISBN 0-201-04558-3

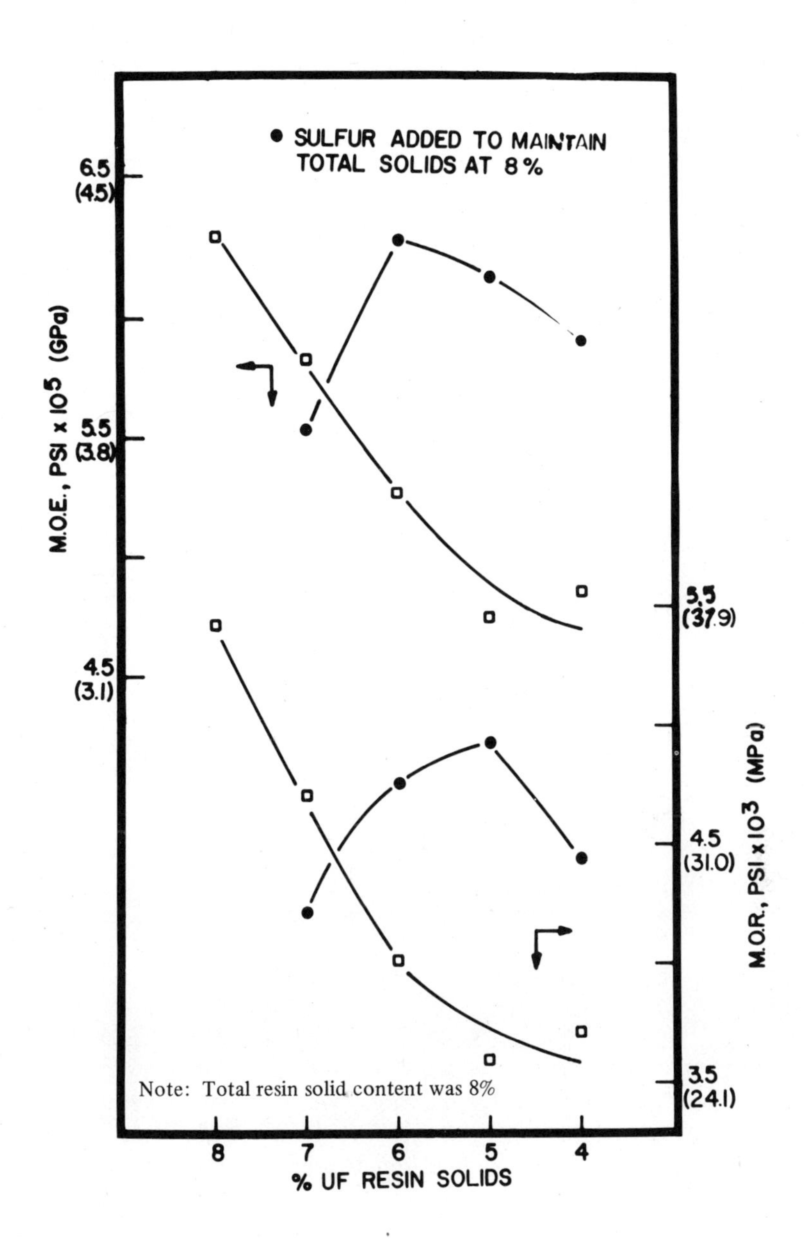

ISBN 0-201-04558-3

Figure 4.11. Correlation Between Sulfur Content of UF-Resin and MOE and MOR of Flakeboard (after Meyer and Johns, 1978)

in the paper industry. However, the reaction also occurs as a side reaction during the curing of resin with ammonium salts:

$$-CO-NH-CH_2O + NH_3 \rightarrow CO-NH-CH_2-NH_2 + H_2O \qquad \text{Eq. 4.10}$$

These reactions have been discussed in Chapter 3. The reaction can continue with formaldehyde and yield methyl derivatives which are disliked in the textile industry because of their characteristic unpleasant odors. Hydroxylamine and ethanolamine can be reacted equally. The first yields ethers. *p*-toluene sulfonamide, $CH_3-C_6H_4-SO_2-NH_2$, also reacts as amine.

Equimolar *p*-toluene sulfonamide and formaldehyde at 110°C yield a semisolid, viscous mass. These resins are compatible with cellulose esters and ethers. They are stable in sunlight and are used in lacquers to increase adhesion, and as plasticizers. In the last five years, amino compounds have been used as a stop-gap to reduce formaldehyde emission from particleboard. The section on odor reduction, Chapter 10, lists some two dozen compounds that are used, usually jointly with the hardener to bind with excess formaldehyde in the press, at 150°C. These compounds are very effective formaldehyde scavengers. However, their behavior during aging remains to be tested. Furthermore, excess ammonia is a well-known softening agent for wood, and thus attacks wood at the resin wood interface, as described in Chapter 6. Aniline makes excellent resins for molding and laminating (Widmer 1965). Melamine resins are described in other parts of this book, including Chapter 3. Such amines are often added as hardener.

ACIDIC CO-CONDENSATION

The best known reactions are those of sodium sulfite or disulfite

$$R-NH-CH_2OH + HSO_3^- \rightarrow R-NH-CH_2SO_3^- + H_2O \qquad \text{Eq. 4.11}$$

To incorporate sulfite or sulfonic groups, the alkaline precondensate is acidified and heated. Sulfite can solubilizes insoluble resins. The solubility is determined by the molar ratio of sulfite. Highly condensed, viscous products are used as paper adhesives and to improve the wet-strength of paper. The acid condensation products can be fixed with aluminum sulfate. Higher fatty alcohols yield anion resins (Aignesberger, Ger. Offen. 2,505,578, 1976) and emulsifiers. The sulfite reactions also apply for sulfamides, phenol sulfonic acid, and amino- or amido-carbonic acids. The reaction of methacrylate and acrylic acid are not yet fully understood.

The reactions of lignosulfonates, the kraft-pulping wastes, are extremely complex and not yet well understood. Lignosulfonates have been periodically used as additives in commercial resins. The products have superior weather resistance, but have to be cured at 180°C for more than 15 minutes and tend to be brittle. The

ISBN 0-201-04558-3

long press time amounts almost to autoclaving the wood, and it has been claimed that any resin would bond well under such conditions. The main application of these resins has been for shaped articles, such as profiles. Sometimes, the board is literally autoclaved at 160°C for two hours at 10 atm. Shen (1974) developed an acid catalyzed cure method, which makes the long heating unnecessary. However, the acid corrodes the press and attacks the wood. Obviously, there is still room for improvement. Kilpelainen (1975) mixed sulfite wastes with PF, a product which is traded in Scandinavia as Karatex. Ammonium lignosulfonate has been used in binders for fiber glass, by Fibre Glass Ltd. (1972) in BP 1,293,744, BP 1,316,911, and BP 1,331,470.

Sulfite is used in some wood adhesives. Sakarada, Kokai 73:51,032, used 2% as additive; Aono, Kokai 74:71,111, used toluene sulfonic acid to prepare crack-resistant overlays for particleboard. Starzynska, Pol. P 74,999 (1975), used disulfite to obtain a water-resistant coating on paper. Szlezyngier, Pol. P 74,060 (1975), incorporated waste sulfite into UF-resin to gain a more flexible and better water-resistant material. Taylor, USP 3,901,752 (1975), used an alpha-olefin; Mita, Kokai 75:25,802 (1975), pulped cellulosic materials with sulfur dioxide and sulfide; El'bert (1972) in USSR 518,363, used lignosulfonate and phosphate as a binder; Nowak (1976), Pol. P 80,377, used sulfite to stabilize resin; and Casebier (1976) in Can. P 986,100 used sulfite to react with bark.

Foam compositions all contain some emulsifiers. These can react with formaldehyde directly, and thus reduce free formaldehyde odor in the finished resin, Chapter 7.

CO-CONDENSATION WITH PHENOL, RESORCINOL

UF-resins with terminal resorcinol have been described by Blommers (1978) USP 4,032,515. Resorcinol has two hydroxyl groups in meta position on an aromatic ring. $(C_6H_4)(OH)_2$. These copolymers form excellent products, but their formulation is still more of an art than a science. Clearly, ^{13}C-NMR will help control and understand these reactions, and encourage more extensive use of the endless combinations of urea-formaldehyde.

Mixed phenol UF-resins have been long used and are well established. Baxter (1963) in USP 3,108,990 added various oxyacids of sulfur as antioxidants, among the latter were bisulfite, disulfite, sulfur dioxide, and sulfides.

Bark and lignin extenders, containing polyphenols, also give good copolymers, if somewhat brittle and often undesirably colored. Trocino (1971) in USP 3,616,201 used Douglas fir bark.

Early resin chemists tried almost any possible addition to resins. Amand, FP 688,269, proposed sugar as additive to molding. Metal oxides, especially zinc, are added as catalysts. Ellis, USP 2,121,076, used starch. It is still used today (Roy 1977). Lactic acid is also suitable, as is cellulose. Cellulose binds as ether, in the manner discussed in Chapter 3. Casein, by itself a wood adhesive, can copolymerize with UF, as was recognized by Ripper. Egg albumin and proteins in general are

ISBN 0-201-04558-3

suitable for reactive interactions. Many of these reagents are commonly used as extenders in plywood, or as gap-filling agents.

Finally, the copolymerization with cyanamide and dicyanamide should be mentioned. H. John, the inventor of the first commercially feasible UF-resin, later worked with cyanamides. Pratt (1944) in USP 2,349,756, assigned to DuPont, described a formaldehyde copolymer. In recent years, isocyanates have become prominent candidates as adhesives. Their properties are superior, but because of their tenacious adhesive prowess, they stick not only to wood, but also to press plattens. Thus, one would need special release agents, or, one would have to restrict their use to the core layer of particleboard. So far, price has limited their application. If spray equipment can be optimized to apply 2 wt % resin solid on dry weight wood, isocyanates will become competitive and popular, because their water resistance is superior. Their use for weather-resistant board has already been approved in Germany (Deppe 1977, Clad 1978, Hutschnecker 1978). The commercial product is diphenyl-methane-isocyanate, $CH_2(-CH_4)-NCO_2$. The reaction product is a urethane bridge: $R-O-CO-NH-C_6H_4-CH_2-C_6H_4-NH-CO-O-R$, where R is either wood cellulose or copolymer or filler. The adhesive is equivalent to PF resins. Koleshnya (1978) described UF-isocyanate copolymers.

ISBN 0-201-04558-3

5. Chemical Analysis and Sampling

This chapter deals first with analytical chemistry. Substantial progress has been achieved during the last few years in this field. It is frequently necessary or desirable to indentify resins and products derived from resins, determine their composition qualitatively or quantitatively, and to determine the effect of resinous products on the environment.

Until five years ago only simple components or resins could be identified by indirect chemical evidence. During the last ten years a truly revolutionary change has occurred in analytical chemistry, because modern electronics, physics, and chemistry have made possible the development of a variety of instruments for rapid, accurate, and *in situ* analysis of elements, compounds, and functional groups with an ease and accuracy which exceeded the wildest predictions. The most powerful example are electron microscopes with semi-quantitative elemental mapping attachments with which cured resin surfaces and interfaces on adhesive substrates can be scanned and photographed; laser Raman spectrometers with which surface chemistry can be followed; Fourier-transform infrared spectrometers; and other electronically enhanced and computer aided instruments, such as nuclear magnetic resonance (NMR) spectrometers with cavity attachments which specifically respond to protons, i.e., hydrogen, carbon, oxygen, and other atoms. NMR spectrometers yield both qualitative and quantitative information, and make it possible to distinguish between carbon atoms in different functional groups, such as alcohols, ethers, and methylene, and in different environments. Further powerful tools are Differential Thermal Analysis, (DTA), Thermal Gravimetric Analysis (TGA), and the various forms of chromatography.

Today it is possible to analyze, identify, and determine the composition of ready-to-use resins and cured resins *in situ* by ^{13}C-NMR. This method is very sensitive to the chemical environment of carbon atoms and thus makes possible to distinguish the functional groups in each molecule, even in a mixture. The method can even be used to measure the concentration of the various species. The method is still expensive and not widely used, because the necessary tools are not every-

Beat Meyer, Urea-Formaldehyde Resins

ISBN 0-201-04558-3

where available, but the situation will undoubtedly quickly change. The NMR method allows simultaneous determination of more than 20 UF-resin components.

This chapter deals first with the conventional analysis of resins, then with sampling techniqes, and finally with the above mentioned new tools. The reader will find most valuable information on theory and practical procedures in the books by Walker (1964), Sawicki and Sawicki (1975), and the references listed in the text below.

Analysis of Raw Materials

UREA

Commercial urea is usually in highly pure form. For the assay of urea, total nitrogen is normally acceptable. The most common procedure is probably still that of Kjeldahl. In this procedure all nitrogen of impurities is included. If urea-nitrogen is to be determined, two procedures are common: the urease method, and the xanthydrol method. In the first method, urea is hydrolyzed by the enzyme urease yielding ammonia, which is acidified with a known amount of excess HCl. The latter is then titrated against methyl red. The method is said to have an accuracy of ± 0.1%. In the xanthydrol method, urea is precipitated and the dixanthylurea ($C_{27}H_{20}N_2O_3$) is determined gravimetrically. The claimed accuracy is ± 0.5%.

The ash content in urea is determined by heating about 50–100 g in a platinum or fused silica dish to 800°C. Typical ash content specifications allow for 30–40 ppm. The ash usually contains iron and copper.

Iron can cause noticeable coloration of resin products. The iron content of ash is determined best by the *o*-phenanthroline method, or with the help of mercapto-acetic acid or potassium thiocyanate. In the *o*-phenanthroline method iron is dissolved in hydrochloric acid, treated with hydroxylamine, and reacted with *o*-phenanthroline. The absorbance of the solution is measured at 508 nm and compared against a calibration curve. Copper can be determined both in urea and urea ashes with the help of diethyl dithiocarbamate. The product absorbs at 450 nm. Water in urea can be determined by the well-known Karl Fischer Titration. Often, it is also desirable to determine the buffer capacity and the pH of urea used in resin synthesis.

FORMALDEHYDE

As indicated in Chapter 3, commercial and reagent grade formaldehyde is either a solid polymer, or an aqueous mixture of methylene glycol and polymethoxy glycol. Usually the latter is stabilized with 1–15% methanol, and contains some formic acid produced by air oxidation.

Specifications for formaldehyde are published in the U.S. Pharmacopoeia (U.S.P.), the American Standards Association (ASA) specifications, the American

ISBN 0-201-04558-3

Chemical Society (ACS) reagent chemical specifications, and the corresponding specifications of other countries. Low methanol solutions contain 35–50% formaldehyde stabilized with 0.3–1.5% methanol. They contain 0.01 to 0.05% formic acid. U.S.P. solutions contain 37.0–37.5% formaldehyde and 6–15% methanol with 0.01–0.03% formic acid. Total solids should be below 0.01%, and iron and other heavy metals are required to be below 5–10 ppm. The pH of the solution is between 2.8 and 4.0. The flash point is 70° to 80°C and the viscosity is 2.1 to 2.7 centipoises at 20°C. The specific gravity is 1.020 to 1.065.

The assay of commercial formaldehyde involves determination of formaldehyde, methanol, and formic acid. The methods for measuring the latter two are described later in this chapter. Formaldehyde is assayed by the alkaline peroxide method or the acidimetric sulfite method.

STANDARD QUANTITATIVE METHODS

Alkaline Peroxide Method

This method was developed by Blank and Finenbeiner in 1898. It converts formaldehyde to formic acid. The reaction is conducted in the presence of a known amount of sodium hydroxide at 60°C. The excess alkali is titrated with hydrochloric acid against bromothymol blue. At the end point the latter turns from blue to green. Neither methanol nor ethanol interferes.

Sodium Sulfite Method

This method was developed by Ripper (1900) who fully recognized the power of this method. While searching for a reagent to determine sulfite in wine, he discovered that sulfite bound to formaldehyde cannot be longer titrated with iodine. Lemme modified the method in 1903 to determine formaldehyde by acid titration. The method is based on the reaction of formaldehyde with sulfite, yielding formylsulfonate, in which the sulfur atom is bonded to the carbon (Meyer 1979, Kurtenacker 1938, Tartar 1948).

$$CH_2(OH)_2 + Na_2SO_3 \rightarrow NaCH_2(OH)SO_3 + NaOH \qquad \text{Eq. 5.1}$$

In this method a one molar solution of sodium sulfite, neutralized against thymolphthalein blue, is reacted with the neutral formaldehyde sample and titrated with hydrochloric acid. One milliliter of a one molar standard acid solution corresponds to 0.03003 g of formaldehyde. As in all acid titrations, one must carefully avoid carbonate impurities. The method is accurate to within ± 0.02%. The method is not very specific for formaldehyde, because sulfite also reacts with other aldehydes and ketones. The sulfite method is the official TAPPI standard method T600os-76.

ISBN 0-201-04558-3

Iodometric Method

This method is based on the reaction of hypoiodite with formaldehyde in alkaline solutions, which yields potassium formate Excess hypoiodite (KIO) undergoes an auto-redox reaction which yields .iodate (KIO_3). Upon acidification the latter reacts with iodide and forms iodine, which can be titrated with thiosulfate. This method is very convenient for determining small concentrations of reasonably pure formaldehyde. It is not suitable for concentrated solutions or for solutions containing other oxidizable organic species.

Ammonium Chloride Method

Ammonium chloride reacts with formaldehyde yielding hexamethylene tetramine. This reaction has been used since 1880 for determining formaldehyde. In this method both a blank ammonium standard and the sample containing excess ammonium chloride are separately titrated with acid against bromothymol blue. The difference between the two results is a measure of the formaldehyde concentration. The method has been refined by Matsumoto (1968).

Mercurymetric Method

Nessler's reagent, alkaline potassium mercury iodide, K_2HgI_2, is used to reduce mercury ion to the element. The latter is determined by iodometry.

Potassium Cyanide Method

Formaldehyde reacts with cyanide forming cyanohydrin, $CN{-}CH_2{-}OH$. Excess cyanide is reacted with silver nitrate and excess silver is titrated with thiocyanate. This method is as reliable as the peroxide or sulfite methods. It has the advantage that acetone, acetaldehyde, ketones, and higher aldehydes interfere less.

Schulek's modification of the cyanide method is even less sensitive to impurities or contaminants in formaldehyde. It involves titration of excess cyanide with iodide.

Hydroxylamine Method

In the presence of phenol or similar resinous materials, formaldehyde can be determined with the help of hydroxylamine hydrochloride which yields formaldoxime, $CH_2{-}NOH$, according to the reaction

$$CH_2(OH)_2 + NH_2OH \cdot HCl \rightarrow CH_2NOH + 2H_2O + HCl \qquad \text{Eq. 5.2}$$

The hydrochloric acid release can be measured by titration against bromophenol blue, or with a pH-meter.

ISBN 0-201-04558-3

Methone or Dimedon Method

Formaldehyde reacts with methone (5,5-dimethyldihydro resorcinol) to form methylene bismethone. The reaction can be followed by volumetric or gravimetric evaluation. The method is selective, because the condensation product of other aliphatic aldehydes convert to cyclic hydroxanthene when treated with glacial acetic acid.

COLORIMETRIC DETECTION OF FORMALDEHYDE

A variety of colorimetric methods have been developed which are useful to identify and estimate small amounts of formaldehyde. Only a few of the currently practical methods are summarized here. Since formaldehyde absorbs light only weakly, with an extinction coefficient of 0.0135 and in the ultraviolet, at a wavelength of 290 nm, the colorimetric method depends on reactions which yield colored species. The most important of these are described here:

Chromotropic Acid, the Eegriwe Method

This method is now widely accepted for assaying formaldehyde in air, in automobile exhausts, in fertilizer, in proteinaceous materials or solutions, and in aqueous solution in general. The chromotropic acid reagent (1.8-dihydroxy naphthalene-3,6-disulfonic acid) reacts with formaldehyde in the presence of concentrated sulfuric acid to yield a violet dibenzo xanthylium salt which absorbs at 580 nm. Under optimized conditions the sensitivity is 0.008 mg per liter. 18 nanograms of formaldehyde in 1 ml solution can be detected with a confidence limit of 90%. The sensitivity range is 0.1 to 2.0 mg/1 in solution, or 0.5 mg/m^3 in air (Clermont 1976).

Optimized Procedure. In this procedure 330 mg of the disodium salt of chromotropic acid is dissolved in 100 ml of concentrated sulfuric acid. A syringe is used to squirt 3.5 ml of the reagent into a 3 ml test solution. The color fully develops in less than the eight minutes. The absorbance is measured at 580 nm.

This method has been extensively tested and reviewed (Sawicki 1975). Sulfur does not interfere. Acrolein, ethanol, and higher molecular weight alcohols and olefins interfere. Phenol gives negative interference, as diethylene and propylene. The method can be made accurate to within ± 5%. Absorbance increases upon standing by 3% after one day.

Schiff Reagent, Denige's Method, p-Rosaniline

In 1866 Schiff discovered that formaldehyde reduces fuchsin sulfite to the leuco form. The method is highly sensitive and simple but optimizing quantitative work is made difficult by a variety of problems. The dye is not easily purified, and other

ISBN 0-201-04558-3

color species must be removed with absorbents before use; absorbance does not change linearly with concentration, and temperature as well as time change absorbance. Part of the problem stems from the fact that *p*-rosaniline has three basic amino groups which can react step-wise. The colored species contains the sequence: $R-NH-CH_2-SO_3^-$. Thus, the molar ratio of sulfite is crucial.

Procedure. Three grams of *p*-rosaniline are dissolved in one liter of water containing 3–4.5 m M disodium disulfite (metabisulfite). The absorbance is measured after two hours at 550 nm. The pH of the solution is about 3–4. The pH influences both color and interferences.

This test is suitable for analysis in the presence of peroxide. Up to 0.002 g/l the absorbance is reasonably linear. With this method 10^{-4} formaldehyde in acetaldehyde can be measured.

Acetylacetone Method, Hantzsch Reaction

This method is highly specific and can be carried out under mild conditions and does not require corrosive chemicals. The reaction was introduced by Nash (1953). Acetylacetone (2,4-pentanedione) yields diacetyldihydro butidiene which absorbs at 412 nm. Acetaldehyde interference is less than 1% at that frequency. The method is suitable for solutions containing up to 8 mg/1 formaldehyde. The reaction is conducted by mixing the sample with a 0.02 molar solution of acetylacetone which is buffered by the addition of 2 molar ammonium acetate and 0.05 molar acetic acid. The solution is developed by heating to 60°C for ten minutes. It is stable for over two days. This test is widely used in industry and research. It is the author's preferred test.

Sulfite interferes with the reaction: 10^{-3} M sulfite causes 90% loss of formaldehyde. Belman (1963) developed a fluorometric modification of the procedure, using excitation at 410 nm and observation at 510 nm, which makes it useful over the range from 0.8 to 2 micrograms. At higher concentrations the method is useless, because the band at 410 nm is saturated and suffers self-inversion. In air analysis, possible interference by nitric oxide, sulfur dioxide, and ozone must be considered.

Procedure. Two milliliters of test solution are added to 2 ml reagent (150 g ammonium acetate, 3 ml glacial acetic acid, and 2 ml 2,4-pentanedione diluted to one liter). The solution is kept for 15 minutes at 60°C, or for one hour at 30°C. The cold sample is tested for absorbance at 412 nm; if formaldedhye is less than one microgram, fluorescence exited at 410 nm is observed at 510 nm. *N*-butanol can be used to extract and enrich the colored species.

Delta-Acid

Delta-acid, 7-amino-4-hydroxy-2-naphthaline sulfonic acid, has the same chromotropic group as chromotropic acid. The detection limit is about 0.0005

ISBN 0-201-04558-3

micrograms. The absorbance is at 470 nm. The color is stable for more than a day. The compound can be used for absorption as well as fluorescent measurements.

MBTH

3-methyl-2-benzothiazolinone hydrazone yields a formazan which absorbs at 670 nm. In this method other aliphatic aldehydes interfere. Ohtomi (1975) and Bitterli (1976) find this method suitable for detecting 30 ppb–1.5 ppm, i.e., it is more sensitive than all other methods except tryptophan (Chrastil 1975). This method has been tentatively accepted by the American Industrial Hygiene Association (AIHA), the American Conference of Government Industrial Hygienists (ACGIH) and eight other societies, for the analysis of air.

Sawicki and Sawicki (1975) described 14 other colorimetric methods and their merits. They, and their detection limits, are summarized in Table 5.1.

Table 5.1

Limit for Identification of Formaldehyde in 19 Colorimetric Tests (after Sawicki & Sawicki, 1975)

Reagent and Procedure	Identification Limit (micrograms)	Color
Chromotropic acid	0·14	Purple
J-acid (A)	0·01	Yellow[a]
J-acid (A; H_2O addn.)	2	Blue
J-acid	0·0005	[a]
J-acid (thermochromic)	0·03	Blue
Phenyl J-acid (B)	0·1	Yellow[a]
Phenyl J-acid (B; with H_2O)	0·5	Blue
Phenyl J-acid (B; steam)	0·05	Blue
2-Hydroxycarbazole (H_2O)	0·004	Blue
2-Hydroxycarbazole (without H_2O)	0·02	Blue
MBTH	0·1	Blue
HBT	0·01	Blue
4-Nitrobenzalhydrazone (C)	0·1	Purple
2-Nitro-1,3-indandione (D)	0·1	Green[a]
2-Nitro-1,3-indandione (E)	0·0003	[a]
1-Ethylquinaldinium iodide (F)	0·5	Blue
1-Ethyl-4-picolinium iodide (F)	2·5	Green
Azulene (G, test tube)	0·1	Blue
Azulene (G, spot paper)	0·1	Blue

[a] Fluorometric procedure.

ISBN 0-201-04558-3

Azulene forms a chromogen which absorbs at 620 nm, but this species is rarely used. 2,7-dihydroxynaphthaline yields a violet color which results from a variety of aldehydes. Gallic acid is known to give a faint luminescent reaction during alkaline oxidation with peroxide and formaldehyde, which can be used to determine formaldehyde. 1-ethylquinaldinium iodide forms a colored species which absorbs at 610 nm. 2-hydroxy-carbazole yields a blue color. 2-naphthol has been proposed for the analysis of phenol containing samples by Ozlorskaya (1971).

The selection of the best analytical method for formaldehyde is not yet conclusive. The choice of the preferred method depends on whether high sensitivity or convenience are desired. At the present time the author prefers the acetyl-acetone method which can be directly correlated with all other methods.

OTHER ANALYTICAL METHODS

The polarographic method for determining formaldehyde has been described by Walker (1964). The method is fairly sensitive and selective. Gas chromatography is another very useful method. A variety of packing materials have been developed for this purpose.

Formaldehyde Oligomers

As explained, formaldehyde solutions contain methylene glycol in equilibrium with a series of linear oligomers. If methanol is present, the polyoxymethylene glycols partly degrade and are converted to hemiacetals. The analysis of these mixtures has been difficult, but Dankelman and Daeman (1976) have developed a method which now makes it possible to determine these mixtures. For this purpose the oligomers are silylated to block the functional groups and freeze the equilibria. The solution is then analyzed by gas-liquid chromatography and the fractions are identified by mass spectroscopy. The silylation is carried out by treating the sample with *N,O*-bis(trimethylsilyl)trifluoroacetamide (BSTFA) and DMF. The reaction is completed in five minutes at room temperature. The sample is then separated on Chromosorb W with 10% OV-1. The same authors also analyzed formaldehyde oligomers by H-NMR and obtained the same results. The data is described in the section on formaldehyde chemistry.

Methanol

The methanol content of formaldehyde solutions can be assayed by several physical and chemical methods. In the standard procedures methanol is determined by subtraction of formaldehyde aliquot from the total chromic acid necessary for total oxidation of the sample. Methanol is oxidized to carbon dioxide. A fairly accurate measure of methanol is also obtained from the specific gravity data (Walker 1964). Specific gravity can be measured to ± 0.01%; 1% methanol in aqueous formaldehyde changes the specific gravity by about ± 0.20%.

ISBN 0-201-04558-3

Formic Acid

Formic acid is slowly produced by air oxidation of formaldehyde and causes commercial solutions to react acidic. The formic acid content can be determined by direct titration with sodium hydroxide against bromothymol blue. Care has to be taken to keep the alkali free of carbonate.

Metal Impurities

The analysis of metal impurities has been described under formaldehyde assay. More details have been described by Walker (1964).

Analysis of Resins

FREE FORMALDEHYDE

Ready-to-use resins and cured resin contain excess formaldehyde. Since resins readily degrade, and thereby release chemically bound formaldehyde, great care must be taken to test for free formaldehyde with nondestructive methods, for example, by sampling the air above the resin. Unfortunately, the term "free formaldehyde" is not well defined, and thus reports on this resin components often differ widely.

The most common methods in use are the acidimetric sulfite method, the hydroxylamine hydrochloride method, and the iodometric sulfite method. Polarography has also been used. The sulfite method is based on de Jong's observation that at room temperature methylol-ureas degrade only slowly in neutral solution. Thus, around pH 7, sulfite first reacts with free formaldehyde, and only slowly shifts the equilibrium. Thus, judicious pH control makes it possible to titrate free formaldehyde at pH 7 and then total formaldehyde at a pH above 9 or below 4 in sequence in any one sample. The procedure has been described by Walker (1964). Reproducible and careful control of time is necessary in this procedure.

TOTAL FORMALDEHYDE

Determination of total formaldehyde is of general importance in the characterization of resins. An excellent review of the procedures has been given by Grad and Dunn (1953). The basic procedures involve acid hydrolysis of the resin. The most common procedure involves digestion with phosphoric acid, followed by distillation of formaldehyde. The latter is usually collected in an aqueous medium and analyzed according to any of the above described methods, such as oxidation with peroxide, etc. Thiourea apparently interferes with the analysis of formaldehyde, but the interference can be minimized by addition of mercuric oxide.

A typical procedure for hydrolysis of resin proceeds as follows: about 1 g of resin solid is placed into a 125 ml distillation flask; 25 ml of 85% phosphoric acid

ISBN 0-201-04558-3

and 25 ml water are added; and the mixture is brought to a boil. The boiling water is replaced until some 200 ml have been distilled.

In the peroxide method, the stillate in the receiver is refluxed one half hour with 50 ml of 0.5 N sodium hydroxide and 60 ml 3% hydrogen peroxide, and the excess alkali is titrated with hydrogen chloride against a methyl red indicator.

TOTAL NITROGEN

The total content of amine and other nitrogen in resin, urea, buffer, and other additives is best determined with the Kjeldahl test; the official ASTM procedure follows.

The unknown containing 0.15–0.25 g of nitrogen equivalent is placed into an 800-ml Kjeldahl flask; 0.5–0.75 g mercury, 10 g potassium sulfate, and 25–35 ml concentrated sulfuric acid are combined and mixed. After frothing ceases the solution is boiled for two hours. Then, 500 ml water is added to the cooled sample which is then added to 80 ml 50 wt % sodium hydroxide and 30 ml thiosulfate. 300 ml of this mixture are distilled into 50 ml 0.5 N hydrochloric acid. The excess acid is determined by titration with hydroxide against methyl red.

UREA

The reaction of urea with aniline yielding crystalline diphenylurea consumes only 80% of total urea. According to Grad and Dunn (1953) urea is best determined by amonolysis with benzylamine. For this about 0.5 g of resin solid is combined with 15 ml benzylamine and refluxed for eight hours. The solution is then cooled to 40°C and acidified with hydrochloric acid against Congo red paper. The resulting oil is then collected by cooling in an ice mixture, until it separates as solid. The dibenzylurea is then dried at 150°C and weighed.

The determination of thiourea, melamine, phenol, and other constituents exceeds the scope of this book.

SOLIDS

The most important factors in determining solid content of a resin are temperature, time, and thickness of layer of the sample during drying. Normally, enough sample is dried in a Petri dish to leave about 0.5 g solid. Samples are either dried at 105°C at normal pressure, or at room temperature *in vacuo* (less than 5 Torr). Sometimes two hours drying at 45°C is preferred. Freeze drying yields slightly higher values. Specific gravity has also been proposed as a measure of solid content.

The PMMA method uses "milk cap" aluminum dishes. These have 58 mm diameter and are 17 mm deep. Five ml water is mixed with enough resin to yield 0.4 to 0.5 g solid. The water is evaporated at 105°C and the sample is kept at that temperature for three hours. The foil method has been described by Averell (1959).

ISBN 0-201-04558-3

WATER AND VOLATILES

Total water and volatiles are determined by weight loss during drying, or by the Karl Fischer method.

pH AND BUFFER

The correct amount of acid catalyst necessary to cure amino resin must be determined on the basis of the resin pH and its buffer capacity. While the pH can be measured with a glass calomel pH meter, the buffer capacity is established by titration with 0.5 hydrochloric acid. The pH versus acid curve reveals not only the total amount of buffer present, but the breaks in the curve indicate the acidity constants of the buffer, and thus its identity. For example, phosphate is readily recognized by a break at pH 6.

FILLERS

Depending on the application, resins might be filled with anything from white, pure alpha-cellulose used for molding articles to wood fibers, ground walnut shells, or asbestos and glass fibers. Analysis of fillers exceeds the scope of this book.

Individual Resin Species and Functional Groups

OLDER METHODS

Amino resins contain carbon, nitrogen, oxygen, and hydrogen in the form of a variety of chemical functional groups, such as amino groups, carbonyls, or alcohols. Each of these groups differs in chemical behavior and imparts the resin different properties and characteristic performance. The relative ratio of the various chemical functions in the cured resin has direct influence on its mechanical properties, stability, and aging. For example, cured resins containing much of the nitrogen as tertiary amides are heavily crosslinked and resist weathering best. In contrast, cured resins containing extensive methylol functions hydrolyze easily, i.e., they lose mechanical strength easily, and, furthermore, they tend to release formaldehyde.

Until very recently it was virtually impossible to analyze ready-to-use resins or cured resins for their chemical functional groups, because the analysis methods were destructive, and altered the resin composition. De Jong developed a sulfite method to identify methylolurea in the presence of formaldehyde. Recently Kaesbauer (1976) and coworkers improved this method and showed time-composition plots for various temperatures. Landqvist (1957) developed a spectrophotometric method for the quantitative determination of urea and methylolurea of the reaction mixture, using absorption at 420 nm. Lee (1972) and Ludlam (1973) developed thin-layer and paper chromatography methods which differentiate

ISBN 0-201-04558-3

between urea, mono-, di- and trimethylolurea, and a total of nine similar species. Duval (1972) described gel-permeation methods for identifying resin components. This paper contains a calibration curve for various different molecules with different molecular dimensions. The method is sophisticated and ingenious, but the products are not fully resolved. Dankelman (1976) discussed gas-liquid chromatography (GLC) and high-pressure liquid chromatography (HPLC) as well as gel-permeation chromatography (GPC). Hope (1973) used GPC to separate both aqueous and nonaqueous UF systems. The shifting of equilibria during chromatography is a serious problem. It can be suppressed by silylation of the resin with *N,O*-bis(trimethylsilyl)trifluoroacetamide (BSTFA) as described by Dankelman (1976). The latter method helps to resolve the fractions of urea, mono- and dimethylolurea by GPC on a Sephadex LH 20 resin in water; likewise, silylated resins can be separated well by gas chromatography. However, nuclear magnetic resonance (NMR) is clearly superior to all earlier methods.

NUCLEAR MAGNETIC RESONANCE (NMR) METHOD

The NMR method, and especially the ^{13}C-NMR method has opened a window for research chemists for direct *in situ* qualitative and quantitative observation of resin composition during manufacture, during resin cure, and in the finished product. When this method reaches the application and quality control laboratories, its application will make it possible to correlate changes in the chemical composition to changes in mechanical behavior and performance of the resin, and it will revolutionize quality control.

The nuclear magnetic resonance analysis of UF-resins was pioneered by Chiavarini and coworker (1975). They succeeded in identifying $-NH$, NH_2, $-CH_2$, H_2O, OCH_3 groups in $NH_2{-}CO{-}NH{-}CH_2OCH_3$, and the same functional groups in six important compounds which occur in UF-resins, Table 5.2 and Figure 5.1. They applied their findings immediately on a series of commercial resins. Since the publication of this work, a large number of papers have appeared in this field, and the NMR spectra of UF-resin components are now well established. The advantage of this method is that the concentration of all functional groups can be directly measured by integration of the spectra. The disadvantage of the proton-NMR method is that the characteristic NMR peaks overlap or are clustered together in mixtures and elaborate solvents, and electronic procedures are necessary to isolate and measure the various peaks. Figures 5.2 and 5.3 show that the data obtained with this method is very valuable. Figure 5.2 is based on the work of Kumlin and Simonson (1978). These figures show that it has become possible during the last years to directly follow the concentration of the main reaction intermediates and products during resin synthesis as a function of U:F ratio. In the example, Figure 5.2. 42 wt % formaldehyde was reacted at 40°C and reached 70°C over a period of 30 minutes. The pH was buffered at pH 8.5. Samples were withdrawn at regular intervals, chromatographed, and analyzed with proton NMR in DMSO-d^6. In

ISBN 0-201-04558-3

Figure 5.3a, after Chiavarini (1978) the relative concentration of primary, secondary, and tertiary amines is shown as a function of U:F. Figure 5.3b shows a plot of three formaldehyde derivatives. Polyoxymethylene drops sharply while methylene increases with increasing U:F. The resin was synthesized at pH 5 and 20°C.

Table 5.2

6 UF Reaction Products and Their Structure
The NMR spectra are shown in Figures 5.1(a) to (g).
(after Chiavarini, 1975)

Name	Formula	NMR Spectrum Figure
Monomethylolurea (MMU)	$H_2N-CO-NH-CH_2OH$	5.1(a)
Dimethylolurea (DMU)	$HOCH_2-NH-CO-NH-CH_2OH$	5.1(b)
Methylene-diurea (MDU)	$H_2N-CO-NH-CH_2-NH-CO-NH_2$	5.1(c)
Dimethyloloxymethylenediurea (DMOMDU)	$HOCH_2-NH-CO-NH-CH_2$ >O $HOCH_2-NH-CO-NH-CH_2$	5.1(d)
Dimethylolureadimethylether (DMUDME)	$CH_3O-CH_2-NH-CO-NH-CH_2-OCH_3$	5.1(e)
Monomethylolurea-methylether (MMUME)	$H_2N-CO-NH-CH_2-OCH_3$	5.1(f)

These figures demonstrate drastic progress in chemical analysis. It took chemists some fifty years of chemical reasoning and many years of elaborate experimentation, synthesis, and indirect wet-analysis to make firm conclusions about the UF-resin chemistry. Today, it takes 30 minutes of NMR analysis to identify the reagent, intermediates, and products and to quantitatively determine their concentration as a function of time. And, all this can be achieved *in situ,* without decomposition.

During the last three years, the NMR method has been perfected by use of natural abundance ^{13}C probes. This method helps resolve and separate peaks of some 60 functional building stones of UF-resin. Table 5.3 and Figure 5.4 show what can be achieved with this method. It has been applied by Dankelman (1978), Slonim (1978), de Breet (1977), Duclairoir (1976), and Tomita (1976–1978).

ISBN 0-201-04558-3

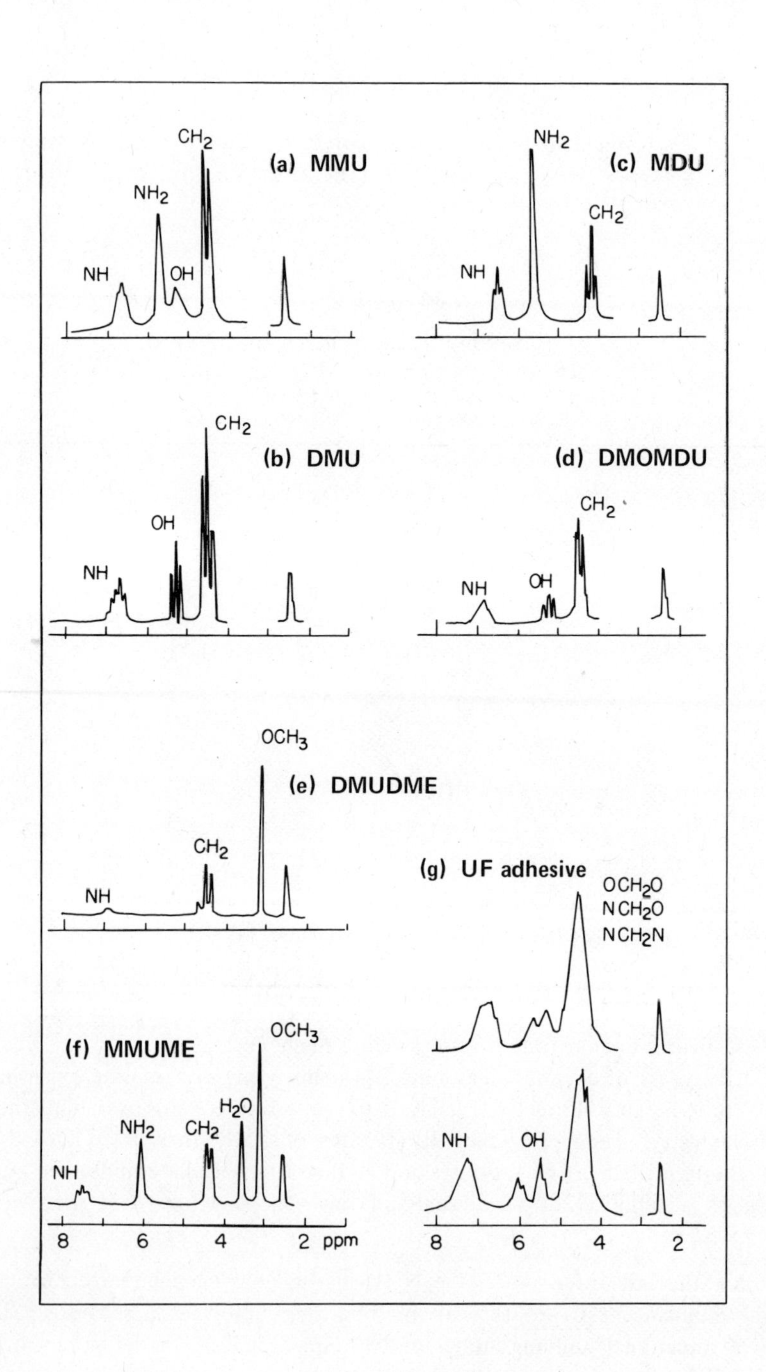

Figure 5.1. H-NMR Spectra of Seven Urea-Formaldehyde Reaction Products in DMSO-d_6/$CaCl_2$ (after Chiavarini, 1970). The reagents are listed in Table 5.2.

ISBN 0-201-04558-3

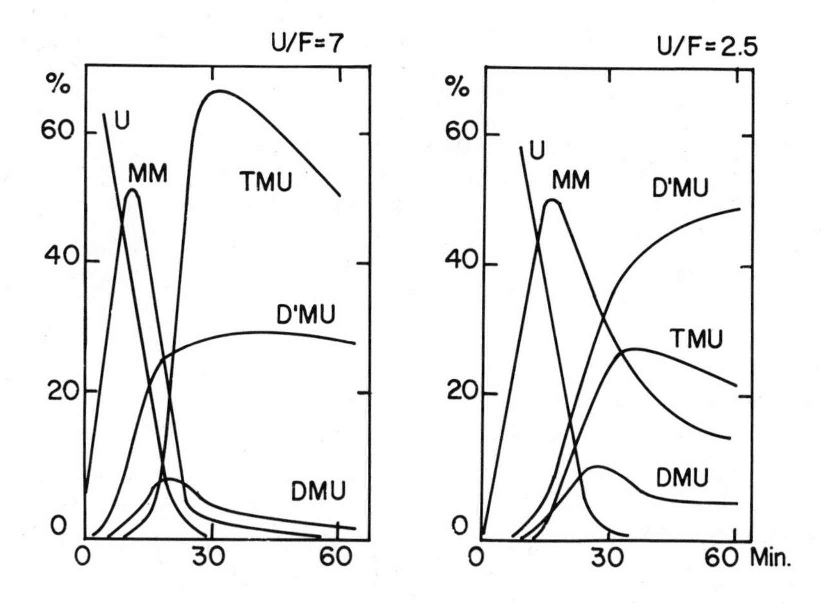

Figure 5.2. Molar Yield of Monourea Compounds as a Function of Reaction Time; a) U/F = 7, b) U/F = 2.5 (after Kumlin, 1978)

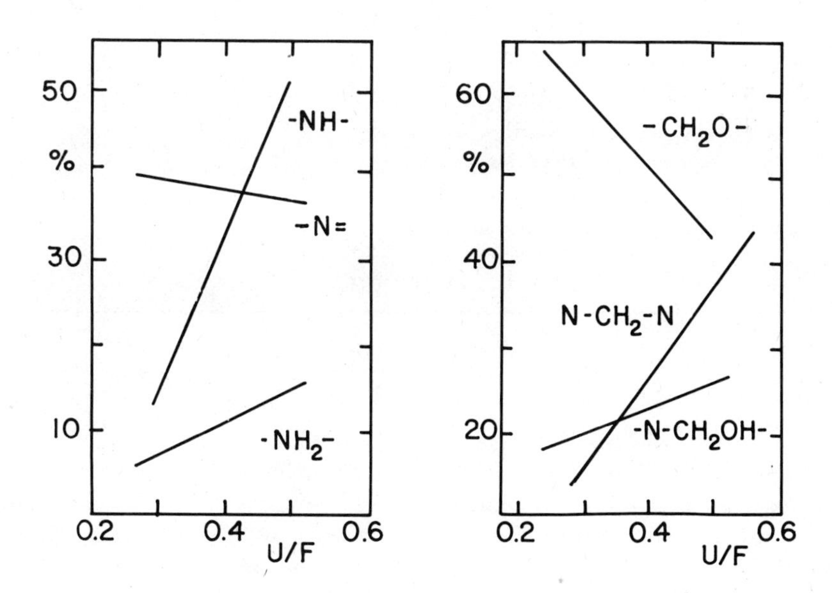

Figure 5.3. Molar Yield of Different Species as a Function of U/F; a) Amide Groups, and b) Formaldehyde Derivatives (after Chiavarini, 1978)

ISBN 0-201-04558-3

Table 5.3

^{13}C-NMR Spectrum of Urea-Formaldehyde Resin[a]
(after Tomita, 1978)
The NMR spectra are shown in Figure 5.4.

Functional Group	Structure	Chemical Shift (ppm)[a]	NMR Signal (Fig. 5.4)
Carbonyl	$=N\underline{C}ON=$	160.2	q
Methylene	$-NH\underline{C}H_2NH-$	47.7	a
	$-N(CH_2-)\underline{C}H_2NH-$	53.8	c
	$-N(CH_2-)\underline{C}H_2N(CH_2-)-$	60.0	e
Methylol	$-NH\underline{C}H_2OH$	65.1	f
	$-N(CH_2-)\underline{C}H_2OH$	71.7	h
Methylether	$-NH\underline{C}H_2OCH_3$	73.2	i
	$-N(CH_2-)\underline{C}H_2OCH_3$	79.7	k
Hemiformals	$-NHCH_2O\underline{C}H_2OH$	87.1	n
	$-N(CH_2-)CH_2O\underline{C}H_2OH$		
Dimethylene ether	$-NH\underline{C}H_2O\underline{C}H_2NH-$	69.4	g
	$-NH\underline{C}H_2OCH_2OH$		
	$-N(CH_2-)\underline{C}H_2OCH_2NH-$	76.0	j
	$-N(CH_2-)\underline{C}H_2OCH_2OH$		
Methanol	$\underline{C}H_3OH$	50.0	b
Methoxy	$-NHCH_2O\underline{C}H_3$	55.6	d
	$-N(CH_2-)CH_2O\underline{C}H_3$		
	$H(OCH_2)_nO\underline{C}H_3$		
Formaldehyde, Methylene glycol	$HO\underline{C}H_2OH$	83.1	l
	$HO\underline{C}H_2O\underline{C}H_2OH$	86.6	m
	$HO\underline{C}H_2OCH_3$	90.7	o
	$H(OCH_2)_nO\underline{C}H_2OCH_3$	95.0	p

[a]Dioxane 67.4 ppm.

Ebdon (1977), Table 5.4, from Tomita's work (1978) showed what type of data can now be deduced. This table lists the combined formaldehyde, the degree of condensation, total methylene, methylol, ether and methoxy concentrations, hemiformal, and total dimethylene ether – a total of 17 components, all as a function of molar U:F ratio, temperature, and history of the samples. It will be only a matter of time before resin synthesis and quality control will rely on ^{13}C-NMR analysis. There is little question that this breakthrough in *in situ* analysis will help tailor new UF-resins and optimize UF-resin properties at will and to a

ISBN 0-201-04558-3

degree that will improve resin performance to a level beyond what is currently possible.

Until recently it was not possible to study NMR spectra of solids, but progress in electronics, and the discovery of the "magic angle," combined with spinning probe holders, has overcome this last hurdle successfully.

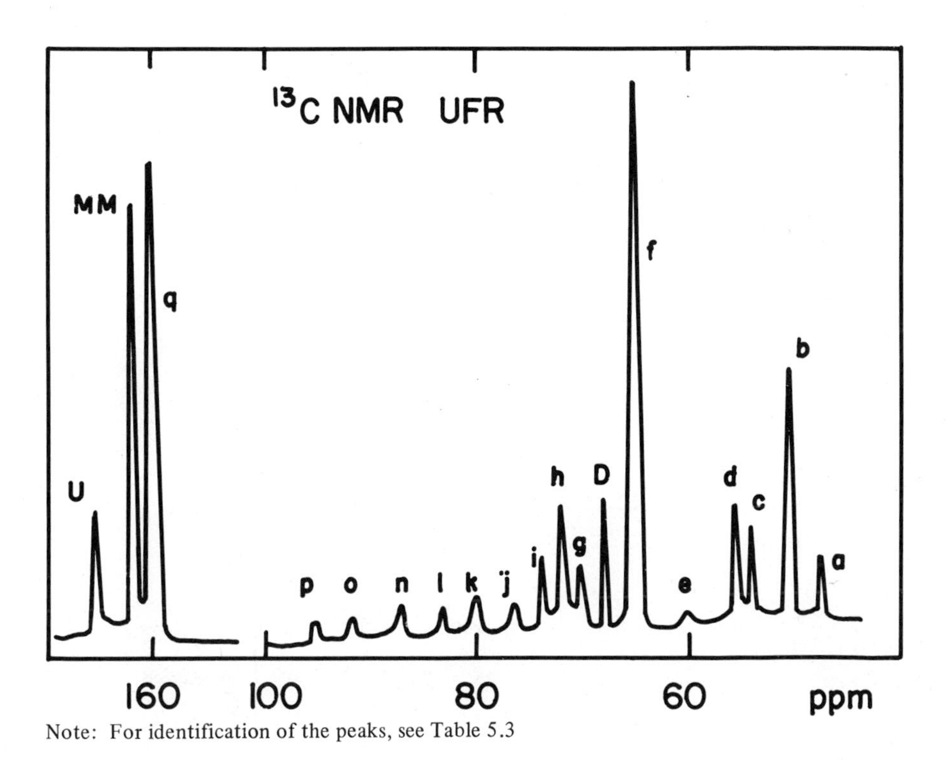

Note: For identification of the peaks, see Table 5.3

Figure 5.4. ^{13}C-NMR Spectrum of Urea-Formaldehyde Resins (after Tomita, 1978)

INFRARED AND RAMAN SPECTROSCOPY

Table 5.5 lists the characteristic vibrational frequencies of some chemical functions found in UF-resins. Infrared spectroscopy (IR) has been successfully used for some time (Myers 1978) to identify and estimate various carbonyl and amine functions (Kozlova 1976). Isotopic substitution can be used to unambiguously identify the active groups in mixtures. The disadvantage of the IR method is that water absorbs so strongly that it often dominates the spectrum and obscures groups of higher interest. Furthermore, samples must be normally transparent, or highly reflective. In contrast, the Raman method is not sensitive to water, and any type of

ISBN 0-201-04558-3

Table 5.4

Analysis of Urea-Formaldehyde Resins by ^{13}C-NMR Spectroscopy[a] (after Tomita, 1978)

Resin No.	1[b]	2	3	4[b]	5	6	7	8	9[c]	10[d]
Synthetic feed ratio F/U	1.4	1.7	1.8	1.8	2.0	2.0	2.2	2.5	–	–
Combined formaldehyde	1.36	1.48	1.50	1.60	1.64	1.70	1.79	1.82	1.60	1.87
Degree of average condensation[e]	0.48	0.68	0.75	0.55	0.63	0.82	0.73	0.80	0.66	0.69
Reaction time (min) at pH 4.9	–	15	60	–	30	60	120	100	–	–
Total methylene group	0.41	0.58	0.68	0.48	0.54	0.71	0.63	0.68	0.54	0.42
$-NHCH_2NH-$	0.14	0.21	0.22	0.16	0.15	0.21	0.16	0.14	0.21	0.11
$-N(CH_2-)\underline{C}H_2NH-$	0.24	0.24	0.40	0.28	0.28	0.34	0.40	0.40	0.26	0.24
$-N(CH_2-)\underline{C}H_2N(CH_2-)-$	0.03	0.12	0.07	0.04	0.12	0.16	0.07	0.15	0.07	0.07
Total methylol group	0.72	0.58	0.51	0.87	0.78	0.69	0.83	0.73	0.63	0.68
$-NH\underline{C}H_2OH$	0.56	0.37	0.28	0.55	0.34	0.31	0.33	0.29	0.46	0.41
$-N(CH_2-)\underline{C}H_2OH$	0.15	0.16	0.21	0.27	0.38	0.32	0.39	0.35	0.17	0.27
$=N\underline{C}H_2OCH_2OH$[f]	0.01	0.05	0.02	0.06	0.06	0.06	0.09	0.09	0.04	0.07
Total methyl ether in methylol group	0.09	0.06	0.17	0.12	0.09	0.06	0.16	0.18	0.22	0.22
$-NH\underline{C}H_2OCH_2$	0.05	0.03	0.10	0.07	0.06	0.03	0.09	0.11	0.15	0.13
$-N(CH_2-)\underline{C}H_2OCH_3$	0.04	0.03	0.07	0.05	0.03	0.03	0.07	0.07	0.07	0.09
Total methoxy group	0.17	0.06	0.23	0.23	0.12	0.09	0.35	0.29	0.35	0.20
Dimethylene ether group and hemiformal[g]	0.14	0.24	0.16	0.18	0.23	0.27	0.28	0.33	0.28	0.54
Total dimethylene ether group[h]	0.13	0.19	0.14	0.13	0.17	0.21	0.19	0.24	0.24	0.47

[a] After Tomita, 1978; values represent the ratio of the listed species to total carbonyl carbons of the urea in resin.
[b] Resins 1 and 4 were synthesized by a second addition of urea to resin 7.
[c] Resin 9 is a commercial plywood adhesive.
[d] Resin 10 is a commercial adhesive for room temperature use.
[e] Increase of carbons in the methylene structure plus half of the carbon in the dimethylene ether structure.
[f] Methylene glycols in hemiformals included.
[g] Sum of peaks g and j in Figure 5.3 and Table 5.1.
[h] Calculated by subtracting n due to hemiformals from the sum of the signal g and j in Figure 5.3.

ISBN 0-201-04558-3

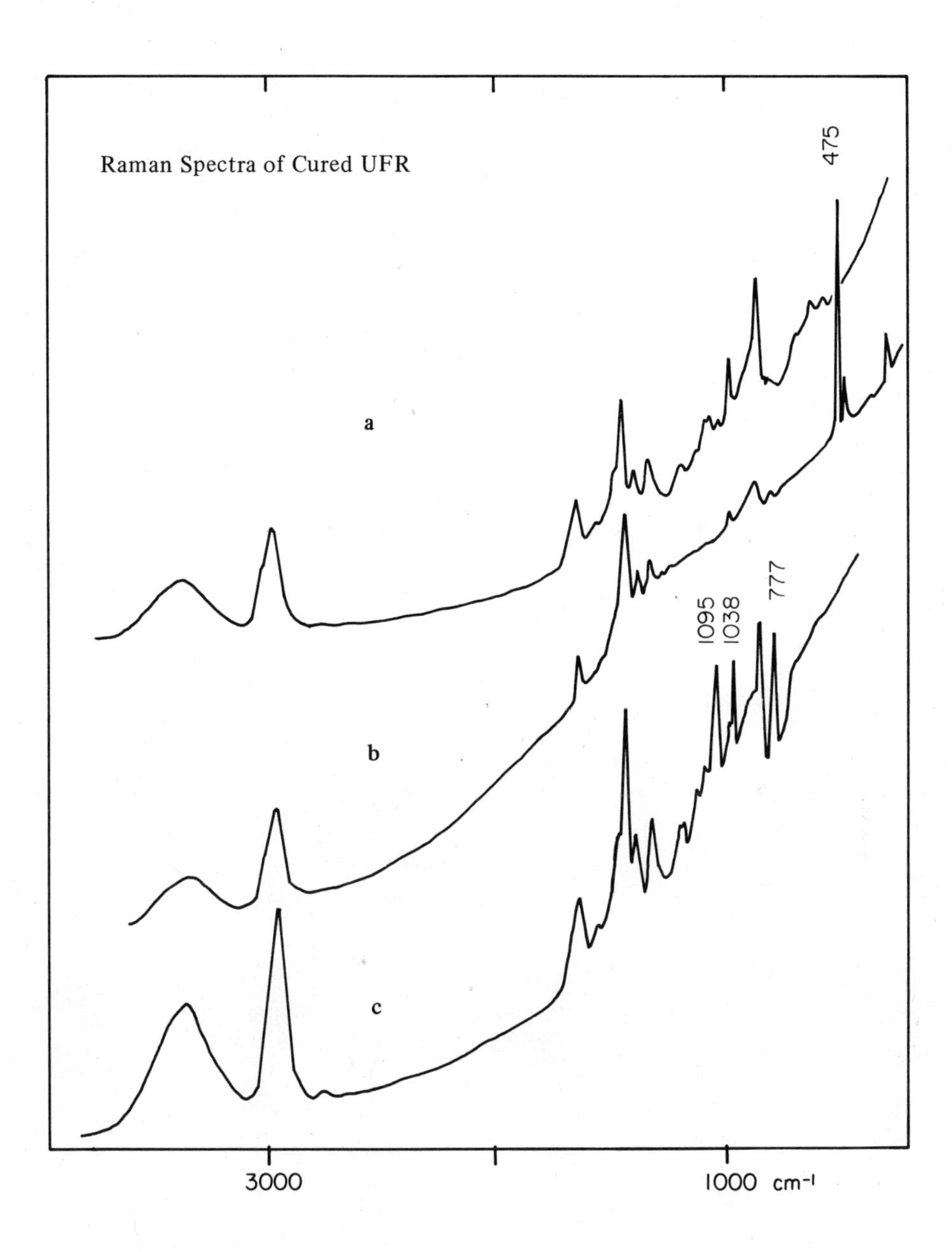

Figure 5.5. Raman Spectrum of a Cured Urea-Formaldehyde Resin; a) Commercial Particleboard Resin, b) Sulfur Modified Resin, and c) Lignosulfonate Modified Resin a) (Meyer, unpublished data, 1978)

ISBN 0-201-04558-3

Table 5.5

Assignment of Infrared and Raman Bands in Urea-Formaldehyde Polymers (after Myers 1978 and Meyer 1979)

Frequency cm^{-1}	Characteristics	Structural Assignment
1650–1670	VS[a]	Amide I; C=O stretch with some CN stretch and NH deformation.
1600–1610	VW; often masked by 1650 and 1550 cm^{-1} bands.	Amide II, NH_2 deformation for primary amide.
1550–1560	S in uncured polymer spectra. Intensity decreases during cure.	Amide II, combined CN stretch and NH deformation; hydrogen bonded secondary, linear amide.
1510–1520	Appears during cure.	Amide II; linear or cyclic tertiary amide and secondary amide in constrained nonhydrogen bonded environment.
1460–1470	VW to W	CH bending in NCH_2N, CH_2O and OCH_3, or NCN vibration.
1380–1400	W	CH mode in CH_2 and CH_3.
1290–1300	W in most UF; more distinct upon cure.	OH deformation in CH_2OH plus Amide III contribution.
1250–1260	M	Amide III of secondary amide; combined CN and NH
1180–1200	W to M	Not identified.
1130–1150	W to M	Aliphatic ether CO stretch and CN stretch in secondary amide.
1000–1050	Decreases during cure and shifts from ~1020 to ~1040 cm^{-1}.	CN or NCN mode.
1030–1050	Broad	NC or NCN mode.
1000–1020	M	CO stretch in methylol.
920	VS	$-CH_2-O-CH_2-$Polymethoxy stretch.

[a] VS = very strong; S = strong; M = medium; W = weak; VW = very weak

sample, solid, liquid opaque, transparent, or translucent, can be studied. Figure 5.5 shows a typical example of a Raman spectra of a cured resin. Cured resin surfaces

ISBN 0-201-04558-3

can be analyzed with equal ease as wood surfaces, fracture surfaces, or glue lines (Meyer 1978; Paul 1976). Thus, it can be expected that the relatively new Raman method will find extensive use, especially as it is far easier and also cheaper to conduct than NMR.

These methods have been successfully used by Myers (1979) to determine the aging behavior of UF-resins. Figure 5.6 shows the effect of temperature on the hydrolysis over a period of 150 hours. Figure 5.7 shows the influence of pH at 40°C over a period of 300 hours. Obviously, these *in situ* methods will yield much direct information on resins in the next few years, and will revolutionize quality control as well as synthesis, and our general understanding of product performance, as is indicated in other chapters.

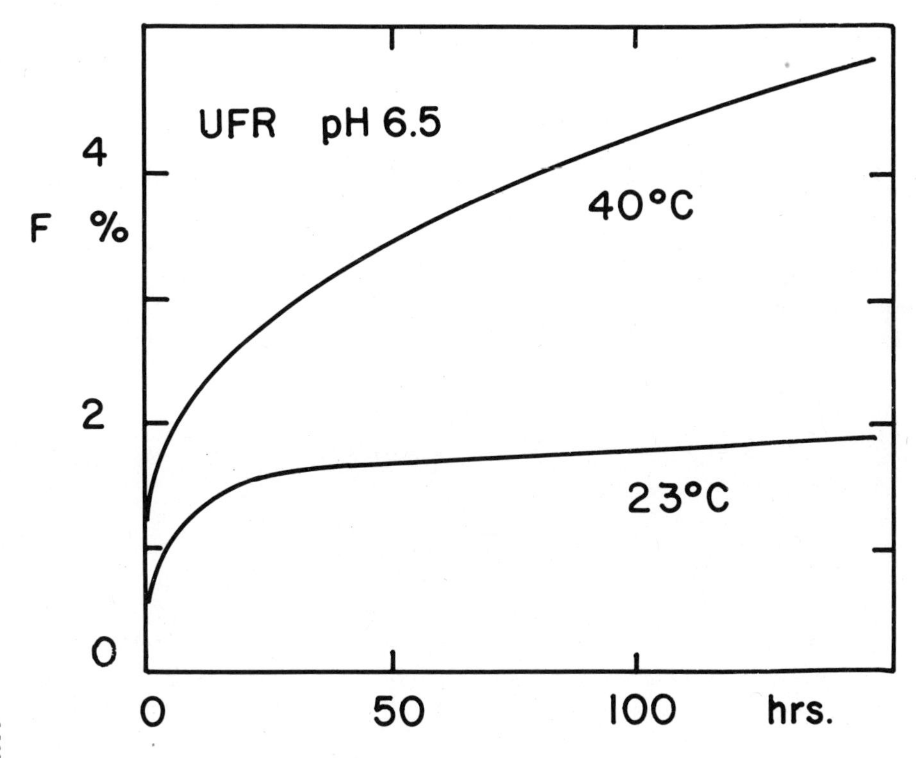

Figure 5.6. Effect of Temperature on Hydrolysis of UF-Resin (after Myers, 1978)

ISBN 0-201-04558-3

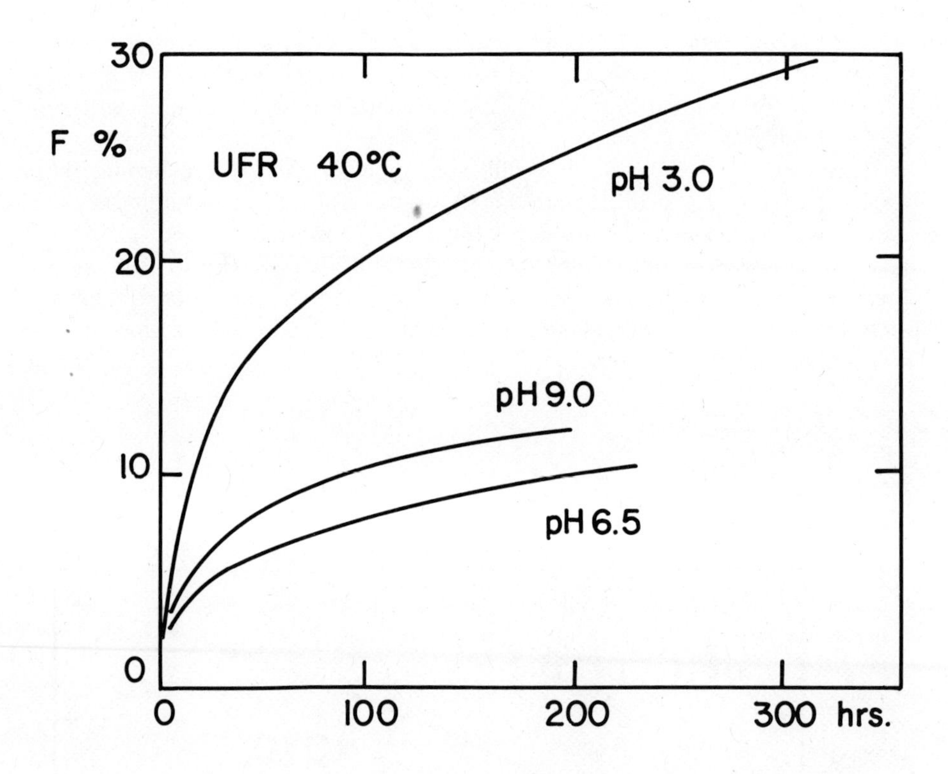

Figure 5.7. Effect of pH on Hydrolysis of Cured Urea-Formaldehyde Resin (after Myers, 1978)

Formaldehyde Sampling

Formaldehyde is difficult to sample, because at room temperature it slowly equilibrates between three phases; the gas, the aqueous, and the solid phase. Sampling can be aimed at (a) ambient air concentration above the source, (b) total free formaldehyde in the source, or (c) release into air space per time and surface area. Unfortunately, most sampling techniques yield a mixture of these types of concentrations and are thus not reproducible. Another difficulty is that the release mechanism from UF-bonded sources changes with time, as indicated in Chapter 9. The difficulty in establishing meaningful formaldehyde readings has hampered every field of research aimed at reducing formaldehyde. The latter is preferentially released from UF-resins because it is the most volatile component. Due to its high vapor pressure, formaldehyde tends to dissipate from the resin surface into adjoining air and onto moist surfaces. Because of its affinity for water in which it dissolves as methylene glycol, formaldehyde accumulates on moist surfaces. Thus,

ISBN 0-201-04558-3

moisture and humidity greatly influence formaldehyde in ambient air. When surface mositure is reduced, formaldehyde vaporizes into air in which it readily diffuses. The combination of these properties causes formaldehyde to spread very unevenly and unpredictably in the environment of UF-resin and in ambient air. In the vicinity of wood (for example, in particleboard bonded with UF) the dissipation of free formaldehyde is further complicated due to the chemical properties of wood cellulose which encourages hydrogen bonding and reversibly absorbs both formaldehyde and water. Obviously, sampling of formaldehyde has to be conducted diligently and with consideration for the chemical affinity of the environment, if air or aqueous sampling is to accurately reflect formaldehyde released from liquid or cured resin. These problems will be discussed further in the chapters dealing with particleboard and health. The next section deals with sampling techniques in air.

AIR SAMPLING

The usefulness of air sampling data depends on the answer to at least two questions: (a) is the sampling method reliable, and (b) is the air specimen representative for the environment. The importance of the first question is widely recognized, but the second question is often overlooked. In the case of formaldehyde, representative sampling is specially difficult, because this chemical tends to accumulate on surfaces where it tends to linger.

AMBIENT VERSUS INDOOR AIR

Most outdoor air tests are ambient air tests. In scientific terms, ambient means open and freely moving; in legal terms it means air which does not belong to anyone in particular and the source of which is not attributable to any one specific source (Rodgers 1978). Ambient air is presumably well mixed and easy to sample. However, indoor air, where most formaldehyde is found, is not ambient, and, furthermore, indoor air is seldom in equilibrium within a building or with the formaldehyde source. And even if the indoor air is in equilibrium with all sources, the equilibrium is highly dynamic. Thus, if formaldehyde is withdrawn, it might not be replenished from the source or from moist surface on to which it might readily have transferred and stayed.

SAMPLING METHODS

The most common air sampling method is that recommended by the U.S. National Institute for Occupational Safety and Health, Physical and Chemical Analysis Branch, as described in the NIOSH-75-12 manual, 1974. It is based on the chromotropic acid technique. Formaldehyde is collected from the air by bubbling the mixture through water. Standard samples are collected over a period of one hour at a flow rate of one liter per minute. Two impingers with 20 ml water are used in sequence as absorbers, Figure 5.8.

ISBN 0-201-04558-3

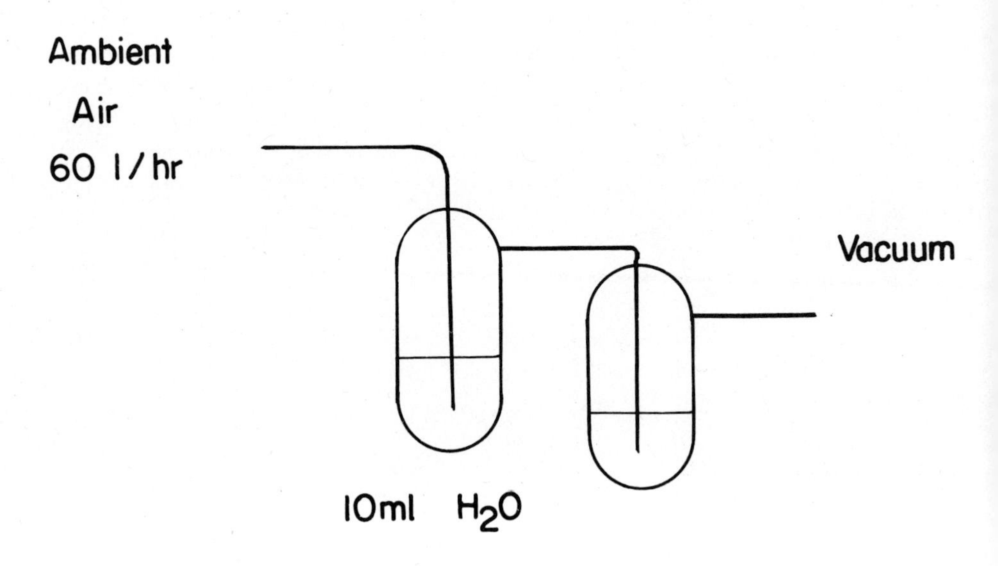

Figure 5.8. Air Sampling Method Using Impingers with Water

Goldman and Yagoda (1943) used a 1% bisulfite solution as an absorber. This method yields the nonvolatile bisulfite complexes of aldehydes which can be determined by iodine titration or other methods described above. Others proposed potassium hydroxide as the absorbing medium.

The European Federation of Associations of Particleboard Manufacturers (FESYP) recommends a procedure, Figure 5.12, similar to that of NIOSH, but the flow rate is increased threefold, to accommodate lower formaldehyde concentration, and expedite sampling collection. Furthermore, FESYP recommends *p*-rosaniline or the iodine titration previously described. Various other modifications have been described by Krikunov (1976), Lehmann (1968), and Masek (1971). Table 5.6 shows an approximate correlation between the formaldehyde concentration in air, and the air volume necessary to yield a given analytical response.

Very often formaldehyde concentrations are very low, and direct analysis is not practical. In such cases, samples must be enriched or concentrated. The basic principles for enrichment of trace contaminants are described in a volume published by Energy Resources Company, Inc. in 1975. A common method consists in stripping gases on a porous polymer (Bertsch 1975, Frankel 1972) on charcoal (Burnett 1974, Kyle 1974, Lee 1965, Levadie 1976, Stakavich 1969), on silica gel (Hughes 1960, Rachmar 1969, Paronyan 1974, Minemura 1973, Snyder 1976) or on any material that has a large surface area. The concentrate is then identified and then measured via gas chromatography (Ustinovakaya 1974, Pellizzari 1975, Smith 1972), with dye lasers (Becker 1975), or by any of the other methods (Gosink

ISBN 0-201-04558-3

1975) described earlier. Of course, the absorber introduces an additional element of potential error, and its reliability must be tested against a concentration curve.

Note: The arrangment is similar to that used in the Conway cell.

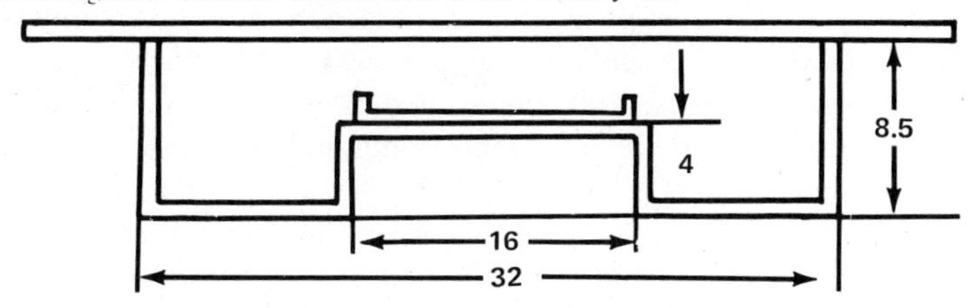

Figure 5.9. Feldman's Sample Cell for Collecting Formaldehyde Above Aqueous Solutions. All dimensions in millimeters.

Table 5.6

Air Volume Necessary to Obtain Analytical Response with the Iodometric Method and the Photometric Method, as a Function of the Formaldehyde Concentration (after FESYP, 1975)

Method		Iodometric		Photometric	
Formaldehyde Concentration in Air; mg/m^3		Titration with 50 ml Solution (x=50/50=1)		20 ml Sample (x=20/50=0.4)	
ppm	(μg/l)	Total Air Volume (m^3)	Na-thiosulfate Consumed (ml)	Total Air Volume (liter)	μg CH_2O in 25 ml Solution
0.1	0.12	10	8	300	14.4
0.5	0.60	3	12	50	12.0
1.0	1.20	1	8	30	14.4
5.0	6.00	0.30	12	5	12.0
10.0	12.00	0.10	8	3	14.4

INDOOR AIR

In most practical situations formaldehyde in ambient air is negligible. High concentrations are found around workplaces and in private residences where formaldehyde- containing materials are prepared or used. Examples in point are the hot press area in particleboard plants, the air space in storage cabinets built from such board, and the air in mobile homes whose walls and flooring are lined with UF-bonded products. In all these cases the air is not normally in equilibrium with the source. Thus, the formaldehyde reading in air depends on the contact time between air and

ISBN 0-201-04558-3

source, the air-to-source surface, the total air volume, and many other factors, including the humidity in the air and on the source surface. Furthermore, the air concentration depends on whether formaldehyde is trapped in air, or can transfer to absorbers. Water is probably one of the best absorbers.

Very often, these processes are ignored, and indoor air is almost always anlayzed as if it were ambient air and could be sampled accordingly. These basic problems are now being considered by several research groups working under the sponsorship of the U.S. Environmental Protection Agency and the U.S. Department of Housing and Urban Development, as a sequel to research programs aimed at increased energy conservation. In the Geomet study Moschandreas (1978) considered indoor air pollution as a function of changing ventilation rates. He recorded the measurements in various parts of residential homes as a function of source and time and derived several equations to describe the pollutant levels of CO, SO_2, NO and NO_2. Formaldehyde was not measured in this study. A similar study was made by Fugii (1973).

A pioneering study on indoor air was conducted by Andersen (1975) at the University of Aarhus. He measured formaldehyde in 25 rooms of 23 Danish dwellings and derived an equation for the air concentration, E:

$$E = (RT + S)(aH + b)/(1 + (nc/z)) \text{ mg } CH_2O/m^3 \text{ air} \qquad \text{Eq. 5.3}$$

where a, b, c, S, and R are constants depending on the nature and surface properties of the emitter, T is the temperature, n the ventilation rate, and z the ratio of surface area to air volume. Details of this work are contained in several reports of 1974. More recently (Andersen 1976, 1978, 1979) the same groups considered the influence of other factors, and other pollutants, including SO_2, on the formaldehyde concentration in air. The sampling method employs 50-liter air drawn continuously by a pump.

A systematic evaluation of formaldehyde at workplaces was conducted by Wayne and Bryan (1976), both at private research institutes, for the National Institute of Occupational Safety and Health. They used two sampling methods. The first employed six-liter plastic bags worn by workers on the job. They were filled with a battery powered portable pump by some 150 intermittent pulses of about 40 ml each over a period of three hours. The sample was then stripped in a bubbler train and the probe developed with chromotropic acid. The second method consisted in continuous sampling at a rate of one liter per min. A similar study was conducted by Flick (1975) in Germany.

Breysse (1977) and his student Carbone (1978) used standard continuous NIOSH procedure for sampling in mobile homes. They interviewed over a dozen associations and manufacturers of UF-resin, including the American Society of Heating, Refrigerating, and Air-Conditioning Engineers, the Synthetic Organic Chemicals Manufacturers Association, and found that all gravitate towards the

ISBN 0-201-04558-3

NIOSH sampling method. A portable analyzer was described by Marcote (1973). The problems of sampling have also been reviewed by Nestler (1977).

AQUEOUS MIXTURES

An excellent and popular nondestructive method for measuring formaldehyde in aqueous mixtures was described by Feldman (Walker 1964). The method is based on vapor diffusion and the fact that the vapor pressure of formaldehyde above a solution is determined by the concentration of free formaldehyde, as described in Chapters 6 and 10. Figure 5.9 shows Feldman's apparatus. It is similar to a Conway dish. A two-milliliter (approximate) sample is placed into the ring surrounding the pedestal. A drop of water, about 0.04 milliliter, is placed on the watch glass, and the cell is sealed and kept at room temperature for 18 hours. The solution in the watch glass is then tested for formaldehyde. The basic assumption for this test is that the formaldehyde diffuses throughout the apparatus, and that equilibrium has been reached after 18 hours. Thus, the concentration in the water probe should be the same as that in the sample.

The collection of formaldehyde vapor in a test solution has been widely accepted and applied in tests for formaldehyde, not only in aqueous resin systems, but also in finished products. However, in some modified Feldman methods, for example, the Fahrni test (Brunner 1978), two factors are usually fundamentally different. First, the sample is not a dilute aqueous solution, and thus the ionic strength of various components influences the vapor pressure of the formaldehyde, the second, the test time is shortened. In such cases, the formaldehyde content of the probe solution is proportionate to, but not identical to that of the sample. Thus, careful calibration curves must be established, and the test conditions must be carefully reproduced.

AQUEOUS UF-RESINS

Aqueous systems of formaldehyde are described in Chapter 3. Sampling is straightforward. Modifications of the Feldman method are most simple and useful for determining total formaldehyde release from ready-to-use resin. The equipment must be calibrated against standard solutions (Meyer and Johns 1979).

UF-BONDED MATERIALS

Cured UF molding releases no perceptible odor. However, filled molded materials, foams, and UF-bonded wood sometimes release substantial formaldehyde. The sampling technique must be matched to the release mechanism. Unfortunately, several different mechanisms can be in operation simultaneously, and the relative importance of these changes with time. Thus, sampling can only yield reproducible results if the method is clearly specific. Three quantities can be

ISBN 0-201-04558-3

measured: (1) equilibrium air composition above the sample, (2) total vaporizable or free formaldehyde in the sample, or (3) formaldehyde release per surface and time unit. Unfortunately, most tests measure a combination of these quantities.

The Japanese Industrial Standard JIS A-5908 1974, updated 1977, the only standard which has required formaldehyde release measurements for several years, employs a modified Feldman method. Since a desiccator is used as a container, this method is also called the desiccator method. The Japanese desiccator method is not related to the German desiccator method. Figure 5.10 shows how the specimens are placed. A surface of about 1800 cm^2 of UF-bonded material is exposed to one liter of air. Formaldehyde is collected in 300 ml of distilled water in a 6-cm deep dish with a 12-cm diameter. The sample container is then sealed and equilibrated for 24 hours at 22.5, ± 2.5°C.

In this method the water volume is larger than that of the sample, and thus most of the formaldehyde transfers from the sample to the probe, and the probe contains the bulk of it. Thus, total free, volatile formaldehyde can be determined with acetylacetone as described in the preceding chapter. A similar procedure is used in the Fahrni method (Brunner 1978), Figure 5.11. However, in this method the sample is only exposed for four hours to the solution. This time is sufficient to equilibrate formaldehyde in the air space, and between air space and probe solution. Furthermore, the probe volume is comparatively small. Thus, this method provides an accurate relative measure for formaldehyde released from the sample, but does not give total free formaldehyde. We have used this method extensively and find it reliable (Meyer 1979).

The European Federation of Particleboard Manufacturers Association (FESYP) recommends two sampling methods. The first, Figure 5.12, collects air from the sample surface. Sixty liters of air are collected in one hour; the air is bubbled through impinger tubes in which formaldehyde is absorbed in water. The formaldehyde is determined iodometrically. The second is the so-called perforator method, Figure 5.13, proposed by Verbestel (1969). In this method all free volatile formaldehyde is extracted from 100 g sample with 600 ml boiling toluene, from which it is transferred into one liter water. Formaldehyde is then determined by iodometric titration. The entire procedure is carefully defined and involves quite large and complicated equipment (Mehdorn 1975). It involves two hours of refluxing and takes a total of about four hours. The method has been widely tested in industry and research institutes. From his own experience, this author finds it reliable but somewhat tedious. Furthermore, the use of toluene creates a potential health and fire hazard. The practical aspects of this test should be improved.

Another popular method is the microdiffusion method proposed by Plath. Her method corresponds to the Fahrni method but employs chromotropic acid as an indicator. Roffael (1975) developed another diffusion method, which he calls the Wilhelm-Klausnitz-Institut (WKI) method which he demonstrated to correlate well to the perforator method. Roffael claims that it is more sensitive and accurate.

ISBN 0-201-04558-3

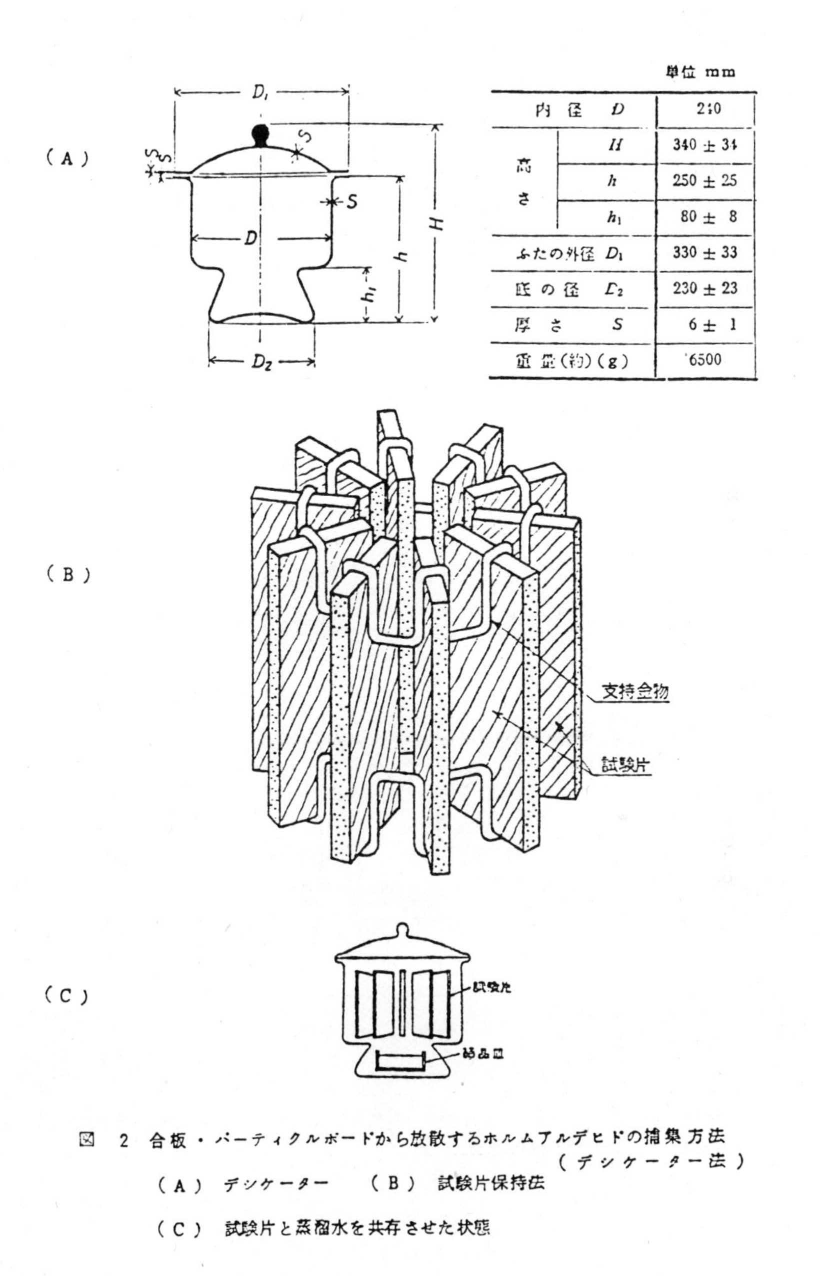

単位 mm

内 径 D		2:0
高さ	H	340 ± 34
	h	250 ± 25
	h_1	80 ± 8
ふたの外径 D_1		330 ± 33
底 の 径 D_2		230 ± 23
厚 さ S		6 ± 1
重 量(約)(g)		6500

図 2 合板・パーティクルボードから放散するホルムアルデヒドの捕集方法
(デシケーター法)
(A) デシケーター (B) 試験片保持法
(C) 試験片と蒸溜水を共存させた状態

Figure 5.10. Japanese Desiccator Method, JIS Standard A5908 – 1977 (A) Glass Desiccator (Dimension listed in Table in mm); (B) Specimen Holder; (C) Loaded Desiccator in Use (after T. Matsumoto, 1975)

ISBN 0-201-04558-3

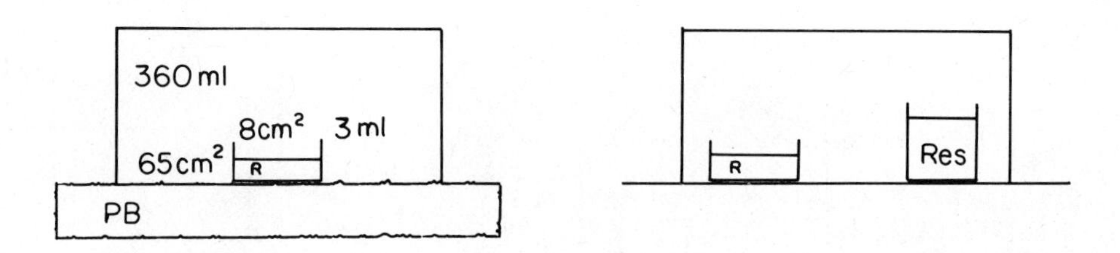

Figure 5.11. Modified Fahrni Method (Brunner, 1978) after Meyer and Johns, 1979

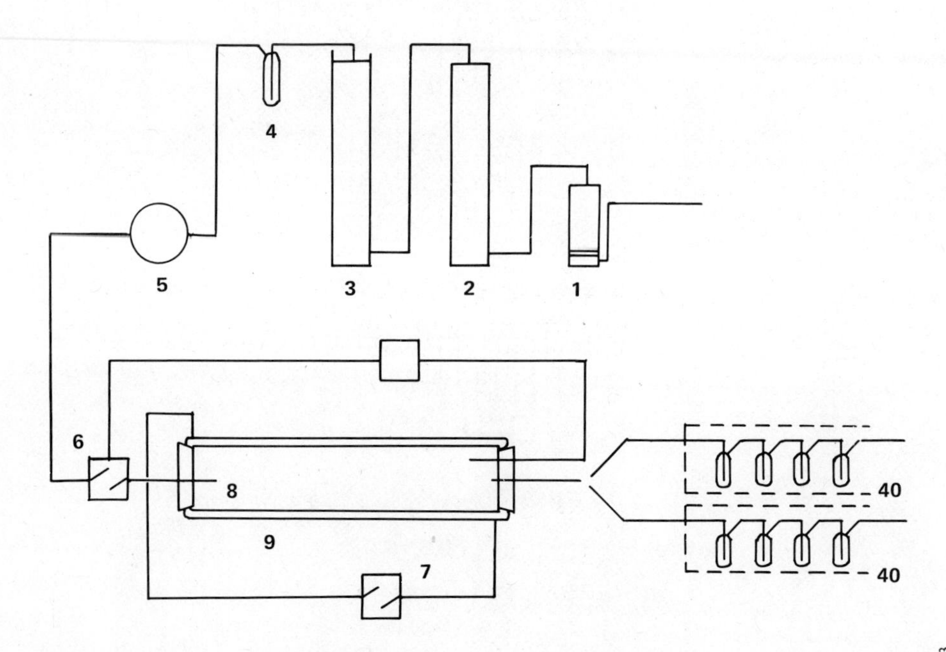

Figure 5.12. Air Sampling Method of FESYP, Using Impingers (FESYP, 1978) (For details see Reference) 1-7: Puritiyev Train; 40: Impingers.

ISBN 0-201-04558-3

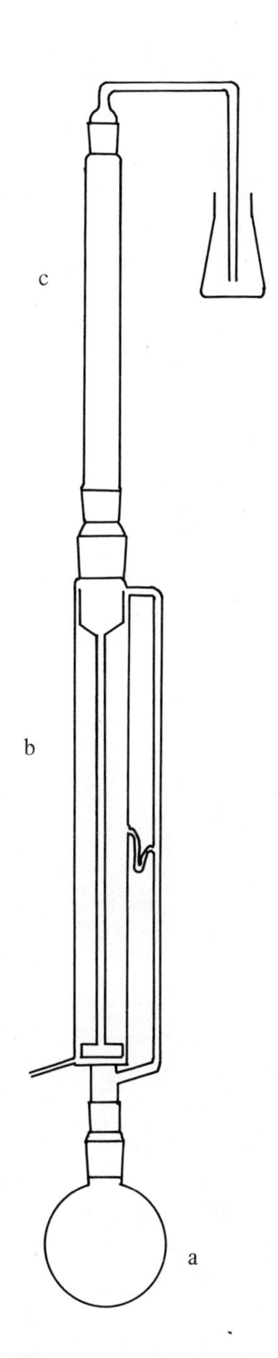

Figure 5.13. Perforator for Determining Free Formaldehyde in Wood Products (FESYP, 1978) (For details see Reference) a: Sample Bulb with Toluene; b: Perforator; c: Condenser.

ISBN 0-201-04558-3

Table 5.7

Comparison of Seven Formaldehyde Determination Methods[a]

Criteria		Perforator (FESYP)	Gas Analysis (FESYP)	Roffael (WKI)	Mohl Suction Method	Fahrni Brunner (1978)	Japan Industrial Standard	Air Analysis
Equipment:	costs ($)	500	3,000	300	1,500	80	300	1,200
	auxiliary	drybox	drybox	drybox	photometer	photometer	photometer	photometer
	chemicals ($)	4	0.50	0.20	0.20	.20	0.20	0.20
	care:	delicate	sturdy	sturdy	normal	sturdy	normal	delicate
Time need:	total	4 hrs	5 hrs	6–24 hrs	1 hr	4 hrs	24 hrs	1 hr
	measurements	15 min	60 min	15 min	15 min	15 min	15 min	15 min
Characteristics and interference:	wood species	sensitive	not known	sensitive	not known	not known	sensitive	–
	board density	none	yes	yes	yes	yes	little	–
	humidity	strongly sensitive	not known	not known	not known	not known	little	little
	formaldehyde in air:	yes	yes	yes	small	none	none	–
	sensitivity	high	medium	high	medium	high	high	–
	board surface	–	possible	possible	yes	yes	small	–
	standardization	yes	yes	yes	possible	yes	yes	yes

[a] After Roffael, 1978

ISBN 0-201-04558-3

In his method the sample is suspended on a string above water in a wide-neck bottle. The bottle is kept at 40°C for one hour. The solution is then cooled and titrated. Painstaking comparisons of this and other methods have been made, Table 5.7.

Several tests are based on the gas analysis of the second FESYP method: for example, Matsumoto (1974) passes four liters of air over a plywood sample, at one liter per 20 minutes, and uses acetylacetone as developer. Mohl (1977, 1978) developed a related method employing a suction bell. Air is either sucked through porous material, such as particleboard, or passed over the sample surface and bubbled through water.

Petersen, Reuther, Eisele, and Wittmann (1974), finally, cured their samples in a closed heat-resistant bag from which air is collected, or through which air can be blown during or after curing.

Roffael (1974–1978) has presented careful comparison of several sampling methods. The difference is sometimes subtle, but it can contain useful information about the nature of the source of formaldehyde or its precursor. For example, the suction methods tend to yield comparatively low values, because formaldehyde does not have sufficient time to fully desorb from the surfaces. In contrast, methods in which an aqueous sampling solution is kept in contact with samples at elevated temperature over a long time period, tend to yield high values, because they increase the humidity of the sample surface. This can cause or enhance hydrolysis of resin, or partly uncured resin components, such as ureamethylol compounds.

Thus, one can expect reproducible measurements only if air humidity, temperature, geometry, and all other test parameters are carefully controlled and monitored, and it is highly advisable to regularly conduct blind tests and test calibration curves.

Concentrated efforts are underway to improve sampling methods and find a better correlation between formaldehyde in air, formaldehyde release rate, and free formaldehyde in the source, and hopefully a correlation between the three and chemical bonding in the UF-bonded product, and, finally, its quality in daily use.

ISBN 0-201-04558-3

6. Adhesives

This chapter consists of six parts. The first deals with general adhesive properties and requirements, the second with lamination of beams, the third with plywood, and the fourth with particleboard and its manufacture. The fifth describes medium-density fiberboard, the sixth deals with wood moldings, from which shaped fittings and furniture can be made. The sixth section reviews the general properties of all UF-wood materials, especially the release of formaldehyde, the abatement of formaldehyde odor, and the emerging indoor standards. The same subject is also touched in Chapter 5 dealing with analytical chemistry, Chapter 9 on product properties, and Chapter 10 dealing with hygiene and health.

In the U.S. currently 75% of all UF-resin is used as wood adhesive, and 85% of this is consumed by the particleboard industry. This chapter deals with UF-resins adhesives, which are popular because of their high quality, their unusual versatility, and their low cost. UF-resins can be formulated to cure at any temperature between room temperature and 150°C. They are resistant to organic solvents, but are hydrolyzed by strong alkali and acids. They are completely resistant to fungi, wood worms, and termites, and they are less sensitive to wood moisture than phenol adhesives. They can be used to satisfy different, and even conflicting needs. For example, they can be designed to produce shaped furniture, or large beams in applications where massive wood would either crack and fail, or in which dimensions and shapes are required which exceed the natural capabilities of wood. They can be used to manufacture plywood, in which gaps must be filled, and a solid glue line must be formed. Finally, they can be used to form particleboard from wood chips or even sawdust.

For the purpose of matching all these needs, resin manufacturers design their resins on the basis of five factors: (1) the molar ratio of urea to formaldehyde, (2) pH, (3) temperature, (4) concentration, and (5) time. The manipulation of these factors, and the sequence in which they are being changed, allows for a very large variety of manufacturing methods. Accordingly, a very great wealth of art is available and is being used. Thus, resin customers can choose from a variety of properties, and each manufacturer supplies the chosen properties by a somewhat different

Beat Meyer, Urea-Formaldehyde Resins

ISBN 0-201-04558-3

method. This variety and diversity is a characteristic property of UF-resins. Accordingly, there is no narrowly defined composition for these resins, or for analyzing them. Recent advances in analytical chemistry, especially gel chromatography (GLC), infrared (IR) and Raman spectroscopy, and ^{13}C-nuclear magnetic resonance (NMR) have made it now much easier to analyze resins. It is now possible for manufacturers as well as users to analyze their resin more fully. This will immediately facilitate quality control and, in the future, when the correlation between chemical molecular composition and the mechanical and other properties of finished products have been established, it will be possible to manufacture far better, and more specifically, formulated resins.

Today, resins come in two forms: spray-dried resin and aqueous resin containing between 35 and 80% resin solid. The dry resins have the advantage of lower shipping cost and better storage stability. The disadvantage is their price. They are used for shipping to isolated plants and countries, or where exacting, reproducible properties are wanted. The solid content of aqueous resins is limited by "water toleration." UF-resins are not water soluble, in the common sense; they are colloidal dispersions. Their water toleration depends on their composition, age, temperature, and many other factors.

Adhesive Properties and Requirements

Adhesion is a complex interdisciplinary science which involves chemical, physical, and mechanical principles. The first adhesive use has been documented in ancient Egypt, which also produced the first plywood. Adhesive properties are often easier appreciated by intuition than by scientific concepts and tests. Thus, adhesion remains an art. The problem of wood adhesion has been reviewed by de Bruyne (1939), Marra (1977), Houwink (1965), Clad (1978), and in many handbooks.

The fact that water exhibits strength normal to the surface when used as film between glass surfaces shows that adhesion and adherence are not the same. Definitions of these terms, as well as cohesion, are found in many specialized books. Marra (1977) analyzed adhesive properties in an excellent chapter which reflects a lifetime of practical and theoretical experience in wood bonding. He explained that "thousands of potentially useful adhesives have died or lie dormant in test tubes because they failed to pass a standard test." The true bond forming potential must include both bond formation and bond performance. The first term includes two steps: the establishing of adhesion and the achievement of strength. The latter is reflected in bond strength and permanence. The conditions under which these factors combine involve four systems: (1) the basic chemical, physical, and mechanical properties of the liquid and solid adhesive; (2) the corresponding properties of the adherends; (3) the consolidation conditions, including pressure, temperature, humidity, and time; and (4) the exposure or service conditions of the finished product. The standard tests involve tack, an elusive quantity which can be established by ASTM D-2979-71; the change of state; the compatibility with

ISBN 0-201-04558-3

adherents, which is reflected in surface tensions; wetting, and the bond forming conditions.

In an adhesive, the bulk properties are only one factor which determines suitability. An adhesive must perform equally well at all points between the pure, bulk glue line and the surface contact between wood and glue. Marra (1977) distinguishes nine links in the adhesive bond, Figure 6.1, between the surfaces of the adherend. Link 1 is the intrinsic cohesive strength of the adhesive within itself. Links 2 and 3 reflect the cohesion between bulk and adhesive absorbed on the substrate surface; 4 and 5 reflect the adherend-adhesive boundary, formed through adsorption or wetting action; 6 and 7 represent the intra-adherend boundary, characterized by checks and incipient ruptures created by surface treatment or degradation of the adherend. Finally, 8 and 9 are the bulk properties of the adherend, which affect bond formation and performance. The quality of the bond is limited by the quality of the weakest link.

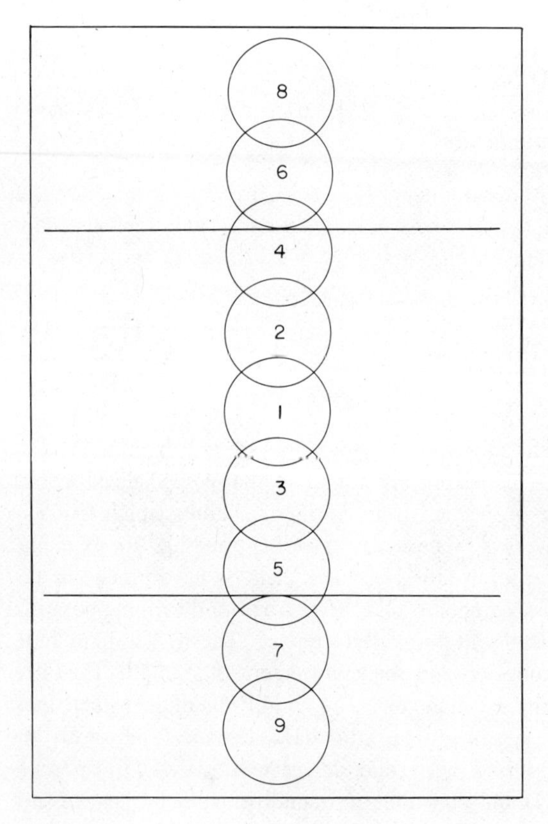

Note:

1) the Adhesive Film, 2) and 3) Intra-Adhesive Boundary Layer, 4) and 5) Adhesive-Adherent Interface (site of adhesion forces), 6) and 7) Adherent Surface, partially fractured during preparation for bonding, 8) and 9) Adherent Bulk

Figure 6.1. Links in the Adhesive Bond between Wood Surfaces (after Marra, 1977)

ISBN 0-201-04558-3

In wood, the bonding process is especially complex, because the adherend is a nonhomogenous multicomponent system containing chemical functions of lignin, cellulose, and water. The surface of the wood is usually weakened by tools, and, thus, a good glue has to penetrate beyond the surface, without soaking totally. In hot-setting resin operations, the situation becomes yet more complex, because the resin viscosity changes, wood extractives emanate, and intrinsic moisture creeps or steams the surface. The action of intrinsic steam is an important factor in the three-layer particleboard design. Fahrni (1944) was the first to recognize the value of higher mositure in the surface layer in the forming of the denser surface. Steam is also a major aid in rapidly transferring heat within the board.

Further complications are caused by the macroscopic and microscopic unevenness of wood surfaces. Glue lines hardly have an even thickness, thus the exothermic and endothermic curing reactions proceed at a different rate in surface films than in thicker layers, or drops, such as those formed in gaps between the adherend. Uneven glue films lead intrinsically to tension within the bonded layer. Another reason for intrinsic strain is that wood is a highly isotropic material, while resins are anisotropic. Thus, the dimensional properties of the two cannot be accurately matched. This problem is further exacerbated by the fact that the macroscopic dimensions of wood are constantly changing because of swelling and drying in response to humidity changes.

Urea-formaldehyde resins have several intrinsic properties that are essential for matching the multitudes of demands in wood bonding. One factor is that resins contain a complex mixture of molecular weight and chemical fractions, each of which reacts differently and fulfills another function in building up Marra's nine links. For example, the high molecular fractions readily coalesce and establish the bulk glue line. The low-weight methylol functions readily diffuse through the wood and establish surface contact with compatible hydroxyl groups in hydrogen-bonded cellulose and intermediate species establish other links. If all goes well, UF-resin provides an almost ideal spectrum of bonds. However, if conditions are wrong, or even only slightly wrong, the resin can totally and stubbornly fail. The margin between failure and success can be small, indeed.

A common misconception among laymen is that the best glues are those with the highest intrinsic strength in layer 1, Figure 6.1. The example of wood bonding with sulfur can illuminate this point. Around 1920 Western Union impregnated some ten thousand wooden telegraph poles across the country with hot liquid sulfur (Meyer 1977). This treatment can also be used to join wood. Such joints have enormous compressive strength, and the tensile strength is such that the wood, rather than the glue joint, breaks. However, this material is utterly useless in pure form, because the joints are stiff, and if used in veneer, the bond readily and reliably fails in Marra's links 6 and 7, the adherend subsurface, because this layer has to absorb all strain. In contrast, in a good glue line, strain is distributed through all layers. To a lesser extent the same can be said of phenol-formaldehyde bonds, which can easily be too strong and too rigid for wood. Good wood glues must be

ISBN 0-201-04558-3

flexible and soft so that they can absorb the swelling of the wood, and the shrinkage during drying, two processes that are part of the unceasing, constant aging cycle. This is shown in Figure 6.2, which shows that wood absorbs and releases up to 60 wt % of moisture during seasonal cycles. The most desirable glue bonding properties are revealed in the practitioner's knife test, during which plywood laminates are forcibly separated. In a good specimen about half of the failure is in the wood, and half in the resin. More or less is equally undesirable.

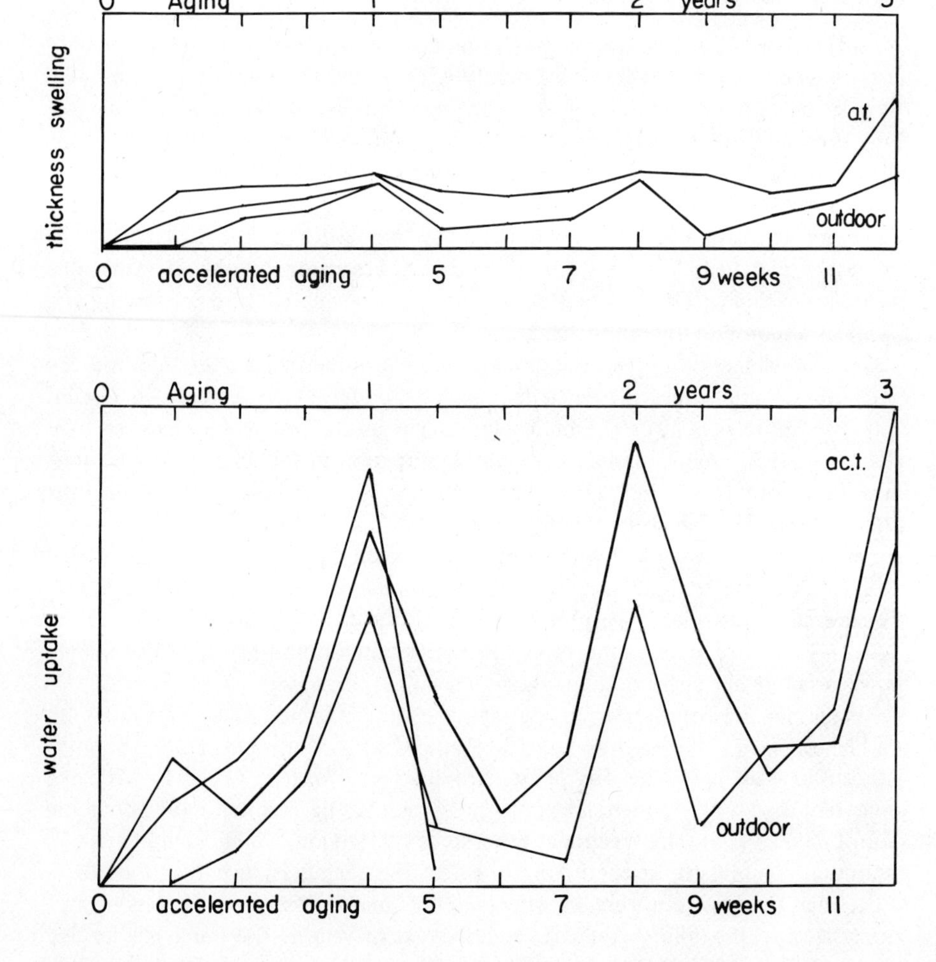

Figure 6.2. Humidity Changes in Open Air Storage of Particleboard a) Thickness Swelling as a Function of Seasonal Weather Changes, and b) Water Uptake (*act.* and *at.*: Accelerated Aging) (after Deppe, 1977)

ISBN 0-201-04558-3

Some problems of adhesive properties and testing have been recently critically reviewed by Clad (1978). Recent trends in various additives to reduce formaldehyde release during and after cure have made the situation difficult. Clad also discusses the problem of establishing standards for testing. Recent trends in establishing absolute adhesive standards might have neglected the synergism of adherend with the bonding agent, and the testing of real-life criteria might occasionally have fallen to an intensive desire to produce tangible optimized numerical accuracy in the laboratory.

Wood Bonding and Lamination

Currently only about 2% of all UF adhesives are used for laminating, but during the last twenty years, architects have widely discovered the attractive properties and looks of super-sized laminated beams. Such "laminated timber" rafters and trusses are produced by cold-bonding of smaller beams or boards. They can be designed to support the roofs of airplane hangars, exhibition halls, sports stadiums, swimming pools, and ice rinks. Already in 1938 a dozen 150-ft beams were used to build an industrial exhibition hall in Basel, Switzerland, which is still in use and in excellent condition. Such beams withstand fires better than unprotected steel. The assembly is usually prepared in 20-m to 60-m sections which are clamped 24 hours for curing, and which are then joined at the final site. In Europe, melamine is widely preferred today for this purpose, but in the U.S. urea-formaldehyde is generally used.

A great wealth of experience in wood bonding was gained during World War II, when torpedo boats and other small and medium craft were built, often from plywood. Likewise, warplanes and even their propellers were made from wood and plywood bonded with formaldehyde resins. Figure 6.3 shows an airplane assembly hall in England in 1943. Several crash programs were conducted to establish the capabilities of these materials, and to formulate military quality standards and specifications. One pioneer in the construction of warships and planes was H. Scott-Paine, a racing enthusiast, who designed and built his own craft and founded the British Power Boat Company which later built torpedo boats 71 feet long and 20 feet wide, cruising at 40 knots (80 km/hr). A forest products professor, who served in the U.S. Navy during World War II, reports that he remembers vividly the heavy vibrations which would build up on the hull and deck of his plywood PT-boat when the 1500-hp engines were fully throttled and he adds that he never understood why the plywood did not delaminate.

In 1946, the workshops of the British Power Boat Company were turned over to the building of prefabricated plywood housing, using the same glues as used during the war. Since then, pleasure boats, gliders, and housing have been made from these materials. In 1954, British television and an adhesive company used the plans and diaries of Sir George Cailey (1773–1857) to build and fly a primitive flying boat, thus giving credibility to the legend that he flew half a century before the Wright

ISBN 0-201-04558-3

brothers assembled their Kitty Hawk. It is, however, not clear how Sir George could assemble his craft, since he hardly could know about UF-resins. The exceptional qualities of UF-resin for bonding wood and other unmodified cellulosic materials, such as paper and cork, are due to the compatibility of their hydroxyl groups. UF does not adhere well to glass, metal, cellophane, or cellulose esters. However, the mechanical strength of bonds is sufficient so that it can be used to bond expanded polystyrene and glass fiber laminates.

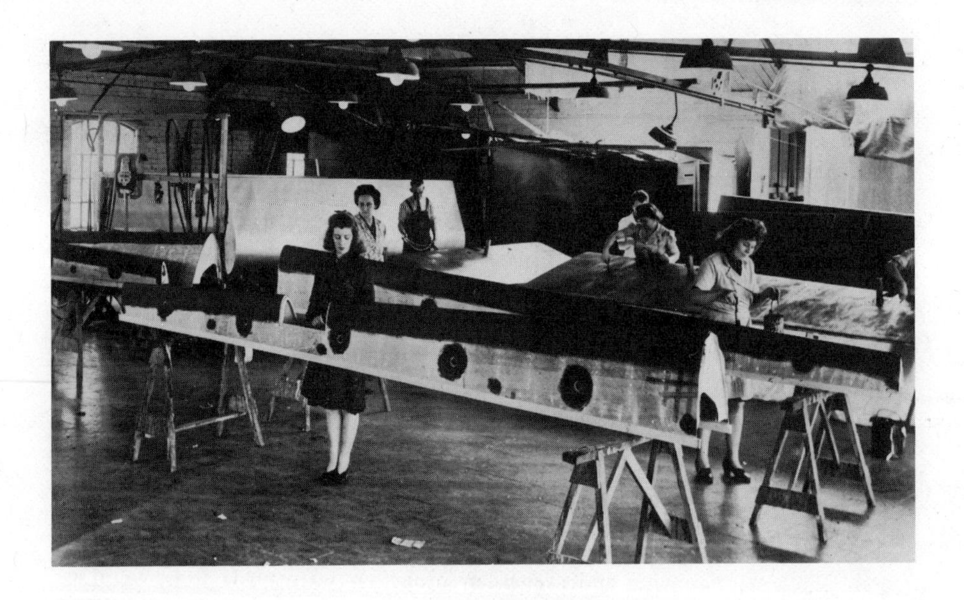

Figure 6.3. Airplane Assembly Line Using Plywood, Cloth, and Aminoresin During World War II (courtesy of CIBA-GEIGY Aerolite Division)

Plywood

Plywood is assembled by cross-stacking 3–5 layers of veneer. The board is often cold-pressed, and then transferred to hot press where it is heated at 90–100°C and 30–300 psi for about 5 minutes. The upper temperature limit must be maintained to prevent blistering of the board due to vaporization of moisture. Hot cured resins usually are reacted by the addition of ammonium chloride or ammonium sulfate catalysts. These salts release acid upon heating. The driving force in the acid release is the formation of hexamethyl tetramine with free formaldehyde:

$$4NH_4Cl + 6CH_2O \rightarrow 4HCl + (CH_2)_6N_4 + 6H_2O \qquad \text{Eq. 6.1}$$

ISBN 0-201-04558-3

The speed of this reaction determines the speed of the acid release. Figure 6.4 shows the rate of four temperatures. Ammonium chloride is far superior to hydrogen chloride. The latter yields weaker joints. If the ammonium chloride shortens the curing period excessively, additional ammonia or hexamethyl tetramine can be used as a retardent. Sodium formate also can serve as a buffer. Many glues have built-in hardeners. Among the suitable salts are pyridine monochloroacetic acid and similar additives.

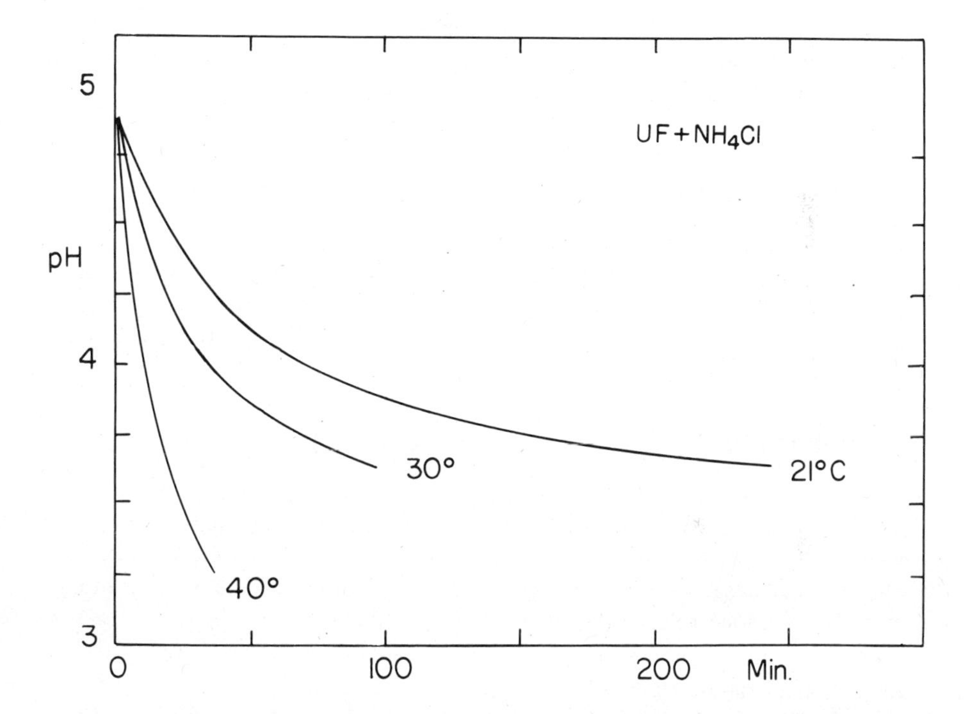

Figure 6.4. pH Change of UF-Resin with Ammonium Chloride Hardener, as a Function of Temperature and Time (after Raynev, 1975)

Formic acid is a popular component of hardener for plywood because it imparts gap-filling properties by preventing craze. Formic acid helps curing at room temperature, and thus is used in exacting applications, such as in parts of air gliders and boats. Formic acid can be applied to the uncoated part of the board. And, it serves as a mild solvent for UF and allows a certain creeping of resin around gaps.

In plywood, fillers are an important part. They produce "body" and prevent the formation of "starved" joints in porous wood. This prevents glue from soaking into the wood, thus starving the glue line and blotching the face of the veneer. Plywood

ISBN 0-201-04558-3

glues contain up to 100 wt % filler. Wheat flour, rye flour, and retch (a weed in cereals) is used. H. Klemm, in the first gap-filling patent (1938), used powdered phenolic molding wastes. Later, cellulosic fillers have been proposed. Today, chemicals such as furfurylaldehyde and dimethyl formaldehyde are also used.

A further additive function is found in so called extenders. Their purpose is to gel and swell during curing. All farinaceous substances, such as cereals or root flours, are suitable. Rye, maize, potato, tapioca or wheat starches are used, as locally available. They rupture between 45–50°C, start gellation at 50–65°C and are fully gelled at 60°C. In times of shortages other materials are used as well, such as coconut or walnut shell flours. In England, gypsum was used for some time. Extenders help mechanical performance, but weaken resistance to moisture.

In plywood applications, UF glues are often fortified to increase water resistance. UF-resins have very good water resistance at room temperature, but hot water rapidly weakens joints and boiling water dissolves them. Normal UF joints pass a three-hour soaking test at 67°C, as prescribed by British standards. Melamine and resorcinol are viable fortifiers. Unfortunately, melamine reduces the shelf-life of resin. Melamine can be blended with UF-resin, it can be added with the hardener, or it can be copolymerized at various ratios, say 1:1. Another additive is polyvinyl acetate. Its mixtures with UF react in an additive way, i.e., no synergism is observed.

Particleboard

Particleboard is a well defined and standardized product with a density of 0.7 gm/cm^3. The quality of wood, the raw material, varies locally and depending on economic conditions. Using the variable resources to make a reliable bond is a task which is divided evenly between the resin manufacturer and the board manufacturer. The division of this task differs in Europe and the U.S.

Particleboard production started during the Second World War. Today it is a complex and mature art. Urea-formaldehyde resins make 6–10 wt % of particleboard and account for a third of the value. The development of particleboard was crucially dependent on the availability of an economic, reliable, and high-performance aqueous resin. The first patent, GP 148,635, describing a wooden board prepared by gluing planer chips to a canvas fleece was filed by Kramer (1889). It was 45 years before Antonia (1933) in FP 749,177 proposed the use of urea-formaldehyde resin, but at that time only dry resins were available. Carson (1936) USP 2,033,411 was the first to propose the use of aqueous UF-resin for preparing a hot-pressed board of an eighth of an inch or greater thickness. Pfrohl, Swiss P 193,139, described the engineering properties of the first "carpenter's board." During the following years, patents continued to rely on dry resins, until the Torfit Company (1938), Swiss P 198,723, proposed the use of liquid sprayed resin, but it was dried before hot pressing. This company was probably the first to produce particleboard commercially. They quickly recognized the problem of blowing during hot pressing,

ISBN 0-201-04558-3

caused by steam trapped in small wood particles. In 1943 F. Fahrni in FP 881,781 recognized the importance of proper humidity control, and in 1945 and 1947, FP 899,487 and FP 926,024, he perfected his principle which consisted in building a three-layer board with higher-density, smooth surfaces made with higher humidity, finer chips, and a coarser core. This engineering design gives the board optimum mechanical properties at a minimum density. The board structure and three steps of manufacturing are shown in Figure 6.5.

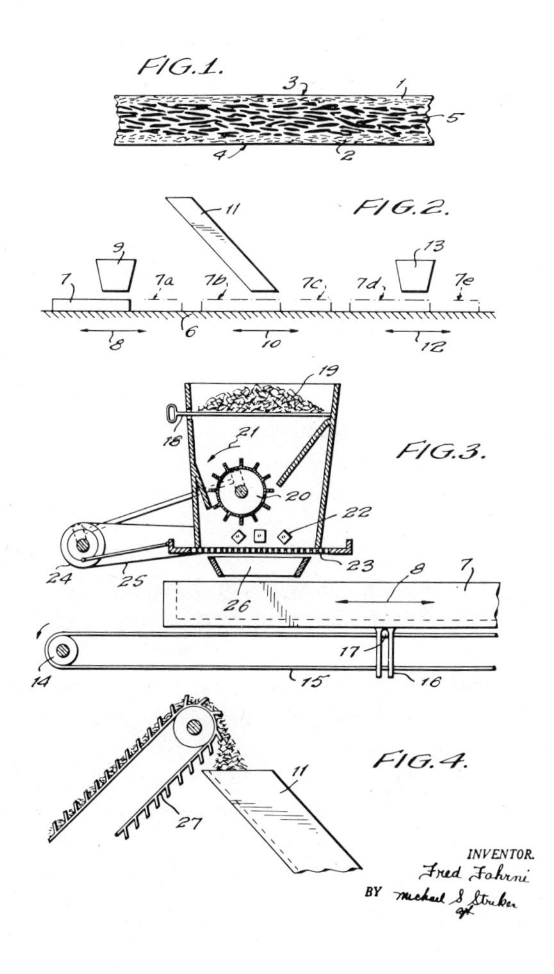

Figure 6.5. Process for Manufacturing Particleboard; USP 2,686,143, F. Fahrni, 1945; Fig. 1 is a cross-section of the board, showing the high density at the surface; Fig. 2 and Fig. 3 explain the mat laying process; Fig. 4 shows how the silo is being loaded

ISBN 0-201-04558-3

A major key in obtaining an economical board is the minimizing of glue use. In a modern board only about 10% of the wood chip surface is wetted by glue. The next proposed improvement, by Roos (1954), consisted of using foamed UF-resin to improve wetting and minimize glue requirements. The latter improvement has not yet found general acceptance.

The history of the particleboard development has been described by Fahrni (1957) and Maloney (1977). Large-scale commercial production of particleboard started during World War II in central Europe. Deppe (1977) lists as pioneers Fahrni, Klauditz, Stegmann, Rackwitz, Himmelheber, Steiner, Fischer, Steeger, Schmitzler, Plath, and Kreibaum. They had to overcome great odds to develop a viable material. Today about 10 million cubic meters of particleboard are produced worldwide; about the same volume as whole wood. Over the last 25 years, U.S. and European production grew at about equal rates and were about equal in size. The history of the technical development is reflected in Figure 6.6, which shows the relative costs for labor, wood, glue, and energy. The data are corrected for constant dollar value, and do not reflect local or short time fluctuations which have plagued the industry. The part of the curve before 1975 is adapted from Deppe (1977). Values for the more recent years have been collected from the proceedings of a symposium series edited by Maloney (1966–1979). Recently, the energy prices have risen to such a degree that wood is now equally valuable as fuel and as board material. The industry has now matured in Europe, where particleboard has almost fully replaced plywood. In the U.S. the transition from plywood to particleboard has lagged, and it currently is hampered by the recent concern for formaldehyde emission. However, the advantages of particleboard are obvious. It can be made from wood wastes, and it has better dimensional stability than wood. Furthermore, bagasse, bamboo, and cotton can be used for particleboard type products. Technical progress is being reported in several journals, and in a series of symposia proceedings (Maloney 1977) and in books by Deppe (1977) and Maloney (1977). The board properties are affected by a variety of interrelated parameters. Figure 6.7 shows 14 of the most important ones. In order to produce standard quality board, the manufacturing process must be able to adjust to local wood species, often to a mix of several. In the U.S. Douglas fir and southern pine, spruce, hemlock and other hard woods are preferred, but any economically available is mixed with these, including sawdust, chopped apple crates, and other wastes. Thus, the mechanical size and properties of wood chips, their surface area and appearance, and the chemical behavior of wood varies greatly. The moisture of board is carefully controlled, either by drying the wood chips and controlled addition of water, or by gradual drying to the desired humidity, usually about 8–12%. Among these changing factors is the pH of wood. This greatly affects its interaction with the adhesive resin, and it can affect the durability of the finished board. Thus, the pH of wood extraction must be carefully measured and matched.

ISBN 0-201-04558-3

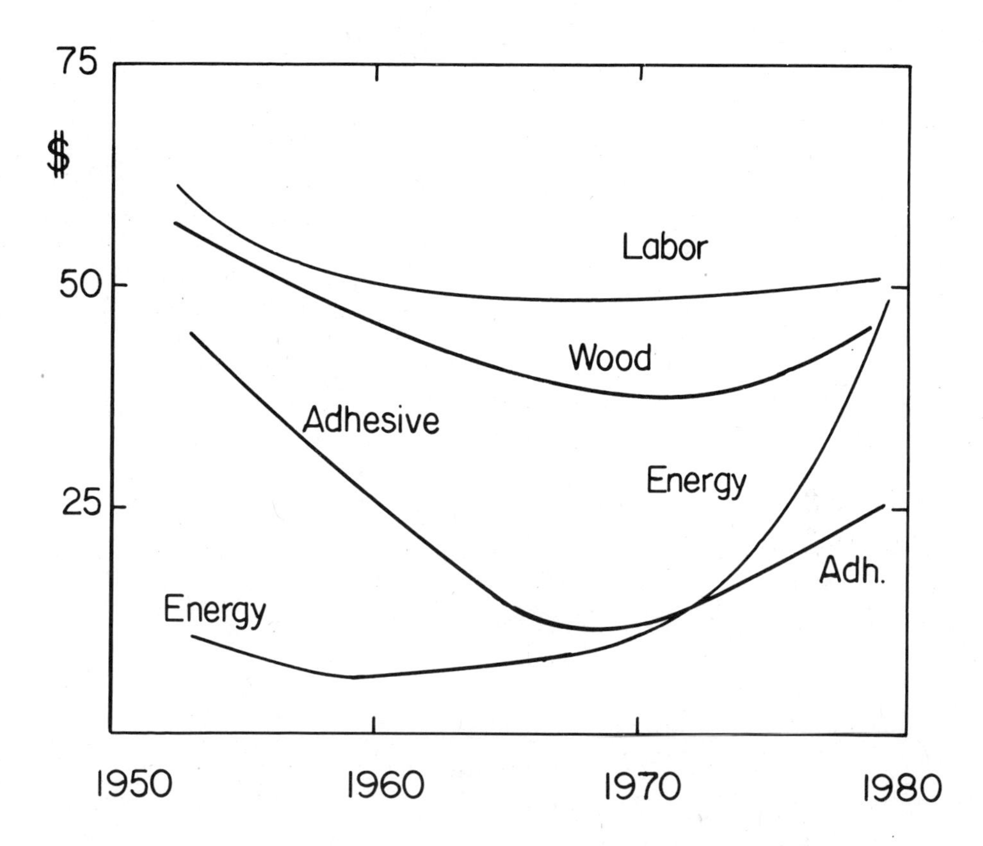

Figure 6.6. Cost Breakdown for Particleboard Production from the Beginning in 1950 to 1979 (after Deppe, 1977 and U.S. Department of Commerce)

Adapting a standard UF-resin for use in a particleboard plant is a sensitive art. While most of the basic UF-resin formulation patents have expired and most resin manufacturers produce lines of basically similar standard adhesives, the formulation of glue from the adhesive is a jealously guarded skill over which the plant manager keeps control and secrecy. In the U.S. it is customary for specialist resin companies to make ready-to-use particleboard adhesive which is trucked to the individual plants and formulated individually according to instructions of the plant supervisor, or the resin company technical representative who regularly and frequently visits plants. Resin batches can be stored for 8–12 weeks. In Europe, the particleboard resin is sold by both bulk manufactureres who produce them continuously, or by smaller jobbers who make batches as needed. It has been proposed for some time

ISBN 0-201-04558-3

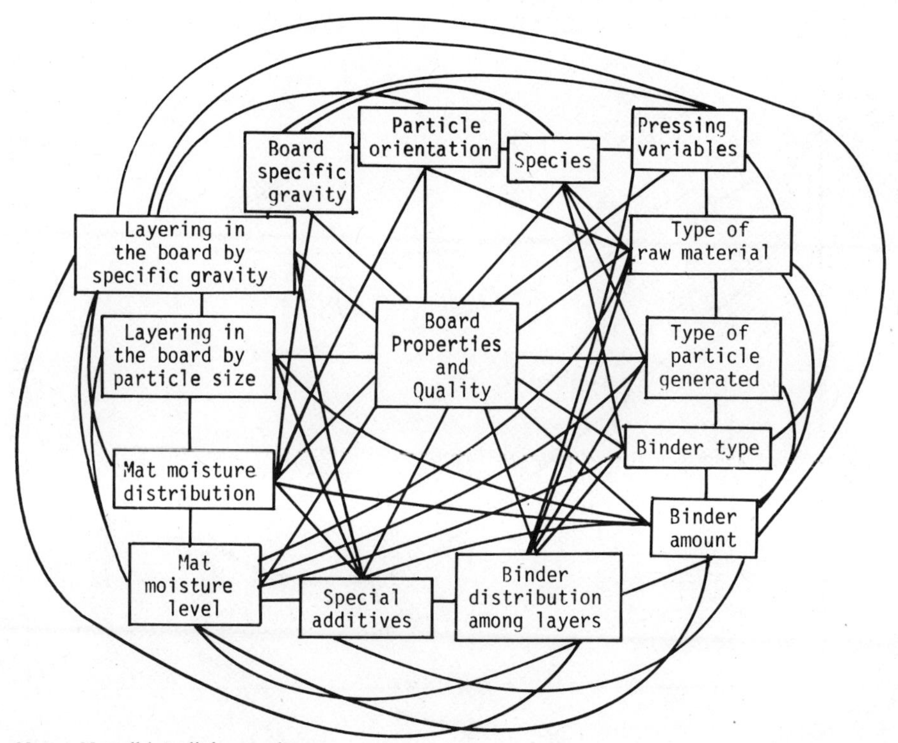

Note: Not all interlinks are shown.

Figure 6.7. Correlations Between the Various Parameters Affecting Particleboard Quality (after Maloney, 1977)

now that on-site resin manufacture directly from urea and methanol by particleboard companies would be feasible and economical, because the modern plant size has increased up to 200 and more tons per day, while the chemical engineering technology makes it possible to design resin reactors that are both small and efficient. However, the economics are still unpredictable, and depend on the availability and price of urea and methanol. The latter fluctuates rapidly and unpredictably. Furthermore, resin manufacturers supply decades of accumulated know-how and other assistance which would be expensive to purchase separately. In South America and other remote countries without an experienced resin industry, much of the resin is purchased as dried resin. This type of resin exhibits better storage stability and more predictable chemistry. Thus, it can be reliably used with lesser need for local chemical skill.

ISBN 0-201-04558-3

GLUE PROPERTIES

Practitioners describe particleboard resin properties with several parameters. The storage life of aqueous resins is normally 8–12 weeks. It is adjusted seasonally, to match the great climatic changes encountered at some plants, in which temperatures are up to 130°F in summer and well below freezing in winter. Modern resins with built-in catalyst and high cure-out rates sometimes suffer from unanticipated instabilities. If a full storage tank accidentally cures, it must be mechanically cleared, a task which might take five or more days. An elusive term is tack, a quality which implies sticking power at room temperature. Practitioners distinguish between (a) the fluid stage, before tack is noted, (b) the tack point at which the glue feels sticky, (c) the tack loss point, at which the resin film has dried sufficiently to lose stickiness, (d) the residual tack level, reached several hours after the resin sets, and finally (e) the dry-out point when no amount of pressure will cause adherence. The viscosity of a good particleboard resin is between 150 and 450 centipoise at 75°F (21°C).

The resin flow is a controversial quantity. It reflects viscosity under press conditions. Resin flow is a decisive factor in making a reliable board. Excessive flow causes soaking of particles and loss in the glue line, insufficient flow means insufficient contact surface. The process during curing is not fully understood. It will be discussed later. All modern resins are buffered. The buffer capacity determines how much acid is necessary to lower the pH and initiate curing, a process which practitioners often call kick-over. Buffering is usually tailored for individual application or wood species, to prevent unwanted pH changes. The pH of particleboard resins is normally between 6.8 and 8 during storage and shipment. Normally, in particleboard extenders are not used. The gellation time, or curing speed is established during the final glue formulation by additives and acid catalysts. Gellation times are measured by recording the time necessary at 100°C for the glue to form a solid plug. The gellation process is not a first-order reaction. Thus, during the test, the resin goes through an incubation period, during which the resin viscosity drops, until the entire volume suddenly coalesces to a solid plug. The gellation times are between 4–12 minutes for face glues, and 30 seconds to 5 minutes for the core, depending on the desired press time and the contemplated press temperature. The actual gellation period is between 1–10 seconds.

While old resins of the 1950s were cured at up to 180°C for up to 15 minutes, modern glues are formulated to fulfill exacting user specifications, providing – or reducing – tack at any point in the manufacturing line, matching any wood or wood extractive species, adapting to quick or slow pressing times, creating any desired viscosity over a wide temperature range, whether sprayed neat, dilute or with wax. All these handling properties have to yield an unerringly strong, durable, and economic bond which resists weather, heat, and aging.

ISBN 0-201-04558-3

ADDITIVES

A wide array of additives are available for treatment of particleboard, as it costs no labor and little more than the value of the additives to incorporate chemicals during the board manufacture. The situation is quite different for the after-treatment of finished board. Several percent of wax is added to UF glues as a sizing agent, to decrease water absorption and penetration. Even though wax merely delays rather than prevents soaking, wax is an important component in most applications. Until recently, high purity petroleum derivatives were used as waxes. A very elaborate assortment of grades and qualities were supplied. Since the recent petroleum product shortage, the situation has drastically changed and most manufacturers have learned how to incorporate lesser wax products, often home formulated mixtures, into their board. Wax emulsions are prepared by blending wax suspensions into aqueous surfactants until the particle size is approximately one micron. Maloney (1977) describes the function and functioning of waxes.

In some European countries, wood preservatives, such as 1.5% of pentachlorophenol is added to prevent decay and termite attack. In some cases fire retardants are added to board. This practice is not yet widely accepted. Boric acid has been used for that purpose, as well as fluorides, urea, and ammonium phosphates. However, incorporation of fire retardants requires experience, because each additive can alter the entire resin chemistry and the synergism between all other components.

PARTICLEBOARD MANUFACTURING

Particleboard plants are often operated side-by-side with plywood plants by the same company. However, modern particleboard manufacture, in contrast to plywood manufacture, is highly automated. Thus, typically four people produce up to 1000 cubic meters (700 tons) of board per day. Figure 6.8 shows an outline of a small-scale, labor-intensive particleboard plant specially designed by Fahrni (1974) for UNESCO for use in developing countries. The process consists of seven steps. (1) Solid wood is hogged, fiberized or flaked, dried to 3–6% moisture content, screened, and stored in bins or silos. (2) The chips are sized in the blender by spraying with wax or petrolatum. (3) Aqueous resin containing 40% solid is atomized to 20–80 micron droplets and sprayed to yield 6–10% resin solid, on the basis of dry wood weight, on the chips. Sometimes operations 2 and 3 are combined. Figure 6.9 shows a typical arrangement for mixing resin, catalyst and wax with a triplex pump. The resin concentration in the face layer is different from that in the core. This process is controlled carefully to provide a reliably uniform resin coverage. A substantial engineering effort is needed for developing the most accurate spray control, such as radioactive meters, to minimize the use of glue, which accounts for 30% of the price of the finished product. (4) The mat is formed on metal cauls which move step-wise from the bottom face to the core to the top face material bin. In some factories the mats are prepared and pressed without a caul. Such caul-less plants can produce board continuously. In such systems cold-tack is

ISBN 0-201-04558-3

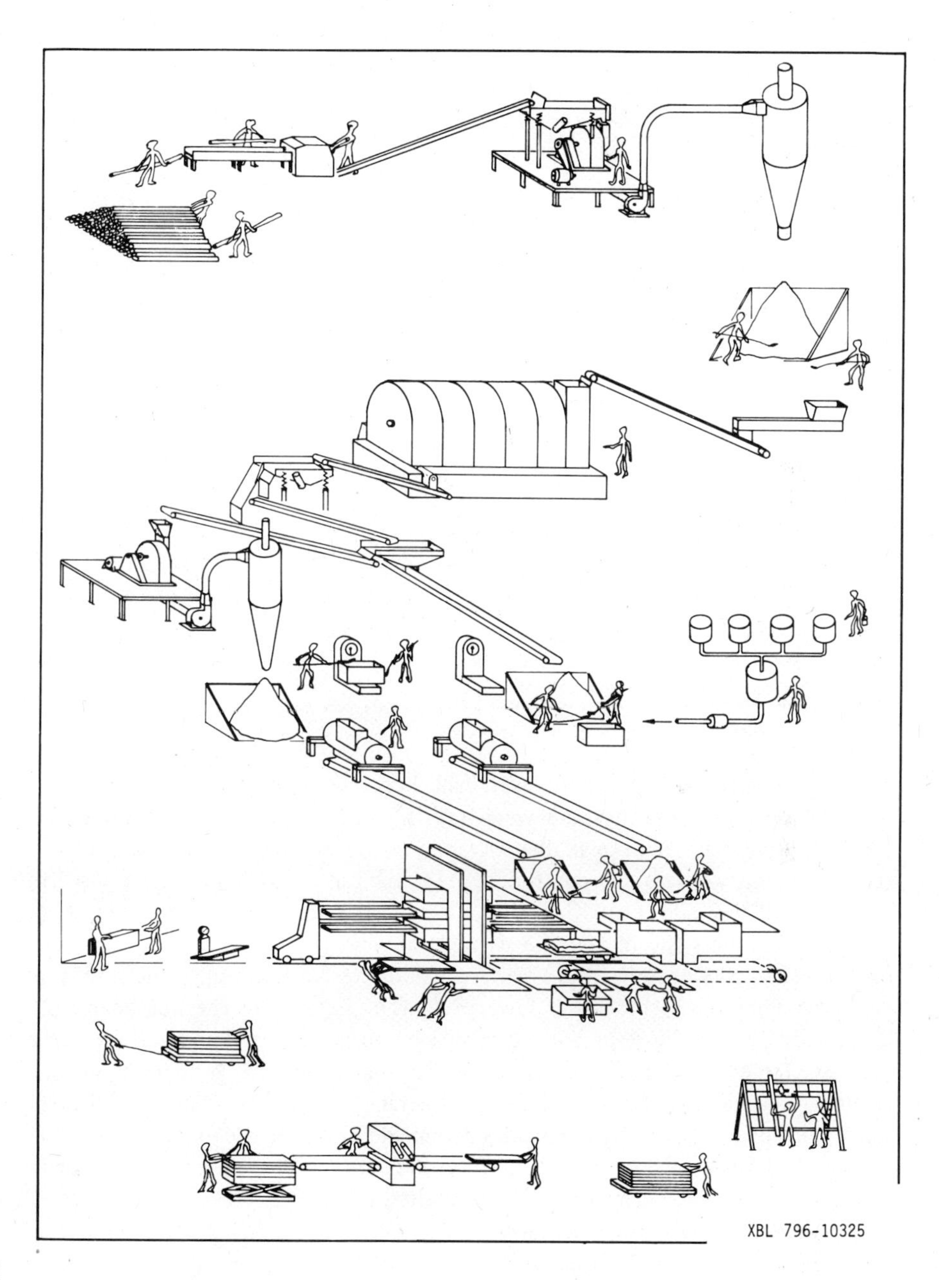

Figure 6.8. Schematic Outline of Small Particleboard Plant for Developing Countries (courtesy of Fahrni Institut; prepared for FAO World Congress New Dehli, 1976, UNESCO)

ISBN 0-201-04558-3

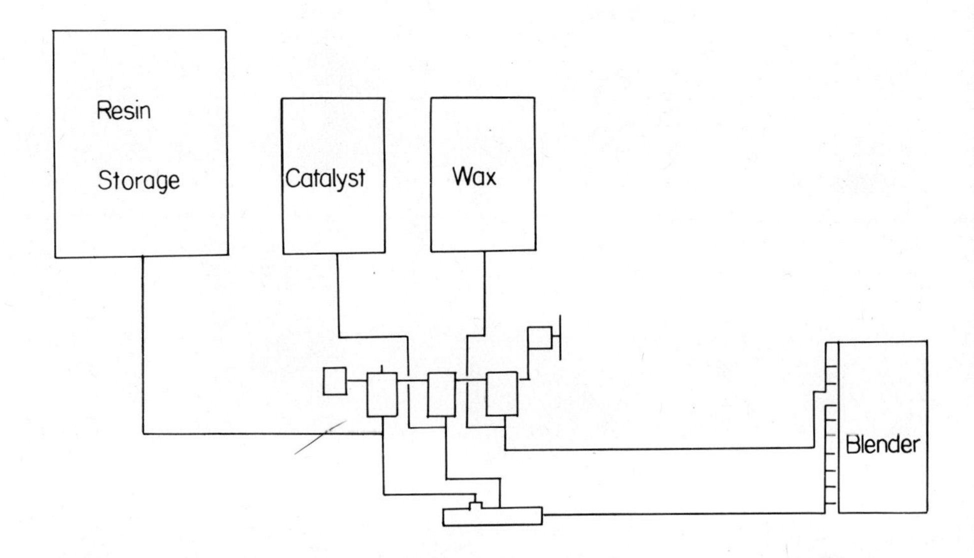

Figure 6.9. Principle Outline of Adhesive Stream in Particleboard Plant (courtesy of Borden Chemical Company, 1979)

necessary. Conventional factories stack cauls in presses with up to 28 layers, Figure 6.10. In older factories the cauls were trays to contain the mat if it did not have sufficient cohesion. Additional glue was sprayed over the redges, as needed. Today, the spray density is carefully controlled. (5) In the hot press the mats are pressed to "stops" at 270–325°F (150°C to 180°C) for 5–12 minutes. The pressure is adjusted to compress the mat at the desired rate and to the desired thickness. The press closure and opening processes are elaborate cycles during which pressure might be temporarily or slowly lowered to vent steam. (6) The hot boards are air-cooled, because UF-resin bonded particleboard is weakened and yellows if the heat is retained too long. Boards are then trimmed and cut. (7) The boards are sanded on both sides bringing thickness tolerance to plus or minus 0.055 in (1 mm). Currently some twelve companies design and build particleboard plants.

The final product and its use in furniture and interior design are shown in Figures 6.11 to 6.13. Particleboard is marketed in standard sizes and qualities. It can be used very much like whole wood. Figure 6.14 shows one of the main uses of particleboard as floor underlayment.

ISBN 0-201-04558-3

Figure 6.10. Opening Press in Particleboard Plant (courtesy of Fahrni Institut)

Methods for making particleboard have been described in several hundred articles and patents. Some important aspects of adhesive formulation are discussed and reviewed in Chapter 5. A few select references must suffice here to indicate areas of more current interest.

Dimter (1970) described air-permeable cellulose acetate overlays. In 1971, Arbuzov, USSR 322,348, described a Russian, three-layered composite board.

ISBN 0-201-04558-3

Figure 6.11. Particleboard Standard Construction (courtesy of National Particleboard Association, 1979)

Vintila, Rom. P 52,770, described fireproofing with sodium fluorosilicate; Seberini, Czech. P 141,245, prepared a calcium chloride polyvinyl acetate mortar board. Trautvetter, East Ger. P 79,962, described a light honeycombed board containing alum and gypsum; Anokhin, USSR 353,846, described a water-resistant board, and in USSR 353,847, a board with U:F between 1.0 and 1.7; Rougier, Fr. Addition 95,499, disclosed a method for fiber or particle panels; Ducorit-Consultants, BP 1,233,396, described the application of an overlay sheet impregnated with etherified melamine; Uchimi, Japan 71:39,479, described a particleboard with a density of 0.68.

In 1972, Kucera, USP 3,668,064, described a lignocellulosic additive containing asphaltites; Brooks, USP 3,668,286, made a board from particles containing up to 25% moisture prior to steaming. Lewin, Israeli P 30,472, added paraformaldehyde or hexamethylene tetramine just prior to pressing to obtain a stronger board; Ball, FP 2,081,925, added diammonium phosphate for door panels; Moralt, BP

ISBN 0-201-04558-3

Figure 6.12. Furniture Assembly Using Particleboard (courtesy of National Particleboard Association)

1,289,465, added asbestos fibers. Temkina, USSR 358,195, made a board with 2–4% hardener, to reduce free formaldehyde to 0.01 to 0.2%; Pahla, USSR 329,897, prepared skis from sawdust layered with wood shavings. The finished skis were oiled and coated with polyethylene. Anokhin, USSR 358,150, made plywood by adding formaldehyde to a 1:1 UF-resin just before pressing. McMinimy, USP 3,674,596, prevented resin bleeding by the addition of alum; Brooks, USP 3,668,286, made board by mixing resin with a particle containing 25% moisture; Senn, Swiss P 516,992, made multilayered board; Peters,USP 3,649,397, used comminuted wood premixed with urea; Marra, USP 3,671,377, made an isotropic building material; Belyi, USSR 346,396, made coatings; Anokhin, USSR 368,150, made waterproof plywood, by late addition of formaldehyde. Sheridan, USP 3,677,808, made moisture-resistant board by using steam-atomized wax.

ISBN 0-201-04558-3

Figure 6.13. Wood Laminates and Composites in Bank in Galesburg, Illinois, USA

In 1973, Purcell, USP 3,761,339, used a paper or fabric impregnated with bis(hydroxymethyl) propionic acid to laminate board, and thereby borrow structural strength from the board. Wilton, USP 3,770,466, made a heat insulating board, incorporating slag or rock wool; Mori, Japan 73:32,418, used bark adhesive as developed later by Steiner and Chow (1974). Hiag Company, FP 2,172,630, applied 70–80 wt % solid content in resin to make board, and in FP 2,161,199 they used 16–20% binder to prepare board suitable to firmly hold screws. Zastanchenko, USSR 395,280, added polyvinyl acetate to make colored, water-resistant particleboard. In USSR 387,847, he made wear-resistant board by using foliate-coniferous chip mixtures. In USSR 390,969, he disclosed colored tiles. Jumncanbhandhu explained the use of para-rubber plants as particleboard material. Lenz, Ger. Offen. 1,570,240, made impregnating resins; Immarino, USP 3,736,275, used crosslinked Xanthomonas colloids; Mori, Japan 73:21,346, added PVC to make water-resistant board; Nakane, USP 3,730,825, employed polyvinyl acetate; Time, USSR 368,067, used the same; and Moroff, Ger. Offen. 2,135,072, made a highly lustrous overlay.

In 1974, Davidovits, S. African P 73:8,363, made sintered composite panels; Domokos, USP 3,784,494, prepared artificial lumber; Fujita, Japan 74:129,753, employed methylated resin; Fujitani, Kokai 74:130,933, added acrylamide; Kato, Kokai 74:131,244, used aluminum hydroxide to fireproof board; Kirsch, Ger. Offen. 2,410,605, used granular vermiculate to make panels with a density of 400 kg/m^3 and a bending strength of 50 kg/cm^2; Isono, Japan 74:4,318 used dicyanamide-U-M-F to make a fire-resistant board. Kobayashi, Kokai 74:71,103, achieved a mildew-resistant plywood by adding benzimidazole; Kurashige, USP 3,804,706, prevented inorganic boards from splitting by using slag wool, asbestos, and

ISBN 0-201-04558-3

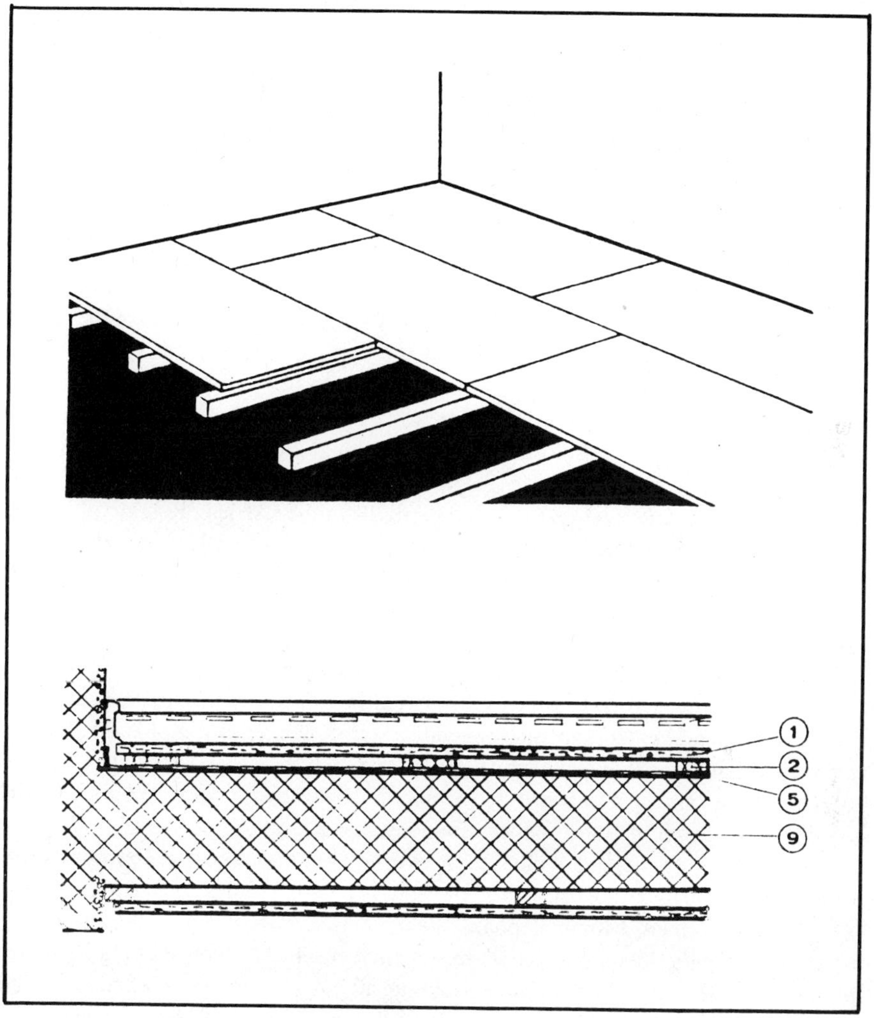

Figure 6.14. Particleboard Floor Underlayment (courtesy of Fahrni Institut)

inorganic filler; Lohenhoffer, USP 3,839,523, precompacted wood particles; Minoji, Kokai 74:38,927 made a top coating melamine layer; Paterson, USP 3,790,417, disclosed a board with improved dimensional stability. Tamura, Kokai 74:130,935, used a curtain-coater to apply resin; Tsubota, Kokai 74:130,977, incorporated bark chips to make insulating board. Shutov, USSR 431,008, impregnated solid wood by vacuum treatment; Baymiller, USP 3,816,236, prepared a cross-banding from a

ISBN 0-201-04558-3

slurry of rag and wood fiber. Deriancourt, FP 2,213,963, made a lignocellulosic fiber mat by a continuous process. Okabe, Japan 74:15,242, disclosed lignin products made from sulfite wastes. Kokai 74:99,775 is an anonymous disclosure of a fiber glass reinforced board; Anokhin, USSR 421,531, described a fire- and bio-resistant board, and in USSR 421,529 an extruded board. His USSR 478,724 disclosed a low-odor parquet panel.

In 1975, Inoue, Japan 75:28,482, reported a fire-resistant plywood; Ishida, Kokai 75:142,703, the same, incorporating phosphate and sorbitol; Iwasa, Kokai 75:43,182, used perlite and amosite to manufacture an inorganic board; his Kokai 75:38,778 incorporates gypsum and diatomaceous earths, asbestos, and fiber glass; Katayama, Japan 75:34,055, used powdered orange peels as an extender for plywood; Kato, Kokai 75:5,471, used volcanic ashes and aluminum hydroxide to make a fire-resistant, lightweight board. Kishi, Kokai 75:45,066, injected freshly pressed board with thermosetting and thermoplastic resins. Ohmura, Japan 75:6,527, applied pigmented nitro cellulose, polyester, or the like to obtain low odor, decorative plywood; Oka, Kokai 75:13,512, made plywood; Sklyarenko, USSR 456,730, used waste veneer and sawdust to make board; Stofko, S. African P 75:495, applied oxidizing agents to wood or veneers; Susicky, Czech. P 157,191, laminated lignocellulosic layers. Tsuruta, Japan 75:6,526, applied bicarbonates as foaming agents; Ueda, Kokai 75:25,716, disclosed fireproof plywood containing aluminum salts, and in Kokai 75:16,753 he used methylated melamine; Utsumi, Kokai 75:21,072, employed large wood chips to obtain a water-resistant board; Webb, BP 1,380,442, disclosed a heat insulating refractory board containing kaolinite clay, bentonite, or fibers which can be worked into shaped articles; Yamamoto, Kokai 75:42,011, applied a titanium oxide pigment as coating; Yoshida, Kokai 75:19,909, applied polyphosphates and pentaerythritol to make a fireproof plywood; Yudina, USSR 485,887, added iron salts to produce low-odor board.

In 1976, Burmester and Wilkie analyzed the swelling of wood; Detushev compared and evaluated the strength of adhesive bond lines. Kurashige, Japan 76:17,598, described a mineral board; Lehnert Ger. Offen. 2,610,286, described the sealing of particleboard edges; Petersen analyzed the influence of gluing conditions over a wide range of conditions, isolating all parameters; Raknes examined the durability of load-bearing timber after ten years; Robitschek analyzed resin degradation; Sundie, USP 3,979,492, described a fiber board; Ueda, Kokai 76:28,148, made fireproof board by using polyphosphoric acid and dipentaerythritol and aluminum hydroxide; Eder, Ger. Offen. 2,716,375, used ammonium persulfate; Hutschneker examined and compared water-resistant adhesives, including isocyanate.

In 1977, Plomley described the industrial application of naturally occurring tannin, i.e., polyphenol, adhesives for UNESCO; Maloney wrote his up-to-date authoritative book on all aspects of particleboard manufacturing; Clad reviewed general problems of adhesive use and established suitable standards; Lein reviewed trends in board production; Meyer and Johns described sulfur-UF formulations;

ISBN 0-201-04558-3

Roffael analyzed the influence of ingredients and manufacturing parameters on product quality; Zenktele conducted a study of UF-bonded particleboard, omitting hardener; and Schedro published a low-odor modification for a Russian resin.

Standard tests for evaluating board have been published in ASTM D1037-72a, annual books of the American Society for Testing Materials. All other countries have similar standards. The German standards are published by DNA, The German Standard Board in the DIN-60 pocket book for particleboard, fiberboard, and plywood.

Medium Density Fiberboard

MDF is a dry formed panel product manufactured from lignocellulosic fibers combined with UF-resin. The panels are compressed to a density of 30–50 pds/cft and hot pressed. The fibers are carefully controlled for geometry. On the U.S. West Coast, sawdust and shavings of spruce, mixed pine, Douglas fir larch, ponderosa pine, and white fir are used. On the East Coast and in the South, chips and round wood of oak, beech, birch, maple, southern pine, and hardwoods are used. The particle size depends on the method chosen for comminuting the wood. A typical particle size distribution is shown in Figure 6.15.

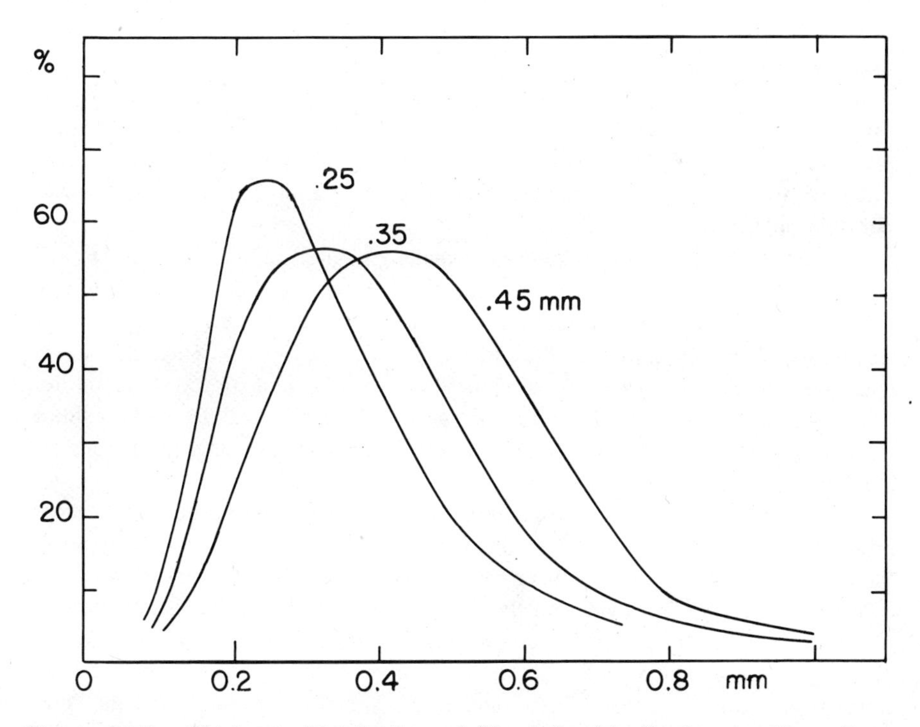

Figure 6.15. Thickness Distribution of Wood Particle Produced with a Comminuter, Model Unichip (after Deppe, 1977)

ISBN 0-201-04558-3

Normally, resins are low viscosity, low tack, and fast curing, with a solid content of 50% to 65%. Resin is applied to make 7.5 to 8.5% of oven-dry fibers. The blended material goes to metering bins. It is metered from 4 to 5 face- and core-forming heads onto a Fourdrinier wire. A 10- to 12-inch mat is precompressed to 4 to 6 inches, cut to length and pressed in a conventional hot platten press or by high frequency. Presses have 20 openings and produce 5 × 18 boards. The press temperature is 300–340°F, the pressure is 750 psi. Conventional press time is 7–9 minutes. High-frequency ovens cure the mat in 2 to 2¼ minutes.

Hardboard is prepared by a similar principle. Hardboard size is 5 × 24 feet with a thickness of 1/12 to 3/8 of an inch. Three qualities are distinguished: general utility service board with a density of 50–60 lb/cft, standard board, and tempered board. The latter contains some 5% oxidizable oil. Only the Asplund process uses synthetic resins. The Masonite process, invented by Mason in 1926, uses natural lignin to bind the board.

Wood Molding

In this process, wood particles are blended with a binder and pressed into a matrix, or mold, to form the desired product using 10–60 mesh wood particles as splinters or flour. Usually an overlay is used. For its application the surface texture and smoothness is critical, because overlays are often thin papers, imprinted with imitation wood designs.

As is the case with particleboard, the choice of the raw wood material depends on local availability, but the end product should fulfill national or international standards. Thus, UF-resins and manufacturing conditions must be locally formulated to adjust. Thus, the forming and pressing is a difficult art which needs constant quality control.

Molded wood products are used for making tables, cabinets, furniture, and coffins. In England, Scandinavia, and in the U.S. mobile home industry, wood molding is used for furnishes and trims, Figure 6.13, and ceiling systems, Figure 8.5, page 212, which are usually finished with a decorative trim. Such ceiling systems are useful for creative architectural design, for acoustic control and for producing special lighting effects. At least part of the art for making wood molding stems from the pre-World War use of UF-resin to make iron castings in the foundry industry.

There is a continuous transition between many of the above mentioned wood products and molded products containing bulk UF-resin with various types of fillers. Periodically, efforts have been made to prepare sandwich panels in which the core is particleboard or foam, and the face is veneer. Such panels lend themselves to factory preassembly of housing components. The main problem with this type of prefabricated housing has not been technical, or with consumer acceptance, but it stems from damage during shipping and during erection. It was found that such components are overly vulnerable to inadvertent or negligent damage along the edges, which cannot be restored economically. However, some assembled com-

ISBN 0-201-04558-3

ponents such as spiral staircases, bathroom assemblies, and kitchens have been well accepted for several decades.

Crocker (1944), in USP 2,327,871, assigned to the Lane Co., used UF-resins to line cedar chests, among other reasons, to reduce the cedar odor. Maksimov (1971) describes UF-resins for bonding colored paper. Kurene (1972) used stains modified with ammonium chloride as a base for an UF finish on furniture panels. Ishin *et al.*, in USSR 332,064 (1972) used UF as a primer. Jakimavicius (9174) used UF-resin to bond imitation design paper. Gupta (1974) experimented with laminated bonding of *Pinus roxburghii* in India; Gusev (1974) used UMF copolymers for making and bonding panels; White (1974) in USP 3,827,995 used UF-MF copolymers to reduce sagging in fiberboard panels. Paszner *et al.* (1974) made UF-bonded veneer-overlaid plywood by cold curing with gamma-rays, and Pastyr (1978) in Czech P 170,983 impregnated cellophane paper with UF.

In 1977 Hubbard and Atkinson, USP 4,035,456, described a method for making an expanded, noncracking, dimensionally stable UF product with a density of 0.2 to 0.8 which would lend itself well to mixing with wood flour or cement.

Properties of UF-Wood Materials

The invention and development of technically viable UF-wood materials, such as particleboard and wood moldings, marked an important breakthrough at the interface between the modern chemical industry and the traditional natural products manufacturers. If properly designed, manufactured and used, these products combine the advantages of natural products with those of the chemical binder. Thus, they are fully environmentally compatible. Particleboard is not more toxic than wood. During fires, its fumes are less toxic than wood, and after use, it degrades readily (as it consists of a mixture of wood chips and a high nitrogen fertilizer, both of which combine to form a desirable mulch and plant nutrient) as described in the section of UF foam. Mechanically, wood-UF products are anisotropic, dimensionally stable, do not crack, and can be formed into bigger, longer, and more creative shapes than natural, massive wood. Chemically, these products resist vermin, mold, moisture, and organic solvent better than either component alone. Accordingly, the potential for the use of these products was quickly recognized and they have been widely accepted by formerly full-wood oriented industries, as well as by industries which never used wood products. The consumer profited from better and cheaper furniture, housing, and other products.

Unfortunately, there seems to be an intrinsic law that outstanding inventions, when recognized, inevitably seed overenthusiasm and overconfidence. Thus, builders, architects, and furniture manufacturers were led to relax their restrictions, and overlooked inherent problems which were well known to those introducing the products.

One major disadvantage of UF-bonded wood products is their susceptibility to hydrolysis. Smith (1979) reports that his long and vigorous efforts to introduce

ISBN 0-201-04558-3

particleboard manufacturing to the U.S. were almost stopped, at the last minute before plant construction, when a competitor showed a piece of soaking and swelling particleboard to his sponsors. This problem has not yet been economically solved, except with coatings. It severely restricts conditions of shipping, storage, and use of particleboard. If this sensitivity is ignored, product performance can be affected. Upon continued exposure to moisture, untreated particleboard loses mechanical strength, blisters, and eventually crumbles. In milder cases, the board releases volatile formaldehyde which exudes an unpleasant, and an obnoxious odor.

Intense efforts have been made to remedy this problem since all other properties have made wood-UF products highly desirable. The problem is composed of many facets, some of which cannot easily be controlled. For one, moisture can penetrate particleboard far quicker than whole wood, and wood inevitably swells upon exposure to water. Another factor has been the problem of resin control. The resin composition changes seasonally, locally, and with age. However, modern analytical chemistry techniques are now available which will undoubtedly help remedy this situation within the next few years. In the meantime, wood panels exposed to weather must be specially impregnated or coated.

It is doubtful that UF-wood products – or for that matter any material – will every be perfected to fit all and any construction needs. However, the development of product classes for particleboard and wood moldings is not yet finished in the U.S., where bulk commodity board is sold and auctioned without consideration for its end use. Thus, particleboard used for manufacturing the interior of mobile homes, which are used as daily living space by its owner, comes often from the same stock as bulk board used, for example, as floor underlayment in agricultural or industrial warehouses.

The periodic attempts to produce materials which combine universally high qualities with lowest possible price, and to formulate all-encompassing product standards covering all possible conditions of use are not practical. They often end in a product which fits all purposes poorly, and none well. It would be easier to introduce graded specialty product lines.

The problem of UF hydrolysis has been analyzed and reviewed by Myers (1978), Roffael (1968–present), Neusser and Zentner (1968). The problem of formaldehyde release from wood products has been reviewed by Nestler (1977). It is directly correlated to water absorption (Roffael 1978, Andersen 1974, Myers 1978).

FORMALDEHYDE STANDARDS

The correlation between the formaldehyde concentration in air and formaldehyde release from building surfaces and components is now recognized, even though it has not yet been quantitatively established. The Japanese Industrial Standards, JIS-A-5980-1977 deals with odor, jointly with mechanical and other performance, and limits formaldehyde to 5 mg/liter. Other countries are contemplating similar standards and the next year will undoubtedly see wide regulation.

ISBN 0-201-04558-3

Trends are discussed in the chapter dealing with health. Thus, control and abatement of formaldehyde are now widely tackled.

FORMALDEHYDE ABATEMENT

Current generation UF-resins were not designed with a view toward indoor use in particleboard or foam. In these applications, UF-resin has a very large surface-to-volume ratio, and the resin has variable thickness, and contact with surfaces which on microscopic scale vary in chemical, physical, and mechanical behavior. The problem can be tackled via different paths. The most sensible approach is to redesign resins chemically from scratch, with the goal to give odor the high priority which it deserves under these conditions. Efforts toward that goal are under way, and are summarized in the chapter on resin manufacture. The other obvious approach, short of suspending the use of particleboard and foam in these applications, is to coat the exposed surfaces with paint or special binders. A variety of treatments are described in other sections. Paper overlays with a weight of less than 80 g/m^2, plastic laminates, veneer and metal faces, and chemical coatings are well-established. This method works well, but costs more than the treatment with additives which can be implemented before or during the forming of the boards or moldings.

Petersen (1977) enumerated the manufacturing factors, listed in Table 6.1, as related to formaldehyde release. He computed graphs in which formaldehyde release, mechanical properties, and water uptake are related to the factors listed in Table 6.1, and he provided data and advice suitable for designing and evaluating experiments to test for these factors using modern statistical methods, based on Yates. He finally provided a procedure for using hand-held programmable computers to conduct such experiments. The problem of producing practical tests was also discussed by Sell (1978).

Table 6.1

Manufacturing Parameters Affecting Formaldehyde Release (after Petersen 1975)

Resin Components	Wood Species
Resin Component Ratio	Wood Moisture
Resin Application	Resin Concentration
Hardener Components	Resin Viscosity
Hardening Quantity	Wood Chip Size
Press Temperature	Wood Chip Geometry
Press Duration	

ISBN 0-201-04558-3

A final, stop-gap approach is to modify resins with additives, to bind free formaldehyde, or absorb it. Among these, a second or third addition of urea to the ready-to-use resin is most common (Petersen 1974). Another proven method is treatment with lignosulfonates. Further improvements can be obtained by addition or fumigation with ammonia which binds free formaldehyde in form of hexamethylene tetramine. Verbestel (1979) developed a machine for this purpose which is described in Chapter 9. The analytical chemistry is described in Chapter 5, and health and hygiene is discussed in Chapter 10.

ISBN 0-201-04558-3

7. Foams

Currently some 75,000 tons of UF-resins are foamed. This is about 5% of all resin. UF foams were developed later than other resin products, but once it was possible to make smooth foams with controllable and uniform cell size, resin chemists and manufacturers recognized that UF foams have versatile applications, ranging from medicinal powders to home insulation and agricultural fertilizers. They have an almost untapped potential in insulation, and their use in agriculture – in plastoponics – opens entirely new vistas. Undoubtedly, it will take many years, or maybe even decades before the utility of foams is fully recongized by their potential users. But, they are so versatile that it is quite possible that foam will account for some 30–50% of all resin consumption in the future. This chapter deals with their chemistry, their manufacture, and their uses.

Chemistry

A. Gams and G. Widmer (1926) patented a porous product with a density of 0.4 g/ml resembling meerschaum GP 523,181, assigned to CIBA. It was prepared by curing with excess concentrated sulfuric acid which was subsequently washed out. In Swiss P 126,827, of the same year, they described the high insulating power of UF. FP 644,648 of 1927 described a highly porous mass. At about the same time Karl Ripper working with Fritz Pollak also worked on insulating compositions. However, it was another eight years before the first practical patent for the preparation of a low-density foam appeared. Finally, in 1933 A. Curs and H. Wolf of I. G. Farbenindustrie in Ludwigshafen first described a foam composition which worked so successfully that it has been basically retained. Their procedure consisted of preparing a separate foam mixture which is then added to the resin and the hardener. Strong acids such as sulfur dioxide, sulfuric acid, or chloralhydrate, a latent acid, are proposed. By example, they formed a foam by stirring 4 ml butylnaphthaline sulfonate in 40 ml water. The foam was then added to 800 ml of a 32% solution of UF-resin which had been acidified. After 3–6 minutes the foam was sufficiently stiff to adhere to a vertical wooden board. It was fully cured after

ISBN 0-201-04558-3

Beat Meyer, Urea-Formaldehyde Resins

two hours. They recommended slow drying and they achieved densities of 0.01 to 0.5 g/ml. This foam can be used to reinforce veneer, wallpaper, wood fibers, and the like. This patent was later modified by use of formic acid and phosphoric acid as hardener, and the foam was dried in a hot, moist air stream at 60°C. The corresponding USP 2,076,295 was assigned to Plaskon. L. S. Meyer of Plaskon, USP 2,273,367, proposed calcium sulfate and gypsum as stabilizers for the foam. He thus gained fifteen minutes between frothing and hardening, time which was useful for transferring the foam into dies, or for shaping it. He added the hardener to the foam before mixing it with the resin, and he used resin that was far advanced. As possible uses, he envisioned linings for automobile tops or airplane fuselages. For strengthening the foam body, he proposed stretchable rubber skins, preferably prepared *in situ* from a latex. The final density was 0.3. In 1943 H. Scheuermann and E. Späth made some important modifications. They prepared a UF-resin foam and a foam catalyst solution separately in pressurized containers and mixed them in a tubular extruder, GP 734,094. In 1949, GP 800,704, H. Scheuermann disclosed a mixing train in which foam could be continuously mixed and extruded. In 1958, W. Bauer, USP 2,860,856, described a portable device with a spiral mixing nozzle. This basic system was adapted widely and was successfully used to make insulating foams from almost any UF-resin.

A high quality foam of low density which remains viable after aging can only be prepared when the foaming conditions are well characterized, and the resin composition is well known. For this, reliable quality control is necessary. There is a considerable art in this field, most of it in the form of trade secrets. The single most effective method for quality control is the use of a spray-dried resin, such as described in USP 3,470,115.

In recent years, substantial efforts have been made to produce foams free of formaldehyde odor. For example, in 1966 F. L. Kelley and R. L. Wells in USP 3,231,525 assigned to Allied Chemical Co., proposed the addition of 0.5 M, preferably 1–2 M for each mole of formaldehyde to the ready-to-use, stabilized resin, preferably that of USP 2,652,377, patented in 1953. The resin contains about 2.5% free formaldehyde and 3.9% *N*-methylol and has a pH of 7. Furthermore, addition of about 4% ammonium carbonate as buffer was proposed, together with a small portion of ethylene glycol which prevents shrinkage. The purpose is to capture not only free formaldehyde, but also formaldehyde in the form of *N*-methylol which might potentially be hydrolysed. The authors claim that 0.1% free formaldehyde is noted by its odor, and that the threshold for *N*-methylol is 1%. In another application, assigned to Allied Chemical Co. (1967), G. H. Justice proposed to fully react *N*-methylol formaldehyde during the cure. He achieved this by heating the resin briefly at pH 4–5 to 135°C, preferably at low pressure. The process is tricky and not fully understood. Apparently, it merely destroys *N*-methylol but does not initiate polymerization. He achieved this by conducting the second stage of resin synthesis at higher temperature than usual. He followed the change of formaldehyde in the resin by analysis. The total of free and *N*-methylol formaldehyde was

ISBN 0-201-04558-3

determined by the sulfite method, as free formaldehyde reacts with sulfite at 0°C, while *N*-methylol reacts at 80°C. For confirmation, free formaldehyde was also measured with the hydroxylamine method, and total formaldehyde by the iodometric method. The drastic drop in formaldehyde above 115°C is shown in Table 7.1. Justice used 1% foaming agent and 0.1 N acid. Triethanolamine alkyl benzene sulfonate with a pH 1.5 was used. He claims that odor is virtually eliminated both during and after curing. In USP 3,312,639 Justice in 1967 also proposed the use of ammonium carbonate to suppress odor. He employs it together with turpentine, camphor, or pinene. These techniques are surprising, because the ammonia salts normally tend to collapse the foam. The resin is prepared according to USP 2,652,377 and starts with a U:F = 4 to 8. 100 parts UF concentrate are mixed with 15 parts urea and 2 parts ammonium carbonate. Ammonium bicarbonate works equally well.

Table 7.1

Free Formaldehyde, Free Methylol, and Total Formaldehyde Vaporizable as a Function of Temperature Resin Mixing (Justice, USP 3,306,861)

Temperature (°C)	Times (min.)	Free Formaldehyde[a] (%)	Total Methylol and Free Formaldehyde[b] (%)	Methylol (%)
80	18	1.85	4.4	2.96
90	10	1.82	4.4	2.86
100	8	1.25	4.0	2.76
105	7	1.00	3.3	2.30
110	5	0.60	1.0	0.40

[a] Bisulfite method at 0°C.
[b] Bisulfite method at 80°C.

In 1971, in FP 2,053,840, Fibreglass Ltd. described a 1-inch-thick thermal insulating mat with a density of 0.1 g/ml prepared from a UF-phenol reinforced resin which is used to soak glass or asbestos fibers. The mats are cured at 250°C under pressure and can be laminated into 6-inch thick boards which do not support combustion up to 260°C. Dicyandiamide and melamine can also be used to reinforce the resin. In 1972 D. Rosenthal, USP 3,651,182 described tough-skinned thermosetting foamed beads. R. Neumann, East GP 96,965, described a hydrophilic open-cell UF copolymer. He used a 2% aqueous solution of polyalkylene glycol fatty ether, 0.3% of a quaternary ammonium salt surfactant, and 5.5% phosphoric

ISBN 0-201-04558-3

acid to a 32% UF resin. The foam gelled in five minutes and dried in two hours at 50°C to a solid containing 20–50% water and a bulk density of about 0.026 g/cm^3. The foam absorbs 100% water in less than 20 seconds and is suitable as oil absorber, insulating materials, etc. British Industrial Plastics Ltd., FP 2,102,406, described a UF foam made from a 40% solid content resin, sodium dodecylbenzene sulfonate and phosphoric acid at 50°C. The density was 0.014 g/ml.

F. Brunnmüller, O. Grabowsky, and J. Lenz of BASF in GP 2,064,102 described a foam of low shrinkage. They used a urea-formaldehyde-triethylene glycol copolymer with a molar ratio of 1:29.0–3.0:0.6–1.5 as additive to a standard UF-resin. Thus, 4.52 kg of 40% formaldehyde plus 0.44 liter water and 0.11 kg 25% aqueous ammonia were precondensed with 2.4 kg of 80% urea at 90–95°C and pH 5.5. The co-condensate was prepared separately by refluxing triethylene glycol, 40% formaldehyde and urea at 90°C. After 30 minutes another 0.6 kg of urea was added, the mixture neutralized, and water added to yield a density of 1.55. Foam was prepared from resorcinol 4,5-sodium dibutyl napthalene sulfonate and phosphoric acid. Linear shrinkage was 1.0%. F. Brunnmüller, GP 2,110,939, also patented a foam of low flammability. For this he added 10 g/kg calcium phosphate, or calcium chloride to a standard foam base as described above. According to German DIN 4102, the foam extinguished within 3 seconds. Shrinkage was 49%, rather than 68% as for a standard foam.

Midwest Research Institute, FP 2,082,270, prepared fire resistant resin from *p*-toluene sulfonamide-formaldehyde copolymerized with a phenol-formaldehyde resin. The foam contained 3% water and 2% unreacted formaldehyde. It was prepared at high pH at 85°C and foamed with polyisocyanate in the presence of dibutyl-tin diacetate. Post-curing at 135°C improved fire resistance. J. Schwab, F. Boehm, and G. Ruthner, in GP 2,219,033, assigned to Sempevil Co., used charcoal or silica gel to increase the active surface for hydrophobic applications. V. D. Valgin, V. A. Novak, and Y. S. Murashov in USSR 328,152 used freons with a boiling point of 20–80°C to foam UF and sulfonphenol resins.

In 1973 H. Humik reviewed volatile and heat-decomposable compounds as blowing agents. K. Oshima, Y. Ito, and S. Suzuki, USP 3,764,428 described the bonding of foam to fibrous mats, such as cotton fleece. The product is cured at 110°C. R. J. Ashall in BP 1,316,911 reported the use of lignosulfonate and urea as extenders for fiber and glass insulation. The resin was a UF phenol-dicyanamide copolymer cured with urea-calcium lye. V. M. Kozin, USSR 403,654 used polyvinyl acetate as resin modifier. S. Suzuki, Japan 73:44,341 likewise used polyvinyl acetate or casein for deodorizing resin. Methyl-cellulose and sodium dodecylbenzene sulfonate were the foaming agents.

In 1974 S. A. Kartashova and S. P. Khainer described the use of UF-resin to bond extra light perlite for building isolation. Diethylene glycol was used, and resorcinol served as stabilizing agent for the foaming agent. V. D. Valgin and V. A. Novak (1974) described a phenol-formaldehyde cellular foam of extremely high flexural strength. V. D. Valgin and N. I. Borodkina described a cellular UF-resin

ISBN 0-201-04558-3

foamed with lauryl triethanol ammonium sulfate. S. A. Druzhinin, E. V. Khylstunova *et al.* (1974) explained that crosslinking directly influenced shrinkage of foams. They proposed the use of dry urea. K. M. Demenkova *et al.* (1974) modified UF with resorcinol, ethylene glycol, or diethyl glycol and compared chemical resistance and shrinkage. H. Nakajima and M. Kubo in Japan 74:15,071 described a UF copolymer containing urethane prepolymer (3–10% free isocyanate) and polypropylene glycol – 2,4-toluene diisocyanate (0.7% free isocyanate) with stannous octoate as foaming agent. They claimed good storability, and uniform cell distribution. D. Blake in GP 2,402,441 claimed a foam useful for removing oil from sea water by reacting 500 g UF copolymer (80% solid) with 1.5 g dicyandiamide, 2 g sodium dioctyl sulfosuccinate, 5 ml tritolyl phosphate, 8 ml silicone solution, 45 ml freon, and 17 ml phosphoric acid. The foam was workable for one minute. N. I. Borodkina in USSR 448,208 described optimum compositions for surfactants. V. S. Borozdin in USSR 431,199 described the foaming of UF-resin at 5 atm. Brown-Boveri Co. in USP 3,853,685 and USP 3,930,066 described a stiff open-cell foam useful as an electrical insulator. It is foamed with sulfur hexafluoride *in vacuo.* Hinteerwaldner in GP 1,745,562 reported a UF-resin reinforced with an elastomer, such as butyl rubber latex. The foam is vulcanized at 150°C. H. Juenger and F. Weissenfels in USP 3,830,894 showed for the example of phenol-formaldehyde resin that wood shavings and other fillers can almost double the mechanical performance of cured foam. H. Petersen, P. Scharwaechter, and J. Lenz in GP 2,241,715, assigned to BASF, disclosed a foam which is both fire- and shrinkage-resistant. The ingredients were 3-dimethyl-phosphono-propyl-carbamate, resorcinol, and phosphoric acid. The density was 0.022 g/cm^3, linear shrinkage 3.8%, burning time 1.8% and volume loss 39.6% in the German DIN test. M. Widmann, Can. P 1,013,499, used polyglycol and chloral in UF foam to fireproof resin.

In 1975 M. K. Frolova in USSR 462,846 proposed the use of sulfophenylurea formaldehyde and phosphoric acid as the foaming-hardening agent. H. Hurst in BP 1,412,559 reported copolymerization with neoprene rubber or acrylic acetate-vinyl-acetate and sodium methyl siliconate to give improved elasticity and water resistance. V. A. Novak *et al.,* FP 2,246,588 and GP 2,447,941, reported their Russian modification of phenol-formaldehyde resin with iron oxide. Transition metal compounds were found to reduce water uptake, and hence corrosion, by the foam to the carrier. S. Okreshy, Japan 75:2,57 disclosed a fireproofed UF cellular resin which contains citric acid, keratin hydrolysate as a binder for a mixture of perlite, and cellular polystyrene. L. Rechner in FP 2,306,225 reported a UF-phenol copolymer to be used as thermal insulator. T. Shiota, T. Sato, and S. Okamura in Japan 75:119,892 claimed that the addition of second urea and starches, or proteins, diatomaceous earths or polysaccharides, or tertiary fatty amines yields open cell foams suitable as microfilter. L. D. Tarasenko in USSR 471,348 summarized the ingredients of a modified UF-resin containing polystyrene and calcium phosphate. G. F. Tobol'skii in FP 2,243,917 described the formulation of UF- and PF-resins foamed with rosin-saporin as binders for glass or asbestos insulating mats.

ISBN 0-201-04558-3

D. A. Yakovlev *et al.* in USSR 481,584 mentioned polyorganohydrosiloxane as stabilizer for UF-resin, and polyvinyl alcohols and basic acids for bonding perlite. J. Zapletal, USP 3,986,988, employed keratin hydrolysate and citric acid plus polystyrene filler to prepare a foam with decreased combustibility. The plastic mass is kneaded, poured into a mold, and allowed to set. The mechanical properties claimed are remarkable, as are alkali and acid resistance. M. Widmann in USP 3,979,341 described a foam assigned to Borden Products of Canada. The foaming composition includes 1.6 to 4.8 mmoles dibutyl nephthalene sulfonic acid as surfactant, 5–20 mmoles phosphoric acid as a catalyst and foam inducing agent, 0–45.5 mmoles of resorcinol to prevent the foam from collapsing, 9.1 to 54.6 mmoles chloral hydrate as a fire retardant, and sufficient water to make a solution of 5% solids content. The resin is made from 1.0–2.4 moles of formaldehyde per mole urea, from 6.5 to 500 mmoles polyethylene glycol as foam charring agent, 0.3 to 1.2 mmoles of calcium chloride to make cell size uniform, 3–310 mmoles dicyanamide to increase resistance against hydrolysis, 2–30 mmoles sodium bromide to improve aging, 0–1.50 mmoles of furfuryl alcohol to mask formaldehyde odor, and sufficient water to obtain 40–45% solids. The two solutions are mixed in a ratio from 0.9 to 1.5 by weight. The Russian literature was equally active during 1976: S. N. Dumov *et al.*, USSR 615,103, described a polystyrene-copolymer foam; V. G.Gorenko, USSR 576,154, a sulfite reinforced foam, M. N. Naidenov *et al.*, USSR 576,327, a PVC-UF composition, and J. Zapletal, Czech. P 162,810, a polystyrene modified UF. Y. Ito *et al.*, Japan 76:6,193, used amine additives. H. Noda, Japan 78:37,767, urea copolymers, A. Jamin, GP 2,537,859, employed polyester-polymethane-UF copolymers and ammonium bisulfate. J. Bravo, Spain P 417,401, added polyglycol to an acid condensate prepared at pH 4.5. W. D. Schoellhorn in GP 2,436,465 employed pentane as a blowing agent, or the decomposition product of hydrogen peroxide. Density was 0.04 g/cm^3. N. Waddleton, BP 1,428,334, mixed keratin hydrolysate with polystyrene and UF-resin. Density of the foam is 0.2 kg/m^3. The material is similar to that of Zapletal. J. J. Cleveland in USP 3,943,994 described UF as a fugitive binder for a ceramic matrix. British Industrial Plastics Ltd., in Japan 76:6,269 used UF-resin cured with phosphoric acid to produce a powder capable of absorbing 5 milliliters water per gram resin. In 1978 W. P. Moore described a method for producing chemically stable foam. He prepared resin at pH 4.4 to 5.5 and gradually added aqueous dialdehyde and additional urea. Glyoxal markedly increased chemical resistance, even at 1%, and removed all formaldehyde odor.

In summary, the art of formulation of UF-resins is now well developed, but the science is not nearly understood. Many UF recipes are suitable for foaming. It is advisable to add several components which are not necessary in adhesive UF-compositions. For example polyglycols help reduce linear shrinkage and improve elasticity. They also serve as foam charring compounds. Salts such as calcium chloride help control uniform cell size. Dicyanamide makes the foam hydrophobic and neutralizes acid formed during aging. Sodium bromide or other halides impart

ISBN 0-201-04558-3

better fire resistance. Odor-masking compounds such as furfuryl alcohol or other chemicals mildly reactive towards formaldehyde, such as ammonium carbonate, are helpful. The most popular foaming agent is clearly dibutyl naphthaline sulfonic acid. It also reacts by itself with formaldehyde, and in fact, the reaction is so strongly exothermic that it forms a porous foam with formaldehyde alone.

Principles of Foam Preparation

RESINS

The basic chemistry of the UF-resin is the same as that for other applications, except that foams must cure at room temperature within the first few minutes, as compared to adhesives which cure within three minutes at 150°C. Thus, the foam resin kinetics must be some thousand fold faster, and the resins must be better precured. Furthermore, the foam has such a large surface area, compared to casts, that resin must be hydrophobic to resist hydrolysis. In this, the surfactant plays an important role.

SURFACTANTS

Almost all surfactants are sulfonic acids, usually aromatic sodium salts. Baumann (1976) has shown that these compounds cannot be recovered from the open foam pores, because they react with the resin. He proposes that alkylnaphthalene sulfonate undergoes alpha and beta reaction leading to three-dimensional structures. During mixing of the foamed surfactant and resin at some 3 atm, the foam collapses to about half its original volume equivalent. Baumann (1978) has succeeded in preparing coplymerized foams in which the sulfonic acid concentration is comparable to that of urea.

CATALYST

In cold-setting foams the pH must be lower than in hot-setting adhesives. A pH of about 1–2 is usually desirable, and most catalysts contain phosphoric acid, great care must be taken to keep acid to a minimum. Excess acid will hydrolyze the finished foam. This is the single most important cause of formaldehyde odor from older, fully cured foam. Excess acid can cause initially normal-looking foam to crumble into powder.

WATER

The water hardness influences shrinkage and surfactant. A good foam contractor will determine the water hardness and adjust it with softener before preparing the final ready-to-foam resin and catalyst mixtures. Most foam manufacturers distribute water hardness measuring kits together with the resin.

ISBN 0-201-04558-3

FOAMING GASES

In bulk insulating foams, air is used as gas, yielding 99% open-cell structures. If freon or other chemicals are employed, the cell structure can change, yielding predominantly closed cells.

ADDITIVES

The copolymerization of UF-resins with foams follows other chemical laws than those for adhesives. Thus, melamine, which improves the performance of adhesives, in contrast, weakens foaming formulations. Mealmine foams lack structural strength. In the application of foam, the compatibility of adjoining materials must be considered. It can be easily pealed from PVC, polyethylene, polyester, polystyrene, polyisobutylene, and does not react chemically. It sticks to butyl rubber foils, but peels off asphalt papers, as used in roofing. Aluminum is discolored, but does not corrode, copper noticeably tarns, zinc plated steel corrodes slightly and steel sheets rust noticeably.

FOAM MANUFACTURE

Today UF foams are made on site in portable foaming equipment, because factory made foam slabs are easily damaged during transportation and occupy excessive and uneconomical shipping volume. In the standard procedure two separate stock solutions are made: (a) a hardener surfactant solution and (b) a resin stock solution. Since the foaming is usually conducted in the field by contractors, the basic chemicals are factory premixed. For this, dry chemicals are preferable because of their lower shipping weight. Furthermore, dry UF-resin stores more reliably and can be used to prepare fresh solutions which have reproducible properties and behavior. The two solutions are usually formulated so that they can be mixed in a 1:1 ratio. The standard procedures consist in pumping the solution in accurately measured proportions from open storage drums, or to expel them from pressure vessels, Figure 7.1. The key tool for producing a good foam is the gun, Figure 7.2. Hardener and resin are combined in the barrel packed with glass beads where the reagents are mixed in turbulent flow. The fully expanded foam emerges from the gun with a consistency of shaving cream. A good foam sets within the first minute sufficiently to be sliced. It hardens fully within a day. Most modern insulating foams have open-cell structure and the residual water vaporizes from the cured foam. If foam dries too quickly, it tends to warp, shrink, or crack. Home insulating foam in closed spaces, on occasion, retains moisture for several years, especially in cold climates.

ISBN 0-201-04558-3

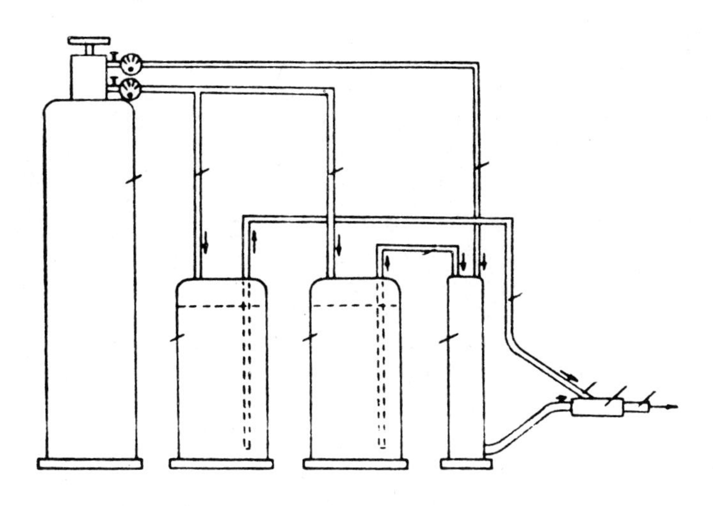

Figure 7.1. Resin and Foam Mixing Set-up for in Field Foam Manufacture (USP 2,860,856, Nov. 18, 1958, W. Bauer, Frankenthal, GFR)

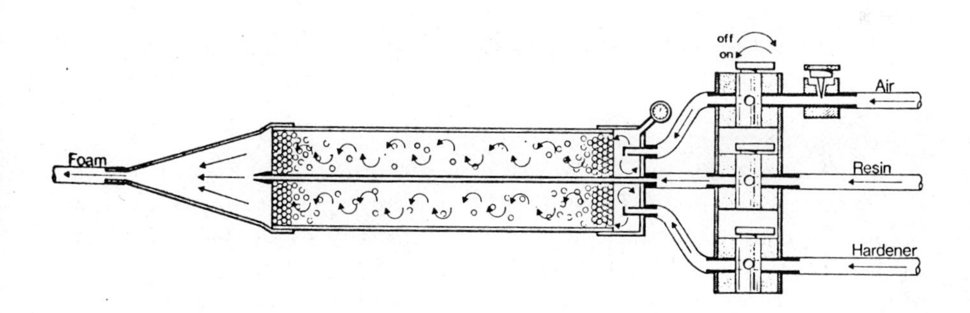

Figure 7.2. UF Foam Gun; Aerolite Model (courtesy of CIBA-GEIGY)

ISBN 0-201-04558-3

Physical Properties

The physical properties depend on the formulation. Some foams are deliberately designed to absorb water and oil, while others are made hydrophobic. Thus, water absorption ranges from 2% of volume to 4000 wt %, i.e., 45% of volume.

The density of typical dry insulating foam, according to ASTM D-1622-63 (1975) is 0.6 to 0.9 lb/ft^3 (10–14 kg/m^3). The wet density of fresh foam is 2.5 lb/ft^3 (40 kg/m^3). Shrinkage is 2.98–4%, depending on formulation. The thermal conductivity of foam, according to ASTM C-177-72 is K = 0.206 BTU·inch/hr·ft^2 (R = 4.85) at 34°F, and K = 0.228 (R=4.39) at 80°F (0.035 W/m·K at 24°C). Astonishingly, the thermal conductivity changes less than 5% over a density range of 10–30 kg/m^3, but it is a function of thickness (Latinerer 1973). If freon is used as blowing agent, conductivity can be further reduced (Dementev 1975). The decomposition temperature is above 392°F (200°C).

The flammability, ASTM E-84-76A of a test speciment comparable to application thickness, rates less than 25 versus 100 for red oak. According to UL 723, smoke developed is 50–80 for a one inch thick specimen, compared with 100 for red oak. The heat of combustion, measured with the Emerson oxygen bomb calorimeter is 7,830 BTU/lb (14 MJ/kg) as compared to 11,000 BTU/lb for polymethane and 17,000 BTU/lb for polystyrene. Smoke density, ASTM D-2843-70 is 1.15%; self ignition occurs at 1150°F (620°C), flash ignition at 755°F (235°C). The toxicity of burning foam vapor is not higher than that of burning wood (Rossiter 1977). The pyrolysis produces 2.7 wt % CO, 1.04 wt % HCN, and 0.35 wt % NH_3 as determined by gas chromatography/masspectroscopy at 800°C.

Toxicity of foam by inhalation, according to U.S. Federal Hazardous Substances Act, Sect. 191(f)(2)-1961 is reated nontoxic. German DIN tests have been reviewed by Baumann (1972 and 1979).

UF foam is an excellent sound absorber, Figure 7.3 (Baumann 1959), but it cannot prevent intrinsic sound transmission by pipes, studs, and walls (Rossiter 1977).

Typically, the foam consists of 99% open cells. Cell size is 5 × 10^{-1} to 10^{-3} mm (2 × 10^{-2} to 4 × 10^{-5} in). Mechanical properties were studied by Y. Sarig *et al.* (1978) and are described in Chapter 9. UF foams are mildly antibacterial. UF foam is stable for at least 10 years (Neumann 1971, Knapp 1966). General properties of UF foam were reviewed by T. Ogiwara (1971) and S. Takashima (1971) and are periodically reviewed at International Symposia of a trade association dealing with foams (Baumann 1978).

Formaldehyde odor varies greatly. Well foamed commercial resin has less than 0.16 ppm free formaldehyde, because the low mechanical requirement makes it possible to formulate the resin for minimum free formaldehyde. Thus, commercial foam exudes no discernible odor, not even during application (Meyer 1978). The U.S. NBS found the following factors responsible for high odor: (a) excessive formaldehyde in resin, (b) excessive catalyst, (c) excess foaming agent, (d) high humidity, (e) cold chemicals, (f) too high density, and (g) applications against

ISBN 0-201-04558-3

recommended practice. Thus, the main cause of odor is lack of quality control. The problem with formaldehyde is that poor quality control might not become apparent for some time, until foam degradation sets in. For example, excess catalyst yields nice looking foam. It might take months before the excess acid hydrolyses foam sufficiently for the smell to become apparent. In the past, quality control was not always recognized nor was it always enforced. In fact, during past years unskilled contractors occasionally used surplus liquid adhesive UF-resins of unknown age and composition, without adequate testing. In some cases, the foam was so poor that it collapsed after application, or crumbled within weeks. The manufacturers have now remedied this situation by establishing certification procedures for their contractors, but the damage done by a few extreme incidences will linger on for some time and it currently haunts even those in the industry who have been always careful and successful. The situation is reminescent of that encountered by polyurethane foams. They are reliably resilient, but the memory of early adverse experiences with collapsing mattresses, some 20 years ago, still negatively influences the image of the material and is responsible for the fact that their price remains lower than that of equal or even of inferior products.

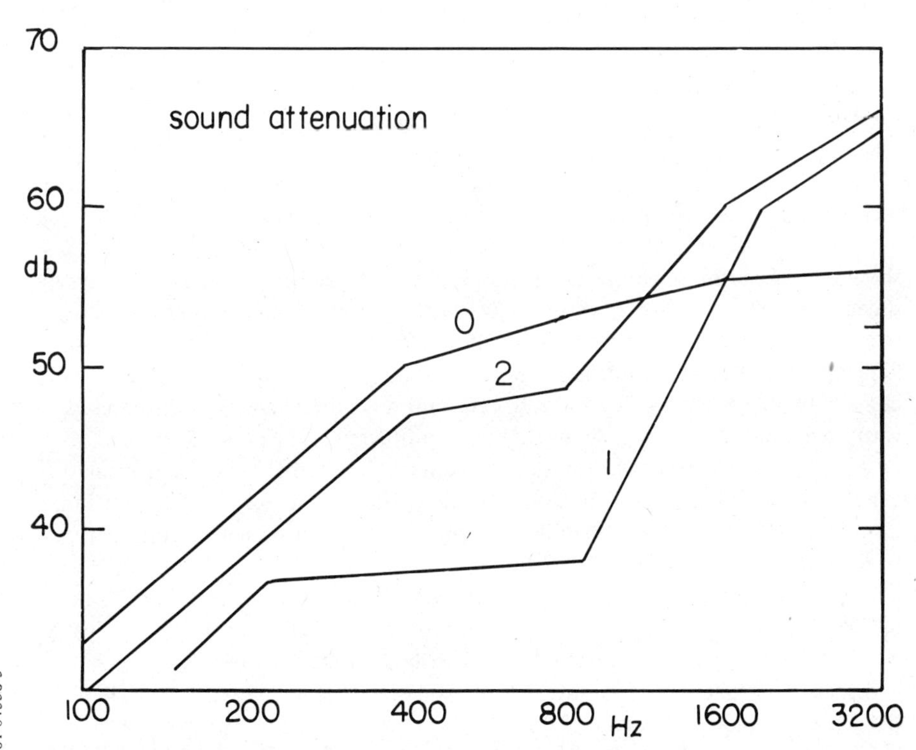

Figure 7.3. Frequency Dependence of Sound Absorption by UF Foam (after Baumann, 1959) 0) Wood; 1) Brick; 2) UFR

ISBN 0-201-04558-3

Foam Applications

UF foam is nontoxic, does not sustain combustion and is economical. These properties, together with its low density and its excellent thermal, sound, and electrical insulating properties determine its applications. Dr. Curs of I. G. Farbenindustrie introduced UF foam as insulation in the smoking room section of Zeppelin's dirigible balloons. Today it is used worldwide for insulating commercial buildings, private residences, ships, and air frames. It has reliably served in mining tunnels (USP 2,835,330) for several decades.

BUILDING AND HOME INSULATION

In the U.S. several large projects such as the Federal Reserve Building in Minneapolis, the Citicorporation Building in New York and the Host of Houston International Airport Hotel, have been insulated with UF foams. New housing developments and retrofitting of older residences are profiting from UF foam insulation, which has several advantages over cellulosic insulation or glass and asbestos mats if properly applied. In the process, it even destroys rodents (USP 3,473,252). The NBS in its recent study recommended (Rossiter 1977) that the following guidelines should be obeyed for successful insulation: the installation should be performed by a skilled and trained applicator, foam should not be used in ceilings and attics, foam should not be exposed; foaming equipment should be kept clean, resins should be labeled with an expiration date; resin and curing agent should be stored at specified storage temperatures; the foaming solutions should be applied at 60–85°F (15–30°C). For cold weather applications, the solutions should be preheated; recommended foaming pressure should be maintained; power lines of more than 220 V should be turned off in cavities to be foamed; the appearance of the foam should be monitored before application. It should be white and fluffy. Freshly sliced foam should have uniform cells, settling time should be between 20–60 seconds; wet density should be about 2.5 lb/ft^3 (40 kg/m^3); vent plugs should be kept open until excess water has escaped; a vapor barrier should be applied in cold climate to prevent blistering of paint, etc.; infrared thermography can be used to verify penetraiton of foam. Metals should be protected against corrosion.

All these guidelines concern quality control. Since the basic foam insulation concept is sound and safe, quality control is the key factor in UF foam home insulation. It is the key problem in all countries where unskilled and unsupervised labor is used in construction.

D. J. Stubblefield (1972) described innovative uses of insulation foam in housing. W. T. Meyer (1972) described general procedure for the construction industry. N. N. Grigoreva and coworker (1973) described formaldehyde release problems, and suggest installation of plastic films or panels as odor and vapor barrier. In 1974 the same group described concrete UF foam-polymer panels and bricks. J. Weizman (1974) reported application of foamed panels in Israel. A. Blaga (1974) described the properties and terminology of foams used in Canada. Baumann (1975)

ISBN 0-201-04558-3

explained the procedure for choosing suitable methods for insulating old houses, and described results from Germany. H. Piechota and H. Roehr (1975) described the principles, manufacture and applications of integral and structural foams.

The potential for use of UF foam in prefabricated building panels has not yet been commercially explored. Since burning UF is less toxic than wood, the foam can be used safely in private as well as commercial construction. Some heat insulating panels and soundproofing materials are now used in Japan (K. Oshima, Y. Ito, and S. Suzuki, USP 3,764,428) (1973).

INSULATION OF SHIPS

Baumann (1959) described the insulation of ocean going vessels as well as river barges. The foam provides heat insulation, reduces transmission of sound and vibration, but adds little to the weight. In fact, USP 3,572,275 describes how foam can be used to raise sunken vessels.

ELECTRICAL INSULATORS

Brown-Boveri Co. in FP 2,193,699 (1974) described the foaming of high dielectric strength materials. Sulfur hexafluoride is used as gas. J. J. Cleveland (1976) in USP 3,943,994 reported the use of UF foam to produce cellular ceramics. Depending on the ceramic to UF ratio, the ceramic will be more porous.

OIL ABSORBERS

UF foam is suitable for collecting oil spilled in ocean water. 12 g foam can absorb 50 ml diesel oil. S. Gruben *et al.* (1971) in Swedish P 336,470 reported a UF-resin cured with tetrapropylene benzene sulfonic acid and phosphoric acid which acts within five minutes. Some foams can collect up to 5 ml oil per g (British Industrial Plastics Ltd. Japan 76:6,269).

MEDICINAL APPLICATIONS

H. Baumann and G. Schmidt (1958) reported the use of medicated UF foam powder on surgical wounds of 20 patients and found that it was superior to cotton or other classic dressings. It appeared to enhance healing. No allergenic reactions were reported. In 1960 W. Schulze and H. Baumann reported further results on a material called Oracid. They implanted UF foam pads in rabbits and found them to be compatible with subcutaneous, muscular, and peritoneal tissues, and reported that they had encountered not a single case of allergy or rejection in six years of human use. According to Baumann (1966) Oracid absorbs 10 m/g water, four times more than silica gel, and 12 times more than talcum powder (USP 3,172,808). Oil absorption was 2.2 ml and again correspondingly greater than that of other customary wound powders.

ISBN 0-201-04558-3

AGRICULTURAL USES, PLASTOPONICS

H. Baumann (1966) summarized agricultural use and future potential of UF foams in a book entitled *Plastoponic*. This term describes the cultivation of plants by placing the roots into synthetic organic materials rather than in soil. The field is a logical extension of hydroponics, the cultivation of plants in salt solutions, first tested by John Woodward (1699) and scientifically explored by Knopf and Sachs in 1860. The original idea for plastoponics stems from F. Döhler, GP 839,944, who looked for a method for disposing of wastes which accured when foam was still marketed in form of slabs. Today foam is placed on site, and agricultural foam can be direcly injected into the soil. UF foam can be used to ameliorate the ground in greenhouses and under plastic covers (T. Geissler and P. Stacke 1972). B. Geyer (1972) described the beneficial effect of UF foams on the physico-chemical properties of soils. P. di Dio (1972) reported biological and physico-chemical modifications of soil by UF foam. B. Geyer and I. Villwock (1973) claimed that UF foam acted synergistically with amide and ammonium, but not with nitrate. The best N:F ratio for potted plants was 0.8 to 1.6. In a second paper the same authors described the microbial and chemical decomposition of foam. They used foam as slow-release nitrogen supplying 80% of nitrogen in their plants. S. Thomas (1973) tested the combined effects of UF foam, mineral fertilizer, and sprinkler irrigation on grass mixtures in sandy soil, and gave quantitative results acquired over several years. S. A. Druzhimin *et al.* (1974) used UF foam to prevent freezing of soil in permafrost regions for strip mining. The mositure content of the soil was found to influence the efficiency of foam strongly. S. Kaniszewski studied the effect of UF foam on sandy soil and observed strongly increased vegetable yields in the second and subsequent years. H. Baumann (1963) described the combination of UF foams with solar energy collectors in Algiers. An aluminum collector with 10 yards diameter can achieve 10% efficiency and delivers about one kW. He calculates that solar energy and foam could help conversion of deserts into fertile soil, USP 3,110,129. Finally UF foam can also be used to cultivate and display plants in pots and other containers (USP 3,833,696) as shown in Figure 8.3 in Chapter 8.

OTHER APPLICATIONS

L. S. Meyer (1943) proposed that foams coated with a rubber bladder could be used to increase the strength of ship and airplane fuselages, and raise sunken vessels. T. Shiota, T. Sato, and S. Okamura describe a UF-resin synthesis which yields foams which are so uniform that they can serve as microfilters. Typical pore diameters are as low as 0.03 micron. Foam can be used to exterminate rodents, as packing material and in a variety of other ways, many of which have been described by Baumann in his publications.

ISBN 0-201-04558-3

8. Resins Applications

In this chapter the most important uses of UF-resins other than adhesives or foam are reviewed. First, the textile treatment with formaldehyde, methylol, or urone-methylol and UF is described; the, UF coatings are mentioned briefly. Next are the paper resins, followed by construction materials, the binders for the metal casting industry, agricultural resins, already mentioned in Chapter 7 dealing with foams, followed by solid resin moldings and a short listing of other current or contemplated applications. Obviously, the combination of all these applications in one short chapter does injustice to many of these uses. For example, textiles and paper would well deserve a more thorough separate discussion, but in these fields several types of amino resins other than UF are important, and thus the reader should anyhow consult the comprehensive review chapters by Widmer (1965) and others whose writings have broader goals than this book.

Textile Treatment

Today, textile treatment is an elaborate art. UF-resins are an important part of the arsenal of chemicals used in this field. Nearly all linen and cotton is now designed to resist creasing, and UF-resins are one of the most powerful tools for imparting this and other desirable behaviors (Marsh 1962, Batty 1967).

The section on formaldehyde cellulose chemistry, Chapter 3, explains that cellulose fibers can be crosslinked with formaldehyde or UF-resins. Eschalier (1906) patented vapor treatment of cotton with formaldehyde to increase strength. He accordingly called it *sthenosage,* using the Greek term. Wagner and Pacsu (1952), Gruntfest and Gagliardi (1948), Steele (1955), Roff (1958), and Woo (1956) reviewed the chemistry. Roff (1959) studied the reaction using textile cotton. Steele used the infrared absorption of the –OH group to follow formaldehyde absorption, Figure 8.1. Roff determined the temperature dependence of formaldehyde uptake by viscose rayon and cotton, Figure 8.2, and studied the influence of pretreatment with acid, alkali, oxidation, and ethylamine. Table 8.1 shows the effect of the latter on cotton. Galil (1972), Hebeish (1976) and others

Beat Meyer, Urea-Formaldehyde Resins

ISBN 0-201-04558-3

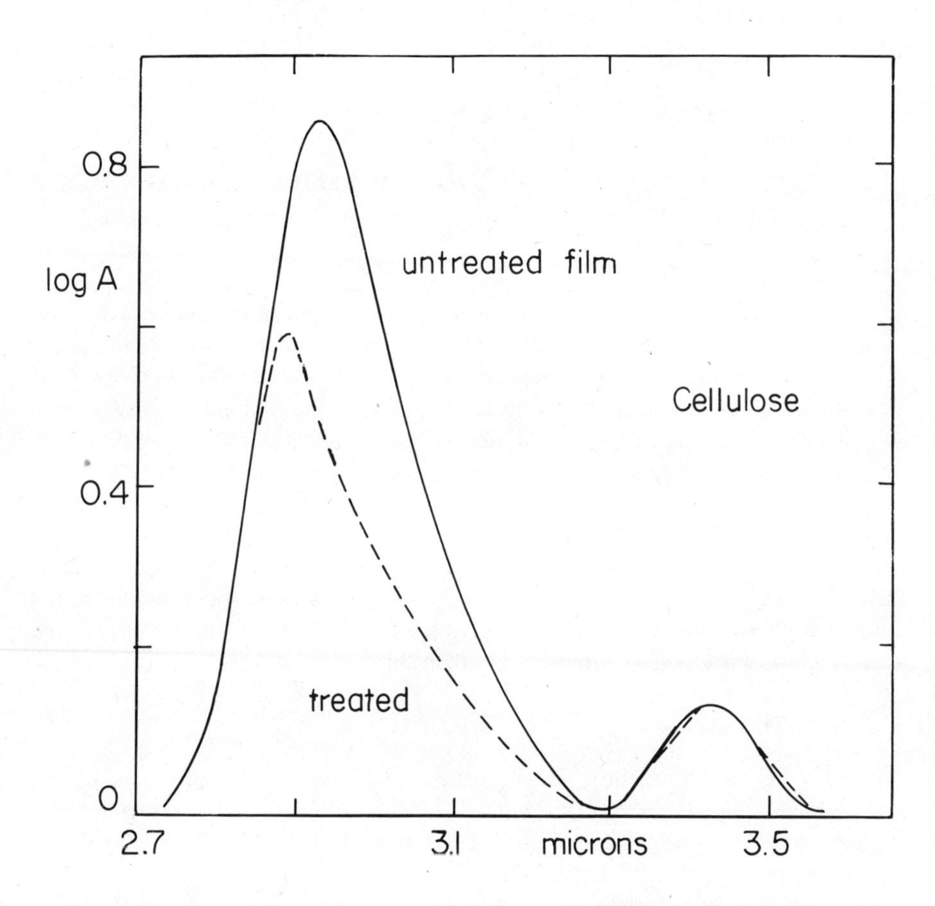

Figure 8.1. Infrared Spectrum of Cellulose Film Before and After Treatment with Formaldehyde (after Steele, 1955)

have described methods to determine UF-resin in textiles. Early treatment was based on formaldehyde alone and rendered textiles excessively hard and brittle. Tootal, Broadhurst, Lee Co. (1926) (Foulds, Marsh, and Wood, BP 291,473, 1926, and USP 1,734,516, 1929) overcame this problem and pioneered the field by use of low-molecular-weight UF. They used a high methylol resin, and correctly analyzed the problem, much to the surprise of both resin and textile researchers, by assuming that starch and high-molecular-weight UF would remain on the fiber surface and act as binder and stiffening agent, rather than by reducing creasing. Their claims were so successfully formulated, that this company became and remained a world leader in this field for many decades and their expired patents are still widely used.

ISBN 0-201-04558-3

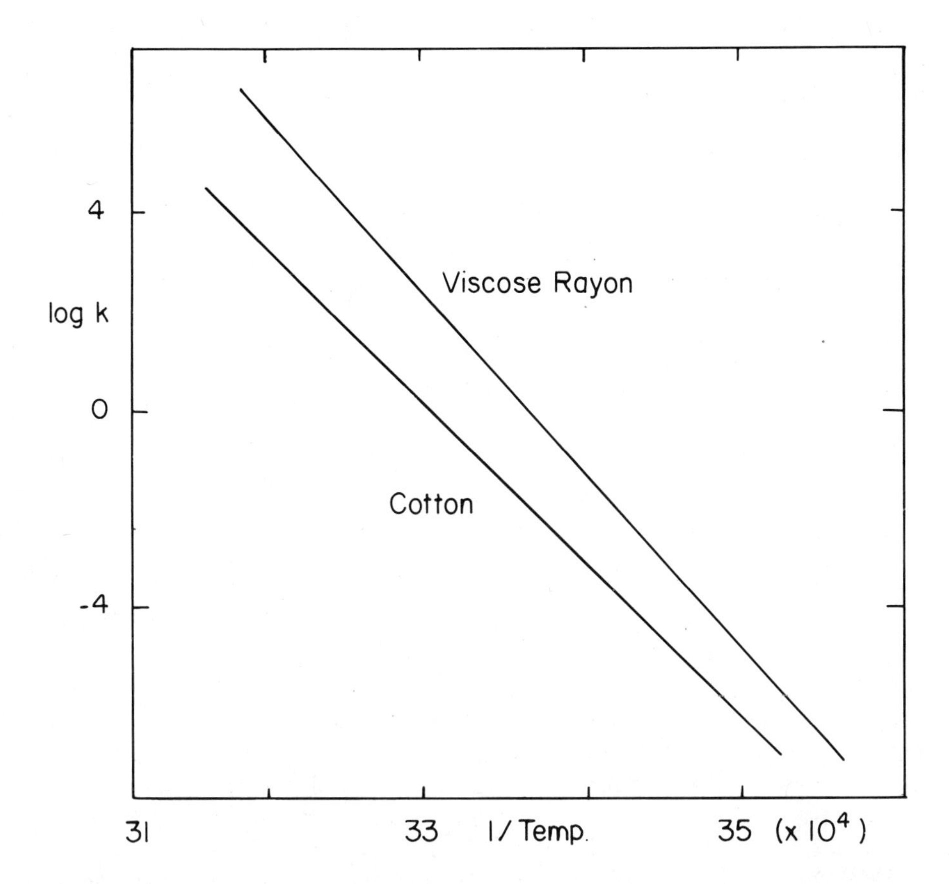

Figure 8.2. Rate of Formaldehyde Absorption as a Function of Temperature on Cellulose (Roff, 1958)

The Tootal patent prescribes only short refluxing of U:F between 1.5 and 2 at high pH. These conditions yield mainly methylol. The resin is quickly quenched to room temperature and diluted to 15% resin solid, aluminum acetate or formic acid is added, and the alkaline fabric is impregnated, squeezed, and pressed to about double its dry weight. The fabric is dried and cured for two minutes at 160°C. The product is scoured, washed, treated with a softener, and dried. Today, dimethylolurea is used as reagent and disodium monohydrogen phosphate buffer is used to reduce the curing temperature to 140°C.

This treatment also reduces shrinkage, doubles wet-strength, and increases dry-strength by 50%. In the last twenty years, resins have been used to add water-repellency, mildew proofing, chintz-effect, stiffening, and drying effects. Wash-and-

ISBN 0-201-04558-3

wear, drip-dry, non-iron, and permanent-press are based on resin treatment. For the latter, methylolurea is added. Ethylene-ureas also overcome the chloride retention, a consequence of bleaching.

Table 8.1

Formaldehyde Absorption on Cotton (after Roff, 1959)

Pretreatment	Absorption (g formaldehyde/ 100 g cellulose)	Moisture regain (%, 65% humidity)	Absorption / Regain	Water retention (%)	Absorption / Retention
None	1.98	6.36	0.31	43	0.05
Ethylamine	2.75	7.90	0.35	45	0.06
Ethylamine (boiled)	2.49	7.32	0.34	47	0.05

Wool can be protected against shrinkage by addition of 7–10% UF-resin. This invention made independently by American Cyanamid Co. (Lanaset) and CIBA (Lyofix) have made it possible to machine wash woolen knit fabrics. Shrinkage can be reduced to one half for tropical worsted, to one fourth in all-wool plaid, and to one fortieth in all-wool skirting. Melamine resins are now widely used for this purpose. They are also used for chintzing, embossing, and fixing pigments. Amino resins are widely used to fix fireproofing agents on curtains. The treatment of textiles is reviewed in a chapter by Heap (1971).

De Bruyne (1973) in BP 470,331 described a method for impregnating stretched fibrous or filamentary materials in fetted, corded, or woven form, imparting them permanent prestress. The patent, assigned to the de Havilland Aircraft Co., was the basis for the construction of the Mosquito and other warplanes which were built from plywood and cloth and bonded with amino resins (see Figure 6.3 in Chapter 6).

Stochel and O'Brien (1973) in BP 1,325,806 described an amino resin carrier for fireproofing cotton and other cellulosic textiles. Date and Fukuoka (1973) in Kokai 73:56,996 describe a similar phosphonium-resin procedure. Nogi *et al.* (1975) in Kokai 75:148,624 proposed treatment of wet-spinning solution with UF in dimethyl-sulfoxide. Dahlberg (1976) in Can. P 991,780 proposed UF with thiourea to fireproof nylon rugs, and reported that a rug with 10% UF retained more than 8.8% after 10 shampoo treatments.

Brandeis, Flory, and Feinauer (1974) in GP 2,259,680 described a UF-M copolymer with sulfite for making cellulosic fabrics chlorine-resistant. Ahmad and Hanif (1975) described the effect of dimethylolurea on crease-recovery; Mints *et al.* (1977) in USSR 550,153 described a UF finish; Getchell (1976) in USP

ISBN 0-201-04558-3

3,944,388 described how abrasion- and wear-resistance can be improved on the edges of durable-press garments; Shelton (1976) in USP 3,951,595 described a vapor phase treatment; Arima and Ichikawa, Kokai 76:55,499, proposed a method for crease proofing; Hermann (1977) in USP 4,039,496 disclosed an etherified and methylolated UF-glyoxal-M resin; Glubish (1977), USSR 584,062, described a soil-repellent finish; and Pusch (1978), Ger. Offen. 2,639,754, formulated a UF-isobutylaldehyde resin for making durable press textiles.

The chemistry of textile resin manufacture has been studied by Slonim (1978, 1979) using ^{13}C-NMR. Usually urone compounds are used for this purpose. Formaldehyde release from textiles has been measured by Galil (1972).

Coatings

Ellis (1935) proposed many applications for UF coatings. He described the excellent properties of UF-nitrocellulose coatings as floor finishes. While nitrocellulose is no longer allowed, UF coatings still make excellent floor finishes (Da Curha FP 2,154,914, 1973). Epoxy is a popular copolymer (Putlyaev 1974). UF is used with sand and carbonundum to make abrasive coatings. Minnesota Mining and Manufacturing (1974) in GP 2,425,887 described a suspension of diamond particles in UF; Supkis (1977) in USP 4,038,046 described a sandpaper; Loehmer (1977) in S. Afr. 77:01,123 disclosed a sanding belt, and Ivanov (1975) made a sandpaper.

Binder *et al.* (1971) reviewed the surface tension and adhesion properties of plastics; Borisyuk (1972) made priming coats. Fiberglass Ltd. (1972 and 1973) in BP 1,333,470, BP 1,316,911, and BP 1,293,744 described fiberglass insulating coatings bound with UF-lignosulfonate resins. Kitayama (1974), Kokai 74:36,732, formulated an adhesive coating for steel sheets; Stene (1974), Ger. Offen. 2,350,820, described adhesion of metal to a plastic film; Akutin (1974) formulated glass fiber UF films, Seeholzer (1974), Ger. Offen. 2,260,186, reported thermosetting UF coatings for wood fiberboards. Putylyaev (1975) described a mastic; Zlodovskii (1977) in USSR 108,579 made a wool-like coating; Antlfinger (1977) in USP 4,018,966 disclosed a coating suitable for aluminum sheet, containing reactive hydroxyl-vinyl chloride; Montasona (1977), USP 4,020,219 made a metal decorating coating, and Smith (1977), Ger. Offen. 2,631,038, formulated a sulfonate UF-lignite and lightener containing kerosene coating for reducing formaldehyde odor as described in Chapter 6. A review of coatings has been provided by Myers (1968).

Paper

Atsuki and Matsuoka (1930) recognized the powerful effect of UF on the strength of paper. In 1940 it was discovered that the wet-strength of sulfate pulp paper was also greatly enhanced. UF-resin is added to the pulp at pH 4.5; 2–4% dry weight of UF is sufficient to improve quality. Kraft paper is most improved,

ISBN 0-201-04558-3

and unbleached sulfite cellulose also increases strength. Rag fiber is not significantly improved. For the latter material, melamine resins work far better. UF paper dissolves in hot water. In order to improve standard UF paper, the resin must be modified. A short chronological review of some select proposals and practices follows.

Fahey (1971) tested accelerated weathering of resin-impregnated paper; Endo (1971), Japan 71:28,785, made corrugated board; Columbus (1972), USP 3,692,185, made filter paper; Schwager (1972), Ger. Offen. 2,209,109, prepared cardboard for gun training targets. Sodium silicate was added to make it brittle; Champion Paper Ltd., BP 1,270,408, in 1972 described encapsulation of opacitying particles in UF, by evaporation; Nakajima (1972) in USP 3,690,976 reported the assembly of decorative panel paper; Obetko (1972) made a starch size for corrugated board. In 1972 the U.S. Food and Drug Administration allowed the use of UF-treated paper in contact with foods, Muresam (1972) added 2–5% UF to kraft paper for manufacturing bags, towel paper, and printing paper. Fife, USP 3,758,377, applied lignosulfonate and isoprene to paper resin. Speiser (1973) in Ger. Offen, 2,233,428 reported a spray-condensation process for encapsulating dyes, fillers, adhesives, etc., in glycol-modified UF; Barzynski in 1974, GP 2,251,433, disclosed solvent-free printing inks which can be hardened by mercury light. Dirska (1974), Pol. P 73,219, described a hardener for photographic gelatin, based on UF-resin at pH 7. Hiyoshi, Japan 74:33,682, made shrinkage-free sheets from cellulose pulp with UF; Vassiliades (1974) in USP 3,886,084 made a microencapsulation system; Belova (1975) made sizing for cardboard boxes; Sander (1975), GP 2,352,190, reduced foaming of fiber solutions, by employing UF-phenolsulfonate with pulp. Kikuchi (1976), Kokai 76:67,335, made waterproof corrugated cardboard. Fiehn (1976) reviewed the brightness stability of bleached paper pulps; Vassiliades (1976), USP 3,993,831, made pressure-sensitive paper; Wojciak (1977) encapsulated metallocenes in UF; Morozova (1977), USSR 559,000 made a saturating composition for textured papers; Knabe (1977) made an emulsion layer for cellulose acetate photographic films; Taylor, USP 4,018,727, described an impregnated tube for *in situ* polymerization. Gavel, Hung P 13,033, prepared hydrophobic formulations; and Petersen (1974) described urons for paper.

Methylolureas are the most economic and simplest textile resin precursors, but melamine resins and urons make better materials. During the early 1960s ethylene urea was one of the most popular resin materials. This resin is prepared by reacting ethylenediamine with urea and formaldehyde. The first step consists of reacting ethylene diamine at 240°C to 2-imidazolidone. A similar procedure also yields propylene urea and triazone resins can be prepared.

Urone resins include a variety of cyclic compounds. Slonim (1978) has analyzed and identified some 50 such building stones in textile resins with the help of ^{13}C-NMR. Gordon, USP 3,962,166, in 1975 prepared a methylol precondensation paste with a storage life of three months.

ISBN 0-201-04558-3

Solvay Co. (1977) in Belg. P 848,093 reported a paste for making electrical insulating paper; and Menashi (1978) discussed the effect of the morphlogoy of UF pigments on paper fillers. Standard paper tests were reviewed by Leneuf (1974).

Construction Materials

There are too many possible applications to review in a short section. UF-resins can be used to prepare a wide variety of construction materials. Plywood and particleboard have been described in Chapter 6 on adhesives. These two materials are also used as forms for pouring concrete, but UF is not a suitable coating for this purpose. UF can be used to bond expanded perlite and other inorganic materials to form nonburning boards, which can be used in tall buildings. This material is prepared with the same equipment, adhesives, and methods as particleboard.

Kuehn (1973), Ger. Offen. 2,261,896, described UF as a fiber glass mat binder. Reeves (1935), USP 1,982,539, described asbestos-UF insulators, Mazur (1970) described UF-reinforced gypsum board; Spirin (1971), USSR 321,505, described a silicate-UF binder. Tobol'skii (1971) made acoustic tiles from slag cotton; Hoedt (1971), FP 2,065,238, made artificial stone using a mineral filler and baking the mixture. Heetman (1972), Swiss P 526,481, made fire-resistant insulators; Ashall (1972), GP 2,161,570, made glass fiber insulation with UF; Knatko (1972), USSR 337,358, added 1–5% UF to asphalt, Pashkov (1972) used UF to achieve well wall consolidation; Ashall (1972), BP 1,273,152, made thermal insulation mat with glass fiber; Jenks (1973), FP 2,161,187, coated various panels against weathering; Smirnov (1975), USSR 487,860, made UF-gypsum boards; Mironov (1973) built agricultural structures with UF-granite and sand mixtures; Znacho-Yavorskii (1973) described Uralite made from asbestos and colloidal quartz.

Heetman (1974), Ger. Offen. 2,227,679, assigned to Novopan-Keller Co., described a lightweight perlite panel with a density of 0.43. The panel contains mainly perlite, some mineral wool fiber, and about 20 wt % of UF-resin (with 60% solid content). The boards are made like particleboard. Katsyuba (1974) made UF-mastics; Shimokai, Japan 74:27,672, made a fiber glass sheet; Blandin (1974) in FP 2,192,552 made a refractory insulator by foaming UF-mineral wool or silica-fillered mixtures; Shearer (1974), USP 3,832,201, added UF to asphalt; Hoener (1974) in GP 2,242,548 described a fire-resistant UF board containing mineral filler. In NP 73:11,929 he described a cork-UF board similar to those made during World War II. Ogden (1974) in Ger. Offen. 2,416,270 binds UF with mineral filler, perlite, etc., in the ratio 1:1 and obtains densities of 0.144 by foaming with alkylarene sulfonates; Materiaux Reunis Co. (1974) in FP 2,219,134 applied asbestos with UF to fireproof its panels.

Laukhin (1974) strengthened friable rock in mining tunnels with UF. This use in the mining industry has been the mainstay for UF foam companies. Volotskoi (1974) stabilized roadbeds with UF; Hayashi, Kokai 74:99,321, made (medium)

ISBN 0-201-04558-3

lightweight gypsum board by bubbling air through the resin. Miller (1975) in USP 3,878,278 made a lightweight plaster board using peroxide as curing agent; Isovolta Co. (1975), Austrian P 324,923, disclosed a perlite board with a density of 0.4 and a bending strength of 40–50 kg/cm^2. Shigeta, Kokai 75:05,421, reported a gypsum board; Towati (1975), FP 1,242,344, described a synthetic stone containing sand, marble or limestone; Lafarge Co. (1976), NP 76,0068, disclosed a gypsum board; Otsuka, Kokai 76:3,355, prepared a composite board using natural cellulosic fillers. Sarkisyan (1977) described a cement which was modified with UF-polyvinyl acetate modified with sulfite liquor. Nelynlin (1977) in USSR 554,344 reported a road building material; Shirokorodyak (1978) in USSR 615,049 used UF as a binder for bitumen clay. Kara (1978) described the use of UF for strengthening the face of mines.

FIREPROOFING AND RESISTANCE

One of the outstanding properties of UF-resin is that they do not sustain burning. If UF is incinerated, the resulting gases are no more toxic than those from wood. Solid moldings resist fire and can be used for forms in iron foundaries; UF foam is permitted in skyscrapers because of its insulating properties. It works best if it is protected with a skin which maintains mechanical strength. The flammability according to ASTM E-84–76A is 25, compared to 100 for red oak. Self ignition occurs at 620°C (1150°F); flash ignition at 235°C (755°F). More details are contained in the section on physical properties of foam. If foam is blown with nitrogen, the properties are yet better. Mineral fillers give UF board a nonburning rating. Plywood and particleboard burn like wood. UF adhesive can be fireproofed by addition of ammonium phosphate buffer. The following section reviews some important references in chronological order.

Boettner (1973) explored combustion of UF as a means for disposal of wastes. Care must be taken not to burn phenol-copolymers, as these react alkaline and attack chimneys. Holmes (1973) described several treatments to fire-retard wood Okada, Kokai 74:131,259, found UF-ammonium phosphate mixture an outstanding fire protection. He applied it by baking it at 140°C for three minutes onto glass wool. Juneja (1974) combined UF-phosphates with dicyanamide. The Dead Sea Bromide Co. (1974) in Neth. P 74:06,625 described a UF-bonded wood or cellulose board containing ammonium bromide. Swire (1975), BP 1,401,960, described fire-resistant pulp moldings; Konishi, Kokai 75:22,493, described fire-resistant composite sheets; Kobori, Japan 75:33,124, used phosphate and dicyanamide to fireproof particleboard and plywood. Konishi, Kokai 75:22,493, used foamed UF-MF copolymers with inorganic filler; Morimoto (1976) studied the combustion product by IR and gas chromatography; Dixon International Ltd., Neth. P 75:11,572, described a fire-resistant sealant; Dolenko (1976) compared various polymer phosphoric compounds; Suzuki, Japan 77:36,899, reported a UF-potassium carbonate fire extinguishing agent; and Drobyazko (1978) formulated a furfural-

ISBN 0-201-04558-3

UF antiscorching paint. Methods for fireproofing particleboard are described in Chapter 6; the fireproofing of buildings by injection of UF foam is covered in Chapter 7.

Binders for Metal Casting

In ancient times, sand cores were bonded with oil, pitch, or starch. Today, sulfite liquor, water glass (hydrated sodium silicate), and formaldehyde resins are used. UF is an excellent binder. It imparts a shiny surface on the metal and can be readily broken. Core resins are a specialty product. They are available as warm sand coating resins, containing 60–70% solid, as hot or cold sand curing resins, as hot box resins to be used at 400°F, as no-bake resins containing furfuryl alcohol, as conventional baking powders or liquids, and they are matched with standard hardeners and release agents. The problem of fumes released during use has been reviewed by Kay (1974), Sato (1976), and Kolotilo (1976).

Agriculture

Urea is a high nitrogen fertilizer. UF foams and mulches are extremely useful as plant nutrients and plant conditioners. In 1967 Baumann published a book in which he presented an extensive summary of the use of UF foam in agriculture, a field which he called plastoponics. He demonstrated by field experiments that UF foams are beneficial as artificial soil for hobby and commercial gardeners, in forestry, in rehabilitating soil, in converting cold, arid or semiarid soils into fertile soils, in greenhouses, and in many dozen other applications. He experimented successfully with 65 plant types from sugar cane to asparagus. UF foam was used to grow tomatoes on Tenerife, bananas in Berlin, and grapevines on the dunes of Wangerooge Island. It enhanced the growth of grass on freeway abutments, and on coal mine refuse; it made possible orchards in northern Germany, and on roof tops, and it provided the basis for cultivating barley, oats, and wheat in schoolrooms. Figure 8.3 shows tomatoes cultivated in UF foam. Figure 8.4 shows the application of foam in road construction where it can be used to protect loose soil, and at the same time as a base for grass or other vegetation.

Malakhova (1967) described the effect of polymers on soil; Karnemaat, Ger. Offen. 2,002,873, described a N–P fertilizer which was made dust-free with formaldehyde. Batyuk (1971) described and compared the use of UF with other polymers; Takashima, Japan 71:09,210, described a slow-release N–P–K fertilizer; Karnemaat (1971), FD 2,074,766, described a slow-release N–P–K material; Toyota (1971), Japan 71:18,570, described a similar mixture; Brunnmüller (1971), Ger. Offen. 2,055,074, described a UF foam for improving soil; the foaming agent was alpha-naphthaline sulfonate; Putseiko (1972) in USSR 361,251 proposed a ligno-sulfonate-reinforced UF; Kuznetsova (1972) in USSR 333,154 disclosed an N–P–K-lignosulfonate UF, neutralized with ammonia gas. Hentschel (1972) used UF to

ISBN 0-201-04558-3

Figure 8.3. UF Culture Growth of Tomatoes (Baumann, 1965)

reduce the pathogenic activity and disinfect soil; Batyuk (1972) described the action of UF foam, and in a second paper he analyzed the effects of surfactants on soil. Kaufmann (1973) described the growth of asparagus in UF foam. Drews (1973) compared UF foam with bark, low moor, and peat; Koberg (1973) in Ger. Offen. 2,164,336 employed UF as a base for fertilizer tablets for *pinus cernbra;* Kramer (1973), USP 3,729,320, fermented chicken manure with UF foam producing N–P–K–Ca–Mg = 15:5:2:5:1. Up to 10% of the product could be added to feed of sheep, ducks, cattle, and horses. Kurshev *et al.* (1973) described the manufacture of UF-fertilizer; Batyuk (1973) used UF foam on rice fields; Davydov (1973), USSR 403,811, used UF to stabilize soil for the installation of foundations, pipelines, etc. Usmanov (1973) analyzed the effect of UF on urea; Almassy (1973)

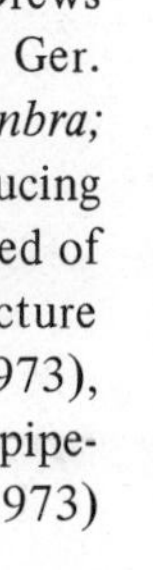
ISBN 0-201-04558-3

Figure 8.4. Road Access Treatment with UF Foam in Switzerland (Baumann, 1965)

ISBN 0-201-04558-3

described UF condensation technology for fertilizers; Nobell (1973), USP 3,759,687, described a UF fertilizer; Takashashi (1973), Japan 73:16,066, formulated a UF slow-release fertilizer with poly(methylene alkyurea). Niedermaier (1973) measured the nitrogen activity of a UF foam hygro mulch. He found an extremely low release rate and questioned the fertilizer action. Vad (1974), Ger. Offen. 2,361,085, used guanidine and dihalo-halogens to decrease surface tension of the soil-water interface. Schulz (1974), Ger. Offen. 2,238,742, described degradable plant pots made from UF with a swellable filler, such as starch. The degradation time of 1.1 m (0.04 in) wall was two growing periods. Saarbergwerke Co., FD 2,270,221 (1971), disclosed a UF-fertilizer synthesis; Kramer (1974), USP 3,800,696, described a display material for flower exhibits. Claes (1974) described the influence of UF on the microbial activity of soil. Schulz (1974), Ger. Offen. 2,238,742, described degradable moldings; Schneider (1974), Ger. Offen. 2,422,238, disclosed a low pH UF-fertilizer; Cropp, BP 378,938, described a slow-release fertilizer. Matsumara (1974) measured the vapor composition above organic soil conditioners, and measured 14.9 g/m^3 ammonia, 96 mg/m^3 formic acid, 1.8 g/m^3 formaldehyde above a freshly gelled UF-sand mixture. Davies (1975) analyzed the effect of UF on the conversion of animal wastes. Gnatowski (1976) injected UF into soil; Kakabadze (1976) studied the influence of UF on tea growth; Lomtadze (1976) used UF-sulfur slow-release fertilizer on tea; Meyer (1977) reviewed the formulation, effect, and potential of formaldehyde bonded N–P–K–S slow-release materials. Pal (1977) fed cattle a straw-UF mixture; Goertz (1977), USP 4,025,329, disclosed another slow-release UF material; Barbera (1977) measured free urea, and An Foras Taluntais (1977) made slurries from UF and manure; Wilkinson (1977) observed Kentucky blue grass; Jacenkow (1977) in Pol. P 92, 502 prepared a soil stabilizer. Gnatowski (1977 and 1978) used UF to seal soil by injection; Jackson (1977) in USP 4,055,974 prepared a degradable tablet; Moore (1977), USP 4,033,745, made a storable, liquid fertilizer; Martirosyan (1978) made a methionine fodder additive; Lubis (1978), USSR 596,561, reduced the cracking of granulated N–P–K; and Yukin (1978), USSR 627,109, prepared another slow-release composition.

Obviously, the potential of UF in agriculture has not yet been explored: UF foam can also be used as frost protection, by foaming blossoms or leaves. Finally, Siegmann (1978) used UF to formulate a slow-release herbicide, and thus demonstrated an entirely different type of use.

Waste Treatment

Golubovskaya (1971) studied the effect of UF wastewater on oxygen and found that it reaches a maximum at the inception. Golyshev (1971) used active carbon to extract UF from water; in contrast, Karnemaat (1972), USP 3,655,395, used UF and nitric acid to treat malodorous industrial and municipal wastes. Orth (1972) in S. Afr. 72:4,876 contemplated clean-up of oil spills with cellulosic fiber and UF. Graham (1973) considered secondary treatment of plywood glue wastes;

ISBN 0-201-04558-3

Kropfhammer (1973), Ger. Offen. 2,224,363, made building materials by mixing refuse and UF in the ratio 5:1. The compressed product has a bulk density of 0.34 and a compressive strength of 28 kg/cm^2. Shank, Ger. Offen. 2,154,630 (1973), patented UF sheathings for glass bottles. The sheathing increased the abrasive strength from 12 to 19 kg/cm^2. Gracheva (1973) studied the burning of UF wastes.

O'Donnell (1975), Ger. Offen. 2,523,483, used UF to convert dewatered sludge into fertilizer. Halama (1975), Czech. P 157,574, used UF foam to protect fermenting molasses; Argo (1977) in USP 4,051,070 described the use of UF on palladium-chromium catalysts to react and absorb nitric oxides by calcination. Finally, Dockal (1977) Czech. P 169,239; Polinek (1978) Czech. P 170,877; Kodo, Kokai 78:42,178; Uchida, Japan 78:34,189, and Uchida, Kokai 78:71677 all published methods for treating adhesive plant wastewaters. Gablin (1977), USP 4,056,362, finally, used UF with sodium bisulfite to bind radioactive wastes.

Solid Resin Moldings

In Ellis' book of 1935 resin moldings were the first and most extensively contemplated uses for UF. At that time, UF-resin was used extensively for making hard panels used as containers for telephones, clocks, radios, and the like. UF has not survived competitively in this field, because melamine resins excel in water resistance and wear, especially as household utensils, dishes, cups, and as furniture coatings. However, UF moldings can be decorative, translucent, and can be dyed in pastel colors.

Molding formulations are often complex and contain up to eight different types of ingredients. UF moldings contain stabilizers, fillers, colorants, mold lubricants, flow promoters, plasticizers, and catalysts or hardeners. A typical older formulation was described by Toledo Synthetic Products (1931) in FP 715,153. A detailed discussion of formulations, machinery, and applications is given in the above-mentioned book of 1935 by Ellis. All these patents have now expired and are free to use.

In recent years some special applications have been reviewed again. UF-resins have retained a firm position as electrical insulation, because of their high electrical resistance and their fire resistance. Dengina (1972) described a Russian formulation. Koromyslova (1972), USSR 338,415, described a molding mixture containing comminuted wood; Gilfrich and Wallhaeusser (1972) described UF products suitable for use at 100 to 200°C for several hundred homes; Greenberg (1973) in USP 3,734,985 described a glass-fiber-reinforced cellular thermoplastic for injection molding. The resin also contains poly(glyceroltriacrylate). Feriday (1973) in USP 3,957,700 uses alpha-cellulose filler and melamine copolymer to make a molding with less than 0.88% shrinkage, a dielectric strength of 1450 V/mm, a bulk resistivity in excess of 10^{13} ohms per centimeter and a flexural strength of 880 kg/cm^2. Lakshminarayanan (1974) described an electric insulator; Ueda, Kokai 74:73,447, made a formulation capable of tolerating embedded metal without cracking. In Kokai 74:73,450 he added polyester to improve impact resistance.

ISBN 0-201-04558-3

Idumura, Kokai 74:112,993, added isophthalic acid and propylene glycol polymers. Bokareva (1975) in USP 476,294 employed glass fibers for injection molding compositions; ICI, Ltd. (1975) in Bel. P 818,541 described a foamy panel that was lightweight and crack resistant. In FP 2,240,254 it disclosed a similar composition. Vinogradov (1975) reviewed the problem of residual stress; Borden Ltd. (1976) in BP 1,421,994 described a traditional mixture; Bonnington (1976) in S. African 76:197 described a fire-resistant insulator; Feriday (1977) in BP 1,483,458 produced a resin for extrusion or injection molding. Sokolova (1977), USSR 580,939, added ligneous substances, Ulbricht (1978), Ger. Offen. 2,639,470 reported an insulation molding, Kuznetsova (1978), USSR 590,318, made stable sulfide emulsions; and Gupta (1978) employed eucalyptus pulp as a high brightness filler.

There is a continuous transition from solid urea moldings to UF-bonded wood moldings, described in Chapter 6. Originally, solid UF-resins had to be filled with cellulose to absorb the water formed during condensation to prevent cracking. The ratio of UF to filler can be adjusted to yield the desired density, hardness, and surface finish. In ceiling panels, Figure 8.5, filler makes a substantial fraction of the product. In the preparation of prefabricated stairs, Figure 8.6, or other applications, the resin makes a larger fraction than the filler.

Figure 8.5. Ceiling Panel Moldings Made from UF Composites (courtesy of CIBA-Geigy's Aerolite Subdivision)

ISBN 0-201-04558-3

Figure 8.6. Amino-Molded Staircase (courtesy Aerolite Division, CIBA-Geigy)

ISBN 0-201-04558-3

Filled UF-resin moldings have still an almost untapped potential as sandwich panels in prefabricated housing. In such panels, the composite can be formulated, layered, foamed, or assembled in any other suitable way.

Other Applications

According to Damier (1978) UF can be used as floculating agent; Aleksandrovich (1978) used it for the flotation of potassium ores; Speyer (1977), USP 4,022,727, formulated a UF hot-melt by incorporating fatty acids. Gibson (1977), USP 4,064,090, produced an electrophoretic crosslinking agent. Wismer (1974), Ger. Offen. 2,415,762, made a metal delustering composition. Seldon (1972), Ger. Offen. 2,157,573, used UF to coat gold and tantalum films on iron to protect the latter against corrosion; Lvovsky (1972), Czech. P 144,662 showed that UF can be used to make wood surfaces sufficiently conductive to electroplate it with metals; Sautin (1972) reported that UF was not sufficiently stable to resist the Siberian climate in footwear. Applications for foamed resins are discussed in Chapter 7.

ISBN 0-201-04558-3

9. Product Properties

This chapter consists of two parts. The first deals with the physical properties and performance of UF-bonded products. The second part deals with formaldehyde release. The latter is given substantial space, even though it represents a transitory phenomenon. There is little doubt that the odor problem can be solved, but the question is whether in the near future solutions can be found which are sufficiently economical to allow particleboard and foam to remain competitive in all markets. Much of the potential use of UF depends on how this problem will be solved.

Performance and Standards

MOLDING

Solid UF-slabs and moldings are widely used as electrical insulators, because their dielectric strength is 1,500 V/mm, and the electrical resistance is in excess of 10^{13} ohms per centimeter, and their flexual strength is 880 kg/cm^2 (Feriday, USP 3,957,700, 1973). However, insulators are usually prepared with fillers which modify the performance of pure resin.

Pure UF-resin slabs were made by Pollak, in the hope of synthesizing an organic glass. The density of such bulk resins is about 1.44 g/cm^3, as compared to 1.77 for ivory, or 2.05 for quartz. The hardness is about 2.95 on the Mohl scale, i.e., comparable to mother of pearl, which has a value of 3.1. The hardness proved to be insufficient for glass. Another problem was presented by slow surface hydrolysis, which amounted to about 0.2% per day at 20°C. This causes surfaces to become opaque. The thermal conductivity is 0.00184 compared to 0.00163 for glass. The refractive index lies between 1.5 and 1.9, exactly like window glass or quartz. The optical absorption, Figure 9.1, is excellent. UF transmits the ultraviolet light better than standard window glass. This fact has been used in greenhouse roofs, verandas, in tuberculosis hospitals in the mountains, and other applications. Most moldings are filled with cellulose, asbestos, and other materials in amounts ranging from 1 to 90 wt %. Furthermore, UF-resins are often chemically modified. Thus, most molded product properties can be adjusted over a wide range.

Beat Meyer, Urea-Formaldehyde Resins

ISBN 0-201-04558-3

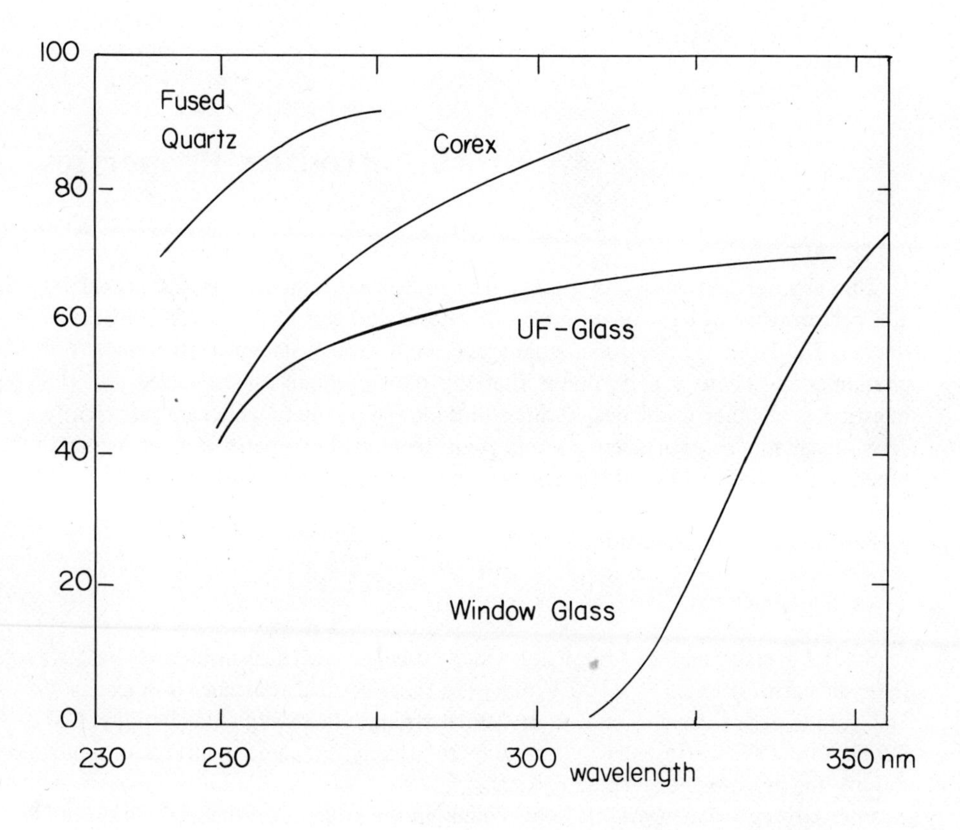

Figure 9.1. UV Transmission of UF-Resin Molding and Window Glass (Crist, 1928)

UF-resins decompose upon extended heating above 200°C. The vapor above UF-resins can contain carbon monoxide. In many ways, it is quite similar to that of wood. Further properties are described in Chapter 9.

FOAMS

The physical properties of foam depend on the formulation density and water content. The density of typical dry insulating foam, according to ASTM D-1622-63 (1975) is 0.6 to 0.9 lb/ft^3 (10–14 kg/m^3). The wet density of fresh foam is 2.5 lb/ft^3 (40 kg/m^3). Shrinkage is 2.98–4% depending on formulation. The thermal conductivity of foam, according to ASTM C-177-72 is K = 0.206 BTU·inch/hr·ft^2 (R = 4.85) at 34°F, and K = 0.228 (R = 4.39) at 80°F (0.035 W/m·K at 24°C). The thermal conductivity changes 5% over a density range of 10–30 kg/m^3. It is a

ISBN 0-201-04558-3

function of thickness (Latinerer 1973). If freon is used as blowing agent, conductivity can be further reduced (Dementev 1975). Sulfur hexafluoride yields highly insulating foams.

The decomposition temperature is above 392°F (200°C). The flammability, ASTM E-84-76A is less than 25 (versus 100 for red oak). Smoke developed according to test UL 723, is 50–80, compared with 100 for red oak. The heat of combustion, measured with the Emerson oxygen bomb calorimeter is 7830 BTU/lb (14 MJ/kg) as compared to 11,000 BTU/lb for polymethane and 17,000 BTU/lb for polystyrene. Smoke density, ASTM D-2843-70 is 1.15%; self ignition occurs at 1150°F (620°C), flash ignition at 755°F (235°C). The toxicity of burning foam vapor is comparable to that of burning wood (Rossiter 1977). The pyrolysis produces 2.7 wt % CO, 1.04 wt % HCN, and 0.35 wt % NH_3 as determined by gas chromatography/mass-spectroscopy at 800°C. Toxicity of foam by inhalation is rated nontoxic, according to U.S. Federal Hazardous Substances Act, Sect. 191 (f) (2) – 1961. German DIN tests have been reported by Baumann (1972).

Sound absorption is shown in Figure 7.3 in Chapter 7. The electrical resistance, ASTM D-257, is 3×10^{12} ohms. Compression strength is 25 lb/in^2 (170 kPa), by ASTM D-1621 it is only 6 lb/in^2 (35 kPa). The flexual strength is 20 lb/in^2 (140 MPa). The flexual modulus is 1600 lb/in^2 (IIMPA). The mechanical strength is changed by fillers, such as wood shavings (Juenger and Weissenfels, USP 3,830,894). Water vapor transmission, ASTM C-355 yields a permeance of 35–40 perms (20–23 $\times 10^{-10}$ kg/Pa·s·m^2).

UF foams are mildly antibacterial and do not serve as nourishment for vermin. Durability of the foam has not been formally established, but several reports indicate no deterioration of UF foam in use in commercial buildings after more than ten years (Neumann 1971 and Knapp 1966).

Physical performance standards have been suggested by the German DIN system (DIN 18 164 and DIN 18 159; Lühr 1979), three Swedish building specifications, the British Agreement Board, the Dutch Bouwcentrum (Wulkan 1979), and the U.S. NBS (Rossiter 1977), the major manufacturers and the many private testing laboratory reports. Baumann (1957, 1959, 1972, and 1979) has carefully reported and summarized German tests and DIN standards. Performance standards exist in Holland, a technical Agreement Certificate in England, and guidelines in Sweden. Standards are being drafted in Germany and Canada.

In the U.S., UF foam is currently approved by New York City, MEA 37-77M, 1977; the City of Los Angeles, Research Report 24,176, April 23, 1978; Metropolitan Dade County, Florida, Acceptance 77-261.1; the U.S. Department of Housing and Urban Development, Materials Bulletin 74 of October 1977; the Southern Building Code Congress; and the International Conference of Building Officials, Report 3,374, December 1977. Formaldehyde odor is normally less than 0.16 ppm; see Chapter 7.

ISBN 0-201-04558-3

UF-BONDED PRODUCTS

Most UF adhesives are used for bonding wood. The most important product categories are laminates, boards, and plywood. The performance and standards are well established for all these products. U.S. test standards are summarized in Table 9.1.

Table 9.1

U.S. Testing Standards for Wood Adhesives[a]

Standard Methods of Test for MOISTURE CONTENT OF WOOD D 2016-74

Standard Method of Testing VENEER, PLYWOOD, AND OTHER GLUED VENEER CONSTRUCTIONS D805-72

This designation covers the determination of the following properties: "Compression, Tension, Toughness, Swelling and Recovery of Compressed Wood Products Due to Moisture Absorption, Moisture Adsorption of Compressed Wood Products, Moisture Content, and Specific Gravity" (in veneer, plywood, and other glued veneer constructions).

Standard Method of Test for TENSILE PROPERTIES OF ADHESIVE BONDS D 897-72

This designation covers the determination of the comparative tensile properties of the adhesive bonds when tested on standard shape specimens and under defined conditions of pretreatment, temperature, and testing machine speed.

Standard Method of Test for APPLIED WEIGHT PER UNIT AREA OF LIQUID ADHESIVE D 899-51

This method covers the determination of the quantity of liquid adhesive applied in a spreading or coating operation.

Standard Method of Test for STRENGTH PROPERTIES OF ADHESIVE BONDS IN SHEAR BY COMPRESSION LOADING D 905-49 (Reapproved 1970)

This method covers the determination of the comparative shear strengths of adhesive, bonds used for bonding wood and other similar materials, when tested on a standard specimen under specified conditions of preparation, conditioning, and loading in compression.

Standard Method of Test for STRENGTH PROPERTIES OF ADHESIVES IN PLYWOOD TYPE CONSTRUCTION IN SHEAR BY TENSION LOADING D 906-64

This method covers the determination of the comparative shear strengths of adhesives in plywood-type construction, when tested on a standard specimen and under specified conditions of preparation, conditioning, and testing.

Standard Definitions of Terms Relating to ADHESIVES D 907-74

This is a listing of definitions of terms relating to adhesives.

ISBN 0-201-04558-3

Table 9.1 (Continued)

Standard Method of Test for IMPACT STRENGTH OF ADHESIVE BONDS D 950-72

This method covers the determination of the comparative impact value of adhesive bonds in shear, when tested on standard specimens under specified conditions of preparation, conditioning, and testing.

Standard Methods of EVALUATING THE PROPERTIES OF WOOD-BASE FIBER AND PARTICLE PANEL MATERIALS D 1037-72a

This designation covers all of the major tests that need to be used in evaluating the properties of particleboard.

Standard Definitions of Terms Relating to VENEER AND PLYWOOD D 1038-52 (Reapproved 1970)

This is a listing of definitions of terms relating to veneer and plywood.

COMMERCIAL STANDARDS:

MAT-FORMED WOOD PARTICLEBOARD Commercial Standard CS 236-66

A recorded voluntary standard of the trade published by the U.S. Department of Commerce (available from the National Particleboard Association).

Standard for Particleboard for Mobile Home Decking, National Particleboard Association NPS 1-73 (available from the National Particleboard Association).

Standard for Particleboard Decking for Factory-Built Housing, National Particleboard Association NPA 2-72 (available from the National Particleboard Association).

U.S. PRODUCT STANDARD PS 1-74 FOR CONSTRUCTION & INDUSTRIAL PLYWOOD Product Standards Section, National Bureau of Standards (available from the National Particleboard Association).

This is the voluntary commercial standard for the plywood industry and contains requirements for producing, marketing, and specifying plywood for construction and industrial uses.

[a]References from 1974 ANNUAL BOOK OF ASTM STANDARDS, Part 22, Wood; Adhesives. American Society for Testing and Materials, Easton, Md., 1974.

Board

Table 9.2 summarizes physical data for softboard, particleboard, and hardboard. All properties are interrelated, often in a complex manner (Maloney 1977, Deppe 1977, Moslerin 1974). Much valuable data is available from FESYP, and many manufacturers. The test performance is described in ASTM Standard D-1037-64. Other countries have similar standards. The German DIN standards are summarized

ISBN 0-201-04558-3

Table 9.2

Classification and Properties of Wood Composition Boads[a]
Based on Data Compiled by Forest Products Laboratory – United States Department of Agriculture

Class of Wood Composition Board	Thickness Generally Available Inches	Specific Gravity	Density Pounds per Cubic Foot	Thermal Conductivity[b]	Modulus of Rupture Pounds per Sq. In.	Modulus of Elasticity in Bending 1000 Pounds per Sq. In.	Tensile Strength		Compression Strength Parallel to Surface Pounds per Sq. In.	Water Absorption 24 hr. Immersion		Maximum[c] Linear Expansion in Percent
							Parallel to Surface Pounds per Sq. In.	Perpendicular to Surface Pounds per Sq. In.		Percent by Weight	Percent by Volume	
Softboard (Insulation)												
Semi-rigid	1/2–1 1/2	0.02–0.15	1.5–9	0.24–0.27	–	–	–	–	–	–	–	–
Rigid	3/8–1[d]	0.15–0.40	9–25	0.27–0.40	200–800	25–125	200–500	10–25	–	–	1–10	0.50
Intermediate-density	3/16–1/2	0.40–0.80	25–50	0.40–0.80	400–4000	90–700	800–2000	–	500–3400	–	–	1.30
Shaving Board[e]	1/4–1 1/2	0.40–1.10	25–60	0.04–0.06[f]	400–8200	75–500	400–4100	55–175	1400–2800	12–88	9–37[g]	0.40
Particle Board												
Low-density	1/4–3/4[h]	0.40–0.80	25–50	0.40–0.80	400–4000	300–700	100–1350	90–400	–	5–50	–	0.40[i]
High-density	1/10–3/8	0.80–1.05	50–70	–	3000–7500	400–1000	300–2500	275–400	3500–4000	15–40	–	0.85
Hardboard (Fibrous)												
Untreated	1/10–5/16	0.88–1.04	50–65	0.80–1.40	3000–7000	400–800	1000–3000	–	1800–6000	3–18	–	0.60
Treated	1/10–5/16	0.95–1.15	60–70	1.50	7500–10,000	800–1000	4000–5500	–	4200–5300	3–13	–	0.40
Special Densified	1/4–2	1.35–1.45	85–89	1.85	10,000–12,500	1250	7700	500	26,500	0.3–1.2	–	–

[a] Values given in table are approximate since they vary among different products depending on materials and manufacturing processes used.
[b] Btu of heat passing through one square foot of one-inch thick material in one hour for each degree Fahrenheit temperature differences between surfaces of material.
[c] Expansion resulting from change in moisture content, from equilibrium at 50 per cent relative humidity to equilibrium at 97 per cent relative humidity.
[d] Specimens up to 3 inches thick are made by the lamination method.
[e] Based primarily on shaving board manufactured in Germany.
[f] Kilo Calories per meter in one hour for each degree Centigrade
[g] Thickness.
[h] Plants using the extrusion process are capable of producing thicker boards.
[i] Does not include extruded board, since it does not have satisfactory strength until a veneer or similar faces are applied.

ISBN 0-201-04558-3

in DIN Vol. 60, 1975. The most exacting standard is currently the Japanese Industrial Standard JIS-A-5908–1977, which limits formaldehyde release to 5 ppm. Japan distinguishes three board classes; the German standards distinguish three types of boards. UF-resins are only allowed in V-20 boards. It is anticipated that four subgrades will be distinguished: (a) those with less than 10 mg formaldehyde per 100 g, as measured by the FESYP perforator test, (b) less than 40 mg, (c) less than 70 mg, and (d) more than 70 mg. The first will be generally admitted, the second will be generally admitted if coated on both faces, the third, only if the edges are sealed, and the last might be excluded altogether.

The moisture uptake, swelling, water diffusion, and similar behavior is discussed in Chapters 6 and 9. Fire resistance is described in Chapter 11. The thermal conductivity of particleboard varies from 0.07 kcal/mh•C at a density of 0.4 to 0.12 kcal/mh•C at a density of 0.7. In comparison, solid oak has a value of 0.3 and cork one of 0.06. Sound alternation is about 0.02 for particleboard, and 0.01 for oak.

Plywood

The physical properties of plywood depend strongly on the wood species of the veneer. Table 9.3 shows the properties of 25 species. Grading of veneer is summarized in Table 9.4. The application and uses of various grade will be shown in

Table 9.3

The Strength Properties of 3/16 in. 3-Ply Woods (after Wood, 1963)

				Static bending					
Name	Species	Weight per 100 sq. ft. lb.	Angle of face grain	Modulus of rupture lb/sq. in.	Modulus of elasticity 1000 lb/sq. in.	Maximum Crushing strength lb/sq. in.	Maximum Tensile strength lb/sq. in.	Panel Impact strength ins.	Panel Shear strength lb/sq. in.
Afara	Terminalia superba	59.3	0	11910	1744	4960	8280	6	1150
			90	2660	127	2830	4800		
Baromalli	Catostemma commune	80.5	0	16240	2959	5380	13130	8½	1300
			90	3390	150	3730	8600		
Berlinia	Berlinia species	72.7	0	15640	1984	6610	11550	9	1730
			90	3250	140	3820	5830		
Canadian Birch	Betula lutea	69.9	0	14380	2036	7360	16080	10½	2850
			90	4460	191	3300	11120		
Red Birch	Bursera simaruba	44.6	0	8320	1074	4030	6360	–	–
			90	1830	98	1890	3880		
Binuang	Octomeles sumatrana	40.7	0	7940	1268	3350	6830	4	730
			90	1540	61	1400	3850		
Daniellia	Daniellia species	40.3	0	6620	1063	2680	5000	3	896
			90	1480	60	1410	2810		
Danta	Cistanthera papaverifera	72.2	0	14200	1902	5940	10640	8	1480
			90	2720	98	3110	6430		
Gaboon	Aucoumea klaineana	45.7	0	8910	1310	4170	7010	5	1530
			90	2220	106	2600	4520		
Greenheart	Ocotea rodiaei	98.0	0	24710	3397	10010	13330	–	1550
			90	4870	187	5790	8720		
African Mahogany	Khaya species	55.1	0	9420	1381	4650	7560	6½	1425
			90	2400	99	2520	4480		

ISBN 0-201-04558-3

Table 9.3 (Continued)

Kokrodua	Aformosia elata	73.5	0	16930	2170	7390	11400	9	1820
			90	3960	163	4150	8300		
Kurokai	Protium species	68.6	0	16300	2309	6830	13080	10	1850
			90	3380	164	3630	7230		
Mabo	Sterculia pruriens	62.1	0	11760	2228	4620	11610	9½	970
			90	2590	120	2080	7090		
Makore	Mimusops heckelii	58.2	0	15080	1930	5790	7920	4	1440
			90	3200	150	3240	5340		
Mora	Mora excelsa	95.2	0	23310	3378	10450	15130	11	2250
			90	5760	216	5340	10020		
Nyankom	Tarrietia utilis	67.4	0	13570	1825	5900	8480	7½	1790
			90	3000	142	3430	5920		
Chilean Pine	Araucaria araucana	54.2	0	10750	1899	4450	8860	5½	1410
			90	2510	104	2360	5590		
Poon	Calophyllum tomentosum	54.4	0	10930	1592	4470	8300	5½	1240
			90	2350	105	2350	5690		
Pterygota	Pterygota kamarunensis	65.9	0	16230	2446	6450	12410	10	1580
			90	3430	148	3610	7520		
Ramin	Gonystylus species	65.5	0	16660	2502	7760	12580	7½	1640
			90	2910	160	4000	6730		
Red Seraya	Shorea species	46.1	0	9100	1605	2990	7900	6½	1190
			90	2290	90	1470	4740		
White Seraya	Parashorea malaanonan	61.7	0	12480	2223	5280	11730	9	1230
			90	2870	116	2240	7280		
Brown Sterculia	Sterculia rhinopetala	84.8	0	17220	2243	7390	13310	10½	1660
			90	3280	126	3290	8550		
Yellow Sterculia	Sterculia oblonga	79.4	0	17640	2488	6450	11980	11	2050
			90	3690	158	3620	8320		

Chapter 11 in Figure 11.6. Grades C and D are largely used in the U.S. The adhesion requirements are shown in Table 9.5. Exterior glues are phenols. UF is used only in interior grades. However, interior grades are frequently used for exterior applications. U.S. standards are summarized in PS 1-74. These standards are voluntary standards established by joint committees of the industry and the Office of Engineering Standard Services of the U.S. National Bureau of Standards. The JIS-K-6801 was established by the Japanese Standards Association by similar procedure.

Information about Japanese standards are available from JSA, 1-24 Akasaka 4, Minato-ku, Tokyo 107, Japan. German standards are published by Fachnormenausschuss Holz (FNHOLZ) DNA, 5 Köln, Kamekestrasse 8, Federal Republic of Germany; European particleboard standards are available from FESYP, Wilhemstrasse 25, D-6300 Giessen, Federal Republic of Germany, and information on almost all standards of all countries can be obtained from the American National Standards Institute, 1430 Broadway, New York, NY 10018.

ISBN 0-201-04558-3

Table 9.4

Grade Description of Plywood Veneers

1. Grade N Veneer (intended for natural finish):

Shall be smoothly cut 100 percent heartwood or 100 percent sapwood, free from knots, knotholes, pitch pockets, open splits, other open defects, and stain; of not more than two pieces in 48 inch widths; not more than three pieces in wider panels; and well matched for color and grain.

Synthetic fillers may be used to fill small cracks or checks not more than 1/32 inch wide.

Patches shall be limited to three "router" patches not exceeding 1 inch in width and 3 1/2 inches in length. There shall be no overlapping. Shims shall not exceed 3/16 inch in width or 12 inches in length and shall occur only at the ends of the panel.

2. Grade A Veneer (suitable for painting):

Shall be firm, smoothly cut, and free of knots, pitch pockets, open splits, and other open defects; and well jointed when of more than one piece.

Synthetic fillers may be used to fill in Exterior type panels small cracks or checks not more than 1/32 inch wide, etc.

Patches shall not be more than 2 1/4 inches in width singly. Synthetic repairs shall not exceed 2 1/4 inches in width. Shims are permitted, except that, they shall not be used over or around patches or as multiple repairs.

3. Grade B Veneer:

Shall be solid and free from open defects and broken grain, except as permitted below. Slightly rough grain is permitted. Minor sanding and patching defects, including sander skips shall not exceed 5 percent of panel area.

Synthetic fillers may be used to fill in Exterior type panels small splits or openings up to 1/16 inch wide if not exceeding 2 inches in length; etc.

Repairs shall be wood or synthetic patching material, neatly made. Wood veneer repairs shall be die cut. Wood panel repairs shall be "router" or "sled" type. Wood repairs shall not exceed 3 in. in width where occurring in multiple repairs, or 4 in. in width where occurring singly. Synthetic veneer repairs shall not exceed 4 in. in width. Shims are permitted.

4. Grade C Veneer:

Sanding defects that will not impair the strength or serviceability of the panel are permitted.

Repairs permitted as for Grade B Veneer.

ISBN 0-201-04558-3

Table 9.4 (Continued)

4.1. C Plugged Veneer

May contain knotholes, worm and borer holes, and other open defects not larger than 1/4 inch by 1/2 inch.

5. Grade D Veneer

Except as otherwise required herein, any number of plugs, patches, shims, worm or borer holes, sanding defects, and other characteristics shall be permitted, provided they do not seriously impair the strength or serviceability of the panels.

Splits measured at a point 8 inches from the end of the panel shall not exceed 1 in. in width. Splits on panel faces and backs shall not exceed 1/4 in. where located within 1 in. of parallel panel edge. Voids due to missing wood on panel backs not otherwise specified above shall not exceed the maximum width of knotholes permitted in the grade and the length of such voids shall not exceed 6 inches.

Table 9.5

Adhesive Bond Requirements for a Panel or a Lot
(American Plywood Association)

General

Any adhesive or bonding system that causes degradation of the wood or latent failure of bond shall not be used.

3.7.1. Interior type bonded with interior glue–A panel shall be considered as meeting the requirements of the Standard if three or more of the five test specimens pass when tested in accordance with 4.3. Panels from lots shall be evaluated for conformence with the Standard in the following manner:

Underlayment, C•D Plugged, and C•D:

A panel shall be classed as failing if more than two of the five test specimens fail. The material represented by the sampling shall be considered as meeting the requirements of this Standard if 90 percent or more of the panels pass the test described in 4.3.

All Other Grades:

A panel shall be classed as failing if more than two of the five test specimens fail. The material represented by the sampling shall be considered as meeting the requirements of this Standard if 85 percent or more of the panels pass, when tested in accordance with 4.3.

3.7.1.1. Mold resistance Underlayment, C•D Plugged and C•D–These grades shall be made with an adhesive possessing a mold resistance equivalent to that created by adding, to plain protein glue, five pounds of pentachlorophenol or its sodium salt per 100 pounds of dry glue base. The equivalency shall be established by testing and evaluation in accordance with *Mold Testing Procedures and Approval Requirements*[3] published by the American Plywood Association, April 9, 1964. This procedure is specifically designed for adhesive qualification approval and is not applicable to inspection and testing, as covered in section 4.

3.7.1.2. Resistance to elevated temperature Underlayment, C•D Plugged and C•D–These grades shall be made with an adhesive possessing resistance to temperatures up to

ISBN 0-201-04558-3

Table 9.5 (Continued)

160°F at least equal to that of plain protein glue. Urea resin glue shall not be used in these grades unless evidence is submitted indicating performance equivalent to that of plain protein glues.

3.7.2. Interior type bonded with intermediate glue–A panel of Interior plywood bonded with intermediate glue shall be considered as meeting the requirements of this Standard if the test specimens taken from a panel average 45 percent wood failure or more when tested in accordance with 4.4.

Lots represented by test panels shall be considered as meeting the requirements of this Standard if all of the following minimum requirements are met:

1. The average wood failure of all test specimens, regardless of the number of panels tested, shall be not less than 45 percent.
2. When more than one panel is tested, at least 90 percent of the panels represented by the test pieces shall have 30 percent wood failure or better.

Specimens cut through localized defects permitted in the grade shall be discarded. Test specimens showing delamination in excess of 1/8 inch deep and 1 inch long shall be rated as 0.0 percent wood failure.

3.7.2.1. Intermediate glue heat durability–Requirements shall be the same as for exterior glue. See 3.7.4.1.

3.7.2.2. Bacteria-mold resistance–Adhesives, in order to qualify as intermediate glue, shall meet the *Bacteria Test* requirements as published by the American Plywood Association, August 13, 1968. This procedure is specifically designed for adhesive qualification and is not applicable to inspection and testing, as covered in section 4.

3.7.3. Interior type bonded with exterior glue–A panel of Interior plywood bonded with exterior glue shall be considered as meeting the requirements of this Standard if the test specimens taken from a panel average 80 percent wood failure or greater when tested in accordance with 4.5.

Lots represented by test panels shall be considered as meeting the requirements of this Standard if all of the following minimum requirements are met:

1. The average wood failure of all test specimens, regardless of the number of panels tested, shall be not less than 80 percent.
2. When more than one panel is tested:
   a. at least 90 percent of the panels represented by the test pieces shall have 60 percent wood failure or better.
   b. at least 95 percent of the panels represented by the test pieces shall have 30 percent wood failure or better.

The above requirements are applicable separately and independently to the results obtained from the vacuum-pressure test and the boiling test as given in section 4. Specimens cut through localized defects permitted in the grade shall be discarded. Test specimens showing delamination in excess of 1/8 inch deep and 1 inch long shall be rated as 0.0 percent failure.

3.7.3.1. Interior type bonded with exterior glue heat durability–Panels shall meet the heat durability requirements for exterior glue (see 3.7.4.1.).

3.7.4. Exterior type–When tested in accordance with 4.5, Exterior type plywood shall be considered as meeting the requirements of this Standard if the test specimens taken from a panel average 85 percent wood failure or greater.

Lots represented by test panels shall be considered as meeting the requirements of this Standard if all of the following minimum requirements are met:

1. The average wood failure of all test specimens, regardless of the number of panels tested, shall be not less than 85 percent.

ISBN 0-201-04558-3

Table 9.5 (Continued)

2. When more than one panel is tested:
   a. at least 75 percent of the panels represented by the test pieces shall have 80 percent wood failure or better.
   b. at least 90 percent of the panels represented by the test pieces shall have 60 percent wood failure or better.
   c. at least 95 percent of the panels represented by the test pieces shall have 30 percent wood failure or better.

The above requirements are applicable separately and independently to the results obtained from the vacuum-pressure test and the boiling test as given in section 4. Specimens cut through localized defects permitted in the grade shall be discarded. Test specimens showing delamination in excess of 1/8 inch deep and 1 inch long shall be rated as 0.0 percent wood failure.

3.7.4.1. Exterior glue heat durability–Exterior type plywood shall be considered as meeting the requirements of the Standard if there is no delamination in any specimen, except when occurring at a localized defect permitted in the grade, when tested in accordance with 4.5.4. When testing overlaid plywood, blisters or bubbles in the surface caused by combustion shall not be considered delamination.

3.7.4.2. Overlaid plywood–The board between veneers of overlaid plywood as well as the bond between the overlay and the base panel shall meet the wood failure requirements described above for exterior. In evaluating specimens for separation of the resin-treated face from the plywood, fiber failure shall be considered the same as wood failure.

Formaldehyde Release

Formaldehyde release is a latent, intrinsic property of UF-resins. This was recognized early, and remedies were summarized in 1935 by Ellis in his book. For all practical purposes, the problem was solved by Rossiter (1924) with his "Beetle" (beat all) inventions, until about 1950 when UF-resins were increasingly used as adhesives in wood products and as foams. In these applications, thin and uneven resin films are formed. They possess a large surface area, which can slowly release formaldehyde if exposed to moisture, or if the resin is not properly cured.

Formaldehyde odor is currently an acute problem, which well deserves a chapter in this book, even though the phenomenon seems transitory. The odor results only under certain conditions, and it is noticed only in some special applications, mainly in closed spaces or in poorly ventilated rooms.

When particleboard was introduced in Europe, around 1950, resins with a much higher formaldehyde content were used. Despite this, the initial odor of the board did not cause any problems, because much of the material was used in furniture which was carefully finished and painted at that time. However, during the last two decades furniture construction has profoundly changed. Pre-coated panels are now cut to size and are assembled with unfinished edges. It is now known that much of the odor emanates from these edges, where coarse wood facilitates formaldehyde release. Furthermore, by virtue of its excellent mechanical properties and low price,

ISBN 0-201-04558-3

particleboard is being used as a bulk construction material, and as floor underlayment covering large surface areas under conditions where inadvertant exposure to weather or moisture can occur.

The odor can be regarded as being primarily either a chemical or an economic problem, or it can be explained as being a problem of adjustment to new products and a new lifestyle.

ECONOMIC FACTORS

Due to their low price and high performance, UF-resins have become attractive in the home construction market under conditions for which they were not designed. In order to enhance penetration of new markets, particleboard must remain competitive in price. These problems are caused by basic market conditions beyond the control of the individual manufacturer or user. Many particleboard manufacturers are unhappy with this situation and claim that an extra penny a pound could double quality. However, this problem is not unique to UF-resins. It is ubiquitous in our current post-industrial economy, which is characterized by fluctuating capacity, mergers, and the fact that all manufacturers seem to compete simultaneously for the same market segment. The odor problem could be changed by factors in the economy alone.

However, the problem can also be reduced by improving the product chemistry, but only to some extent, and only with cooperation from the user and designer of products. In order to succeed the user must know the product that he plans to use, he must know the work conditions, and he must match the two. In the case of UF-bonded wood, a major problem is rooted in insufficient understanding of the capabilities and limits of the materials. Everybody knows that glass is brittle, that bricks have very poor shear strength, and that wood burns or swells. Nobody would ignore the adverse properties of these materials, nor would one consider them a sufficient reason for outlawing their general use in construction. Yet, with particleboard the situation is different, and many users and consumers have unrealistic expectations and treat it as if it were superior in every way to any type of plywood.

The current problem of formaldehyde release is also related to the fact that the art of UF-resin chemistry was perfected before particleboard was invented, and that particleboard became immensely popular before formaldehyde release was considered a serious drawback. This sequence of events is not unique to formaldehyde. It is typical for all truly successful inventions. In fact, there almost seems to be an intrinsic law that important inventions, by virtue of their self-evident reputation, will promote their use to such a degree, so quickly and so widely, that they inevitably reach wide use in fields for which they were not designed nor suitable. Their failure under conditions of misuse then tends to turn overenthusiasm into disappointment, causing a backlash which often threatens the reputation of the entire invention. This situation certainly holds today for UF-foam and for UF-bonded particleboard in the U.S.

ISBN 0-201-04558-3

CURRENT PUBLIC ATTITUDES

If the limit of the products suitability is exceeded, failure may result. Failure provokes basic human defense instincts and readily activates insecurity or fear towards all product applications or even the product as such. Such feelings are deeply rooted in basic and viable human emotions and involve vital survival instincts, and breed sudden distrust which ignores all earlier evidence. It ignores the fact that UF-resins were successfully used during World War II to bond components in the British Mosquito and other warplanes, and torpedo boats; that plywood was used in the "super constellation" planes which were flown commercially for more than twenty years, and that modern aircraft, including the wings of jumbo-generation planes are held together with man-made epoxy and similar adhesives. The specific problem of our time is that particleboard is not the only new product available. For example, in the case of mobile homes, the entire concept, as well as many of the construction materials, are all new, and the individual parameters responsible for the odor problem are intertwined with the psychological and physical problems of the owner's adjustment to a different lifestyle.

In fact, the consumer is being deluged with new products with new properties, both superior and inferior, often in an irrational combination. To use a drastic example from the electronics field, a variety of modern hand-held electronic computers are available, often with no rational correlation between price and capability among the cheaper models. Amazingly, many models are cheaper than their carrying cases, even though the materials, and the design, development, and manufacturing costs cannot possibly be even comparable. To make the situation even more extreme, the durability of the case is often lower than that of the calculator. Every reader surely must have his own favorite example reflecting upon our confused time. And economics only reflect the depth of the revolution in which we live. The response to the overwhelming difficulties in coping with the multitude of rational judgments and decisions which we are called upon in daily life is again emotion, touches at our herd instinct and favors a trend towards stampeding into drives for product standards and legislation, nourished by an urge to find universal safety and protection. However, the development of such standards requires extensive practice and experience, which is not easily developed in quickly changing times.

No matter how advanced modern education might be, and how well skilled we might be professionally, it is only natural that man has an urge to respond irrationally to such an exposure, especially in his private life. The author believes that the lack of product familiarity explains many of the current intense consumer reactions to products ranging from plastic toys to formaldehyde odor in schools and residential and mobile homes which are meant to serve as emotional refuges in the industrial world. This situation can also explain the equally serious current quality control crisis: architects, engineers, and laborers are constantly being exposed to new materials with new properties which not only pose new challenges, but negate

ISBN 0-201-04558-3

earlier experience and make many skills and procedures obsolete. This causes a higher tendency toward an almost manic-depressive oscillation between overconfidence and total lack of trust; this fosters indifference or a half-hearted attitude and leads to failure which is countered by overreaction. By way of practical example, particleboard was recently used as a structural component of a roof of an airline pilot training center. When a leak developed, the roof showed blisters and the rooms below smelled from formaldehyde. Because of poor workmanship in the vapor barrier and the tar roof, the entire concept was questioned, and the roof rebuilt with other materials. A similar combination of poor workmanship, lacking design and insecure expectations is responsible for a large fraction of the current problems with UF-resins. Thus, modern furniture and cabinet makers feel their way into the market with products in which novel finishes and unfinished UF-bonded products are used in a manner for which the original resin was never tested or designed.

PERFORMANCE AND PROPERTIES

Product properties and product performance are not the same. Marra (1977) discussed the difficulties in designing laboratory tests for UF adhesive which adequately predict its behavior in the contemplated applications. This problem is common. For example, it is well known that steel, for many a synonym for strong materials, structurally fails in fires. Thus, the use of unprotected steel is not permitted in the construction of tall buildings, while UF foams, consisting of organic materials, have successfully weathered fires in tall structures and are thus allowed.

The end-product performance depends on the intrinsic chemical, physical and mechanical suitability of the material, on the design of the product, quality control during manufacture, and the nature of the use.

The strength of UF-resins is their low cost, high strength, low toxicity, and the wide compatibility which gives a wide range of possible uses. Their weakness is their general sensitivity to extended exposure to humidity or weather.

Table 9.6 shows a comparison of UF with similar artificial, polymeric materials. Bulk UF-resin castings resist acute water exposure well. In fact, Ellis (1935) and many early pioneers specifically praised its water resistance, and predicted a range of further applications based on this property. Today, UF is indeed used in many exacting applications, including wrappers for meat and other moist food. It is good enough to make paper cups, but the early expectations have not been fulfilled, because unprotected organic polymers, including UF-resins, are not resistant to prolonged extended outdoor exposure. A detailed analysis of the theory and field experiments on polymer degradation is found in a book series edited by Grassie (1978). Such polymers are vulnerable to UV, singlet oxygen, and thermal decay. UF excels among these materials in most respects. For example, UF-resins are hard, they are transparent to sunlight in the near UV and make excellent greenhouse roofs. The use of UF as a glass substitute was explored for many years, and Pollak,

ISBN 0-201-04558-3

an untiring inventor and skillful businessman, was willing to stake two decades of his career on the pioneering of this product. He overcame incredible odds in fighting many difficult problems, such as the casting of large, strain- and fault-free panes, but he and others eventually failed to produce a viable UF glass because they could not solve the seemingly minor and innocuous problem of surface tarnishing, caused by slow degradation upon prolonged exposure to humidity. The same problem still haunts UF-resins in many of its applications in durable products. Today, UF-resins are not normally used unprotected in outdoor use. Exceptions are UF foams used in agriculture, and fertilizer coatings which are deliberately formulated to facilitate degradation, so that the plant nutrient value becomes slowly released to the soil exactly at the rate required by the plants. In all other outdoor applications, UF-resins, or UF-resin-bonded products are coated, painted, treated, or protected otherwise. If the protection is properly designed and applied, UF foams and UF-bonded particleboard or plywood can resist rough weather and climates in such uses as entrance doors, window case work, and floor underlayment. Problems also arise if these materials are stored unprotected in transit or on construction sites. In such exposures, the wood component in particleboard absorbs water and swells, regardless whether bonded with UF, MF, PF, or other resins, and it never recovers it original shape or state.

Table 9.6

Physical Properties of UF Resins and Their Resistance Against Inorganic and Organic Solvents and Materials (Ullman, 1959)

Property	UF	PF	MF
Density (g/ml)	1.46	1.35	1.48
Refractive index	1.54–1.56	1.6	1.6
Water uptake (wt%/24 hrs)	0.5	0.5	0.5
Warping temperature (°C)	130	80	140
Softening point (°C)	60	70	80
Upper temperature for use	60	70	80
Thermal expansion coefficient	0.3	0.2	0.15
Specific heat (cal/g. deg.)	0.4	0.4	0.4
Tensile strength (kg/mm^2)	4–6	3–8	1–5
Modulus of rupture (%)	0.8	1.5	0.8
Compressive strength (kg/mm^2)	20	10	20
Modulus of elasticity (kg/mm^2)	800	250	900
Hardness (Rockwell scale)	120	90	120

ISBN 0-201-04558-3

Construction and Indoor Uses

Plywood and particleboard have been successfully used in Europe for several decades under the most exacting conditions. For example, aluminum sheet coated particleboard is widely used in commercial and residential kitchens, dairy factories, and similar applications where surfaces must be scrubbed, washed, or hosed daily. If properly planned and installed, particleboard can serve as a structural component of furniture, of cabinets, and even of residential and commercial housing. Thus, particleboard because of its dimensional stability is eminently suitable for walls and floors, and in some European countries, where the roof construction is basically different, particleboard can even be safely used as underlayment for roofing. Accordingly, extensive tests have been made in such applications, and design data is available to architects and builders. Furthermore, exploratory efforts have been made, especially in France, to construct prefabricated, fully integrated wall components comprising sandwiched insulation and particleboard layers up to 5 inches (12 cm) thick. However, quality control problems have arisen in the transition from exploratory to routine use.

Despite these virtues and the extended experience in many practical uses, a host of problems have arisen during the last fifteen years. On first sight these complaints are astonishing, if viewed against the background of historic experience. However, they are real and are obviously justified. They fall into two categories. The first category deals with mechanical damage and the second with odor.

Weathering

Particleboard is artificial wood. At its best, it excels because it neither warps, or cracks. At its worst, it can combine all the disadvantages of wood with all those of UF-resin. All untreated particleboard swells in water, regardless of resin, Figure 6.2. Wood particles are in contact with glue on 10% of their surface; 90% of the surface is free wood in contact with air. Thus, air and moisture can readily penetrate, migrate, or emanate from the finished board. The swelling of wood irreparably damages board, regardless of resin, because it strains the resin-wood interface, as described in Chapter 6. Particleboard is normally prepared from particles pretreated with waxes which impart some moisture resistance. Thus, standard particleboard resists short-time water exposure sufficiently to be used in window case work and similar applications where short-term acute exposure occurs. If problems arise in such uses, the reason is almost invariably that moisture is trapped by faulty design. Damage is then not due to acute exposure to water, but due to chronic moisture, collected by leaks, or by improper vapor barriers. Extended exposure to water leads to the swelling described above, and slowly hydrolyzes the resin at the wood-resin interface, Figure 6.1, Chapter 6.

ISBN 0-201-04558-3

Moisture eventually also hydrolyzes cured resin. If the resin contains excess hardener, or the wood is acidic, the influence of moisture is enhanced because the speed of degradation is strongly dependent on pH. Recent data on hydrolysis are summarized in Figure 5.6 and Figure 5.7. The pH is determined by the amount of acid per water volume, i.e., concentration. Thus, small amounts of moisture cause the most serious degradation and the damage can begin as soon as a microscopic water film forms on the surface of the wood particle forms. Temperatures above 40°C greatly increase the reaction. At 70°C UF-bonded plywood is stable for more than three days in water; at 100°C, the bond dissolves within a few minutes. Different components of the UF-resin are vulnerable to a different degree. Table 9.7 shows how the stability of the various resin components is correlated. The weakest link is in the cellulose-resin link, the hemiacetals, ethers, and methylols. The oxygen-free methylene linkage is most resistant. The degradation of the amine group increases with the number of protons, i.e., hydrogens. Thus, tertiary, cross-linked N-bonds are strongest. The correlation between chemical structure and resin performance has been reviewed by Myers (1977, 1978) and can now be directly monitored with the help of ^{13}C-NMR and similar techniques. If the resin degrades significantly, the mechanical performance is reduced quickly. Thus, a loss of 5% of the resin leads typically to a 50% loss in strength (Myers 1978).

Table 9.7

Relative Stability of Bonds in Cured UF-Resin
(in order of increasing stability)

Bond Type	Structure
C–O in cellulose to resin	R–O – $CH_2NHCONH$–
C–O in methylene ether (formal link)	–$NHCONHCH_2$ – $OCH_2NHCONH$–
N–C in amido methylol	–NHCONH – CH_2OH
C–O in polyformal	–$NHCONHCH_2O$ – $CH_2OCH_2NHCONH$–
C–N in primary amide	–$NHCONHCH_2NHC$ – NH_2
C–N in secondary amide	–$NHCONHCH_2NH$ – CNH–
C–N in tertiary amide	–$NHCONHCH_2NCONH$– – CH_2 – –$NHCH_2NCONH$–

The control of this problem is conceptually simple. Particleboard must be either sealed or not used in exposed applications. The practical problem is mainly one of economics design, quality control, and workmanship.

ISBN 0-201-04558-3

Odor

Normal, cured, and seasoned UF products are virtually odor-free. Thus, Crocker (1944) in USP 2,327,871 proposed coating cedar chests with UF-resin to reduce the cedar aroma. However, strong UF odors are emanated during the curing process; for example, from floor parquets which are being sealed with UF-resin, from particleboard or plywood presses, and from porous articles which have been in contact with curing resins, such as textiles, paper, or wood. Chronic odor is a sign of incompletely cured resin, improperly formulated resin, or resin decay. Unfortunately, odor vanishes only slowly. The decay time in a finished home is about two years (Mattson 1977) unless the board or space is fumigated or otherwise treated. Odor can be a symptom of acute or chronic release, or of both simultaneously. Pre-1970 resins were often formulated with a large excess of formaldehyde. This caused acute initial release, but after some aging such high U:F ratio resins were very stable and little chronic odor was reported. Today, resins contain far less free formaldehyde. Thus, initial odor is greatly reduced. Ironically, the reduction of initial U:F ratio to reduce initial odor produces chemically less stable resins, more susceptible to slow hydrolysis and chronic odor.

FORMALDEHYDE RELEASE MECHANISM

UF odor is caused by the high volatility of formaldehyde; the cause for free formaldehyde is complex. It can stem from a variety of partly related sources such as free, unreacted formaldehyde in the resin, from formaldehyde dissolved in moisture on the product surface, where it readily dissolves, and its vapor pressure and its release rate change with changes in air and product humidity. In particleboard it can come from free formaldehyde which was bound to wood cellulose during the hot press cycle, and which slowly hydrolyzes under the influence of the wood humidity acidity. It can result from degradation of incompletely cured resin, or resin components, such as methylolurea, and finally it can result from bulk resin degradation. As expected, the latter is more rapid in products in which thin and nonhomogenous resin films with a comparatively large surface area are present. This is the case in foams and in wood composites. Particleboard is a poor source for analyzing the origin of free formaldehyde, because the wood is a heterogenous, complex chemical system, the resin-wood interface is uneven, and contains many weak links, for example, the particle surface where the resin components individually interact differently, almost like on a chromatographic separation column, and finally, because wood absorbs moisture, adsorbs formaldehyde, and concelas, or delays, the release of decay symptoms. Kobuta (1977) observed that untreated plywood veneer could conceal up to 100 mg/sq ft, Figure 9.2. Foam surfaces are simpler and formaldehyde is more directly related to moisture content. Further-

ISBN 0-201-04558-3

more, the sulfonate surfactants are known to react with much of the free resin formaldehyde during the curing.

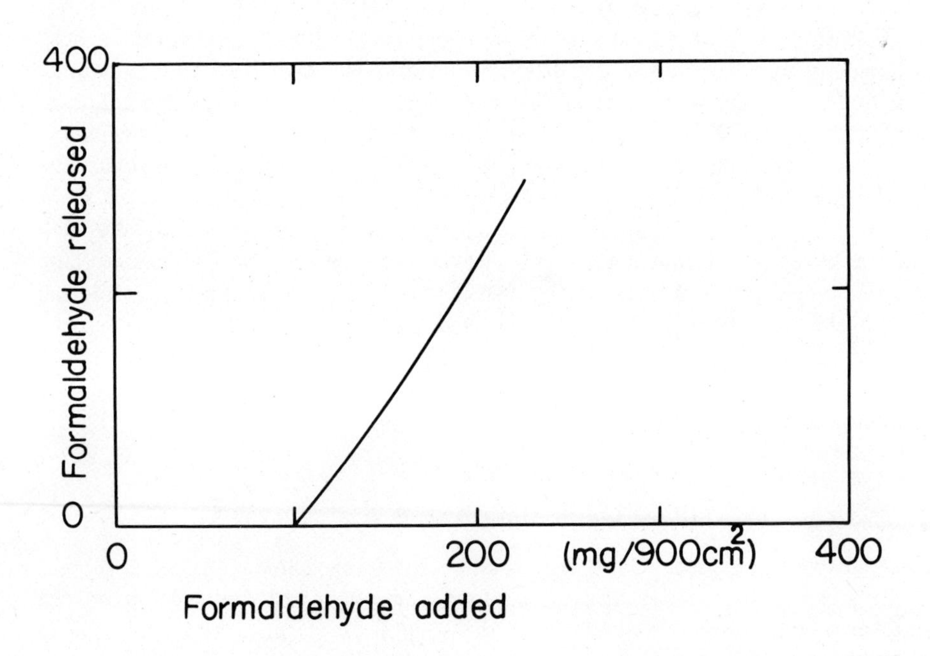

Figure 9.2. Formaldehyde Retention by Plywood Pressed without Adhesive (Kobuta, 1977)

A further problem in dealing with formaldehyde odor is the fact that formaldehyde in air and formaldehyde in the product are not always related by the same mechanism. Table 9.8 shows an approximate correlation between formaldehyde in particleboard, as measured with the FESYP perforator test, and formaldehyde in air, in a room containing 40 m^2 of board at 23°C and 45% humidity, in an air volume of 34 m^3 (Barghoorn 1979). However, a given formaldehyde release rate does not always lead to the same odor level in a room, because ventilation, absorption by other substances and the human body, and the size of the air space above the UF surface all influence airborne concentration, as discussed in Chapter 5.

The potential odor problem has been recognized since the very beginning, in the 1920s. Neusser and Zentner (1968) analyzed the problem in particleboard in a landmark paper which is still valid. They correlated air concentrations to ventilation rates, and established a dose-response relationship. They measured and defined free formaldehyde in UF-resin, and correlated it to storage time, U:F molar ratio,

ISBN 0-201-04558-3

hardener concentration, filler, extender, and the influence of wood. Verbestel (1965) showed that birch causes 30% more free formaldehyde than fir. Wild (1964) predicted a relationship between free formaldehyde release in the press and the resin formulation, the influence of wood moisture content, press temperature, and press time. Neusser and Zentner (1968) report on several earlier studies. It was suggested that the difference between formaldehyde release from standard UF, as compared with PF or modified lignosulfonate UF-resins was mainly a question of press temperature or time. They summarized the work of Wittmann, Chalmin, Deppe, Stolger, Plath, and Verbestel in seven rules.

1. Formaldehyde release is proportionate to free formaldehyde in the resin.
2. The total resin solid content is not related to free formaldehyde.
3. Additives of hardener reduce free formaldehyde.
4. The formaldehyde release is directly related to the humidity of the face wood particles.
5. Formaldehyde release is reduced by extended cure time, and is inversely proportionate to the moisture released during the pressing. Thus, some excess formaldehyde is expelled with the steam. Standard manufacturing practice only expels about half of the formaldehyde available for varporization.
6. The influence of the post-press cooling cycle is not clear.
7. Moisture and himidity changes influencc the release from finished boards.

Table 9.8

Correlation Between Free Formaldehyde in Particleboard and Air Concentration (Barghoorn 1979)

Perforator Value of Particleboard mg/100 g board	Air Concentration of Formaldehyde Average ppm (± 30%)
(10)	0.1
20	0.5
30	0.9
40	1.3
50	1.75
60	2.2
70	2.6
80	3.0
90	3.4

ISBN 0-201-04558-3

Neusser and Zentner collected between 0.03 and 0.3 wt % of formaldehyde per total wood weight from the finished board. They correlated the release to age and plotted the result as the sum of total formaldehyde released and as release rate. They determined the temperature coefficient for the release, compared different analytical methods, and built test chambers and boxes. Finally, they recommended and tested ammonia as an abatement agent, and gave advice on the handling of consumer complaints. Among the pioneers in this field were Wild (1964), Wittmann (1962) and Noguchi (1963), Chalmin (1965), Karahasanovich (1966), Kopitovic (1965), Mildenstein (1965), Plath (1966–68), Stoger (1965), and Verbestel (1965).

Their work was followed by many unpublished studies in industry and research associations. The problem has been reviewed by Kelley (1970), Nestler (1977), Sundin (1978), Hanetho (1978), Winkler (1972), Mølhave (1977), and Roffael (1969–78). Roffael, in several important papers, invented a new testing method (1975), correlated strength and formaldehyde release in oak (1975), measured the influence of free formaldehyde on reactivity (1976), and in 1977 he observed that wood gives off formaldehyde upon drying, and discussed possible lignin and cellulose precursors, and measured levels of release. Beech gave off 2 mg/kg wood at 160°C, 8 mg/kg at 220°C; fir 1 mg/kg and 5 mg/kg, and pine 0.8 and 3 mg/kg, respectively. In 1976 he simplified his analytical method; in 1976 he studied the effect of bark on formaldehyde in air, using ultraviolet spectroscopy of aqueous products and found that formaldehyde is released most for fir and pine and least for beech. In 1977 he further developed his Wilhelm-Klausnitz-Institut (WKI) method, using test rooms; in 1978 he compared his method with that of others, Table 5.5. In another paper he evaluated the perforator method of FESYP in another evaluated the effect of aging; in another explained how one can differentiate between small, but significant differences in release. Petersen (1974–78) collected formaldehyde in heat-resistant bags containing the board during the curing in the hot press, and showed that formaldehyde release increases 50% if the wood humidity is increased from 10 to 15%. He measured formaldehyde release as a function of aging, as a function of U:F, as a function of press time and temperature, during and after forming, and finally, as a function of hardener, urea, ammonia, phosphate, magnesium chloride, zinc nitrate, and pyrrolidone. These papers establish a standard of experimental care which is hard to match. Andersen measured the air levels in Danish homes (1974), developed an equation to correlate values for air humidity, Figure 9.3, and board release; in 1976 he observed that SO_2 as copollutant significantly reduces formaldehyde in air above wood, as shown in Figure 9.4, and in 1979 reported human exposure experiments as a basis for recommended ambient indoor standards of 0.15 ppm (Andersen 1974–1977). Rehn and Peterson (1977) in an unpublished report used 64 volunteers to establish the detection limit for formaldehyde in air (see Figures 10.1 and 10.2 page 251; 253). In Russia studies were done by Sukovatov (1968), Kravchenko (1974). At the Kiev Hygiene and Forest Products Institute, the latter compared formaldehyde release as a function of temperature and humidity, and further found that sawdust retains

ISBN 0-201-04558-3

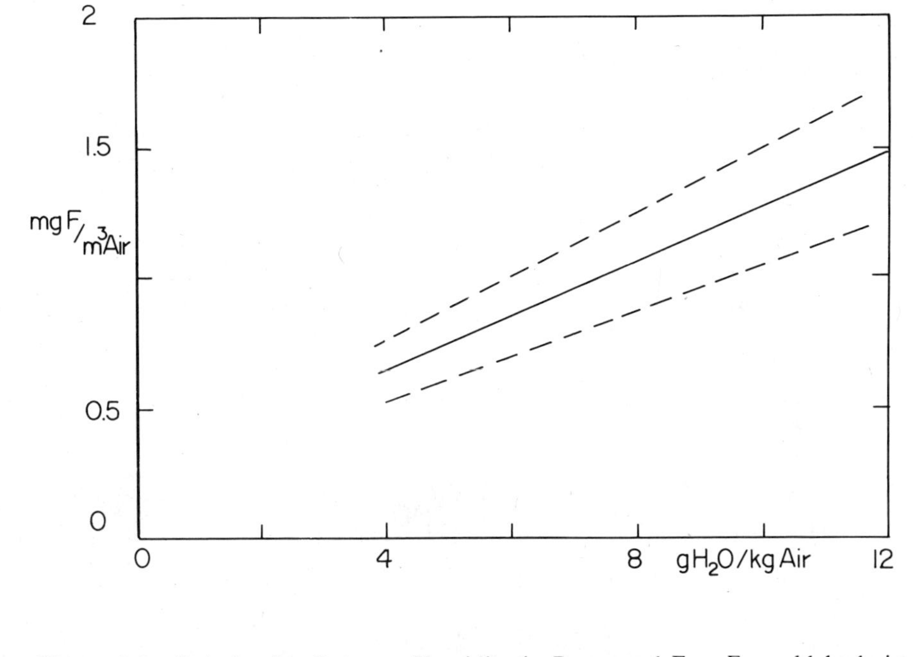

Figure 9.3. Relationship Between Humidity in Room and Free Formaldehyde in Air, with Constant Particleboard Exposure (Andersen, 1976)

more formaldehyde than larger chips. He compared boards made in eight different factories from Kiev to Kostopol by measuring formaldehyde in air above the surface. In 1977 he measured formaldehyde release from fiber glass coatings as a function of time. Malygina (1974) made laboratory experiments and monitored boards for many months. He tested post-cure heating, floor varnish, and other methods for reducting formaldehyde. Mikhaylova *et al.* of the central plywood research and development institute in Leningrad studied the influence of the U:F ratio, and amonia, and compared English, Austrian, Hungarian, Polish, Japanese, and Russian resins. Similar work was done by Tendrovskaya (1973) and in Japan by Matsumoto (1974), Kubota (1977), Fuji (1973), and others. Matsumoto (1972) studied the release from UF, MUF, and PF bonded plywood in a test room, and correlated the free formaldehyde in the board with that in air; he developed a suction test, and measured the influence of temperature and humidity. In 1975 he correlated the Japanese desiccator test with air concentrations, Figure 9.5, and explained the Japanese industrial standards. Fuji (1973) explained the reasoning behind the introduction of the formaldehyde standard, and the choice of a value of

ISBN 0-201-04558-3

5 ppm based on the desiccator test, by developing a theoretical equation for release:

$$C' = \int_0^t dC' = \frac{C_\infty a S'}{W}\left[t - \frac{V}{a'S+aS'} + \frac{V}{a'S+aS'}\, c - \left(\frac{a'S+aS'}{V}\, t\right)\right]$$

where C' is the concentration of formaldehyde in the collector dish, S is the surface are, a' the constant (mg/h·ppm m^2), a the factor (mg/h·ppm m^2), C the formaldehyde concentration in air, V the air volume, w the water volume, and t is time. The correlation between formaldehyde released from the sample surface (m) versus that collected in the water is linear, as shown in Figure 9.6, for the 24-hour period chosen in the Japanese test.

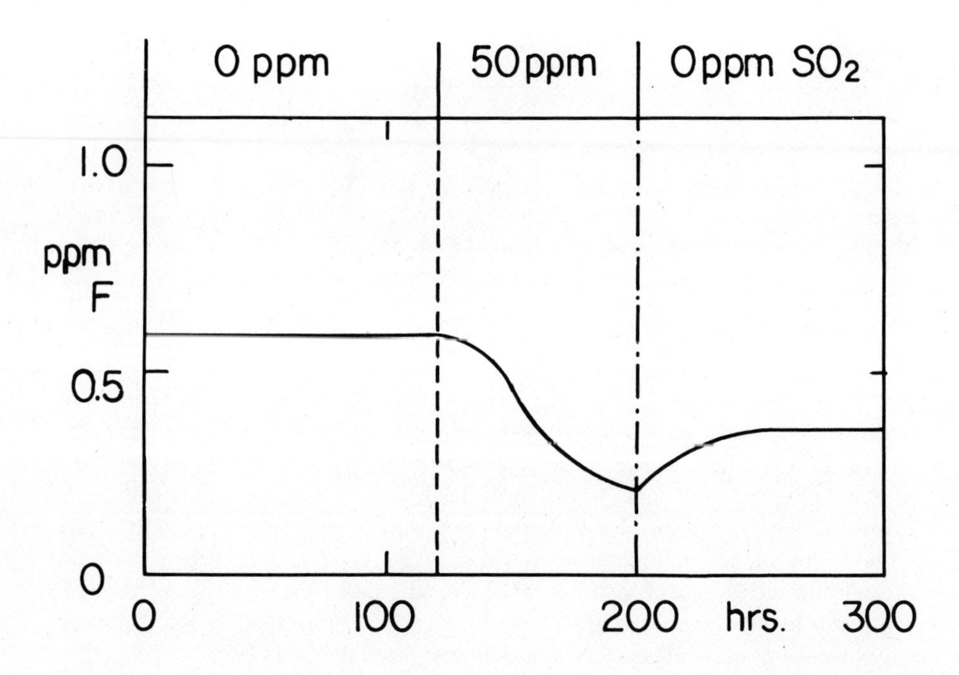

Figure 9.4. Variation of Formaldehyde Concentration During and After Exposure to SO_2 (50 ppm at 22°C; humidity 5g H_2O/kg air)

ISBN 0-201-04558-3

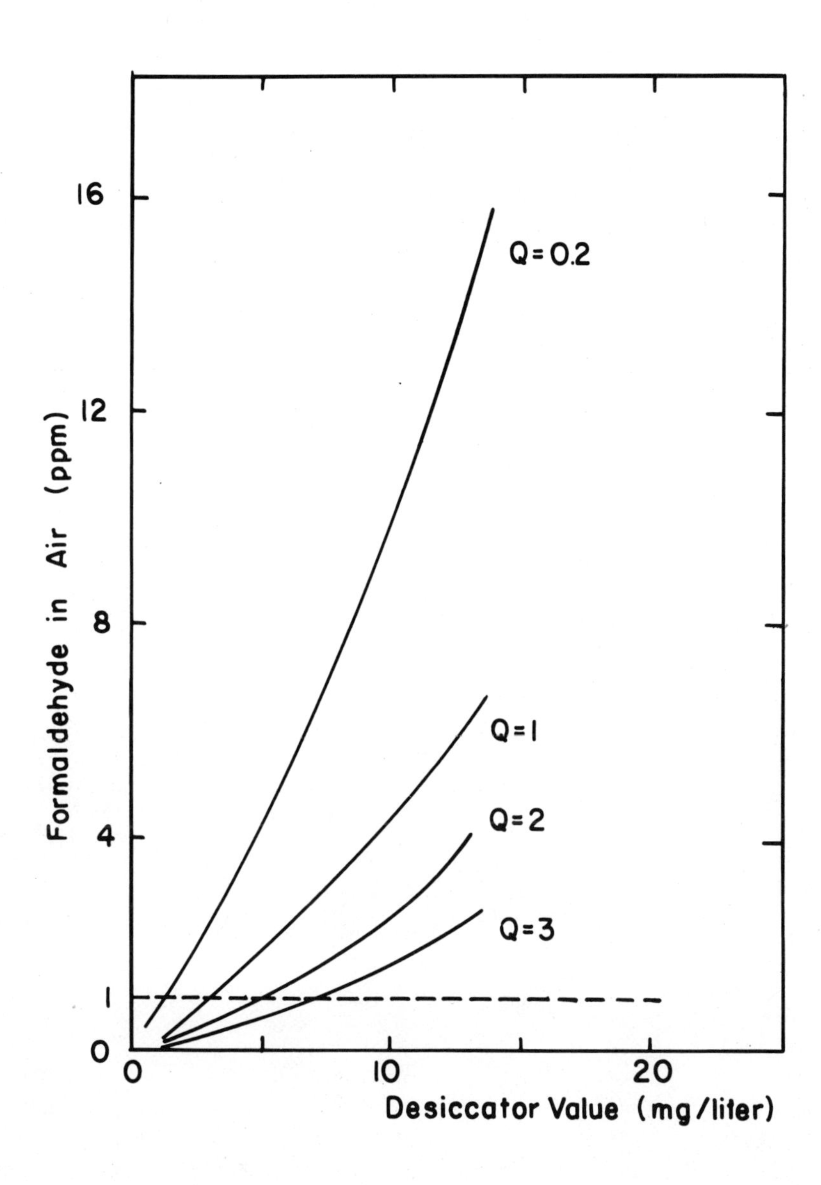

Figure 9.5. Relationship Between Formaldehyde Release (Japanese Desiccator Method) and Formaldehyde in Air (Fuji, 1973) (after 24 hours at 80% humidity, 30°C)

ISBN 0-201-04558-3

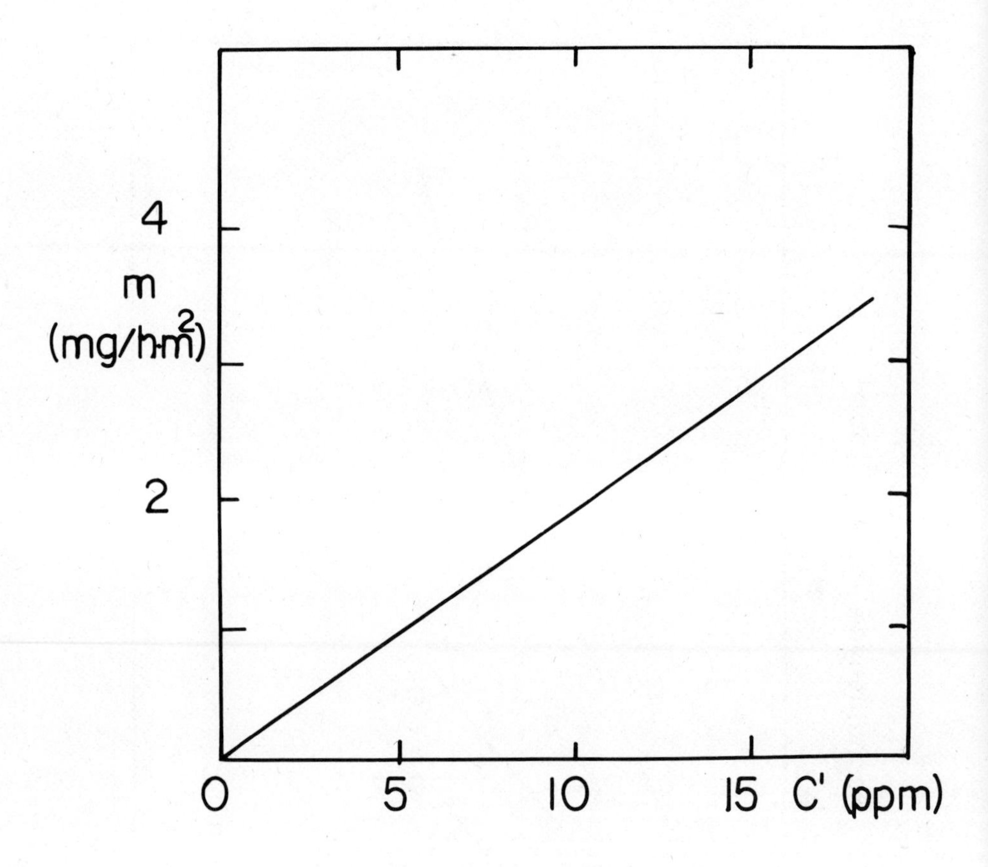

Figure 9.6. Relationship Between Formaldehyde in Air and in Water, in the Japanese Industrial Test JIS-A-5908-1977 (Fuji, 1973)

The effect of temperature on hydrolysis rate (Myers 1978) was shown in Chapter 5 in Figure 5.6; that of cure temperature was in Figure 5.7. Hanetho (1978) studied the release from different layers with the board, using UF and PF layers, to determine migration within the board. Almost all forest products laboratories currently conduct work in this field (Myers 1977–1978). Analytical methods for measuring free formaldehyde in board and those for air sampling are described in Chapter 5. Cherubim (1976) observed hydrolysis of UF-resins under a variety of conditions, and reviewed earlier literature. Mølhave (1977) related formaldehyde in air to formaldehyde in wood, the method used by Andersen's group.

A detailed report on the status of knowledge in Scandinavia was edited by Mattsson, Sundin, and Astrom (1977). He found a direct correlation between free

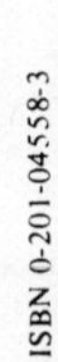
ISBN 0-201-04558-3

formaldehyde in board and in the air, and found the half life of formaldehyde in a newly erected house to be about two years. The sources were particleboard and a new hard fiber board–honeycomb ceiling structure–introduced widely in 1974.

The Houtinstituut, TNO, Delft, Holland (1978) evaluated the formaldehyde release problem. They proposed limits for release at 1.3 g/m^2 over a 9 X 24 hour period, according to Roffael; Mohl (1978) described a suction method for surface and air mentioned which he had tested in Karlsruhe for several years.

Formaldehyde Abatement

COATINGS

The odor can be reduced by different means. The easiest, but not the most economical, is to coat or paint the finished product. Papers, plastic sheets and sheet metal are all used for this purpose. Barghoorn (1979), Table 9.9 compares seven different types of treatments with and without edge finish; the data refers to a surface to volume ratio of 0.6, one air change per hour and an original FESYP perforator value of 0.027%. The table shows that the edge finish alone can reduce emission by a factor of ten. Table 9.10 shows the effect of age.

Table 9.9

Formaldehyde Values (Perforator) for Various Types of Coatings and Finishes (Barghoorn 1979)[a]

Surface Finish	Formaldehyde Concentration in Test Room (ppm)
Untreated (0-value)	1.20
Wallpaper with sealed edge	0.81
Falima coating with sealed edge	0.03
Macore over-lay with sealed edge	0.19
Melamine coating with sealed edge	0.02
Melamine coating with exposed edge	0.10
Varnished surface with coated edge	0.02
Varnished surface with uncoated edge	0.09
Overlay paper with sealed edge	0.08

[a]Perforator value of board: 0.027%; 0.6 m^2/m^3; one air change per hour; board dimension: 1x2 m.

ISBN 0-201-04558-3

Table 9.10

Effect of Surface Treatment on Formaldehyde in Air[a]

Surface Treatment	Formaldehyde Concentration in Test Room (ppm)
Untreated	3.1
With coating, fresh	0.33
With coating, after 6 months	0.24
With coating, after 1 year	0.47

[a]Perforator value not known; one air change per 2 hours; humidity 63%; 0.8 m^2/m^3.

CHEMICAL AFTER-TREATMENT

Studies with plywood were conducted by Matsumoto (1974) in a test room. Minemura, Hirata, Imura, and Takahashi (1976) at the Hokkaido Institute published a 35-page paper on formaldehyde release from plywood and methods for reducing it, correlating press temperature, and the effect of viscosity, melamine, urea, potato starch, and other fillers and extenders, the moisture in the veneer, the effect of foaming of the resin with 2% alkyl benzene sulfonate, the effect of wood species, of phosphate, iron chloride, disulfite, and iron or manganese chloride hardener, the effect of the additives, Table 9.4, the collection of formaldehyde in cotton. Kobuta, Hirata, and Takashashi, also at the Hokkaido Institute (1977) measured total formaldehyde released during and after pressing of plywood, and the formaldehyde retention from veneer under manufacturing conditions, Figure 9.2. Haui (1962) published a book on wood bonding, and Mitsua and Matsunaga (1978) one on UF- and MF-resins, which gives an excellent review of UF-resin chemistry and analysis, containing many very valuable tables and figures, but it is in Japanese. Imura and Minemura (1977) discussed the effect of sulfur oxyacids on the formaldehyde release. Ten weight % aqueous solutions were painted on 900 cm^2 (1 ft^2) of 3-ply board, four millimeters thick and dried five minutes at 105°C. They also coated particleboard with polyurethane paint or with nitrocellulose varnish, giving a coating of 100 g/m^2. The latter was odor free. Figure 9.7 shows the reductions that were achieved, Table 9.11 summarizes the shear strength and internal bond strength, both fresh and after exposure to light. This treatment is chemically interesting, but not very practical, because of problems wtih pollution

ISBN 0-201-04558-3

and corrosion. A much better result is achieved if the compounds are added to the resin before curing (Meyer and Johns 1979). Figure 9.8 shows the effect of Verbestel's commercial ammonia treatment as a function of treatment time.

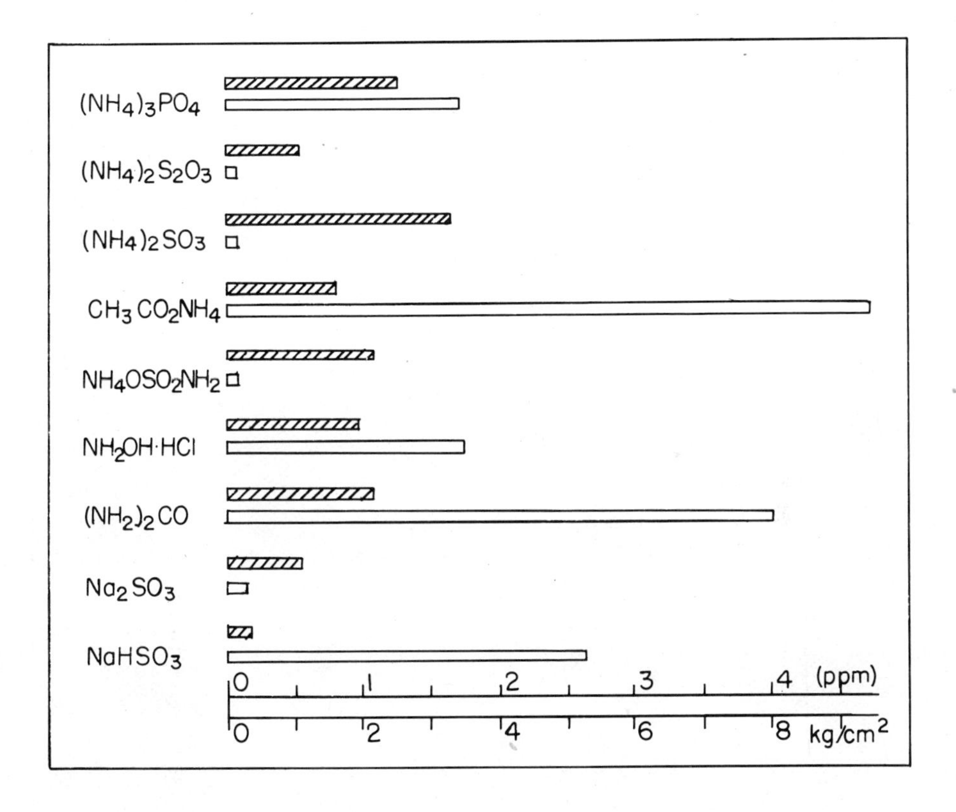

Figure 9.7. Plywood Strength and Formaldehyde Release Following Aftertreatment with Sulfur Oxyacids (after Minemura, 1978) [hatched bar] Formaldehyde Release; [open bar] Bond Strength

ISBN 0-201-04558-3

Table 9.11

Physical Properties of the Plywood Impregnated with Different Scavengers (after Minemura, 1977)

Scavenger	CH_2O emission JIS desiccator method	Shear strength ( ) = wood failure	Coating strength ( ) = separating portion between wood and coating			After irradiation by carbon arc light for 125 hours	
			Untreated	After carbon arc irradiation	After recycle test at various temperatures	Color change	Loss of brightness
	Ppm	Kg/cm^2 (%)	Kg/cm^2 (%)	Kg/cm^2 (%)	Kg/cm^2 (%)	NBS	%
Control no painting	13.20	11.2 (16)	–	–	–	7.2	2.8
Polyurethane painting	.50	11.5 (8)	8.0 (66)	10.1 (0)	9.9 (61)	12.6	10.5
Urea	.07	11.6 (27)	8.8 (39)	9.1 (23)	8.4 (43)	12.5	9.8
$(NH_4)_2SO_3$	.15	10.3 (17)	7.4 (55)	7.8 (10)	7.8 (30)	15.1	12.1
Na_2SO_3	.00	12.5 (26)	4.3 (73)	5.5 (18)	8.3 (43)	18.5	9.6
$NaHSO_3$	.25	11.0 (17)	8.4 (20)	7.9 (5)	7.6 (56)	17.3	8.7
$Na_2S_2O_4$	.34	11.7 (34)	6.1 (32)	5.6 (10)	8.8 (68)	19.7	9.8
$Na_2S_2O_5$	.26	10.4 (14)	12.9 (55)	6.1 (0)	5.5 (50)	16.6	9.7
NH_4HCO_3	.37	13.2 (18)	9.1 (27)	7.6 (5)	4.5 (24)	13.7	11.0
$(NH_4)_2S_2O_3$	.01	12.2 (18)	4.5 (45)	7.1 (0)	4.7 (40)	27.7	16.1
$NH_4OSO_2NH_2$	.64	11.2 (16)	7.3 (19)	9.0 (20)	4.8 (29)	30.8	12.1
H_2NCONH_2	.06	11.2 (16)	5.6 (100)	5.6 (95)	5.7 (100)	16.1	7.6

ISBN 0-201-04558-3

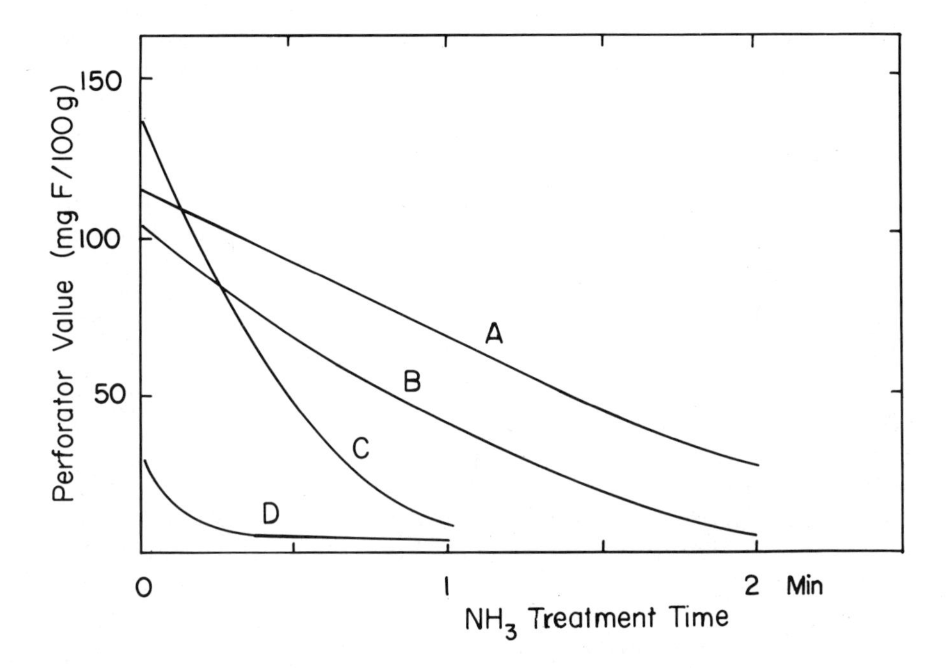

Figure 9.8. Free Formaldehyde in Particleboard after Ammonia Treatment
1) Particleboard (2 cm thick; 600 kg/m^3); 2) Hemp Shive Board (2 cm thick; 500 kg/m^3); 3) Hemp Shive Board (2 cm thick; 400 kg/m^3); and 4) Particleboard, (0.6 cm thick, density 750 kg/m^3) after Verbestel, 1979

RESIN ADDITIVES

Tanaka (1971), Japan 71:28,784, added ethylenediamine, Przybylak (1971) added lactic acid to a UF cold resin; Neumann (1971) in Ger. Offen. 2,130,577 used ammonium sulfide, and polysulfide during condensation at pH 5 to produce tougher resin products, Shvartsman (1972) and Varlacheva (1972) tested formaldehyde reduction; Davidson (1972) in USP 3,642,042 described a method for bleaching and impregnating wood with ammonia and sulfur dioxide; Shastri (1972) in Ger. Offen. 2,041,566 bleached pulp with dithionite, a treatment which also prevents self-ignition. Nishino (1972) in Kokai 73:92,541 employed guanidine, and its derivatives; Ernst (1972) in Ger. Offen. 1,920,283 used aluminum phosphate to fireproof UF-mica board, Ludwig and Stout (1972) in USP 3,658,638 reacted ligno sulfonate with PF and UF; Yoshimitsu (1973) in Kokai 73:103,705 disclosed treatment of plywood with 2% ammonia in air at 80°C; Higuchi, Kokai 73:60,737, used ammonia and phosphate and additional urea to reduce formaldehyde; Kawahara,

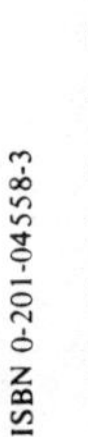
ISBN 0-201-04558-3

Kokai 73:72,309, used excess urea and excess ammonium chloride; Kato, Kokai 73:17,755, used malonic or acetoacetic acid and butylated melamine with triethanolamine at pH 5; Starzynski *et al.*, Pol. P 68,454 (1973), disclosed a three-step alkaline condensation; Kitakado, Kokai 73:58,109, impregnated plywood with 30% aqueous urea; Kawahara, Kokai 73:72,309, used urea and flour; Suzuki *et al.*, Kokai 73:14,747, added alpha cellulose and zinc stearate and ammonium chloride to a molding composition; Ferber (1973) contemplated polyester alkyd resins and other known plasticizers; Starzynska (1973) added furfuryl alcohols; Szlezyngier (1973) used *m*-phenylene-diamine, dicyanamide and resorcinol, as well as thiourea, albumin, lignosulfonates, caprolactanin, and melamine; Kubitzky (1973) in Ger. Offen. 2,206,696 added aliphatic amides, thioamides, imines, or amines to the resin. Sodium rhodamide was also effective. Zeigerson (1973) in Ger. Offen. 2,247,036 used brominated lignosulfonate and thereby also improved flame resistance; Ohno, Japan 74:43,126, used ammonium acetate, sodium sulfite, calcium hydroxide and ammonium carbonate; Hasegawa, Kokai 74:99,191, employed nitrates, sulfate, carbonate, and chloride at pH 7; Ikeda, Kokai 74:18,926, added urea and melamine; Inui, Kokai 74:44,060, added enzymes during or after resin manufacture; Ikeda, Kokai 74:18,925, added urea; Miwa, Kokai 74:25,108, added inorganic peroxide and reduced formaldehyde release from 150 to 0.8 ppm; Miwa, Kokai 74:124,207, wrapped finished plywood in a paper impregnated with ammonium sulfamate reducing free formaldehyde from 30 to 3 ppm. Miwa, Kokai 74:124,203, then impregnated paper with sodium bisulfite and after seven days observed reduction from 25 to 3 ppm. Kawashima, Kokai 74:71,118, used urea, and calcium lignosulfonate at 30 g/900 cm^2, and reduced formaldehyde from 30 to 0.3 ppm; Kitakado, Kokai 74:80,210, added wheat flour and soybean flour; Kitakado, Kokai 74:66,808, used 30% aqueous urea; Sasaki, Kokai 74:125,504, used ammonium thiosulfate, sulfite, and acetate baths to treat plywood, finding that formaldehyde release decreased with increasing pH; Ohara, Kokai 74:131,243, reduced formaldehyde tenfold by addition of ammonia or amines. Enkvist (1974) in Ger. Offen. 2,334,540 used lignosulfonates and chalk with a PF copolymer; Lenz (1974) in Ger. Offen. 2,324,440 added sodium trisulfite to increase the shelf life. Ohashi, Kokai 74:119,948, made acetaldehyde copolymers. Ohara, Kokai 74:131,243, used isobutylene-maleric anhydride and wheat flour and reduced free formaldehyde from 20 to 2 ppm; Shiota, Kokai 74:48,782, used alkaline calcium salts and ammonium sulfate; Kotani, Kokai 74:112,952, used bisulfite-acetic acid as modifier and found 1.4 ppm after 24 hours of soaking in water; Wirpsza (1974) in Pol. P 72,885 added acetone and acetic acid to reduce formaldehyde to 0.4%; Tashiro, Koaki 74:101,418, added 2-amino-*w* methyl 1,*e*-propanediol; Tsuge, Kokai 74:261,072, added urea and cellulose with ammonium persulfate yielding 1.9 ppm formaldehyde two days after curing. Tange, Kokai 74:92,160, used magnesium, calcium, zinc, or barium carbonate and ammonium salts; Kotani, Kokai 74:35,511, used aliphatic beta-diketones, beta-ketone acid esters, malonic acid and methyl acetoacetate and found 2 ppm free formaldehyde. Dashkovskaya (1975) in

ISBN 0-201-04558-3

USSR 480,555 added mineral oil and sodium silicate. Teodorescu (1975), Rom. P 57,015, added pulping bisulfite and sodium pyrosulfite. Schmidt-Hellerau (1975) in Ger. Offen. 2,354,928 added up to 10% sulfite waste; Higuchi, Kokai 75:34,332, used polyethylene glycol, phenolic resin and lignosulfonic acid; Suematsu, Kokai 75:24,404, added urea and amino compounds; Takahashi (1975) added urea and melamine; Szlezyngier (1975), Pol. P 74,060, added lignosulfonate; Kawahara, Kokai 75:43,181, used urea and amino compounds; Coyle (1976) in USP 3,931,072 used phenol and lignosulfonate; Minoji, Kokai 76:20,225, used acid amide polymers, such as polyacrylamide with *p*-toluene sulfonic acid; Fukuda, Kokai 76:41,031, added 1–10% 2(3,4-dihydroxy phenyl) 3,6,8-trihydroxy-4H-1-bezopyran (quebracho); Kamachi, Kokai 77:95,745, used guanidine; Kitamma, Kokai 77:74,015, used isopropanol; Nizhni, Kokai 77:155,697, added isooctyl-phenol; Takahashi, Japan 77:20,516, added dicyandiamine; Edler (1977) in Swedish P 397,834 employed lignosulfonate. Meyer and Johns (1979) incorporated phenol, urea, lignosulfonate and various oxyacids of sulfur into ready-to use commerical resin and observed in some cases substantial reduction of formaldehyde release during curing and in the finished product. Figure 9.9 shows the formaldehyde release as a function of additive concentration. Similar results are achieved with the more conventional ammonia or urea treatment.

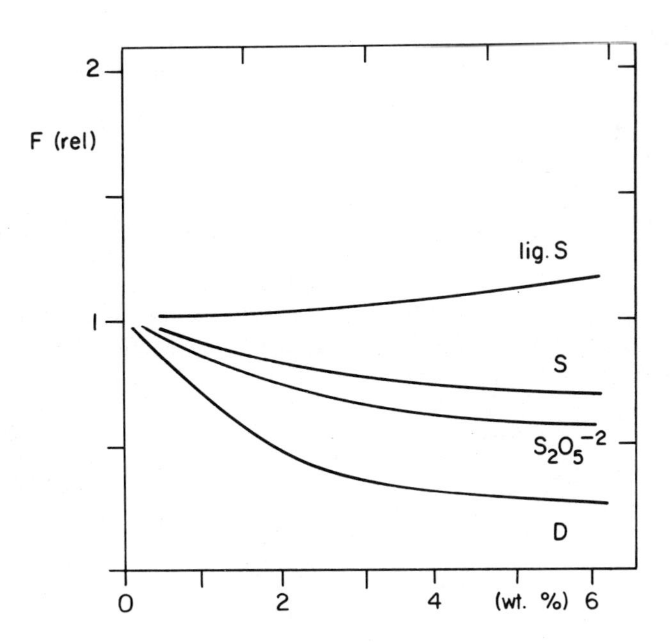

Figure 9.9. Formaldehyde Release from UF Particleboard Resin containing 0–6 wt % of Lignosulfonate (lig S), Sulfur (S), Sodium Disulfite ($S_2O_5^{-2}$) and Dithionite (D) (Meyer and Johns, 1979)

ISBN 0-201-04558-3

NEW RESIN FORMULATIONS

Almost any conceivable resin mixture was already tested by the early pioneers (Ellis 1935, Scheiber 1943), but comparatively little research has been conducted during the last decades. However, a new generation of low-odor resin is currently appearing on the market in which traditional reagents are used, but the synthesis is more carefully controlled. For example, Willegger and Thiel (1976) in USP 3,994,850 describe an improved lignosulfonate UF; O'Neill and Steiger (1976) in USP 3,996,190 describe a four stage UF-resin in which the third stage increases formaldehyde, and a fourth stage brings an additional 10% urea; Brunnmüller, Schatz, Mayer, and Grabowsky (1974) in USP 3,816,376 assigned to BASF describe a continuous process for making UF-resin with backmixing; Eisele, Petersen, Mayer, and Wittmann (1977) in USP 4,021,413, assigned to BASF, describe a wood adhesive using U:F values between 2.5 and 4.5 in the initial mix; Hubbard and Atkinson (1977) in USP 4,035,456 assigned to ICI described an expanded UF panel material; Renner (1977), USP 4,064,088, prepares fillers, adsorbents, carriers, and thickeners from UF and naphthalene sulfonic and with ammonium bisulfites as catalyst; Blommers (1977) in USP 4,032,515 assigned to Koppers uses resorcinol to terminate UF-resin sulfite and dimethylolurea are used in some examples. The resin can be used for bonding plywood or laminated timber; and Moore (1978) in USP 4,097,419 describes a UF-resin suitable for foam, particleboard or plywood in which 1–5% dialdehyde, preferably glyoxal, is added, and lignosulfonate can be used, as well as naphthalene sulfonic acid. Many other methods are described in Chapter 4.

FOAMS

As indicated above, foams are easier to modify than particleboard, especially if foam is used for insulation, and no structural strength is required. All commercial foam resins are modified to reduce odor and shrinkage. Foam modifications are explained in Chapter 7. The formaldehyde release in foam is complex. Figure 9.10 shows that formaldehyde emission goes through a maximum after several weeks. This cannot be solely due to drying, but must be due to post-curing (Wulkan 1979). The figure shows that the release drops quickly after about four weeks. Foam can be modified with all reagents suitable for adhesives, except those which interfere with the foaming agent.

ISBN 0-201-04558-3

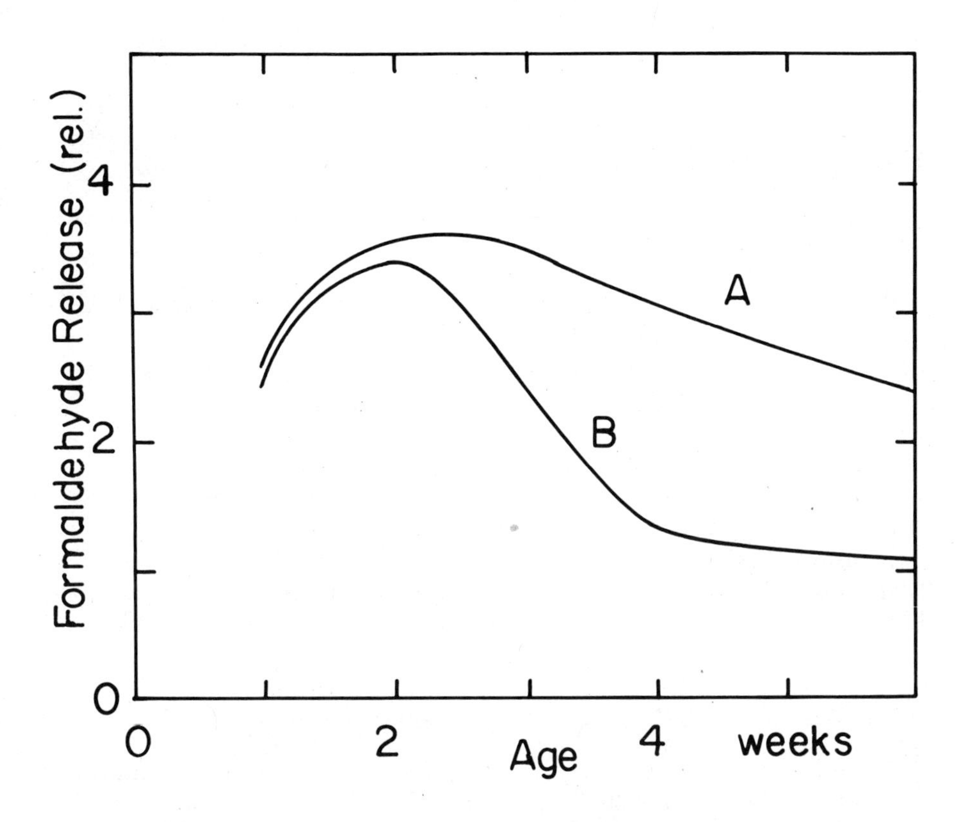

Figure 9.10. Formaldehyde Release from UF Foam as a Function of Time (Meyer, unpublished work; see also Wulkan, 1979) (Foam Sample: A stored in close place; B exposed to air.)

ISBN 0-201-04558-3

10. Health and Environment

This chapter deals with health, toxicity, and environment in nine sections. First the toxicity of urea and formaldehyde is reviewed. Then, the importance of acute exposure is explored and compared to that of chronic exposure. The formaldehyde levels found in air, the public attitude, regulations, and recommended limits, especially the new concept of indoor air quality, are discussed and it is pointed out that the formaldehyde problem intrinsically is determined by the nature of the vapor's source, which causes it to linger on surfaces. Furthermore, the reported health effects are reviewed and the current standards are discussed. Finally, the formaldehyde levels in water and in food are discussed. It will be shown that cured UF-resin is essentially inert, but that the slow but persistent release of low levels of formaldehyde from incompletely cured resin or improperly designed products into poorly ventilated rooms can cause odor, discomfort, or irritation.

Toxicity

UF-resins are made from urea, formaldehyde, and water. The finished resin contains the reagents in chemically bound form, and is inert. The main hazards of UF are those of the free reagents into which the resin can revert by hydrolysis. Thus, for the users of finished products, and for workers who deal with aqueous resins, knowledge of the properties of urea and formaldehyde are important. For those who conduct the chemical resin synthesis, proper precautions must also be taken against the toxicity of lesser, but more dangerous ingredients, such as acids, ammonia, organic amines and other ingredients which are consumed, spent, or converted during the chemical manufacture of resin.

UREA

Urea is a solid which readily dissolves in water. For all practical purposes, urea, a natural product, is inert. The human body excretes daily some 50 g which is discharged with urine. One can ingest at least 500 g per kg body weight without poisoning. Dogs tolerate up to 3 g/kg subcutaneously, and the same quantity can

Beat Meyer, Urea-Formaldehyde Resins

ISBN 0-201-04558-3

also be introduced intravenously. Aqueous urea is not toxic up to 1000 ppm. Urea is used as a fodder supplement for cows and other ruminant animals, and it serves as a fertilizer and plant nutrient. Upon heating, urea eventually decomposes into carbon dioxide and ammonia, Chapter 3.

FORMALDEHYDE

Ambient air contains 0.15 to 120 ppb formaldehyde as an intermediate in the oxidation of methane. One ppm formaldehyde is equivalent to 1.248 mg/m^3; 1 mg is 0.801 ppm. Formaldehyde is a volatile and reactive aldehyde which readily hydrolyses. This property has been used for many decades in hospitals to desinfect tools and working surfaces.

Acute Effects

Rehn (1978), Figures 10.1 and 10.2, conducted human volunteer tests and established a dose-response curve for the odor threshold. Of the 64 subjects, nobody noticed less than 11 ppb, and everybody noticed 3 ppm. It is generally believed that most people become desensitized by high formaldehyde levels.

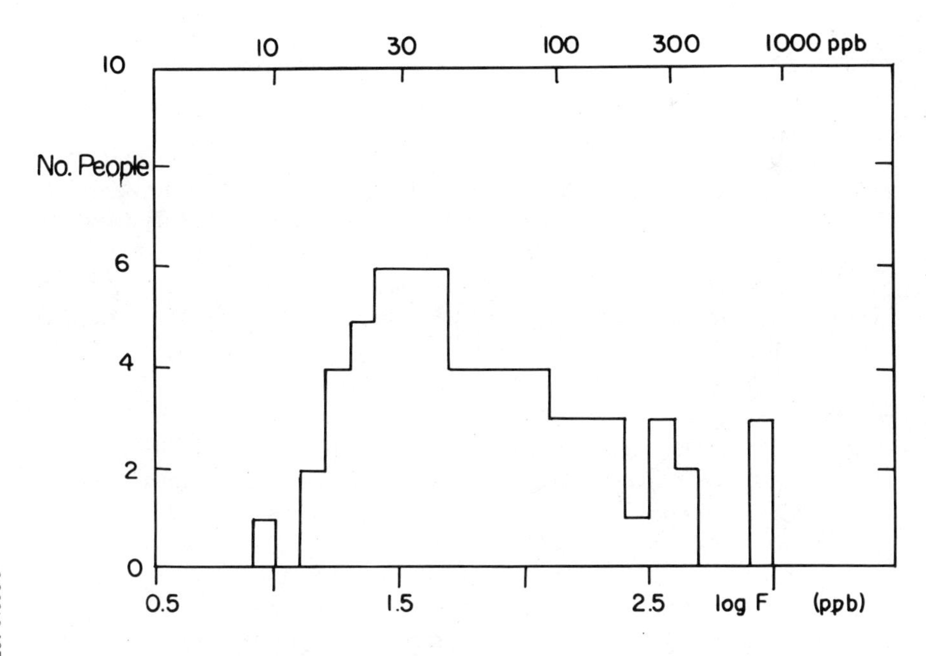

Figure 10.1. Recognition of Formaldehyde by Human Subjects (Rehn, 1977)

ISBN 0-201-04558-3

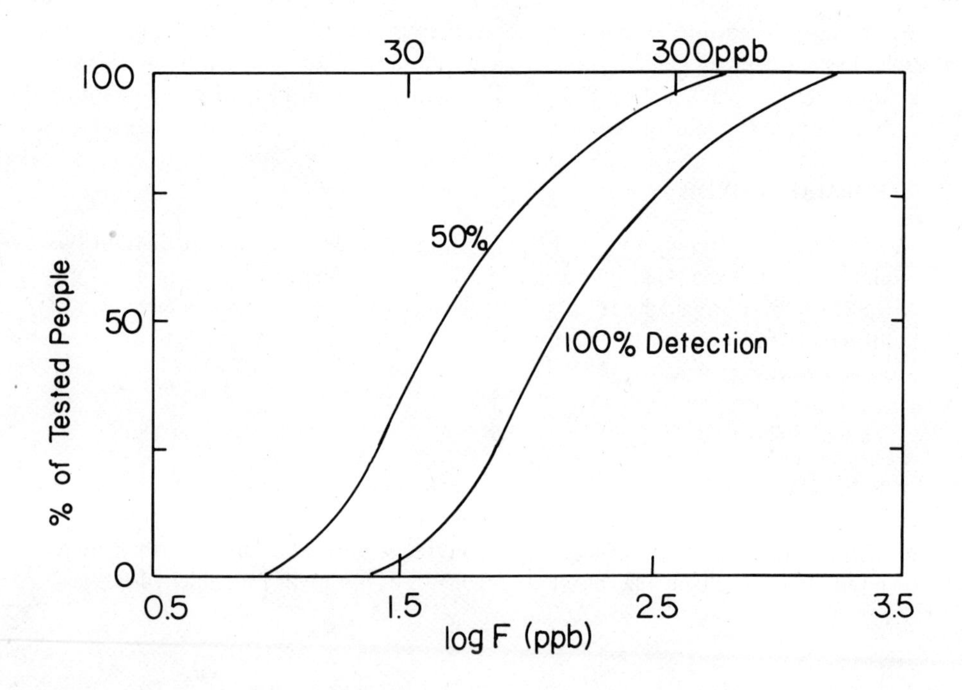

Figure 10.2. Dose-Response Curve for Formaldehyde Odor (Rehn, 1977)

Andersen (1979) tested 16 healthy young volunteers over a period of four hours, Figure 10.3, and found that the time response curve depended on the concentration. Formaldehyde is almost always in equilibrium with several of its derivatives and precursors. UF-resins contain two formaldehyde sources: (a) excess formaldehyde dissolved, trapped or absorbed in UF-resins or UF-bonded products, and (b) hydrolysis products, formed by aging of UF, especially unevenly cured resin, such as uneven, thin films.

With water formaldehyde converts almost quantitatively, but reversibly, to methylene glycol, as described in Chapter 3. At room temperature, it spontaneously polymerizes to polyoxymethylene glycol or solid paraformaldehyde. The hydrolysis of paraformaldehyde to methylene glycol and the vaporization to formaldehyde gas is slow. The equilibrium vapor pressure of formaldehyde at room temperature depends on the chemical form of its precursor, but it is sufficient to cause discomfort. Solid and aqueous formaldehyde can cause skin allergies. Japanese doctors, for example, have established that UF-treated cotton underwear, especially underpants, can release up to 1336 ppm. The allergies disappear if the source is abolished (Yoshihara 1974, Horima 1974, Kojima 1975).

ISBN 0-201-04558-3

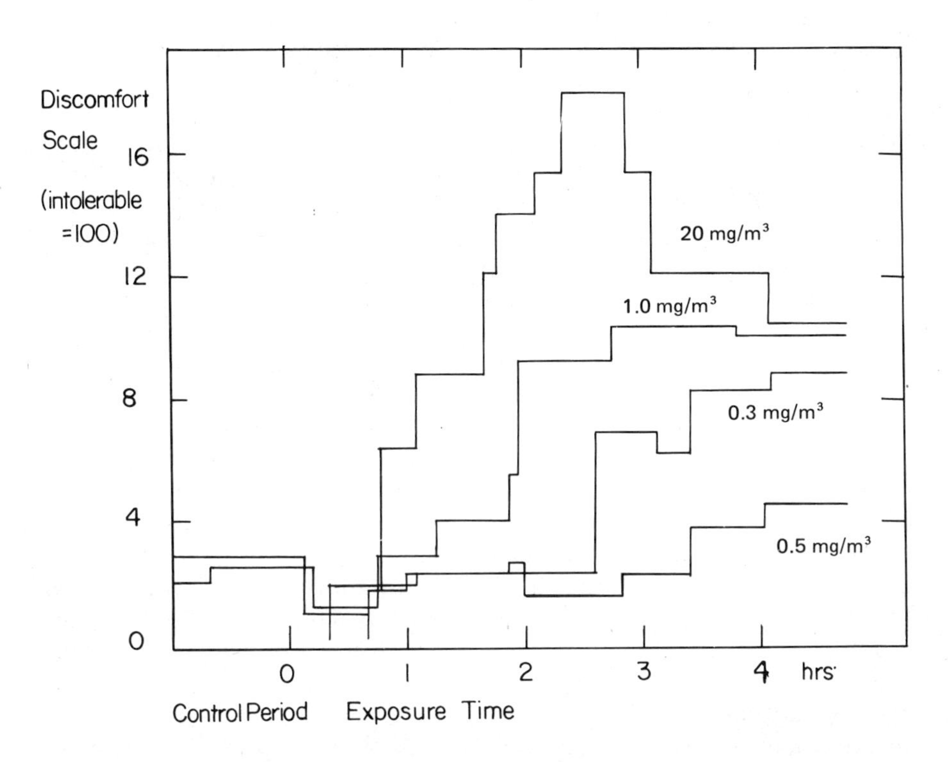

Figure 10.3. Discomfort Response as a Function of Exposure Time for Four Levels of Formaldehyde (Andersen, 1979)

Guinea pigs can ingest up to 800 mg/kg. Cats, dogs, and rats tolerate about 400 mg/kg subcutaneous formaldehyde. Rats can inhale some 300 ppm for four hours.

The result of extended exposure is very poorly known. Current U.S. Department of Health, Education, and Welfare recommendations by the National Institute for Occupational Safety and Health (NIOSH) give a ceiling of 1.2 mg/m^3 for a thirty-minute period, and OSHA workplace standards of December 30, 1976 are 3 ppm over an eight-hour workday, with a 5 ppm acceptable ceiling and a 10 ppm maximum for any period of not more than 30 minutes. Standards in other countries differ greatly. Russian and the Scandinavian countries are more strict, and central European countries contemplate similar standards. Recommendations and regulations are currently in flux (Andersen 1979).

Chronic Exposure

Formaldehyde has been in regular use for over a hundred years in mortuaries, in hospitals, and in dissecting rooms where air concentrations up to 6 ppm have been

ISBN 0-201-04558-3

measured (E. Strofft 1971). During the last thirty years, workers in plywood and particleboard plants have been exposed to similar concentrations. So far, no chronic, work-related complaints have been confirmed. While there is full agreement that such concentrations are unpleasant, there is no agreement whether there is any chemical or physiological hazard connected with such work exposures. Some people seem to become rapidly desensitized without adverse effects. Those who develop allergic reactions, quickly avoid it and seem to suffer no long range ill effects. The problems with chronic low-level exposure studies are many. For one, it is not clear whether this type of exposure has a cumulative effect, as is known for silicosis, caused by dust. Such an effect is unlikely because the human body produces enzymes which convert formaldehyde, while silicosis is due to the fact that asbestos and other particulates are inert, and therefore cannot be dissolved and removed from the lung. Furthermore, it is not clear whether the effect of low-level exposure can be predicted by extrapolation of higher levels of acute exposure, especially if the latter are not established either. Obviously, the study of such effects is a difficult task which requires that the toxicity or complaints in response to formaldehyde are clearly specific to that substance and well defined, or that very large population groups are tested. During the last decade, substantial progress has been made in the statistical analysis of epidemiology. For example, Finklea, of the U.S. HEW has studied asthma and croup among children and other sections of the population, and found significant trends around certain types of smelters, partly due to heavy metals, possibly also due to sulfur dioxide (Meyer 1977). Similar studies cannot be done with formaldehyde, because formaldehyde does not occur in geographically isolated areas. Furthermore, the ill effects of formaldehyde are not known, not even at acutely uncomfortable levels.

Strangely, the main thrust of current interest in the potential health hazard does not originate from work related, high-level exposure which can reach 15 ppm and more, but from the much lower, but quite persistant, nuisance noted by consumers who share houses with products which were either improperly designed or manufactured under conditions of improper quality control. Such UF products can exude formaldehyde at the 0.2 to 3 ppm level for many months. The latter can cause irritation to sensitive individuals, and it adds a stress factor which is noted by the many who suffer from other acute or latent complaints. Chapter 9 explains that such psychological consumer reaction is understandable and justified towards unsolicited intrusion into the private home. Furthermore, it is not astonishing that formaldehyde should be perceived to be unpleasant. Formaldehyde reacts readily with amino groups and with hydroxyl groups. Such groups are characteristic functions of DNA, proteins, sugars, and almost all other biochemical systems. The reactions are so vigorous that formaldehyde has been employed in one of the oldest and longest used test of proteins, first developed by Van Slyke (1910). Accordingly, formaldehyde is a known mutagen (L. Fishbein 1970) and a possible carcinogen (H. S. Rosenkranz 1972) in some animals, but Andersen (1979) believes that such reactions are unlikely in humans, because the liver produces an enzyme which can destroy formaldehyde before it reaches its chemical target.

ISBN 0-201-04558-3

Formaldehyde Levels in Air

In remote, ambient air formaldehyde varies between 0.1 to 0.4 ppb (Warneck 1978). In Los Angeles it is 50 ppb to 180 ppb (Altschuller 1963) depending on weather. In Europe concentrations of 140 ppb, i.e., 0.4 ppm, are regularly observed in urban areas. Above a saturated boiling solutions, the concentration is 0.3 atm, i.e., about 3×10^5 ppm as indicated in Figure 3.4. The high vapor pressure of formaldehyde facilitates spreading in the work environment of 1–20 ppm and is responsible for the unpleasant odor. Three types of exposure must be distinguished:

1. The formaldehyde concentration is highest in the industrial resin factory, where chemical hazards are so significant that they are well recognized and carefully controlled. Thus, problems are well defined, reactions take place in closed vessels, and injury occurs only when quality control is temporarily lacking.
2. The formaldehyde concentration in the facilities of professional resin users, such as in textile factories and particleboard plants, often exhibit high formaldehyde levels which cause acute discomfort to workers.
3. Concentrations around UF-resin bonded products are usually low, but in enclosed spaces the vapor can accumulate and reach sufficient levels to constitute for the consumer a noticeable long-range nuisance.

The conditions (1) and (2) are covered by occupational health laws, the conditions of (3) fall under general indoor air quality. While occupational health standards have been widely formulated and are being widely accepted, the ambient indoor air quality standards are still in a formative stage and are hotly disputed for theoretical and practical reasons.

Public Attitude

In the early days, U:F ratios of up to 8 were employed. In factories and in the air surrounding fresh products the formaldehyde smell was fierce. However, high formaldehyde ratios produced tough products from which the odor quickly dissipated. Ellis (1935) describes several procedures to reduce free formaldehyde, for example by a second addition of urea or thiourea, by purging the resin with air, and by the addition of ammonia or of cellulose. His main concern was not odor or safety, but the product quality. Until very recently, the workers' health and comfort were not adequately protected in all countries. For example, those who impregnated felt hats with UF-resins according to Pollak's formulation were not only exposed to formaldehyde foams, but traditonally worked with mercury which was used to soften felt. This caused notoriously debilitating work-related illnesses which led to the coining of the term "mad hatter."

As recently as during the 1940s there were amazingly few complaints about UF-resin conditions. In fact, the author has been unable to collect any information about complaints connected with the use of some 4000 tons per year of UF foam for home insulation during World War II. Even today, workers in the textile and

ISBN 0-201-04558-3

forest products industry are quite indifferent to the odor, partly because they are desensitized, partly because of inertia, and possibly for fear of losing job security. However, the users of final products, especially the home user, the consumer, has changed his attitude towards living quality. An interesting example is the sensitivity towards contact allergies caused by formaldehyde. Thus, according to Baer (1964 and 1973) the index of sensitivity to formaldehyde increased by 290% during the period from 1937 to 1962. This is only partly due to changes in textile finishing techniques. Today, formaldehyde in textiles is considered one of the ten most common contact allergens.

In regard to airborne formaldehyde, Neusser and Zentner (1968) in an excellent technical paper, advised handling consumer complaints by ventilation or by using stalling tactics, because aging of resin products would solve most problems in time. Today, the situation is quite different. In the U.S. the indoor ambient air question is intrinsically connected with the development of basic changes in lifestyles. Economic pressures and government coercion towards energy savings in the private sector have caused homeowners to seek tax credits by improving insulation. This has created ventilation problems, which caused measurements of 0.5–1.5 ppm of formaldehyde in mobile homes. At the same time, the current urge to foster a less materialistic value system in a more environmentally oriented framework has enhanced sensitivity to ambient air quality. The result is that questions of safety and quality are being debated simultaneously, and often jointly. This situation is not unique to formaldehyde, but is reminescent of the outdoor pollution by automobiles, power plants, and chemical plants (Meyer 1977). It is important to remember that pollution, according to Webster, means an "action which makes or renders unclean; defiles, desecrates, or makes profane." This implies mystic as well as religious forces and involvement. Since pollution and waste dumping have been tools for social struggles in establishing class structure since antiquity, it is not astonishing that covert elements of class sensitivity can be triggered. For example, a substantial thrust towards regulation of indoor quality in Germany has come from parents and teachers, because during the record heat of summer 1976, in a new school building in Karlsruhe, formaldehyde from ceilings and school furniture reached irritating levels. The problem had been further enhanced by the accumulation during vacation, during which the building had not been vented. Similar concerns were reasons for earlier actions in Russia (Gebtishchevra 1969). In the U.S. public concern has been stimulated by isolated incidences of hasty or plainly shoddy workmanship during installation of UF insulating foams, and by the reccuring problem with mobile homes. In many cases, mobile homes are intended to provide the cheapest possible shelter. A tempting factor in optimizing this goal is to reduce general quality control standards. Furthermore, economic factors have made it common practice to use unfinished and uncoated industrial bulk quality particleboard which was never designed for such applications. The same procedure caused many problems in the furniture industry where complaints increased greatly when manufacturers ceased to coat and finish the side surfaces on doors of drawers

ISBN 0-201-04558-3

and cabinets. These complaints, together with the general concern about energy and environment led to the current studies, such as the U.S. Department of Energy's "Geomet" study by Morschauser (1978).

Regulations and Recommended Limits

Neuss and Zentner (1968) reviewed the status of knowledge and regulations of their time on airborne formaldehyde. In 1966 occupational regulations permitted 12 ppm in England, 6 ppm in the U.S. and Yugoslavia, 5 ppm in Czechoslovakia, and 1 ppm in the USSR. In Germany the VDI-2306 guidelines of 1966 proposed a maximum workplace concentration of 6 ppm, a maximum 30-minute celing of 0.07 ppm and a continuous load factor of 0.03 ppm. The corresponding U.S. recommended concentrations of 1965 were: 5–10 ppm as ceiling, 10–20 ppm peak. The values were formulated with the stated intent to be flexible to match diverging needs and situations. It was already well recognized that ventilation rates could greatly alter workplace concentrations. For example, in the dissecting rooms of medical schools, the formaldehyde concentrations varied from 6 ppm in the the middle of the hall to 20 ppm immediately above the bodies, and up to 100 ppm in the storage vaults. The concentrations were also a function of the number of students present in the rooms (Stofft 1971).

STANDARD AMBIENT INDOOR AIR

Concern for clean air in nonindustrial areas has been highest in the USSR, where E. A. Geltishcheva, V. Khrustaleva, and S. K. Osokrisa proposed in Moscow in 1969 that particleboard and fiberboard not be used in children's schools and institutions, unless they were coated with paper laminates. Even then, their recommendation suggested that all rooms containing UF-bonded materials should be aired twice daily for at least one hour. In 1969 Y. G. Nefedov *et al.* discovered that formaldehyde was exhaled by healthy people as normal metabolites and that such formaldehyde could be identified in human breath. I. K. Pushkina (1969) established that quality control in particleboard factories reflected itself directly in the level of formaldehyde in air, and that such levels readily fluctuated by a factor of three or four between different factories. In 1971 V. K. Golyshew observed that UF-resin in industrial wastewater could cause excessive formaldehyde concentrations. In 1973 A. N. Bokov *et al.* in Rostov, Russia established that old melamine resins emitted 45–75 times the allowable formaldehyde concentration and should be replaced by the new generation of UF-resin. UF was found permissible for bonding plywood, coating parquets and molded fixtures. In 1974 J. Michal and J. Mitera claimed that burning UF-resin at 550°C generated HCN, but this was not confirmed by others, including K. Takeyama (1974). In 1975 K. Flueck of the state occupational health office of Bochum, Germany, measured formaldehyde in a particleboard plant and measured 0.6 to 3.2 ppm at the press console panel, and 0.7 to 6.5 ppm at the press

ISBN 0-201-04558-3

exit and the resin mixing stand. The variations were explained by differences in ventilation rates.

In 1974 I. Andersen, a medical doctor, G. R. Lundqvist and L. Mølhave measured up to 0.8 ppm in Danish homes at 22°C, 35% humidity, and an air exchange rate of one half per hour. This work started a long series of excellent papers. In 1975 they reported data from 25 rooms in 23 Danish dwellings and found a range of 0.08–2.24 ppm, with an average of 0.62 ppm. They correlated air concentration to surface area and ventilation rate. In the U.S. official concern remained long restricted to workplaces. In 1976 L. G. Wayne and R. J. Bryan visited 18 U.S. industrial plants and reported results of a study on a shirt factory, a plywood plant, and two particleboard factories to NIOSH. In the shirt factory, they found 25 workers in the cutting room exposed to 0.1 to 0.8 ppm, with an average of 0.35 ppm. The values for the 30 workers in the pressing room ranged from 0.0 to 0.4 ppm and average 0.17 ppm. In the sewing room 250 people were exposed to 0.0 to 0.2 ppm at an average of 0.08 ppm. Thus, the total average for the entire factory was 0.21 ppm. In the plywood plant, 40 workers were exposed to 0.4–1.1 ppm with an average of 0.60 ppm; in the particleboard plant, 4 people experienced 0.5–1.4 ppm with an average of 0.92 ppm; and in finishing and the warehouse, 22 people were exposed to 0.7–1.5 ppm, again with an average of 0.92 ppm. The total, overall exposure ranged from 0.4–1.5 ppm, with an average of 0.81 ppm. In another plant, values were higher, averaging 1.28 ppm, and were up to 2.7 in the assembly line and almost every other part of the plant. During the daytime, hourly variations were up to 200%. The authors considered fluctuations uncontrolled.

In 1977 Breysse, a scientist at the public health school of the University of Washington, in a study going back to 1961 reported formaldehyde levels measured in 74 mobile homes, with concentrations up to 2.11 ppm, and normally around 0.4 ppm. All measurements were the result of complaints from occupants, all of whom noticed the odor and many of whom reported health complaints. Thus, this study constitutes a summary of extreme cases. In 1976 K. A. Akhmedzhanov *et al.* measured urea emission from school desks and children's furniture. In 1978 Carbone sampled 91 homes over a period of 13 months and found that 80, or 88%, had concentrations levels above 0.1 ppm, the ambient air level recommended by the American Industrial Hygiene Association. Five, or 5%, exceeded the proposed industrial 8-hour standard of 1.0 ppm of NIOSH, Table 10.1. The concentration levels were followed as a function of age, atmospheric temperature, climatic factors (including daily precipitation), size of the home, and humidity. Several homes were revisited. The only clear correlations found were those with age and temperature. In 1978 Sundin reported measurements from 319 Swedish homes with an average of 0.58 ppm, with the frequency distribution shown in Table 10.2. This value corresponds to a concentration of 0.37 ppm in homes where 5 people smoked cigarettes each within 30 minutes in a room with 45 cubic meter volume.

ISBN 0-201-04558-3

Table 10.1

Current Tentative Air Standards in Four Countries, 1979 (ppm)

	USA[f]	Germany[f]	Russia	Sweden
8 Hr Time Weighted Average[a] (TWA)	1.2[r]–3	0.07	1	2
30 Min. Threshold Limit Values[b] (TLV)	3[r]–10	6	1	1
Ambient[c]	0.1[r]	0.1[r]	0.01[d]	0.3
Indoor	–	–	0.01	0.13[e]

In Germany: [a]MAK, [b]MIK_K, [c]MIK_D
[d] Ambient city air has a background up to 0.12 ppm
[e] Action taken above 0.7 ppm
[f] r: recommended

Table 10.2

Formaldehyde Level in 319 Swedish Homes (Sundin, 1978)

Formaldehyde Values in Houses Sept. 1975 – Oct. 1977			
ppm interval	Frequency	%	Cumulative %
<0.3	72	22.6	22.6
⩾0.3–<0.4	60	18.8	41.4
⩾0.4–<0.7	100	31.3	72.7
⩾0.7–<1.0	49	15.4	88.1
⩾1.0	38	11.9	100.0

Average for 319 homes: 0.58 ppm

Nature of Source

Formaldehyde has two properties which influence health and environment. First, it is vaporizable. Second, it is readily chemically absorbed by water and thereby transforms reversibly into highly soluble species. The first allows formaldehyde to readily enter air; the second causes it to linger on surfaces and resist removal by ventilation.

ISBN 0-201-04558-3

TOTAL VAPORIZABLE FORMALDEHYDE

It is frequently overlooked that airborne formaldehyde concentrations are not a reliable measure for the total formaldehyde pool. First,formaldehyde does not totally evaporate, but air concentrations are in equilibrium, or below equilibrium, with the formaldehyde source. Especially on moist surfaces a considerable quantity of formaldehyde remains absorbed and chemically dissolved. Second, the vaporization process is normally slow compared to the normal ventilation rates of homes. Only closed furniture cabinets and specially designed energy conservation homes reach an air composition which corresponds to equilibrium.

Thus, only in the case of short and acute formaldehyde exposure, does the level of exposure correspond to, and directly correlate to, airborne concentrations. In contrast, during the long-range and low-level exposure in mobile and other homes, the air concentration is not steady, and it is not necessarily a measure of integrated formaldehyde exposure. Especially when the recipient surface is moist and strongly absorbs the gas, the gas concentration merely indicates the transfer rate between the source and the recipient. The hazard is then no longer necessarily primarily due to air, but it can be caused by accumulation of formaldehyde transferred from the source. Thus, it is conceivable that formaldehyde distills from a source surface, say a particleboard, to the recipient's clothing, especially if it is humid or sweaty. Such clothing, flowerpots, and the like could become sinks and could serve as secondary formaldehyde sources. Accordingly, modern product standards measure and restrict total free or vaporizable formaldehyde. The definition of these terms, and the problem in determining them are discuseed in the chapter on particleboard.

RELEASE MECHANISM

Hot-cured molded and cast UF articles exude very little formaldehyde. Likewise, cold-cured UF floor varnishes rapidly lose their odor. Properly treated and aged UF foam is odorless, and even upon exposure to moisture, exudes so little formaldehyde that it can be used as wound powder, without causing the well-known formaldehyde skin allergies. UF-bonded plywood normally smells from veneer, and not from the glue, and particleboard does not release noticeable concentrations of formaldehyde if properly aged, used, and coated. In fact, UF-resin has been proposed as coating for cedar, to reduce its odor (Crocker 1977, USP 2,327,871). The problem of formaldehyde arises from two sources: (1) improper quality control during manufacture, or (2) improper design or application. Examples of the first are excess formaldehyde, excess methylol, or excess acid catalysis in resin or foam. Examples of the latter are exposure to improperly designed mositure barriers or the use of unfinished particleboard. More subtle problem areas arise from the use of improperly applied or furmulated insulation foam, from improperly selected or used particleboard, or from freshly treated textiles, all of which share the problems of having large and uneven surface areas. Large surface-to-volume ratios not only

ISBN 0-201-04558-3

increase the vaporization rate, but they are difficult to cure, because of uneven heat dissipation in the vicinity of surface anomalies.

Normally, the formaldehyde vapor pressure above solid molded UF is comparable to the ambient air background. The same is true for pure medicinal UF foam. The concentration above commercial home insulation foam is about 0.1 ppm, and decreases with age (Bowser-Mower 1978). Even during the foaming operation in closed spaces, the concentration can be kept below 1 ppm (Meyer 1978). In foam as well as in particleboard, the formaldehyde control is also a question of economics. Progress since 1970 shows that it can and will be significantly reduced by improvement in resin formulations, such as discussed in Chapter 9.

ABATEMENT

In chemical industry, formaldehyde containing gases are incinerated. Properly manufactured and installed UF products are normally free of odor. If necessary, the formaldehyde odor can be abated by several methods. The manufacturers can age and air particleboard at the plant. They can coat or impregnate the finished board. Builders can paint, impregnate, or coat the board or install a vapor barrier. If desired, they can fumigate the board with ammonia (Verbestel 1979) or sulfur dioxide (Andersen 1978). The consumer, finally, can improve ventilation. The effect of ventilation is shown in Figure 10.4 for fresh, untreated particleboard.

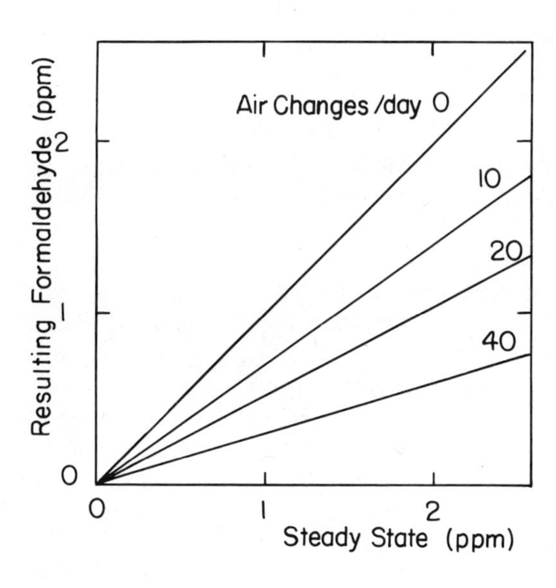

Figure 10.4. Effect of Ventilation Rate on Formaldehyde Level as a Function of Air Change and Steady State Level (Sundin, 1978)

ISBN 0-201-04558-3

Normally, the formaldehyde odor has a half-life of several weeks, or even months, and thus, the problem cannot be solved by ventilation alone, or by aging.

Health Effects and Dose Response

Response to formaldehyde varies greatly among individuals. Thus, the dose-response curve has a wider spread than for many other irritants. Furthermore, while some individuals become increasingly allergic, others apparently become desensitized. Every visitor of workplaces with high formaldehyde concentrations is surprised that people seem to go about their jobs unhampered, while newcomers rub their eyes. Whether such work conditions are safe is currently a matter of study and dispute.

A. K. Sgibnev (1978) observed orientation reactions and olfactory sensation. Above 0.83 ppm (= 1·mg/m^3), all subjects responded. At 0.3 ppm half of the subjects noted olfactory sensation. H. Trinkler (1968) did similar work. J. Akabane (1970) reviewed 190 papers dealing with metabolism, toxicity, and pharmacological effects of lower aldehydes. Dalham (1971) studied the effect of aldehydes on tracheal mucus. Y. G. Feldman (1971) the inventor of an excellent sampling test, studied the long-range effect of 0.1 ppm ambient formaldehyde on rats and reported delayed nerve responses. Pavlenko (1973) found strong adaptation to inhalation. Conte and Devitofrancesco (1973) believe that irritation is limited to the upper respiratory tract. Takhirov (1974) observed that formaldehyde reacted additively with hydrogen chlorid, acetone, phenol, and acetic acid. Kulle (1975) observed that the trigeminal nasal sensory system is desensitized. Anokhin (1976) and Burakinskaya (1976) studied the toxicity of UF-bonded materials. Breysse (1977) and Carbone (1978) reported health complaints of mobile homeowners. Wayne and Bryan (1976) made a very thorough study of irritant effects of industrial formaldehyde at 3 plants selected from 18 factories visited, and conducted and evaluated ophthalmologic tests on 83 workers during and after work. They found no correlation between chronic formaldehyde exposure which averaged 0.4 ppm and eye symptoms or disorders. Visual acuity, depth perception, peripheral vision, accommodiation, fixation, and color vision were measured. They prepared a list of factors for NIOSH to suggest for use by ophthalmologist in obtaining a medical history of workers exposed to formaldehyde. They concluded that workers who were exposed to 1 ppm and more performed worse on all tests during exposure, but no adverse effects remained noticeable after work. However, exposed workers had a noticeable tendency to blame their eye problems on their work atmosphere, even though the frequency of their symptoms were not greater among exposed workers than among controls.

Finally, in 1979, I. Anderson, a medical doctor, reviewed the biological effects of formaldehyde and conducted a careful study with 16 healthy young adults to determine sensitivity. They were exposed to 2 ppm for four hours. No change in airway resistance was observed, but the threshold towards the smell of ethyl valeriate was somewhat increased. A small decrease in nasal mucus flow was found.

ISBN 0-201-04558-3

Eye irritation and dryness in the nose and throat was experienced by three subjects (20%) at 0.3 ppm, and by 15 subjects (90%) at 1 ppm. He rejects claims of sensitivity to 0.01 ppm, because ambient urban air has a background of about 0.07 ppm.

I. A. El-Sayad (1972) reported that the dermatitis induced by phenol resins was far lower in urea resins. Ivanova (1973) found that the skin of humans and animals was sensitized. In 1973 Kharchenko found that resins irritate the skin. Sevliakov (1974) observed allergic dermatitis; Dueva (1974) observed desensitization upon second intracardial injection in guinea pigs. Schorr (1974) in his clinical practice observed extreme allergies and measured in concentrations of 1 to 3500 ppm in women's fabrics. In Japan similar dermatitis has been repeatedly reported from underwear.

Formaldehyde was suspected to be a potential carcinogen by Rosenkranz (1972). Shumilina (1975) suspected a correlation between formaldehyde exposure and menstrual disorders. Traganos (1975) observed *in situ* changes in DNA denaturation. Vologodskii (1975) explored theoretically the unwinding of DNA; Cowan (1975) lists formaldehyde as a mutagen for yeast. Faichney (1975) studied the effect of formaldehyde on the gastrointestinal tract of sheep. Kitchens (1976) claimed that it is neither mutagenic nor carcinogenic on mammals. Andersen (1979) pointed out that the human liver and the erythrocytes have enzymes which are capable of detoxifying formaldehyde, by conversion to formic acid. He thus rejects a cumulative effect, and doubts that it acts as a human mutagen or carcinogen. He accepted the well-established skin dermatitis, but stated that he is unaware of any reports of asthma below 2 ppm. He reported discomfort responses as a function of time, in response to four exposure levels ranging from 0.5 ppm to 2 ppm, Figure 10.3, page 253, and suggests a TLV standard of 0.15 ppm, corresponding to about three times the background value of ambient urban air.

Table 10.3

Human Response to Formaldehyde
(U.S. Department of Health, Education and Welfare)

Dose-Response Following Human Exposure to Airborne Formaldehyde			
Concentration (ppm HCHO)	Duration of Exposure	Frequency N	Responses
0.9–2.7	Hour	Several	Tearing of eyes and irritation of nasal passages and throat (Irritant effects were greatest at very beginning of work
0.9–3.3	Hour	Several	Mild eye irritation, objectionable odor

ISBN 0-201-04558-3

Table 10.3 (Continued)

Dose-Response Following Human Exposure to Airborne Formaldehyde

Concentration (ppm HCHO)	Duration of Exposure	Frequency N	Responses
0.9–1.6	8 hr/d	2	Itching eyes, dry & sore throat, disturbed sleep, and unusual thirst upon awakening in morning
1.0	Odor panel	4	Odor threshold
1.4	Min	12	Eye sensitivity to light lowered in unacclimated group
0.06–1.3	"	12	Optical chronaxy changes in unacclimated group
+ 1.0	"	Several	Increased worker complaints
0.8	"	12	Altered functional state of cerebral cortex
0.8	Daily	?	Equilibrium and olfactory sensation shifts; irritation of upper respiratory tract and eyes in most sensitive individuals; enhancement of alpha-rhythms
0.3–0.5 (smog chamber)	5 min (eye only)	12	Increased blink rate, rate proportional to formaldehyde concentration
0.05–0.5 (smog chamber)	"	12	Eye irritation range in unacclimated group
0.13–0.45	?	Several	Complaints of temporary eye and upper respiratory tract irritation
0.07	Min	15	Odor perception threshold for group
0.06	Min	12	
0.05	"	5	No alteration of cerebral electrical activity in subjects most sensitive to odor

ISBN 0-201-04558-3

Table 10.3 (Continued)

Dose-Response Following Human Exposure to Airborne Formaldehyde			
Concentration (ppm HCHO)	Duration of Exposure	Frequency N	Responses
41./	10 min	1	Bilateral vesicle reaction on hands of hypersensitive person
16–30	8 hr/d	60	Eye and throat irritation, skin reaction
10–30	Min	1	Skin and eye tingling in hypersensitized worker, progression to generalized skin reaction
13.8	30 min	12	Nose and eye irritation subsiding after 10 min in chamber
1–11	8 hr/d	+50	Eye, nose and throat irritation
4.2–10.9	Min		Unbearable without respiratory protection
10.–	10 min	1	Bilateral vesicle reaction on hands of hypersensitive person
0.5–7.3	Daily	278	Increased occurrence of upper respiratory irritation
4	?	Several	Complaints of irritation of conjunctiva, nasopharynx, and skin; increased incidence of caterrhal conjunctivitis, slight reddening and drying of the skin
0.9–5.26 (with paraformaldehyde)	Hr	"	Eye and upper respiratory irritation; lessened during day, returned after lunch or next day
0.3–217	8 hr/d	"	Annoying odor, constant prickling irritation of the mucous membranes, disturbed sleep, thirst, heavy tearing

ISBN 0-201-04558-3

A summary of human dose-response reactions is given in Table 10.3; recent measurements are shown in Figures 10.1 to 10.3.

Health Standards

The formulation of standards is a complex process. Van Atta (1976) described the current OSHA mechanism for establishing standards in a short article entitled "Developing Occupational Health Standards." Toxic substances lists, of the U.S. HEW are published in the Federal Register, and guidelines for working with formaldehyde are available from HEW, and many private and public firms and associations. The U.S. Government released three reports dealing with formaldehyde, all in 1976: "Criteria for Recommended Standards: Occupational Exposure to Formaldehyde," published by NIOSH; "Irritant Effects of Industrial Chemicals: Formaldehyde," also by NIOSH; and "Investigation for Selected Potential Environmental Contaminants: Formaldehyde," by the U.S. Environmental Protection Agency. A comparative evaluation of the toxicity of formaldehyde with other dangerous industrial materials is available in the book by Sax (1975). Toxic effects of chemical substances have been reviewed by Christensen (1976) for NIOSH; suspected carcinogens are listed and compared in the files prepared and published by Christensen (1976). The USSR standard is 0.35 mg/m^3. No official standards have yet been set for formaldehyde in Europe or the U.S. The tentatively suggested value has been 0.12 mg/m^3. If one accepts the proposal of the American Society of Heating, Refrigerating, and Air-Conditioning Engineers (ASHRAE) that air used for indoor ventilation purposes should not contain more than 10% of the TLV exposure, then the current U.S. TLV standard of 3 mg would suggest a value of 0.3 ppm. With the proposed NIOSH value at 1.2 ppm, the home air standards would be 0.12 mg/m^3. This corresponds to the value suggested for spacecraft missions of 1000 days by NRC-NAS (1968). The current values are summarized in Table 10.1. In the U.S. the Occupational Safety and Health Administration (OSHA) and NIOSH, the Department of Housing and Urban Development (HUD), the Environmental Protection Agency (EPA), the Consumer Product Safety Commission, and the Department of Energy (DOE) have all task forces or research projects aimed at monitoring formaldehyde levels and their impact on health. Several industries have formed a coordinating group, the Formaldehyde Institute, for conducting their own work. Among its members is the Manufacturing Chemists' Association and other trade groups. It was shown in Chapter 9 that various product safety guidelines and standards are being developed to enforce formaldehyde emission limits. The implicit assumption in such product standards is that emission is directly proportional to air concentrations. This is unfortunately not always the case. Furthermore, emission rate measurements cannot be easily standardized, because they depend not only on a variety of connected parameters, such as humidity and temperature, but also on the age and the history of the products. The main reason for these difficulties is the complex chemistry of the formaldehyde

ISBN 0-201-04558-3

precursors. Other problems are rooted in sampling and the analytical chemistry, both described in other chapters.

Japan is the only country in which formaldehyde is an integral part of the official product standard. JIS-A-59087.8 of July 1973 and JIS-K-6801 provide emission limits of 5 mg for all classes of board, including 100, 150, and 200, with exactly prescribed measuring methods. This standard can be obtained from the American National Standards Institute. In Germany, four product qualities are being defined for particleboard to meet the anticipated TLV of 0.1 ppm:

1. Emission below 10 ppm/100 g is to be admitted without restructions.
2. Emission below 40 ppm/100 g is to be admitted with face treatment or coating.
3. Emission below 70 mg/100 g is to be admitted only if all surfaces are coated or treated.
4. Emission above 70 mg/100 g is not tolerable.

No decision has been made about the analytical method, but it is likely that the FESYP perforator method, described on page 153, will be used as a reference standard (Nestler 1977).

FORMALDEHYDE IN WATER AND FOOD

Until recently, formaldehyde was considered harmless, if not beneficial, and was freely dumped. Today, formaldehyde and resin containing waste waters are carefully treated (Diem 1974, Brunnmuller 1974). Formaldehyde is a powerful preservative, but it is not permitted in food. Its use in milk has been illegal for some time, and several tests have been designed to test for formaldehyde in milk. However, formaldehyde is permitted in some cheeses. Traces are found in smoked products, especially fish. The U.S. Food and Drug Administration recognizes that some formaldehyde might enter food from containers or utensils made from formaldehyde-containing resins, and from wrapping containing resins as a wet-strength improving agent. They limit free formaldehyde to 5 mg/kg, i.e., 5 ppm, however in some food additives, for example dimethylpolysyloxane up to 1 mg/kg, i.e., 1000 ppm are tolerated. The qualitative tests include the Hehner's test, used primarily for milk; Schryver's test; and Shrewbury and Knapp's test and the chromotropic acid test. The determination, i.e., quantitative measurement, is based on hexamethylene tetramines (Pearson 1976).

ISBN 0-201-04558-3

11. Future Trends

This chapter offers a brief view of what the author considers current and possible future developments. This chapter consists of ten short sections. First, the production statistics are analyzed, then chemical development will be reviewed, and finally a short section is devoted to summarizing or highlighting select developments in the field of each preceding chapter in the books.

Production Trends

GENERAL TRENDS

Urea-formaldehyde resins consume currently some 8% of the U.S. production of urea and 25% of the U.S. production of formaldehyde. Of these, urea as a raw material for resin, represents only 4% of the total U.S. ammonia production and only 8% of the U.S. urea consumption. The chemical urea market is a stabilizing influence for manufacturers, because it constitutes a steady market compensating for the highly seasonal fertilizers. Formaldehyde is marketed mainly as 37% aqueous solution. It is stabilized with methanol, from which it is produced. About half of the U.S. formaldehyde production of three billion pounds, i.e., about one million tons, is used for resins. Half of this is used for urea resins. Figure 11.1 shows the relationship between ammonia use and urea consumption (Gutheil 1978). Figure 11.2 shows the uses of methanol and formaldehyde demand. Forty-three percent methanol is converted to formaldehyde. Half of the latter goes into resins. Figure 11.3a lists the data for UF-resin, consumption, and Figure 11.3b that for phenol resins. Comparison shows how dominant adhesive UF-resins are. A more accurate distribution among the different use categories is contained in Table 11.1. Historic trends are excerpted in Table 11.2. Table 11.3 shows some typical 1975 prices for various commercial UF foams. These prices were twice those of 1971. Prices stabilized thereafter, but are expected to double again by 1982, due to general inflation. At the same time, the price gap between UF and PF will likely double, too, because of price increases for petrochemicals. A comparative melamine resin would cost about twice as much, a phenol resin perhaps three times as much.

Beat Meyer, Urea-Formaldehyde Resins

ISBN 0-201-04558-3

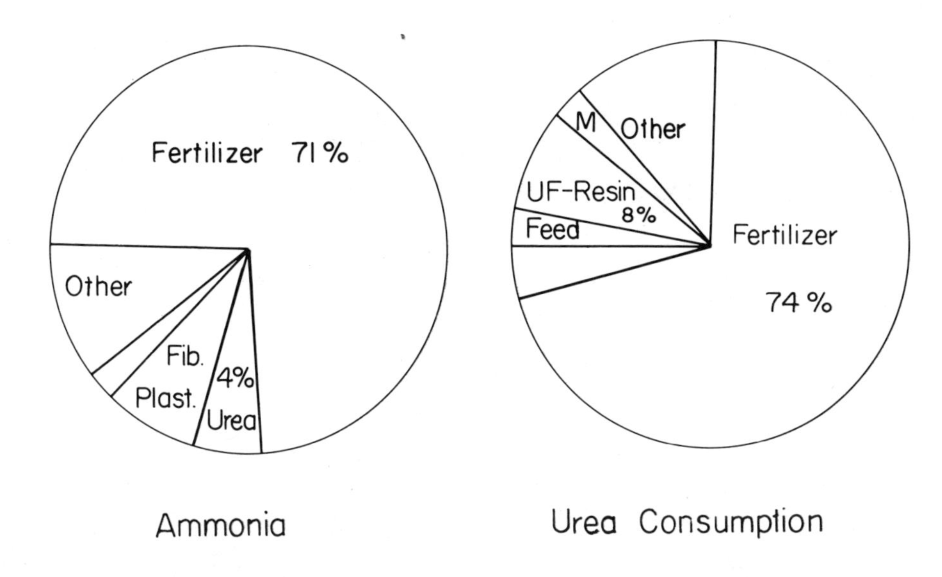

Figure 11.1. U.S. Consumption of a) Ammonia and b) Urea

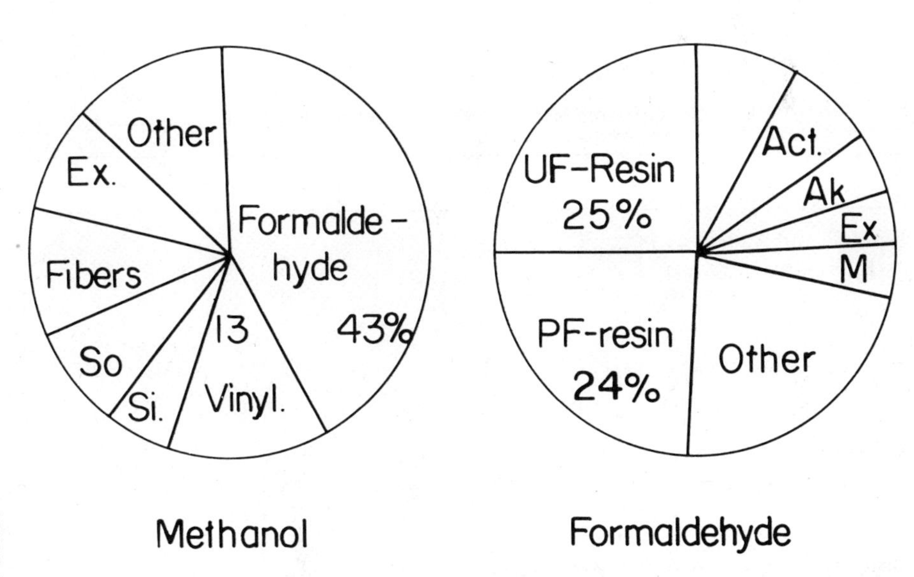

Figure 11.2. U.S. Consumption of a) Methanol and b) Formaldehyde (after Gutheil, 1978)

ISBN 0-201-04558-3

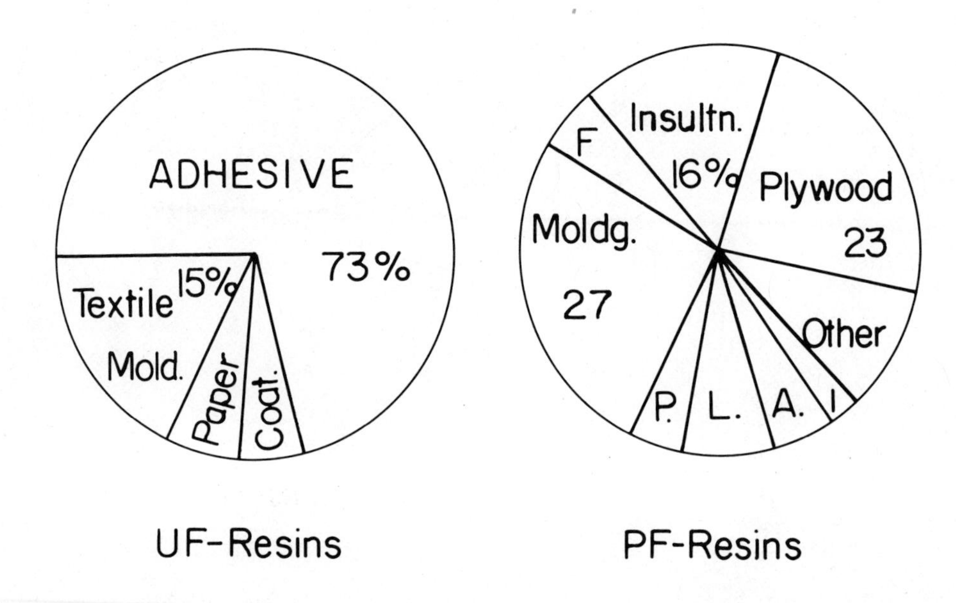

Figure 11.3. U.S. Resin Consumption a) Urea-Formaldehyde and b) Urea-Phenol (After Gutheil, 1978)

Table 11.1

U.S. Uses of Urea Resins, 1975 (after Updegraff, 1978)

Resin	Amount, kt	% of Total
Bonding and Adhesive		
Fibrous and Granulated Wood	274	60
Laminating	14	3
Plywood	33	7
Molding Compounds	43	9.5
Paper Treating and Coating Resins	24	5.5
Protective Coatings	35	8
Textile Treating and Coating Resins	14	3
Exports	15	3
Other	2	1
Total	454	100

ISBN 0-201-04558-3

ISBN 0-201-04558-3

Table 11.2

U.S. Consumption and Production of Aminoplastics and Aminoresins from 1930 to 1965 (after Widmer 1965)[a]

Year	All Plastics	Urea Resins[b]	Melamine Resins	Total Amino Resins	Consumption of Amino Resins for Various Uses: Adhesives and Plywood		Lami-nates	Molding and Misc.	Protective Coating	Paper Treatment	Textile Treatment
1930		(2.0)		(2.0)							
1931	77.3										
1933	103.3	3.2		3.2							
1935	183.2	4.2		4.2							
1938	236.7	8.2		8.2							
1940	423.4	21.5		21.5							
1942	(622.9)	37.5		37.5							
1944	999.0	18.1	(18.0)	66.1	27.1		5.5[c]	24.0	5.9	1.2	2.4
1946	1301.2	77.3	14.1	91.4	37.5		5.9	24.7	9.4	4.8	9.1
1948	(1686.5)	125.3	24.4	149.7	50.0		10.0	36.2	23.6	10.8	19.6
1950	(2431.6)	178.6	40.5	219.1	85.6		14.8	58.9	29.8	12.1	28.0
1952	2600.2	175.6	52.2	227.8	79.8		12.6	45.7	24.3	23.4	42.1
					Adhesives	Plywood					
1954	3209.9	197.5	67.6	265.1	6.4	79.8	18.9	67.3	27.8	20.6	44.4
1956	4408.4	245.2	96.3	341.5	7.6	107.6	22.6	93.7	37.7	26.1	46.2
1958	4978.3	238.7	110.6	349.3	11.6	101.6	32.0	96.1	32.5	24.7	50.8
1959	6344.5	280.5	143.1	423.6	27.7	106.5	36.4	138.0	38.6	30.4	46.1
1960	6300	268.2	130.8	399.0	31.0		35.4	122.2	41.2		44.8
1961	6857	305.5	134.8	440.0	56.1	99.3	38.5	110.0	40.9	42.6	52.6
1962[e]	7942			448.9	11.8[d]	95.7	51.0	121.5	42.0	51.7	51.8
1963[f]	8959			516.6	15.8[d]	120.3	42.9	119.8	50.9	48.3	51.4

[a] All figures in millions of pounds.
[b] Composition board consumption: 1962, 63.4; and 1963, 67.2.
[c] Underlined figures, estimates based on Tariff Commission data.
[d] Adhesives or other bonding uses.
[e] Chem. Eng. News, Sept. 2, 1963, p. 105.
[f] Chem. Eng. News, Sept. 7, 1964, p. 136.

Table 11.3

Commercial Urea-Formaldehyde Resins and 1975 Prices
(after Updegraff, 1978)

UF Resin	US $
Particleboard Syrup, 65 wt % solids	0.176
Plywood Syrup, 60 wt % solids, tank cars	0.185
Spray-dried Powder, water soluble, 100% resin, in 113 kg drums, such as Borden's Casconite 151 for wood adhesives	0.86
General Purpose Coating Resin, 55 wt % solids	0.72
General Purpose Molding Compound, 90.7 kg drum of granules, such as Allied Chemical's S-447, alpha-cellulose filled	0.97
Type A Butylated UF Resin, 208 L (55 gal) drums of liquid such as Cyanamid's Beetle 227-8, 50 wt % solids, for mixing with alkyds for baked surface coatings	0.80
Powdered UF Adhesive, 31.8 kg bags of powder, such as Cyanamid's Urac 110	0.48
Adhesive Syrup, 250 kg steel drums, chemically modified UF resin in aqueous dispersion, such as Cyanamid's Urac 185	0.73
Syrup for Paper Treating in Tank Cars, 35 wt % solids in water, such as Borden's Casco PR-335	0.20

Table 11.4 shows an approximate comparison of prices in six countries. The per-capita use of amino resins in 15 geographic areas is given in Table 11.5. The same table also lists the same data for all plastics combined.

The historic trend in plastics is shown in Figure 11.4. The growth is almost exponential and reflects the growth of GNP in the leading industrial nations. A similar growth has been observed for wood products, Figure 11.5. The data for U.S. particleboard production are shown in Table 11.6 for the period between 1965 and 1977.

It is likely that the European and Japanese markets are now reaching maturity. In the U.S. the shift from plywood to particleboard will continue. Thus, the particleboard could almost double in the next twenty years. Whether UF-resins will profit from this increase will depend on whether isocyanates will become cheaper, and whether technological problems in their application can be solved. Furthermore, it depends on legislative control of adhesives. Legislation and quality control will determine whether UF foam will find wide acceptance as a home insulation material. In the long range, UF foams are bound to become much more important.

ISBN 0-201-04558-3

Table 11.4

Price Trends for Amino Resins and Raw Materials in Six European Countries (after Widmer, 1965) (U.S. cents per pound)

	England	France	Germany	Italy	Belgium	U.S.
Urea-formaldehyde molding material	25.7	25.4	21.0	16.0	19.6	32.0
Melamine-formaldehyde, top-grade molding material	35.1	35.6	34.0	35.6	30.1	47.3
Urea, 46% N	4.1		4.8	4.4	3.7	5.0
Formalin	3.7	4.1	4.4	4.0	3.2	3.75

Table 11.5

Approximate Per Capita Consumption of Plastics and Aminoresins in 15 Industrial Nations (after Widmer, 1965)

		All Plastics			Amino Resins
Country	Population 1960/61	Consumption[a]	Consumption per Capita, kg	Sales per Capita, kg	Sales per Capita, kg
Austria	7.0	73.0	10.44	6.1	1.40
Belgium	9.0	97.4	10.8	5.1	
France	43.8	379.5	8.65	8.6	0.73
West Germany	53.8	853.2	15.9	20.6	0.37
Italy	48.4	326.3	6.7	8.3	0.98
Netherlands	11.1	115.6	10.4	8.5	
Norway	3.45	36.1	10.5	9.7	4.75
Portugal	8.8	6.5	0.74	0.12	0.04
Spain	29.4	69.1	2.35	1.8	0.19
Sweden	7.34	91.1	12.4	9.6	2.59
Switzerland	5.04	70.7	14.0	6.4	
United Kingdom	51.6	483.3	9.35	12.0	1.18
United States	172.0	2658.1	15.4	17.8	1.16
Japan	90.0	(511.3)	(5.7)	(6.1)	1.55
Western Europe	296.0	2672.9	9.0	9.6	0.71

[a] In 1000 tons per year

ISBN 0-201-04558-3

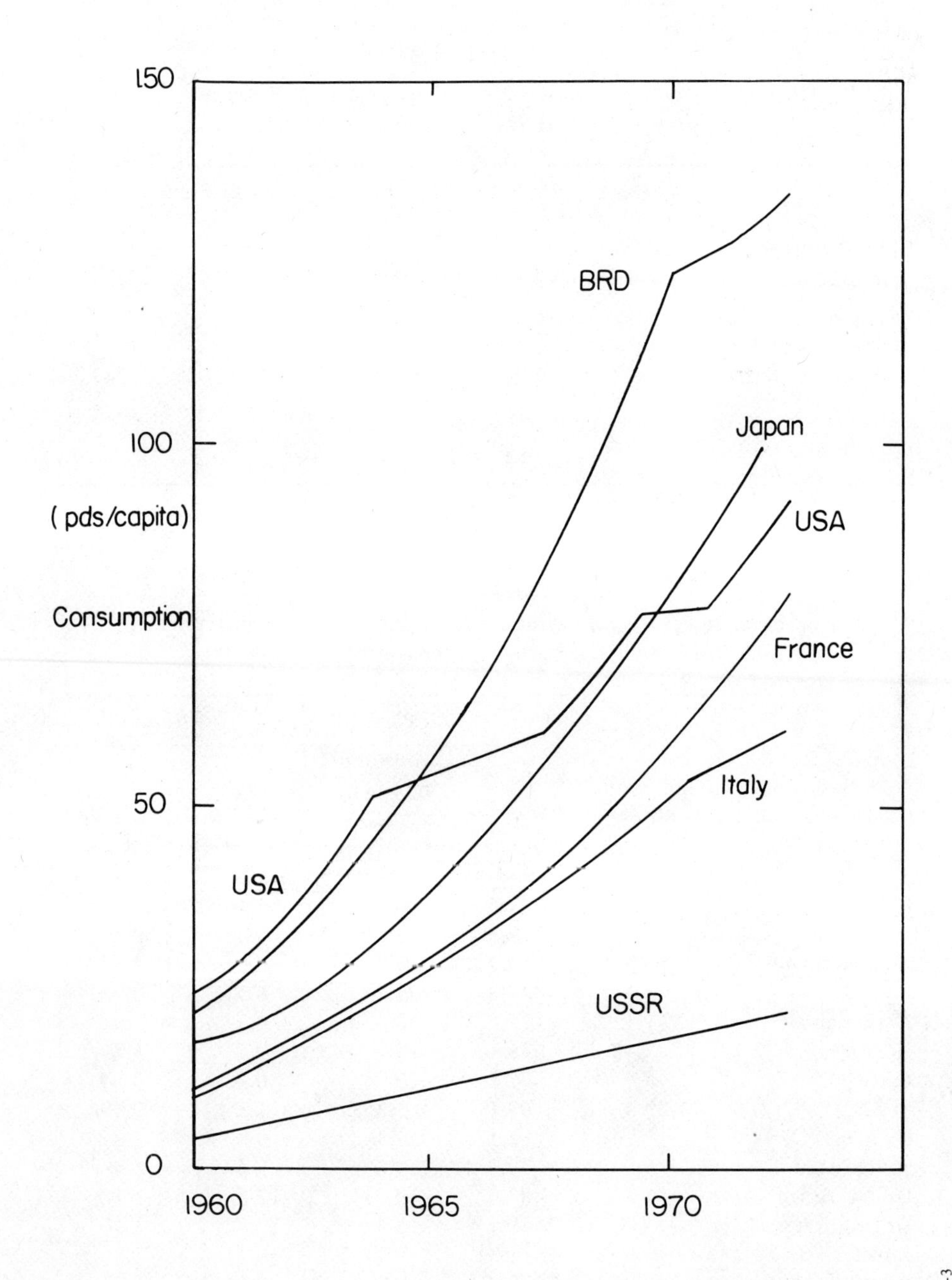

Figure 11.4. Plastic Consumption in Six Nations Between 1960 and 1975 (after Platzer, 1974)

ISBN 0-201-04558-3

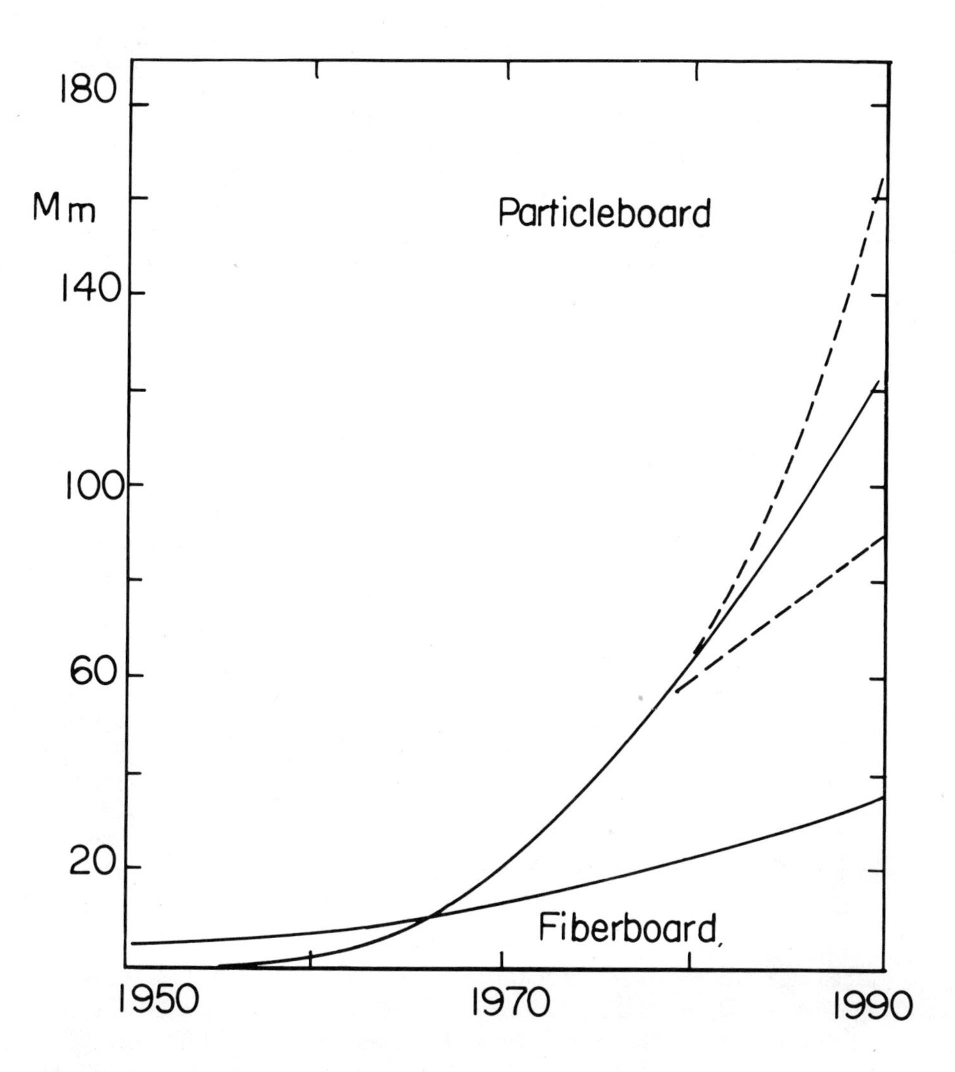

Figure 11.5. Anticipated World Consumption of Wood Products, 1970 to 1990 (after FAO, Proceedings New Delhi Conference, UNESCO, Brussels 1976)

ISBN 0-201-04558-3

UF foam production is currently about 75,000 tons. The insulation market could increase tenfold. Plastoponics are an entirely new field, and only imagination limits the growth in new agricultural uses. If their potential is realized, as much as one half of all fertilizer urea could be applied as slow-release fertilizer, and much of it as foam. Thus, the agricultural fertilizer market could increase demand for UF-resins fivefold. Other new uses could open new markets tenfold, again, compared to current size.

Table 11.6

U.S. Production of Particleboard between 1965 and 1977 (after U.S. Bureau of Census and National Particleboard Association; Morschauser 1979)

Year	Total[a]	Industrial	Medium Density Fiberboard	Mobile Home Decking	Floor Underlayment
1965	780,180	448,347	–	–	282,556
1966	994,361	573,020	–	–	371,801
1967	1,115,155	617,664	–	–	456,531
1968	1,424,988	843,727	–	–	547,450
1969	1,716,059	1,130,078	–	–	551,855
1970	1,763,548	1,135,024	–	–	596,427
1971	2,393,760	1,605,810	–	–	753,411
1972	3,116,944	2,137,770	–	–	941,353
1973	3,493,648	2,351,772	–	461,983	870,935
1974	3,103,825	1,859,624	–	313,558	901,253
1975	2,533,934	1,492,032	215,496	226,541	784,007
1976	3,202,225	1,535,713	280,036	276,998	1,092,321
1977	3,592,910	1,825,460	441,354	289,154	1,096,564

[a] Total includes extruded particleboard.

Japan currently produces and consumes more UF-resin than any other country. The U.S. is second, Russia third, and the Federal Republic of Germany, France, and Great Britain follow in that order. The market distribution could shift if new foam uses open markets in developing countries. Among the manufacturers BASF in Ludwigshafen, the successor to I. G. Farbenindustrie, is probably again one of the largest single producers. Casco, a member of the Swedish Nobel group, has very large worldwide markets, and licenses its technology as do most other large manufacturers. In the U.S. Allied Chemicals, American Cyanamide, Borden, Celanese, Georgia-Pacific, Pacific Resin, and Reichhold Chemicals are among the largest producers, to name only a few.

The future of UF-resins depends on how their performance will be improved, as compared to that of competing materials; how their price will be, and what profits manufacturers will be able to make. Currently, the last factor clearly limits expansion, because the low price of the raw materials does not allow the mark-up

ISBN 0-201-04558-3

necessary to pay for research and development of new products. This, in turn, prevents these materials from becoming more useful and more valuable. The future depends on whether this thirty-year old deadlock can be broken.

WOOD ADHESIVES

Seventy-three percent of all UF goes into adhesives. Of these, some 80% are used for particleboard. Obviously, the fate of UF-resin is closely tied to the forest products industry. Thus, a short review of this industry is in order. Table 11.7 lists the countries with the largest forest areas. Russia, Canada, Brazil, the U.S., and Australia have the largest holdings. Russia, the U.S., China, Brazil, Canada, and Indonesia harvest more than a hundred million cubic meters per annum, each.

Table 11.7

Thirty Countries with the Largest Forest Areas (FAD Yearbook, 1970)

Country	Area 1 000 ha	ha per capita	Estimated yearly removal 1 000 m^3	Estimated yearly removal from 1 ha
U.S.S.R.	738 117	3.1	380 400	0.52 m^3
Canada	420 328	19.1	111 872	0.26 m^3
Brazil	335 100	3.8	167 080	0.50 m^3
U.S.A.	292 721	1.5	336 158	1.15 m^3
Australia	207 267	17.2	13 757	0.06 m^3
Congo Dem. Rep.	129 141	7.7	11 588	0.08 m^3
Indonesia	121 177	1.1	100 994	0.83 m^3
China	96 380	0.1	169 000	1.75 m^3
Angola	72 000	13.4	6 459	0.89 m^3
Colombia	69 400	3.5	25 235	0.36 m^3
Peru	65 300	5.1	3 227	0.05 m^3
India	57 393	0.1	23 340	0.40 m^3
Sudan	55 100	3.7	20 982	0.38 m^3
Venezuela	47 970	4.6	6 824	0.14 m^3
Burma	45 274	1.7	4 900	0.11 m^3
Mexico	39 747	0.8	6 534	0.16 m^3
Zambia	37 631	9.2	4 022	0.11 m^3
New Guinea and Papua	36 422	15.8	4 101	0.11 m^3
Ecuador	34 711	6.0	2 285	0.07 m^3
Tanzania	34 015	2.7	30 525	0.89 m^3
Nigeria	31 592	0.5	55 235	1.74 m^3
Thailland	28 053	0.8	3 600	0.13 m^3
Cameroon	24 081	4.3	7 070	0.29 m^3
Rhodesia	23 570	4.7	3 799	0.16 m^3
Japan	23 556	0.2	51 588	2.19 m^3
Sweden	21 948	2.8	53 700	2.44 m^3
Finland	21 157	4.5	43 200	2.04 m^3
Paraguay	20 900	0.9	3 413	0.16 m^3
Gaboon	20 000	5.0	2 906	0.14 m^3

1 ha = 2.471 acres

ISBN 0-201-04558-3

Compared to its own holdings, the largest rate of use takes place in Sweden, Japan, Finland, Nigeria, China, and the U.S. The world consumption of processed wood is shown in Table 11.8. A more detailed breakdown of categories is shown in Table 11.9. Table 11.9a lists data for plywood, 11.9b for particleboard, 11.9c for fiber board, and 11.9d for veneer sheet production. The leading nations in each field are shown in Table 11.10. The data for the U.S. is further illustrated by a graphic comparison of the various uses. Figure 11.6a shows the distribution of timber products, Figure 11.6b compares lumber end uses. New housing and residential upkeep account for about 50%.

Table 11.8

Processed Wood Products (1970 FAO Yearbook of Forest Products)

			Estimated Shipping Cost		
	Volume (m^3)		per m^3		Total Value
Lumber-coniferous	261,744,000	x	$ 8.00	=	$2,093,952,000.00
Lumber-nonconiferous	89,205,000	x	9.50	=	847,447,000.00
Railway ties	9,117,200	x	4.50	=	41,026,000.00
Plywood	28,112,400	x	13.00	=	365,461,000.00
Particleboard	16,095,300	x	10.00	=	160,953,000.00
Veneers	2,690,000	x	15.00	=	40,350,000.00
Hardboard	4,368,100	x	16.00	=	69,896,000.00
Softboard	2,029,800	x	8.00	=	16,238,000.00
	total cost		$3,635,323,000.00		

Table 11.9

1973 Production, Import, and Export Statistics for Plywood, Particleboard, Fiber Board, and Veneer Sheet (FAO Yearbook 1970)

a) 1973 Plywood Production, Import, and Export, by Region

	Production	Import		Export	
	($10^3 m^3$)	($10^3 m^3$)	mil $	($10^3 m^3$)	mil $
World	42,656	6,940	1,460	6,443	1,362
N. C. America	20,669	2,636	460	801	1,368
Asia	13,702	1,161	253	3,673	60
Europe	4,604	2,900	701	1,419	366
U.S.S.R.	2,200	41	7	316	52
Canada	2,417	346	56	417	60
U.S.	18,049	2,200	390	308	73

ISBN 0-201-04558-3

Table 11.9 (continued)
1973 Production, Import, and Export Statistics for Plywood, Particleboard, Fiber Board, and Veneer Sheet (FAO Yearbook 1970)

b) 1973 Particleboard Production, Import, and Export, by Region

	Production	Import		Export	
	(10^3 m^3)	(10^3 m^3)	mil $	(10^3 m^3)	mil $
World	31,671	3,831	364	3,658	356
N. C. America	6,959	242	16	167	11
Asia	1,079	116	13	24	2.3
Europe	19,337	3,430	327	3,271	330
U.S.S.R.	3,120	–	–	168	11
Canada	586	197	13	4	–
U.S.	6,283	35	1.6	10	1.0

c) 1973 Fiber Board Production, Import, and Export, by Region

	Production	Import		Export	
	(10^3 m^3)	(10^3 m^3)	mil $	(10^3 m^3)	mil $
World	18,134	2,490	243	2,415	241
N. C. America	9,231	658	53	307	34
Asia	1,293	77	8.3	20	3
Europe	4,825	1,662	171	1,616	164
U.S.S.R.	1,912	–	0.1	1,889	140
Canada	1,137	105	7	150	15
U.S.	8,001	545	44	156	18

d) 1973 Veneer Sheet Production, Import, and Export, by Region

	Production	Import		Export	
	(10^3 m^3)	(10^3 m^3)	mil $	(10^3 m^3)	mil $
World	3,685	1,519	486	1,489	465
N. C. America	203	722	115	314	87
Asia	849	141	26	578	74
Europe	1,567	507	312	292	220
U.S.S.R.	476	63	12	8	.9
Canada	197	65	31	165	51
U.S.		656	83	147	36

ISBN 0-201-04558-3

Table 11.10

Leading Trade Nations

Import		Export	
Country	Volume ($10^3 m^3$)	Country	Volume ($10^3 m^3$)
Plywood			
U.S.	2127	Korea	1320
U.K.	1480	China	1213
Japan	779	Finland	600
Germany	377	Singapore	500
Canada	346	Canada	471
Particle Board			
U.K.	1281	Belgium	760
Germany	461	Finland	480
France	240	Austria	400
Canada	200	Germany	360
Denmark (& Neth.)	180	Sweden	330
Fiber Board			
U.S.	367	Sweden	370
U.K.	289	U.S.S.R.	172
Germany	211	Finland	154
Netherlands	127	So. Africa	102
Denmark (& Poland)	70	Poland	99
Veneer Sheet			
U.S.	656	Phillipines	215
Germany	81	Canada	173
France	75	U.S.	110
Canada	65	Congo	99
U.K.	63	Germany	93

Plywood

Figure 11.6c shows the plywood grades currently marketed in the U.S. Much of the production is CD-X and almost 75% is grade C-6, a grade which is not manufactured outside the U.S. Figure 11.6d shows the U.S. end-uses distribution.

Plywood is the oldest of the commercial bulk board products. Forty years ago, it had no competition from other products. Today, the situation has changed, because plywood is made from veneer which has become scarce. In Europe, plywood has long been replaced by particleboard. Recently, the U.S. Plywood industry has

ISBN 0-201-04558-3

conducted efforts to export plywood, but the increasing transportation costs will hamper this effort. The goal of these exports was not to regain the market lost to particleboard, but to penetrate the construction market to test whether it would be possible to alter European home construction methods, and replace masonry, brick, and gypsum board with less labor-intensive plywood panels. However, the current trend towards a 10-year forest harvesting cycle will soon drive up the price of veneer, and plywood will undoubtedly become a high priced, luxury product everywhere.

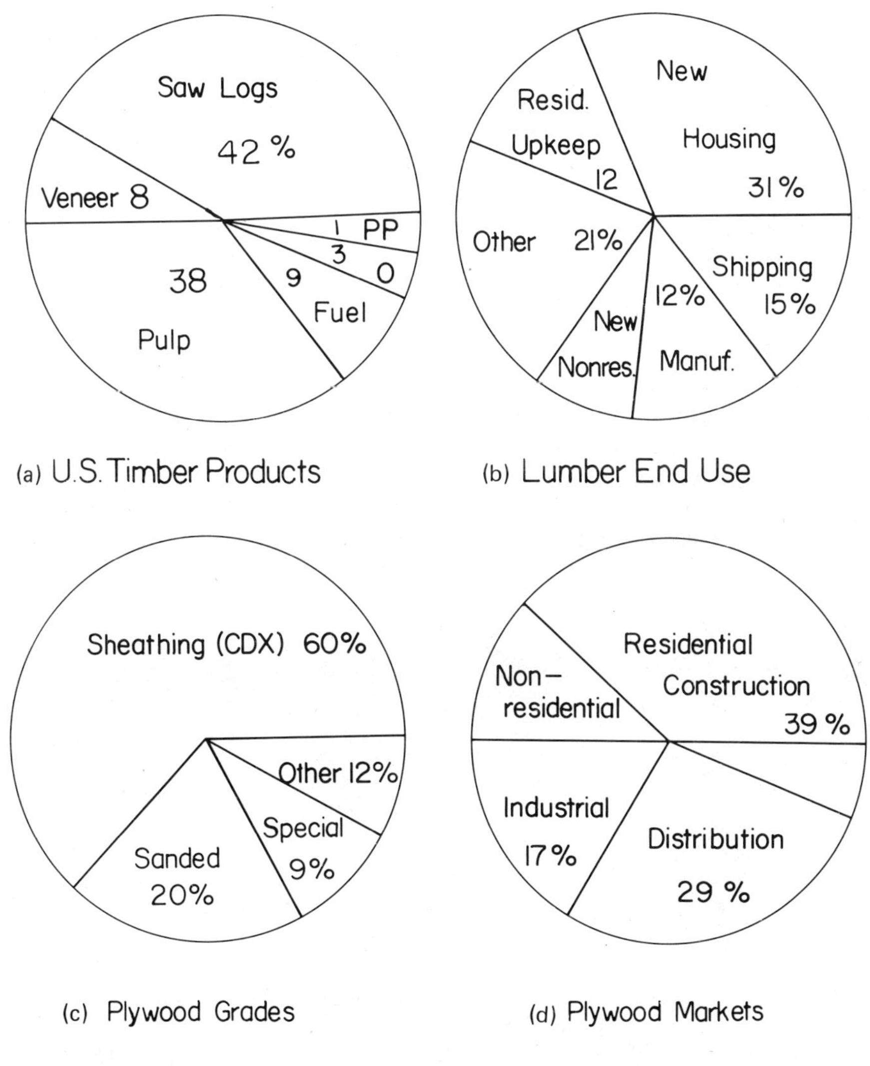

Figure 11.6.

ISBN 0-201-04558-3

Particleboard

80% of all UF-resins go into particleboard. From an environmental viewpoint, particleboard is one of the most modern and desirable inventions, because it converts local wastes, such as sawdust, or even used produce boxes, into a product which is superior to the waste's original precursors! Figure 11.7a shows the various uses for particleboard. Industrial uses still account for 60%. This figure could change quickly if the trend from plywood to particleboard continues. Figure 11.7b shows the data for building board consumption. Table 11.11 shows the panel board consumption per housing unit. The large emphasis on insulation board in mobile homes is striking. It readily explains why the quality control in the particleboard manufacture is crucial in this field. All the 3.5×10^9 sq. ft. is produced in only 70 plants. Thus, small changes in the large market can have a strong influence on the producers. Accordingly, the forest product market has traditionally oscillated greatly from a large oversupply to a serious shortage. Either particleboard plants cannot produce enough or they suffer from overcapacity. Thus, for plant managers the problem periodically changes from searching for a cheap adhesive with the shortest possible curing time (so that the plant capacity can be fully used), to finding the cheapest possible adhesive which still will yield a tolerable final product (so that profits stay tolerable, despite insufficient use of the plant capacity). In both situations, the foremost consideration is price, and this remains the main advantage of UF-resins. This situation will likely continue, and if anything, the gap between the price of particleboard and competing materials could increase. The reason for this is that the wood need not be high-quality, aged whole wood, since wood waste from many sources can be readily converted into comminuted wood, and that the trend of prices for the UF-resin components indicates that they will remain cheaper than other adhesive raw materials which depend on petrochemicals.

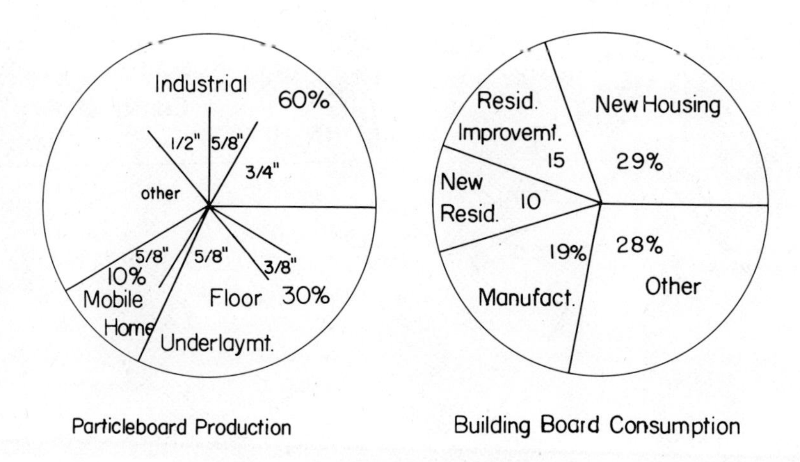

Figure 11.7.

ISBN 0-201-04558-3

Ironically, the main problem facing particleboard is due to exactly the same factors. The above described developments will narrow profits for the manufacturers and they will be increasingly tempted to turn to manufacturing more expensive products. This will put the future of UR-resins into the hands of UF-users. Thus, the future will depend on whether users will take the initiative to conducting UF-resin research. In summary, UF-resins have as excellent prospects and as clouded a future as any other material used in the attractive but turbulent construction industry.

Table 11.11

Panel Board Consumed per Housing Unit
(Sq. Ft., 1970)

	1–2 Families	Multifamilies	Mobile Homes
Hardboard (1/8")	1000	40	170
Insulation bd. (1/2")	935	40	7110
Particle bd. (3/4")	250	55	580

Resin Chemistry

As has been pointed out before, all fields of chemistry are currently profiting from a veritable revolution in analytical techniques. While resin chemistry has stagnated since the first generation of patents expired, a flurry of new activity is now commencing. The easy control of complex chemical reactions, even in mixtures, will greatly help these efforts. It is now possible to use *in situ* techniques to observe quickly and quantitatively the path of many simultaneous reactions taking place in polymer mixtures at all stages – from the first addition of reagent, through the various stages of resin synthesis and the resin cure, all the way to the aging, weathering, and degradation reaction which occur during the use and decay of the final UF-bonded products.

COMPUTER AIDED ANALYSIS

Electronically assisted chemical analysis is tailored for dealing with real-life complex mixtures which were out of bounds for traditional basic chemists. The newly available methods include infrared spectra, which now can be guided with computers over long periods of time. Computers can subtract reagents or product spectra in mixtures, and they can sum spectra of individual components to predict and simulate product spectra. This makes it possible to determine whether filler or extender mixtures are chemically reactive or whether they remain inert. Furthermore, Fourier-transform techniques allow scanning a spectrum within 20 seconds. Thus, curing and thermal degradation kinetics can be followed readily. Computers

ISBN 0-201-04558-3

can be programmed to compare not only total absorption changes, but also relative changes between peaks that change simultaneously. For example, during resin cooking, one can now observe not only the rate of change of primary amine, but one can follow simultaneously the ratio of the disappearance of amine and the appearance of secondary or tertiary amines. Thus, the rate of crosslinkage versus chain growth can be observed and guided.

Infrared can also be used to compare the spectra of unsupported UF-resin films with those of films on wood surfaces. This helps elucidate the chemistry of adhesion.

Raman spectra, like IR spectra, record molecular vibrations, but the selection rules, and thus the intensities, are different. The Raman peaks of water and hydroxyls are far weaker than in the IR. Thus, the Raman method is ideal for aqueous chemistry, cellulose chemistry, and all chemistry involving hydrogen bonds. Laser Raman spectrometer can yield high sensitivity, which can be further enhanced by multiple scanning, and storage of reference or intermediate spectra. Raman spectra are not dependent on light transmittance.

Nuclear magnetic resonance (NMR) is a well established field, but until recently it did not hold much promise for UF chemistry, because spectra of carbon chains were poorly resolved, and all spectra were dominated by protons of hydroxyl. This has radically changed with ^{13}C-NMR, which can identify and distinguish carbon atoms within chains, even if their identity differs little. This method is now so sensitive that the natural abundance of ^{13}C, i.e., 1.11%, is sufficient for work in real-life systems. The discovery of the "magic angle" of 54.70 opens up study of solids in addition to the liquids to which the method was restricted in the past. Finally, some relaxation studies indicate that it might become possible to find a correlation between ratios of relaxation and the impact strength of chemical structures.

ESCA, electron spectroscopy, makes it possible to observe surface effects, for example, UF-resin surface degradation (Grassie 1978). Ion spattering can be used to study lower, subfaces of resin surfaces. Fluorescence can be used to study polymer blending, if energy donors and acceptors are separately mixed with components. Neutron scattering, finally, makes it possible to identify the orientation of polymer chains.

The application of these instrumental methods to UF-resins are comparatively new, and it will probably last several years before they become commonly used in industry, but they offer so many advantages for direct quality control that it seems inevitable that they will be widely used. This will cause several changes. First, UF-resin synthesis can be continuously monitored, evaluated and altered, step-by-step, during the entire manufacturing process. Thus, complex copolymerization reactions between UF and resorcinol, phenol or isocyanates will become manageable and controllable, and thus will become useful for synthesis on a large, industrial scale. Second, the modern methods will make it possible to carefully control the resin synthesis, thus making available more reliable resins. Thus, resins can be standard-

ISBN 0-201-04558-3

ized, graded (for example, according to free formaldehyde content), and upgraded. Finally, chemical analysis will make it possible to conduct routine, nondestructive production tests to identify the quality of finished UF-bonded products. Therefore, independent of other resin chemistry improvements, it will become feasible to take standard production quality particleboard, eliminate the few substandard boards which might later cause complaints, select the top 10% of boards with truly superior performance, and guarantee their performance for special applications, at a permium price which more than compensates for the elimination of subpar quality.

Resin Manufacture

The possibility of identifying resin reactions *in situ* will increasingly lead to continuous manufacture of resin in backflow reactors with full electronic control. Only a minimum of modifiers will be needed to obtain maximum resin quality. This should make it possible to use more modifiers than now common. The entire development suggests that resin manufacture will become yet more complex, and that large-scale producers will be able to offer higher quality at a discount. Thus, smaller manufacturers might temporarily suffer. On the other hand, cheaper microelectronics will make it undoubtedly feasible to implement resin control on a local scale, maybe even at the individual plant. Thus, the day might come when individual resin users can manufacture resin as needed, on the premises, giving them independence, as well as access to simpler resins which need not have the extended storage capability which requires properties which are often in conflict with the chemistry of optimum cure.

Local control would have the added advantage that the forest products industry could formulate the resin which fits its individual, local needs imposed by local wood species or other raw material demands, rather than buy the resin qualities which the chemical industry finds most convenient or profitable to market.

Wood Adhesives

These adhesives constitute currently the predominant resin use and particleboard is the main product. Particleboard is an invention of the 1940s and its manufacture is highly automated. Because of the high capital costs of plants, the trend towards large plants will continue, except in smaller, less developed areas, where highly labor intensive plants might be feasible (Fahrni 1976 for FAO). The basic technology has matured, and the optimum plant size is probably 300,000 m^3 per annum partly because of shipping costs. The largest growth potential for particleboard, in absolute tonnage, is probably in the U.S. where conversion from plywood to particleboard has been slow, because of the high availability of veneer.

The key factors determining new potential markets for particleboard are density, weather resistance, fire resistance, and odor. For shipping and for optimizing its properties, particleboard should be light. For technical reasons, modern particle-

ISBN 0-201-04558-3

board is made denser than it was. This trend should be reversed. To be weather-resistant the board should swell less than 5% after 10 days of soaking. Figure 11.8 shows the mechanical properties of UF- and PF-bonded particleboard, as a function of natural and accelerated aging tests. This figure shows the superiority of fresh UF-bonded board; it also shows the loss of quality upon exposure to weather, and shows incidentally, that current test methods for accelerated testing are not yet very reliable. In regard to fire resistance, particleboard is essentially equivalent to wood, Figure 11.9. Fire resistant boards can be made if wood is replaced by vermiculite, asbestos, or fiber glass in organic materials. Fireproofing of the wood particle is feasible, but it causes higher water absorption, and possible poorer adhesion (Maloney 1977). The future success in market penetration will depend also on odor reduction. If housing units are to contain 7110 sq. ft., as they do in mobile homes, rather than 1000 as is currently the case in conventional homes, odor must be proportionately reduced. For this purpose, the wood particles must be coated to reduce moisture retention (Burmester 1976), one of the causes of odor, the resin must be made more water-resistant or the finished board must be treated. Currently, several approaches are used. The cheapest, the bulk treatment of wood before assembly of the caul, has been neglected because of the drastic influence of the energy crisis on the price of common waxes. There is little doubt that eventually better resins will be available, but the development of a new resin product takes time. Several steps have been taken by the forest products industry to develop stop-gap techniques. In Europe and Japan, particleboard is quite frequently upgraded by coating. Boehm (1977) in East Ger. P 124,312 describes coating of board with films. Roffael (1974) developed a coating called Falima, i.e., Verbestel (1979) introduced equipment to fumigate board with ammonia; Maslinkovskii (1977) in USSR 554,233 described a material for finishing walls. Paper laminates have been long used to give board an imitation woodgrain look. Iwasaki (1972), Japan 72:32,422, described a vein-like overlay. Minoji (1974) in Kokai 74:38, 927 disclosed a method for producing lustrous finishes, and Moroff (1973) in Ger. Offen. 2,135,072, and Lee (1972) in USP 3,686,021 described overlay papers.

If formaldehyde odor is to be significantly reduced, the edge of the panel should be sealed. Lehnert (1976) in Ger. Offen. 2,610,286 explained methods for doing so. Barghoorn (1979) showed that sealing the edge reduces the release by a factor of ten; while coating the surface reduces odor only to about one-half, Table 9.9, page 241. If these problems can be reliably reduced, particleboard could immediately penetrate the building market. The use of board in ceilings and roofs, Figure 11.10, is currently only possible if great precautions are taken. These precautions negate much of the price advantage inherent in the potential use of board.

ISBN 0-201-04558-3

While the plywood industry is dependent on ample supplies of veneer, composites have hardly started to show their potential. Their use, currently largely restricted to furniture, could easily encompass laminated sandwich panels, with foam cores, for prefabricated housing (Gfeller 1977). Various chemical pretreatments will allow mechanical modification. So far, the main problem with these products is that they are far too vulnerable to damage in transit, and that they have a slight tendency to creep when used as self-supporting part of walls or structures.

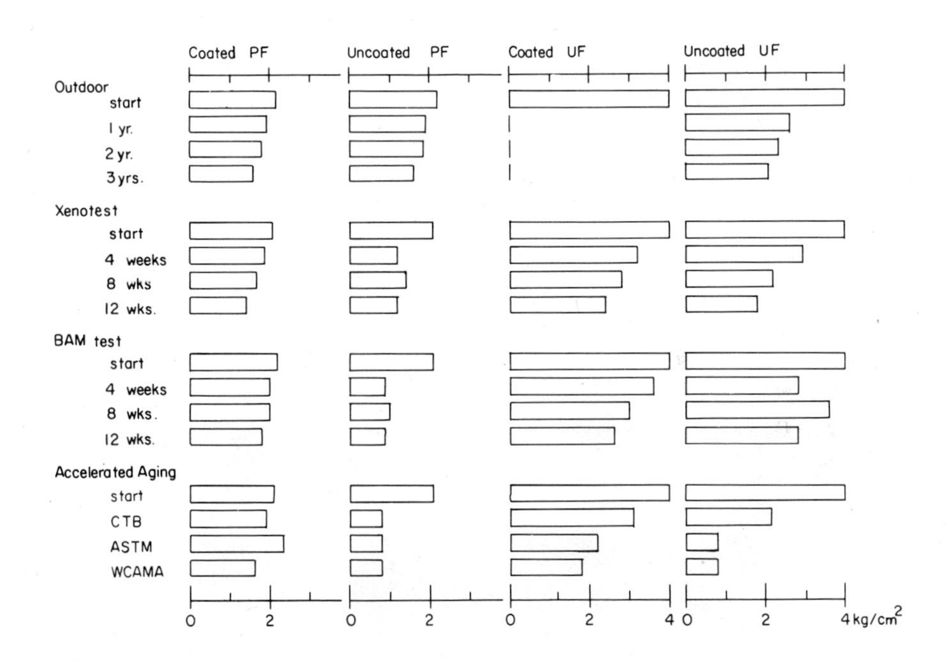

Figure 11.8. Weather Resistance of Particleboard; a) Phenol, Coated, b) Phenol, Uncoated, c) Urea-Formaldehyde, Coated, and d) Urea-Formaldehyde, Uncoated (after Deppe, 1977)

ISBN 0-201-04558-3

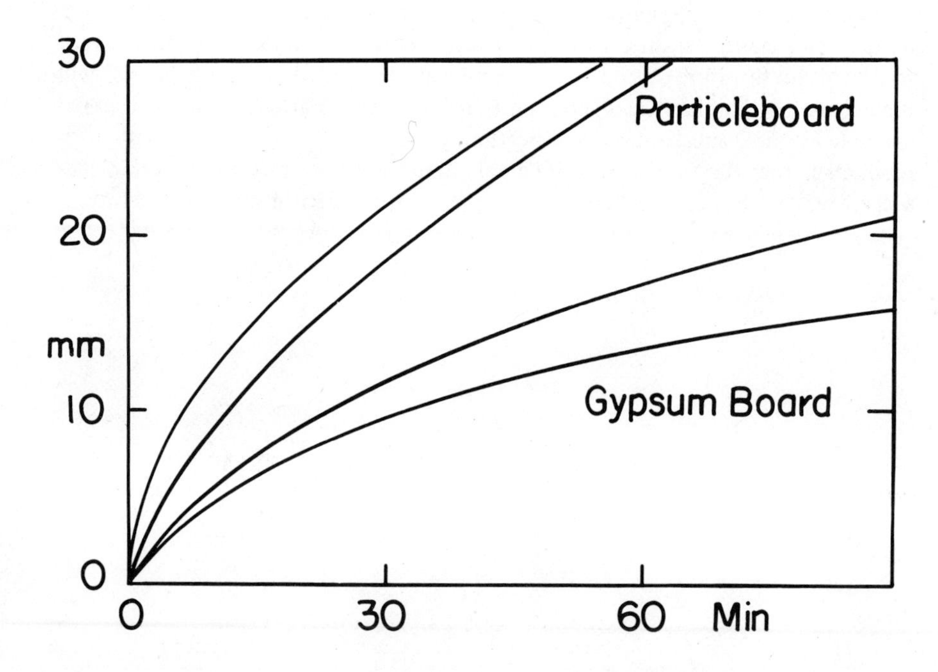

Figure 11.9. Fire Resistance of Particleboard and Gypsum Board According to German Test DIN 4102 (after Deppe, 1977)

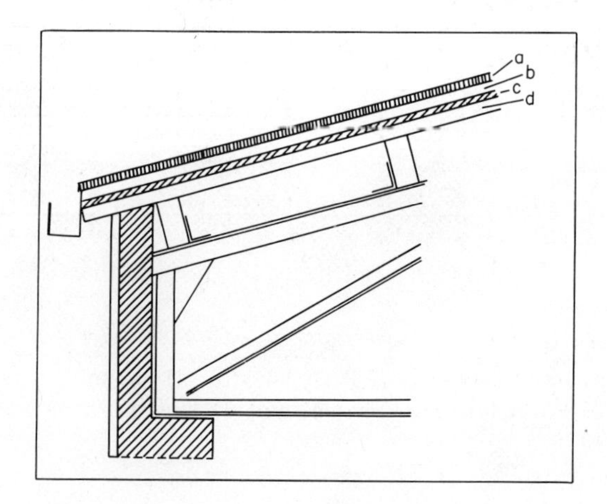

Figure 11.10. Particleboard Roofing (after Fahrni, 1975), a) Roof Skin, b) Insulating Layer, c) Vapor Barrier, d) Particleboard

ISBN 0-201-04558-3

UF Foams

The potential for use of UF foam was early recognized by the pioneers, but only few of the users who could profit from this potential are yet aware of the advantages of this material! If properly developed and introduced, the materials described in Chapter 7 will undoubtedly find increasing markets. In fact, properly introduced foams could easily dominate all other resin uses, within the next twenty years. The problem with foam is primarily one of communcation with the potential users who still widely holds a conservative stance towards any chemical innovation except fertilizers. However, foams could well be introduced jointly with the slow-release fertilizers (Meyer 1977) which are now widely tested and commercialized.

The plastoponic and plant-nutrient use of UF foam demonstrates the intrinsic environmental compatibility of UF-resins. However, for increased use as insulating material, resin manufacturers must be able to successfully control the quality of foam application by contractors. This is difficult in large markets where traditionally unskilled labor is used. Unfortunately, the few rare cases of UF foam abuse are sufficient to cause public back-lashing. This could cause a delay in the new use of this insulating material and it could cause inflexible government regulations. This would be regrettable, because UF foams have few rivals in fire resistance and insulating capacity and hence are uniquely useful in tall buildings.

Other Applications

In the textile field, UF-resins, mainly urons, are in a mature market dominated by urons and melamines. It is possible, but not likely, that resin improvement will increase use. In contrast, coatings are a field of wide potential and UF could gain wider acceptance due to its low price. In paper UF-resin use also faces a mature market but new specialized markets are opening in book binding (Ditmev 1970, East Ger. P 77,033).

The construction market for UF-resin is currently determined by particleboard. UF gypsum board, vermiculite board, perlite, and other low density boards have a great potential in outdoor use and in tall buildings if the manufacturer's interest is sincere rather than a stop-gap activity to dispose of unused particleboard plant capacity. In this application the fire resistance is a vital factor.

In agriculture, UF foams are the preferred product. However, Kirkeby, Neth. P. 24,479 (1979), described sprayed coatings of UF useful for seeding grass. Likewise, UF-resins are excellent materials to bind wastes or as oil-spill collectors on oceans as described in Chapters 7 and 8.

In molding, UF has a firm place as an electrical insulator. Efforts have been made to use foams in the same application. Among the more exotic uses, UF-resins are now used for encapsulating ink and prepare storageable, pressure sensitive paper, and the like.

ISBN 0-201-04558-3

Formaldehyde Release

Chapter 6 describes the formaldehyde problem. The problem appeared when UF-resins were introduced in new, high-volume markets in which unexpectedly high quality standards were necessary. In fact, these standards are higher than those encountered in any traditional UF-resin use. The odor problem can be solved by a variety of technical means explained in Chapter 9.

There is still some dispute whether the odor problem needs to be tackled. For example the president of the Manufactured Housing Institute has been recently quoted in the Seattle Times (May 17, 1979, page G-8), a metropolitan newspaper with a circulation of 500,000, as saying: "The (federal government) agencies would be quick to tell us if there is a danger in any aspect of our product. They see no need at this time for any change in our method of construction," and "a correlation might be drawn between formaldehyde and aspirin. Aspirin is the most common pill used by almost everyone in the country, except for those very few which are bothered by this miracle drug." Most manufacturers would not agree with this statement and feel that the odor problem is serious and can be solved.

Superficially read, the above quote seems to reflect more on legality than on concern for consumer acceptance and product quality. Some readers will see the statement as the result of a manager caught in the struggle for day-to-day survival. Consumer advocates will be tempted to blame the entire industry for holding a calloused attitude towards product performance. The author believes that this quote again reflects the unique situation and structure in the UF-resin application field.

The manufacturer recognizes the odor problem and he can correct it, but he would do so rather by switching to a higher priced product. The furniture and construction companies recognize the almost untapped and unlimited potential for particleboard, but it is not in a position to abate odor cheaply, and they recognize that the size of the market depends directly on price. Thus, odor problem is handed on to the consumer and it is hoped that he will adjust to the odor, or that the problem will blow away. One is tempted to solve the problem with stop-gap procedures such as air filters for mobile homes, by fumigation, or perfuming. However, the current trend towards reducing air circulation to preserve energy will exacerbate the problem. The only long range solution will be to redesign UF-resins from scratch and to consider formaldehyde release a factor in the product formulation. This was not necessary when the UF-resins were first introduced.

Health

In recent years a new field of public health epidemiology has developed to deal with the large-scale chronic exposure to chemicals which have been entering the environment at levels above those found in nature. Formaldehyde is currently not high on the list of such studies, because it belongs to a group of substances which have been used in industry for many decades, without any acute permanent damage, and without obvious chronic effects. However, because of the well-known

ISBN 0-201-04558-3

skin allergy problem, and because of the chemical fact that formaldehyde can react with amino groups, formaldehyde has been labeled a "suspect" by several researchers. Morin (1978) for example, showed that DNA can react with formaldehyde. Andersen (1979), also a medical doctor, points out that liver and other enzymes convert formaldehyde quantitatively before it can enter the metabolism. However, many studies, especially in Russia, indicate more serious effects. Volkova (1971) tested blood levels, Kratochvil (1971) reviewed health hazards from textiles, Vinogradov (1974) found formaldehyde in automobile exhaust, Kolesnikov (1974) studied the effect of formaldehyde on radiation sickness. Basmadzhieva (1974) studied inhalation, Ostapovich (1975) looked at the synergism for formaldehyde with sulfur dioxide, Kuzmenski (1975) tested for allergies. None of these studies arrived at alarming results, but there is a definite need for objective testing of the type initiated by the newly formed Formaldehyde Institute, a subsidiary of the Synthetic Organic Chemical Manufacturers Organization. This will take many years and involve statistical evaluations (Tendrovskaya 1973). It currently seems that formaldehyde might well prove to be mainly a nuisance rather than a danger, but such nuisances add stress which can aggravate existing other health problems.

Regulations and Standards

Since consumers cannot easily propose chemical or manufacturing improvements, the dissatisfied consumer is left in a defensive stance. If he shifts the complaint to politicians or government agencies, the same situation prevails. This leads invariably to stricter regulations. This topic has been explored in Chapter 10. In transition times, such as the current, the daily new experiences can test interhuman trust in a drive for increased regulation to supercede individual judgment. However, such regulations are inflexible, and especially in a transition time, they tend to be ineffective.

In regard to air pollution, formaldehyde will likely fall into a group of indoor chemicals which will probably be limited to a value between 0.1 to 0.2 ppm in indoor air, as discussed in Chapters 9 and 10. These values are not founded on health and safety considerations, but are necessary if freedom from odor is to be considered a factor in living quality. If these values are to be attained, the release from UF-bonded materials should be less than 5 ppm per m^2 per hour. Such standards are not yet routinely met by untreated standard particleboard, but they are highly preferable to regulations which restrict the chemical compositions which are admissable as adhesives. The regulating of chemicals rather than the properties of products ignores chemical progress which might well be possible and which might already be underway (Clad 1978).

In the U.S. the following five agencies are possibly involved in regulating indoor air quality: the Occupational Safety and Health Administration (OSHA, NIOSH), the Department of Housing and Urban Development (HUD), the Environmental Protection Agency (EPA), the Department of Energy (DOE), and the Consumer Product Safety Commission. In addition, the individual States' Attorney General

ISBN 0-201-04558-3

Conference can challenge licensing of product sales, and several other state, county, and city agencies can exercise various controls. Almost all federal agencies have initiated some research program to study the formaldehyde problem. However, the problem ranks low among their priorities, and their decision will probably be based on outside expertise. It is to be hoped that such advice will remain objective, and not be influenced by competitive considerations.

Conclusion

The purpose of this book is to make it possible for a wide group of readers to gain an overview of the current status of UF-resins and their applications. UF-resins have immense new potential. The author has stressed two areas: particleboard and foam. In these areas UF-resins have demonstrated outstanding performance, and have found quick acceptance. However, in recent years it has become obvious that in order to remain competitive UF-resins need improvement to become more weather resistant and to reduce initial odor. Whether UF-resins can find the potential acceptance in the construction market depends on how the weathering and odor problems are solved. Industry has already taken several successful stop-gap measures, and there is no question whether these problems can be technically solved. The most important of the remaining questions is whether innovative solutions can be found to formulate resins that allow particleboard and foam to remain sufficiently economical to penetrate the construction market and to remain competitive with other new materials, and whether such solutions can be found before over-reaction to early problems causes rejection of the product before its potential is recognized. The author believes that UF-resins have great, untapped intrinsic potential, and that the future of these resins depends primarily on the interest among people who hold the tools and know how to conduct novel and inventive work. These people are spread over many different fields of basic and applied science. The problems can only be solved if the specialists maintain close communications. It is a goal of this book to assist such interdisciplinary work.

ISBN 0-201-04558-3

Appendix

Appendix

RESIN FORMULATION

The resin formulation is explained in Chapter 4. Here, a few practical examples are given for preparing typical resins.

Water Soluble UF-resin Containing Excess Formaldehyde in Acidic Solution.

Acidify 930 g of 30% formaldehyde solution to yield pH 2 and heat it to 90°C. Stir 120 g urea into the solution. Keep the mixture at 60°C for 30 minutes and then it is neutralized. Pump off water at 60°C and 20 Torr until 225 g highly viscous, clear condensation product remains. It is water soluble in any proportion.

Monomethylol Urea

Combine 60 g urea with 60 g water and 1 g barium hydroxide. Bring the temperature to 3°C. Slowly add 100 g of a 30% formaldehyde solution over a period of two hours. Immediately afterwards, the pH is corrected with carbon dioxide gas at 20°C. The barium carbonate solid is removed, and water is evaporated below 40°C *in vacuo*. The yield is about 86 g of MMU which can be recrystalized from ethanol.

Dimethylol Urea

Adjust 210 g of a 30% formaldehyde solution to pH 8. At −10°C add 60 g urea as well as 0.8 g monosodium phosphate. Dimethylolurea precipitates slowly. The reaction is complete after one day. The product is filtrated and recrystalized with alcohol.

Alkaline UF-resin Wood Adhesive

Neutralize 126 g of a 37% formaldehyde solution with concentrated NaOH. Add 60 g paraformaldehyde and 120 g urea. Adjust the pH to 8 and bring the temperature to 95°C. The pH slowly drops to 6.8; after 30 minutes it is readjusted to pH 8 for another 10 minutes. Concentrated phosphoric acid is used to adjust the pH to 5.8. After some 45 minutes the viscosity has increased 2.5 times. The mixture is cooled, and the pH corrected to 8. To 100 g of this resin add 7 g concentrated ammonia (25%) with 3 g urea and a buffer consisting of 0.5 g sodium and ammonium acetate. The pH should be 7.

ISBN 0-201-04558-3

UF-sulfite Adhesive

Mix 650 g of a 30% formaldehyde solution at pH 7 with 150 g urea and bring to a boil. The pH is adjusted to 4.7 with 10% acetic acid. The mixture is refluxed for 30 minutes. Add 156 g sodium disulfite and reflux mixture for 30 minutes. Then, add 50 g urea and reflux one more hour. The product is neutralized with NaOH and vacuum concentrated to the deisred viscosity.

Solubilizing UF-resin Gel with Sulfite

Neutralize 77 g of a 30% formaldehyde solution and heat to 80°C. Add 21 g urea and bring the mixture to 90°C. The pH is brought to 4 with oxalic acid, and the mixture is heated for three hours. Upon cooling a gel is obtained. It can be thickened by vacuum evaporation. The insoluble resin can be solubilized by bringing the pH to 7–8, and adding water to make a product with a solid content of 40%. The mixture is brought to 95°C and a 15 g solution (made by combining 6 g sodium bisulfite, 0.1 g sodium sulfite, and 1.1 ml NaOH in water) is added. NaOH is added as necessary to maintain the pH at 7.5. The resulting adhesive is water soluble.

ISBN 0-201-04558-3

Glossary

Glossary

ACS American Chemical Society.

Adhesive A substance capable of holding materials together by surface attachment. The term is synonymous with glue.

AIHA American Industrial Hygiene Association.

APA American Plywood Association.

ASA American Standards Association.

ASTM American Standard Testing Method.

Binder A bonding agent used to bond together the particles in a particleboard, or a material added during manufacture to form bonds.

Buffer Regulates the acidity (pH), and thus the reactivity of a resin.

Catalyst *See* ***Hardener.***

Caul A flat metal plate on which wood particles are conveyed and pressed.

CBU Cyclobutylene urea.

Comminuted Wood Wood reduced to small or fine particles.

Composition Board Board made of comminuted wood or other lignocellulosic material, bound with adhesive and made under heat and pressure.

Condensation A chemical reaction in which two or more molecules combine with the separation of water or some other simple substance. If a polymer is formed, the process is called polycondensation.

Core (Middle) The center layer in a composition board panel.

Cure The physicochemical change brought about in thermosetting synthetic resins under the influence of heat and catalysts (polymerization).

Desiccator Test Formaldehyde determination method. The German and Japanese methods differ significantly.

DIN German industrial standard.

DMU Dimethylol urea.

DTA Differential thermal analysis.

EPA (U.S.) Environmental Protection Agency.

ESCA Electron spectroscopy for chemical analysis.

EU Ethylene urea.

Extender A low-cost additive used with an adhesive or binder. Cereal flours are often used for this purpose.

F; Formaldehyde This term is often also used for latent formaldehyde.

ISBN 0-201-04558-3

FESYP European Federation of Particleboard Manufacturers.

Fiberboard A sheet of material made from fibers of wood or other lignocellulosic material and manufactured by an interfelting of the fibers followed by compacting between rolls or in a platen press.

Filler An additive used with adhesive or binder to reduce excessive penetration of the adhesive into the wood and to improve other characteristics such as gap-filling. Finely ground minerals, and wood and nut shell flour are commonly used.

Free Formaldehyde Excess formaldehyde in UF resin. This conceptually simple term is ill-defined and cannot be unambiguously measured. Different test methods respond to different types of chemical and physical states of formaldehyde.

GLC Gas liquid chromatography.

GPC Gel permeation chromatography.

HEW (U.S.) Department of Health, Education and Welfare.

Hardboard Usually a fiberboard of 0.50 to 1.20 g/cm^3 (31.5–75 lbs/ft^3) density.

High-Density Particleboard Particleboard with a minimum density of 0.80 g/cm^3 (50 lbs/ft^3).

HUD (U.S.) Department of Housing and Urban Development.

Industrial Board Nonstructural board used for furniture.

Internal Bond Tensile strength perpendicular to surfaces. A measure of the bonding force between particles.

IR Infrared spectrum.

JIS Japanese industrial standard.

Mat Particles loosely laid together in preparation for being pressed into a board.

MDF Medium density fiberboard.

MDU Methylene diurea.

Medium-Density Fiberboard Normally a compressed fiberboard of about 0.40 to 0.80 g/cm^3 (25–50 lbs/ft^3) density, manufactured for a wide range of uses such as panel material for use in building, furniture, and similar constructions.

Medium-Density Particleboard Particleboard with a density of 0.60 to 0.80 g/cm^3 (37.5–50 lbs/ft^3).

MF Melamine formaldehyde resin.

MHI Manufactured Housing Institute.

MMU Monomethylol urea.

Modifier Any chemically inert ingredient added to an adhesive formulation that changes its properties.

MOE Modulus of elasticity.

Moisture Content Amount of moisture normally expressed as percentage of dry weight.

MOR Modulus of rupture.

MPU Monopropylene urea.

MUF Melamine-urea-formaldehyde resin.

ISBN 0-201-04558-3

NAS (U.S.) National Academy of Sciences.

NMR Nuclear magnetic resonance.

Novolak A phenolic-aldehydic resin which, unless a source of methylene groups is added, remains permanently thermoplastic.

NPA National Particleboard Association.

NRC (U.S.) National Research Council.

OSHA (U.S.) Occupational Safety and Health Administration.

Overlay A thin layer of paper, plastic, film, metal foil, or other material bonded to one or both faces of a panel.

Panel A sheet of plywood, fiberboard, particleboard, etc., of any type.

Particle Fraction of wood or other lignocellulosic material produced mechanically. Used as the aggregate for a particleboard. Types of particles include:

Curl. Long, thin flakes manufactured by the cutting action of a knife.

Flake. Specially generated thin, flat particles, with the grain of the wood essentially parallel to the surface of the flake.

Flax shives. Fine rectangular-shaped particles of lignocellulosic material obtained by longitudinal division of the stalk of the flax plant.

Granule. A particle in which length, width, and thickness are approximately equal, such as a sawdust particle.

Shaving. A small wood particle of indefinite dimensions developed incidental to certain woodworking operations involving rotary cutterheads, producing a thin chip of varying thickness, usually feathered along at least one edge and thick at another.

Sliver. A particle whose length is parallel to the grain of the wood and at least four times the thickness.

Splinter. An alternative form for *sliver.*

Strand. A relatively long shaving consisting of flat, long bundles of fibers having parallel surfaces.

Wood wool (excelsior). Curly, slender strands of wood made usually by scoring and cutting knives moving along the grain of a block of wood.

PC Paper chromatography.

Perforator Extractor used to determine formaldehyde content in particleboard according to FESYP test.

PF Phenol formaldehyde resin.

pH Measure of acidity. Negative logarithm of proton concentration.

Platen A part of a press consisting of a rigid metal plate, usually heated, for exerting pressure on the mat. Intermediate platens in a multi-opening press are also called plates.

PU Propylene urea.

Resin Content The amount of dry solids in the resin usually expressed as a percentage of the dry weight of the wood in a finished board.

Set To convert an adhesive into a fixed or hardened state by chemical or physical action, such as condensation, polymerization, oxidation, vulcanization, gelation, hydration, or evaporation of volatile constituents.

ISBN 0-201-04558-3

Size Alum, wax, petrolatum, asphalt, resin, or other additive introduced to the stock for fiberboard or to the agglomerate for a particleboard prior to forming, primarily to improve water resistance.

Storage Life The period of time during which a packaged adhesive can be stored under specified temperature conditions and remain suitable for use. Sometimes called shelf life or ***Working Life.***

Structural Board A board of high strength and stiffness suitable for use in load-bearing structures.

Sulfite-Bonded Board Particleboard made with sulfite spent liquor as a binder. Sulfite spent liquor contains the lignin removed from wood in the sulfite pulping process.

Tack Stickiness of a resin in the uncured state.

Tempered Hardboard Hardboard which has been specially treated in manufacture to improve its physical properties.

Thermoplastic A material which will repeatedly soften when heated and harden when cooled.

Thermoset A material which will undergo or has undergone a chemical reaction by the action of heat, catalysts, ultraviolet light, etc., leading to a relatively infusible state.

THF Tetrahydro furane.

Three-Layer Particleboard A particleboard in which the core particles and face particles are of different qualities.

TLC Thin layer chromatography.

TMU Trimethylol urea.

UF Urea-formaldehyde resin.

UV Ultraviolet spectrum.

Working Life The period of time during which an adhesive, after mixing with catalyst, solvent, or other compounding ingredients, remains suitable for use.

ISBN 0-201-04558-3

Bibliography

Bibliography

Introduction

This section contains all references to earlier chapters. Patents, journal articles, and government reports are all integrated alphabetically; according to the first author's name. For each author, the listing is chronological with the newest paper first. If a reference is missing, the article is quoted in another paper by the same author, or in a review quoted in context with the chapter. Anonymous articles, or patents without authors are listed under the company name. Government documents are listed according to the government agency's name; the countries are indexed in the author's index. U.S. Government reports can be ordered from the National Technical Information Service (NTIS), Department of Commerce, Springfield, VA 22161.

U.S. Patents can be ordered from Box 9, Patent and Trademark Office, Washington, D.C. 20231. The correlation between U.S. and foreign patents often can be established with the help of the Patent Concordance, published yearly by the Chemical Abstract Service of the American Chemical Society. The abbreviations used for foreign patents are self-evident, with exception indicated below. However, the international patent law situation is currently in flux. Japan has two systems, the Japanese patents (Japan) and the Japanese Kokai (Kokai). German patents (GP) are now first published as Offenlegung (Ger. Offen.) and only later converted into GP. The same system is followed in several other countries. The date given in this bibliography is the year during which a patent was issued. BP stands for British Patent, Belg. P for Belgian Patent, Neth. P for Netherland Patent, and so forth.

ISBN 0-201-04558-3

Adamca, M., Drevo **31**, 7 (1976). "Evaluation of Adhesives with Regard to Dielectric Heating."

Adamca, M., Drev. Vysk. **18**(4), 251 (1973). "Investigation on Electric Conductivity of Adhesive in Bonding of Wood by Radio Frequency Heating."

Aero Research Ltd., Brit. P 545,409, May 26, 1942. "Urea-Formaldehyde Adhesives."

Afanas'ev, N. V., B. V. Yarlykov, M. S. Akutin, and I. K. Sanin, Tr. Mosk. Khim.-Tekhnol. Inst. **86**, 174 (1975) (Russ). "Characteristics of the Viscous Properties of Reactive Oligomers."

Ahmad, S., N. A. Jamil, A. H. Hanif, and N. Ahmad, Pak. J. Sci. Ind. Res. **18**(6), 277 (1975). "Effect of Crease-Recovery Reagents on Physical Properties of Yarn Made from AC-134 Cotton."

Aignesberger, A., and H. G. Rosenbauer, Ger. Offen. 2,505,578, Apr 1, 1976. "Production of Anionic Sulfonic Acid Group Containing Melamine-Formaldehyde Condensation Products in Aqueous Solution with High Solids Content."

Akabane, J., Int. Encycl. Pharmacol. Ther. **20**, 523 (1970). "Aldehydes and Related Compounds."

Akhmedzhanov, K. A., A. M. Muminov, S. L. Dubrovskii, A. M. Orlovskaya, and M. N. Vladimirova, Probl. Gigieny i Organiz. Zdravookhr. v Uzbekistane (Russ) **1976**(4), 1974 (1976). "Hygienic Appraisal of Adhesives MF-60 and MF-17 Based on a Urea-Formaldehyde Resin and Used in the Manufacture of School Desks and Child's Furniture."

Akutin, M. S., V. Kafarov, V. A. Reutskii, V. V. Polukhin, S. S. Pozdnyakova, V. I. Vatamanyuk, and V. Z. Afanas'ev, USSR P 567,730, Aug 5, 1977. "Urea-Formaldehyde Oligomers."

Akutin, M. S., V. A. Sharkovskii, and M. L. Kerber, Mekh. Polim. (Russ) **1974**(3), 442 (1974). "Adhesion and Wetting in Urea Glass-Fiber Reinforced Plastics."

Akutin, M. S., L. G. Bekkert, N. B. Tagil'tseva, and O. A. Tarakhtunov, USSR P 384,843, May 29, 1973. "Urea-Formaldehyde Oligomers."

Akutin, M. S., N. B. Tagil'tseva, O. A. Tarakhtunov, and V. P. Khviler, Plast. Massy (Russ) **1972**(10), 4 (1972). "Polymeric Catalysts for Poly-condensation Processes."

Aleksandrovich, Kh. M., A. D. Markin, and I. B. Kolochinskaya, Khim. Prom-st (Moscow) (Russ), **1977**(4), 286 (1977). "Effect of Urea-Formaldehyde Resins on the Flotation of Potassium Ores."

Alekseev, A. V., Nauch. Tr., Leningrad. Lesotekh. Akad. **1972**(146), 3 (1972). "Properties of Hardened Adhesives and Laminated Wood Affecting the Wear of a Cutting Tool."

Allgemeine Elektrizitaets-Ges., Brit. P 316, 194, July 24, 1928. "Adhesives Comprising Organic Condensation Products."

Allyn, C. L., USP 4,065,421, Dec 27, 1977. "Continuous Process for the Production of Aqueous Urea-Formaldehyde Solutions."

Almassy, G., Magy. Kem. Lapja (Hung) **28**(9), 452 (1973). "Technology of Urea-Formaldehyde Condensates Used as Fertilizers."

American Reinforced Paper, Fr. P 821,840, 1937. "Wet Strength Paper."

American Society for Testing and Materials, in Annual Book, ASTM Standards, part 22, ASTM designation D1037-72a, Philadelphia, PA, 1974. "Standard Methods of Evaluating the Properties of Wood-Based Fiber and Particle Panel Materials."

ISBN 0-201-04558-3

American Society for Testing and Materials, ASTM Standard Designation D1391-57, Philadelphia, PA, 1971. "Standard Method for Measurement of Odor in Atmospheres (Dilution Method)."

American Society of Heating, Refrigerating, and Air-Conditioning Engineers, Inc., Handbook and Product Director: 1973 Systems, Ch. 33, p. 33-1 to 33-12, ASHRAE, New York. "Odor Control."

American Society of Heating, Refrigerating, and Air-Conditioning Engineers, Inc., Guide and Data Book: Systems. Ch. 19, p. 287, ASHRAE, New York, 1970. "Ventilation of the Industrial Environment."

Andersen, I., G. R. Lundqvist, and L. Mølhave, Ugeskr. Laeg. **141**, 6287 (1979). "Formaldehyde in the Home Atmosphere. Proposed Introduction of Limits in Housing."

Andersen, I., unpublished work, 1978. "Human Dose Response Curve to Formaldehyde Odor."

Andersen, I., G. R. Lundqvist, and L. Mølhave, Holzfor. Holzverwert. **28**(5), 120 (1976). "Effect of Air Humidity and Sulfur Dioxide on Formaldehyde Emission from a Construction Material (Chipboard)."

Andersen, I., G. R. Lundqvist, and L. Mølhave, Atmos. Environ. **9**, 1121 (1975). "Indoor Air Pollution Due to Chipboard Used as a Construction Material."

Andersen, I., G. R. Lundqvist, and L. Mølhave, Ugeskr. Laeg. (Dan), **136**(38), 2140 (1974). "Liberation of Formaldehyde from Particleboard under Controlled Conditions in a Climate Chamber."

Andersen, I., G. R. Lundqvist, and L. Mølhave, Ugeskr. Laeg. (Dan) **136**(38), 2145 (1974). "Liberation of Formaldehyde from Particleboard: Mathematical Model."

Andersen, I., G. R. Lundqvist, and L. Mølhave, Ugeskr. Laeg. (Dan), **136**(38) 2133 (1974). "Formaldehyde in the Atmosphere in Danish Homes."

Anderson, R. W., and R. C. McAfee, Ger. Offen. 2,354,607, May 9, 1974. "Aqueous Coating Composition Containing Polyepoxides and Polycarboxylic Acid Monoanhydrides."

Ando, S., H. Akino, and S. Miura, Japan 74 44,581, Nov 29, 1974. "Composition for Treating Plastic Surfaces."

Andrezen, L. M., L. A. Mazurova, L. F. Orlova, T. A. Taskina, and A. M. Khomyakov, Elektrofiz. Apparatura (Russ), **1977**(15), 145 (1977). "Hardening of Urea-Formaldehyde Resins in the Presence of Ionizing Radiation."

Andrezen, L. M., L. A. Mazurova, L. F. Orlova, T. A. Taskina, and A. M. Khomyakov, Dokl. 2-go Vses. Soveshch. po Primeneniyu Vskoritelei Zapyazh. Chastits v Nar. Kh-ve, 1975 (Russ), **1976**(1), 279 (1976). "Hardening of Urea-Formaldehyde Resins Exposed to Ionizing Radiation."

An Foras Taluntais Co, Brit. P 1,483,150, Aug 17, 1977. "A Method of Treating Human and Animal Waste Products and Products so Obtained."

Anokhin, A. E., Sb. Tr. – Tsentr. Nauchno-Issled. Inst. Fanery, **1976**(6), 107 (1976). "Toxicity of Polymeric Products Based on Wood and Urea Resins."

Anokhin, A. E., N. N. Kostin, A. N. Bokov, and A. N. Zabrodkin, Sb. nauch. rabot. Rostov. med. in-t (Russ), **1973**(2), 174 (1973). "Effects of Hardener Concentration and Hardening Time on Toxicity of Chip Boards and Adhesive Properties of SK-75 Urea Resin."

Anokhin, A. E., A. N. Bokov, and N. N. Kostin, USSR P 478,724, July 30, 1975. "Parquet Panel."

ISBN 0-201-04558-3

Anokhin, A. E., A. N. Bokov, A. G. Zabrodkin, and N. N. Kostin, USSR P 421,531, Mar 30, 1974. "Particle Boards."

Anokhin, A. E., A. N. Bokov, A. G. Zabrodkin, N. N. Kostin, and G. P. Trubitskaya, USSR P 421,529, Mar 30, 1974, "Extrusion Particle Boards."

Anokhin, A. E., A. G. Zabrodkin, and T. V. Zakharova, USSR P 364,426, Dec 28, 1972. "Adhesive Composition."

Anokhin, A. E., A. N. Bokov, A. G. Zabrodkin, and N. N. Kostin, USSR P 358,150, Nov 3, 1972. "Plywood."

Anokhin, A. E., A. N. Bokov, A. G. Zabrodkin, and N. N. Kostin, USSR P 353,847, Oct 9, 1972. "Particle Boards."

Anokhin, A. E., A. N. Bokov, A. G. Zabrodkin, and N. N. Kostin, USSR P 353,846, Oct 9, 1972. "Waterproof Particle Boards."

Anonymous, Japan. Kokai 74 99,775, Sept 20, 1974. "Glass Fiber-Reinforced Wood Substitutes."

Anonymous (Food Drug Adm., Washington, D.C.), Fed. Regist. **37**(90), 9317 (1972). "Food Additives. Components of Paper and Paper Board in Contact with Dry Food."

Antlfinger, G. J., and R. F. Reinhart, USP 4,018,966, Apr 19, 1977. "Compositions Containing a Reactive Hydroxyl-Containing Vinyl Chloride Polymer."

Antoni, K. A., Fr. P 749,177, July 20, 1933. "Procédé de Fabrication de Bois Artificiel et d'Objects en Bois Artificiel."

Aono, I., and T. Yoshinaga, Japan. Kokai 74 71,111, July 10, 1974. "Crack-Resistant Decorative Boards with Overlays."

Arbuzov, V. V., USSR P 322,348, Nov 30, 1971. "Particle Boards."

Argo, W. B., J. W. Edwards, A. P. Little, and W. M. Lumb, USP 4,051,070, Sept 27, 1977. "Catalyst Preparation Utilizing Metal Nitrates and Nitrogen Oxide Scavenger."

Aries, R., Fr. Demande 2,230,348, Dec 20, 1974. "Sulfamoylanthranilic Acid Derivatives."

Arima, Y., and M. Ichikawa, Japan. Kokai 76 55,499, May 15, 1976. "Creaseproofing of Textiles."

Ashall, R. J., Brit. P 1,316,911, May 16, 1973. "Lignosulfate and Urea Extenders for Binders for Glass Fiber Insulation."

Ashall, R. J., Brit. P 1,273,152, June 6, 1968. "Thermal Insulation Based on Resin-Bonded Mineral Fibers."

Ashall, R. J., Ger. Offen. 2,161,570, June 29, 1972. "Binders for Glass Fiber Insulation."

Auer, L., Fr. P 674,703, May 7, 1929. "Artificial Substances."

Averell, P. R., in *Analytical Chemistry of Polymers,* Part I., G. M. Kline, ed. Vol. 12 "High Polymers," Interscience, New York, 1959. "Amino Resins."

Azarov, V. I., A. N. Oblivin, and V. I. Semenov, Izv. Vyssh. Uchebn. Zaved., Lesn. Zh. (Russ) **18**(4), 103 (1975). "Hydrolytic and Thermal Stability of Urea Resins under Conditions of Chip Board Molding, Acclimatizing, and Thermal Hardening."

Baer, R. L., D. L. Ramsey, and E. Biondi, Arch. Dermatol. **108**(1), 74 (1973). "The Most Common Contact Allergens: 1968–1970."

ISBN 0-201-04558-3

Baer, R. L., G. Lipkin, N. B. Kanof, and E. Biondi, Arch. Dermatol. **89**(1), 63 (1964).

Bagley, J. M., Brit. P 1,380,919, Jan 15, 1975. "Hardening Agent Compositions for Synthetic Resins."

Baldwin, J. T., USP 1,793,666, Feb 24, 1931. "Floor Covering."

Ball, G., and E. A. Boettner, Nuova Chim. (Ital) **50**(5), 76 (1974). "Combustion Products of Nitrogen-Containing Polymers."

Ball, G. R., and F. Grantham, Fr. Demande 2,081,925, Jan 14, 1972. "Bonding Panels with Adhesives."

Baranowski, P., and D. Lupa, Zesz. Nauk. – Akad. Ekon. Poznaniu, Ser. 1 (Pol) **58**, 127 (1974). "Infrared Spectroscopic Study of Chemically Hardenable Lacquers, II."

Baranowski, P., and D. Lupa, Pr. Zakresu Towarozn. Chem., Wyzsza Szk. Ekon. Poznaniu, Zesz. Nauk., Ser. 1, No. 46, 93 (1972). "Infrared Spectroscopy of Chemically Cured Undercoatings and Enamels."

Barbera, A., and D. Canepa, J. Assoc. Off. Anal. Chem. **60**, 708 (1977). "Determination of Free Urea in Fertilizers Containing Urea Formaldehyde Polymers."

Barghoorn, A. W., Holz-Zentralblatt **104**, 994 (1978). "Spanplatten und Formaldehyd."

Bartashevich, A. A., and P. N. Markevich, Derevoobrab. Prom-st. (Russ) **1974**(11), 3 (1974). "Coating Furniture Panels with Films."

Barthelemy, H., USP 1,691,427, Nov 13, 1929. "Condensation Products of Urea and Formaldehyde."

Barzynski, H., G. Heil, and G. Storck, Ger. Offen. 2,251,433, Apr 25, 1974. "Solvent-Free, Light-Hardenable Printing Inks."

BASF, Fr. Demande 2,163,578, Aug 31, 1973. "Improving Stability of Amino Plastic Adhesives."

Basmadzhieva, K., T. Burkova, M. Argirova, and others, Khig. Zdraveopaz. (Bulg) **17**, 480 (1974). "Biological Effect of Threshold Concentrations of Hydrochloric Acid and Formaldehyde Inhaled into the Organism."

Basterfield, S., and E. C. Powell, Can. J. Research **1**, 261 (1929). "Isoureas and Isoureides. I. Some New Isoureas, Salts and Acyl Derivatives."

Basterfield, S., and E. C. Powell, Can. J. Research **1**, 285 (1929). "Isoureas and Isoureides. II. Condensation of Isoureas with Diketons and Ketonic Esters."

Batty, J. W. "Textile Auxiliaries.", Pergamon Press, New York, 1967.

Batyuk, V. Ts., Zh. V. Malakhova, and L. M. Fenenko, Probl. Genezisa Melior. Oroshaemykh Pochv (Russ) **3**, 54 (1973). "Improvement of Physicochemical Properties of Soils of Rice Fields."

Batyuk, V. P., S. K. Samokhvalenko, Nauch. Osn. Melior. Pochov (Russ) **1972**, 184 (1972). "Effect of Surfactants on the Aqueous-Physical Properties of Soil and on the Physiological-Biochemical Features of Plants."

Batyuk, V. P., Melior. Pochv SSSR (Russ) **1971**, 175 (1971). "Chemical Bases for the Improvement of Solonetzes with Polymers and Use of Surfactants for Increasing Soil Fertility."

Baum, E., Bauwirtschaft, Heft 35 vom 24. Aug 1972. "Brandverhalten von UF-Schaumkunststoffen."

Baumann, H., "UF-Foam Bibliography", Schaum-Chemie W. Baur, Co; Frankenthal, GFR, 1979.

ISBN 0-201-04558-3

Baumann, H., Angew. Makromol. Chem. **71**, 221 (1978). "Treatment of Urea-Formaldehyde Resins with Concentrated Solutions of Alkaline Earth Hydroxides."

Baumann, H., ed., *FSK Schaumkunststoffe 1978,* Vorträge anlässlich der 8. Int Congress Plastic Foams, Mai 1978 Düsseldorf, Fachverband Schaumkunststoffe e.V., Frankfurt, 1978.

Baumann, H., Japan Plastics, Aug-Sept, pages 13–17, 1976. "Preparation and Processing of UF-Foam Polymers."

Baumann, H., Kunststoffe im Bau **11**(5+6), 35 (1976). "Formaldehyd in UF-Schaumkunststoffen, Teil 1."

Baumann, H., HR-Zeitschrift für Versorgungstechnik (7), 328 (1975). "UF-Schaumkunststoffe zur Sanierung von Altbauten."

Baumann, H., Plasticonstruction **5**(3), 115 (1975). "Harnstoffharz-Schaumkunststoffe–sie kommen doch wieder."

Baumann, H., Chemiker Zeitung-Chemische Apparatur **87**(8), 267 (1963). "Die Ausnutzung der Sonnenenergie."

Baumann, H., Kunststoff-Rundschau **9**(4), 169 (1962). "Plastoponik–ein neues Kulturverfahren fur aride Gebiete."

Baumann, H., *Leime und Kontakleber,* Springer, 1967. Chapter 15: "Chemisch bindende (härtende) Leime (wasserabspaltend)."

Baumann, H., *Plastoponik: Schaumkunststoffe in der Agrarwirtschaft,* Dr. Alfred Hüthig Verlag, Heidelberg, 1967.

Baumann, H., Chemiker-Ztg./Chem. Apparatur **90**(13), 449 (1966). "Apparatur nach Baumann–Graf zur Ermittlung der Schaumbeständigkeit und des Schäummaximums von Lösungen schaumbildender Substanzen."

Baumann, H., Fetter, Seifen, Anstrichmittel die Ernährungsindustrie **68**, 741 (1966). "Apparatur nach Baumann zur Bestimmung der Flüssigkeitsaufnahme von pulvrigen Substanzen."

Baumann, H., Parfümerie u. Kosmetik **40**, 287 (1959). "Pudergrundlage aus Kunstharzschaumstoff."

Baumann, H., Wärme-, Kälte-, Schall-, und Feuchtigkeitsschutz **3**(16), 217 (1959). "Isoschaum, Prüfteste und Erfahrungshinweise."

Baumann, H., Hansa (1), 2 (1959). "Isolierung von Schiffen mit spritzbarem Schaumkunststoff."

Baumann, H., Kunststoffe **48**(8) 362 and (9) 406 (1958). "Properties of Foamed Cold-Setting Urea Formaldehyde Resins, Part I and II."

Baumann, H., and G. Schmidt, Fortschritte der Medizin **76**(2), 59 (1958). "Überprüfung der Reizlosigkeit eines Puders und einer neuen Wundauflage aus Kunstharzschaumstoff ('OBC-Puder')."

Baxter, G. F., USP 3,108,990, Oct 29, 1963. "Stabilizing Phenol-Formaldehyde Resins."

Bayersdorf, F., "Separation and Structure of Low Molecular Weight UF-Condensation Products." Ph.D. Thesis, Univ. Darmstadt, GFR, 1977.

Baymiller, J. W., USP 3,816,236, June 11, 1974. "Cross-banding Material for Wood Veneers."

Beck, K. M., USP 3,725,323, Apr 3, 1973. "Cyclohexylammonium *N*-Cyclohexylsulfamate as Resin Catalysts."

ISBN 0-201-04558-3

Becker, K. H., V. Schurath, and T. Tatarczyk, Appl. Opt. **14**(2), 310 (1975). "Fluorescence Determination of Low Formaldehyde Concentrations in Air by Dye Laser Excitation."

Becker und Van Huellen Niederrheinische Maschinenfabrik K.-G., FP 2,117,512, Aug 25, 1972. "Nonflammable Wood-Based Components."

Behar, R., and A. Perrier, Ger. Offen. 2,347,401, Apr 4, 1974. "Urea-Formaldehyde Resin."

Belova, T. A., T. N. Egorova, E. S. Batalina, and A. E. Anokhin, Tr. Vses. Nauchno-Issled. Inst. Tsellyul.-Bum. Prom-sti. (Russ) **68**, 107 (1975). "Studies on the Development of Technology for the Production of Packaging Boxboard by Its Surface Treatment in a Sizing Press."

Belyi, V. A., N. I. Egorenkov, N. M. Klimashevich, and Yu. M. Pleskachevskii, N. V. Syrov, USSR P 346, 396, July 28, 1972. "Coating of Products from Thermoplastic Material."

Benning, C. J., "Plastic Foams", J. Wiley, New York, 1969.

Benteli, A., Swiss P 161,052, June 16, 1933. "Adhesive."

Bercovici, T., J. King, and R. S. Becker, J. Chem. Phys. **56**, 3956 (1972). "Formaldehyde. Comprehensive Spectral Investigation as a Function of Solvent and Temperature."

Bergin, E. G., and V. Godin, Can. Forest. Serv., Publ. No. 1296 (1971). "Durability of Wood Adhesives in Birch Plywood After 15 Years Exposure to the Weather."

Bergsund, K., Norw. P 131,891, May 12, 1975. "Adhering with Hard Adhesive and Use of Solid Hardener."

Bernert, J., Wasser, Luft Betr. **20**(2), 75 (1976). "Emissions from Particle Board Factories."

Bertsch, W., E. Anderson, and G. Holzer, J. Chromatogr. **112**, 701 (1975). "Trace Analysis of Organic Volatiles in Water by Gas Chromatography-Mass Spectrometry with Glass Capillary Columns."

Bezverkhaya, L. M., K. Ya. Martynov, and V. M. Khrulev, Izv. Vyssh. Uchebn. Zaved., Stroit. Arkhit **17**(12), 75 (1974). "Use of Resin 248 as a Binder for Particle Boards."

Billingsfors Bruks AB, Neth. Appl. 74 06,753, Nov 25, 1974. "Film with Finished Surface."

Binder, K., R. Soboll, and K. Lukesch, Mitt. Chem. Forschunginst. Wirt. Oesterr. Oesterr. Kunststoffinst. **25**(3), 121 (1971). "Surface Tension and Adhesion Properties of Plastics."

Bitterli, W., and J. M. Sire, Textilveredlung **11**(8), 345 (1976). "Detection of Trace Amounts of Formaldehyde in Aqueous Solutions–Comparison of two Spectral Colorimetric Methods for the Quantitative Determination of Formaldehyde in Textiles."

Black, W. R., USP 3,905,847, Sept 16, 1975. "Urea-Aldehyde Resin Binders."

Black, W. R., USP 3,697,355, Oct 10, 1972. "Urea-Aldehyde Resin Adhesive for Bonding Water-Penetrable Cellulosic Substrates."

Black, W. R., FP 2,098,692, Apr 14, 1972. "Urea-Aldehyde Adhesive for Agglomerating Water-Impregnable Cellulosic Substrates."

Black, W. R., S. African 70 05,055, Mar 8, 1971. "Urea-Aldehyde Resin Adhesives for Bonding Water-Penetrable Cellulosic Substrates."

ISBN 0-201-04558-3

Blaga, A., Can. Build. Dig., CBD 166 (1974). "Plastic Foams."

Blais, J. F., *Amino Resins,* Reinhold Publishing Co., New York, 1959.

Blake, A. D., Ger. Offen. 2,402,411, July 25, 1974. "Aminoplast Foam."

Blandin, H. M. F. F., M. R. A. M. Blandin, and P. P. P. C. Blandin, FP 2,192,552, Feb 8, 1974. "Refractory Insulator."

Blank, W. J., USP 3,907,740, Sept 23, 1975. "Stable, High Solids, Methylated Urea-Formaldehyde Coating Compositions."

Blommers, E. A., and R. H. Moult, USP 4,032,515, June 28, 1977. "Curable Resorcinol Terminated Urea-Formaldehyde Resins."

Boehme, P., G. Dahl, H. Damme, S. Dammer, P. Froehlich, H. Konrad, M. Scholz *et al.,* E. Ger. P 124,312, Feb 16, 1977. "Finishing Film Based on Urea-Formaldehyde Resins."

Boehm, W., H. Drechsel, and M. Mueller, Chem. Prum. (Czech.) **25**(10), 542 (1975). "Continuous Production of Urea-Formaldehyde Resins."

Boehm, W., H. Drechsel, J. Frydrych *et al.,* E. Ger. P 88,196, Feb 2, 1972. "Continuous Preparation of Urea-Formaldehyde Resins."

Boehm, W., H. Drechsel *et al.,* Ger. Offen. 2,202,202, Aug 24, 1972. "Urea-Formaldehyde Resin."

Boettner, E. A., G. L. Ball, and B. Weise, U.S. Nat. Tech. Inform. Serv., PB Rep. No. 222001/0, 154 pp (1973). "Combustion Products from the Incineration of Plastics."

Bögemann, M., S. Petersen, O. E. Schultz, and H. Söll, "Preparation and Reactions of Sulfur Containing Carbonicacid Derivatives," p. 160 in "Houben-Weyl, Methoden der Organischen Chemie." 4th edition, Thieme Verlag, Stuttgart, 1955, E. Müller, editor. Volume 9, "Sulfur-, Selenium and Tellurium Compounds."

Bogosavljevic, M., and M. Popovic, Hem. Ind. (Croat) **26**(12), 489 (1972). "Stability of a Continuous Process of Formaldehyde Addition with Urea."

Bohlen, J. C., U.S. Nat. Tech. Inform. Serv., PB Rep. No. 204981, 34 pp (1971). "Shear Strength of Douglas-Fir Lumber Laminated at High Temperatures."

Bokareva, E. Z., L. V. Kraineva *et al.,* USSR P 476,294, July 5, 1975. "Injection-Molding Composition with a Glass-Fiber Filler."

Bokov, A., and I. Shamshura, Sb. Nauch. Rabot Rostov. Med. In-T (Russ) (2), 212 (1973). "Comparative Sanitary-Hygienic Evaluation of Particle Board Facing Tiles Produced from Autoclaved and Nonautoclaved Sawdust Based on Phenol-Formaldehyde Resin SBS-1."

Bokov, A., *et al.,* Gig. Primen. Polim. Mater. Stroit., Mater. Vses. Soveshch., 1st 1972 (Russ) p. 84 (1973). "Hygienic Aspects of the Production and Use in Construction and the Furniture Industry of Synthetic Resin-Based Materials Manufactured from Wood."

Bondi, E., and A. Larcher, GP 1,645,016, Oct 4, 1973. "Stable, Aqueous, Concentrated Solutions of Urea/Formaldehyde Condensation Products."

Bonner, W. H., Jr., USP 3,645,981, Feb 29, 1972. "Poly(thioureas)."

Bonnington, M. G., S. African P 76 00197, Nov 5, 1976. "Improvements in Thermosetting Molding Materials."

Borden (UK) Ltd., BP 1,421,994, Jan 21, 1976. "Improvements in or Relating to Urea-Formaldehyde Condensates."

Borisyuk, I. D., N. D. Ishin, A. G. Yakhno, and V. I. Papeshina, USSR P 362,855, Dec 20, 1972. "Composition for Prime Coats and Mastics."

ISBN 0-201-04558-3

Borodkina, N. I., *et al.*, USSR P 448,208, Oct 30, 1974. "Urea-Formaldehyde Foam Plastic."

Borozdin, V. S., and G. S. Filippov, USSR P 431,199, June 5, 1974. "Cellular Plastic Based on an Urea-Formaldehyde Resin."

Brancato, J. J., and D. F. Herman, USP 3,928,272, Dec 23, 1975. "Urea-Formaldehyde Solid Microspheres."

Brandani, V., G. Di Giacomo, and P. U. Foscolo, La Chimica e L'Industria **58**, 99 (1976). "Termodinamica delle soluzioni acquose di formaldeide: equilibri in fase liquida."

Brandeis, J., K. Flory, and A. Feinauer, Ger. Offen. 2,259,680, July 4, 1974. "Chlorine-Resistant, Finishing Treatment for Textiles."

Braswell, J. R., D. R. Spiner, and R. K. Hoffman, Appl. Microbiol. **20**, 765 (1970). "Adsorption of Formaldehyde by Various Surfaces During Gaseous Decontamination."

Braum, D., and F. Bayersdorf, Chem.-Ztg. **96**(6), 352 (1972). "Amino Plastics. 1. Silylation of Urea-Formaldehyde Condensates."

Breysse, P., "Formaldehyde Levels in Mobile Homes." University of Washington, Environmental Health and Safety Newsletter, Volume 26, Jan-Jun 1977.

Brial, J. C., and C. Duclairoir, Fr. Demande 2,338,956, Aug 19, 1977. "Urea-Formaldehyde Polycondensate with Great Storage Stability."

Brial, J. C., and C. Duclairoir, Ger. Offen. 2,702,421, July 28, 1977. "Urea-Formaldehyde Polycondensates."

British Cyanides Co., The, FP 714,101, Mar 31, 1931. "Plastic Compositions."

British Cyanides Co., The, BP 359,498, July 23, 1930. "Urea-Formaldehyde Condensation Products."

British Industrial Plastics Ltd., Neth Appl. 76 00373, July 19, 1977. "Thermosetting Urea-Formaldehyde or Melamine-Formaldehyde Resin Molding Compositions."

British Industrial Plastics Ltd., Kokai 76:06,269, Jan 19, 1976. "Water-Adsorbent Urea-Formaldehyde Resin Foams."

British Industrial Plastics Ltd., FP 2,102,406, May 12, 1972. "Manufacture of Urea-Formaldehyde Resin Foam."

Brooks, S. H. W., and H. A. Raddin, USP 3,668,286, June 6, 1972. "Fiberboard Produced from Wood Particles Having a 5 to 25 Percent Moisture Content Prior to Steaming and Mechanical Reduction in the Formation Process."

Brose, H., D. Hentschel, and H. Pelshenke, Ger. Offen. 1,966,796, June 20, 1974. "Ultraviolet Light-Hardenable Pigmented Alkyd Resin Lacquers."

Brown Boveri und Cie, Fr. Demande 2,193,699, Feb 22, 1974. "Stiff Foam with Open Cells."

Brunner, K., Holz-Zentralblatt, **111**, 1661, (1978), "Wechselwirkung zwischen der Formaldehydabgabe von Spanplatten und der Formaldehyd konzentration in der Umgebungsluft."

Brunnmüller, F., H. Schatz, J. Mayer, and O. Grabowsky, Ger. Offen 2,241,995, Apr 4, 1974. "Continuous Preparation of Aminoplast Solutions."

Brünnmüller, F., H. Schatz, J. Mayer, and O. Grabowsky, USP 3,816,376, June 11, 1974. "Continuous Manufacture of Amino Resin Solutions."

Brunnmüller, F., J. Mayer, *et al.*, GP 2,215,947, Apr 1, 1972. "Phenol Resin Adhesives."

ISBN 0-201-04558-3

Brunnmüller, F., O. Grabowsky, and J. Lenz, Ger. Offen. 2,110,939, Sept 14, 1972. "Difficultly Flammable Aminoplast Foam."

Brunnmüller, F., H. Schatz, J. Mayer, and O. Grabowsky, Ger. Offen. 2,109,754, Sept 14, 1972. "Continuous Manufacture of Formaldehyde-Urea Resin Solutions."

Brunnmüller, F., O. Grabowsky, and J. Lenz, Ger. Offen. 2,064,102, July 13, 1972. "Hardenable Aminoplast Foams of Low Shrinkage."

Brunnmüller, F., J. Mayer, and J. Lenz, Ger. Offen. 2,055,074, May 18, 1972. "Soil-Improving Urea-Formaldehyde Foams."

Brunnmüller, F., "Aminoplaste" in "Ullmanns Encylopädie der technischen Chemie." Volume 7, page 403; Verlag Chemie, Weinheim, 1972.

Bulygin, B. M., and Z. V. Malafeeva, Plast. Massy (Russ) (12) 48 (1978). "Isolation of Volatile Products in the Preparation and Use of Phenol-Formaldehyde Cellular Plastics."

Burakinskaya, A. V., V. Ramanauskiene, and R. S. Taubkina, Vopr. Epid Epidemiol. i Gigieny v LitSSR. Puti Uluchsheniya Uslovii Truda v S.-kh, Proiz-ve, (Russ), pgs 188-90 (1976). "Toxicological and Hygienic Evaluation of Spacklings Using a Urea-Formaldehyde Resin."

Burmester, A., and W. E. Willie, Holz als Roh- und Werkstoff **34**, 67 (1976). "Swelling Reduction of Wood Over Partial Relative Humidity Ranges. (1) Incorporation of Materials in Cell Walls."

Burnett, R. D., P. Diamond, M. P. Anderson, and M. L. Swigart, USAF Environmental Health Lab, EHL-74M-13, pg 34, McClellan AFB, CA (1974). "Evaluation of Charcoal Sampling Tubes."

Buschfeld, A., Ger. Offen. 2,337,233, Dec 19, 1974. "Phenolic Resin Solution from a Storage-Stable Precondensate."

Cammerer, W. F., Bull. Inst. Int. Froid, Annexe **1973**(4), 189 (1973). "Thermal Conductivity as a Function of the Thickness of Insulating Materials."

Campbell, C. C., USP 3,869,414, Mar 4, 1975. "Wax Emulsions Containing Proteinaceous Emulsifiers."

Carbone, R. D., Masters Thesis, University of Washington, Seattle, 1978. "Formaldehyde Exposure in Mobile Homes."

Carson, F., USP 2,033,411, Oct 3, 1936. "Manufacture of Artificial Lumber and Pressed Molded Products."

Casebier, R. L., and K. D. Sears, Can P 986,100, Mar 23, 1976. "Aminated Sulfite Derivatives of Coniferous Barks."

Catoni, M., U. Massa, and P. Pellegrini, Cellulosa Carta **27**(2), 27 (1976). "Urea-Formaldehyde Resin for Stock Suspensions–Examination of Application Factors."

Central Lab. Phys. Chem. Chem., Sofia, Brit P 1,499,285, Jan 25, 1978. "Improvements in or Relating to the Curing of Aminoplastic Plastics Materials."

Champion Paper Co. Ltd., Brit P 1,270,408, Apr 12, 1972. "Opacifying Particles Providing High Opacity in Fibrous and Nonfibrous Substrates."

Chand, R., Ger. Offen. 2,354,149, 1974. "Apparatus and Process for Electrochemical Gas Determination."

ISBN 0-201-04558-3

Chang, T., Tai-Wan Sheng Lin Yeh Shih Yen So Lin Wu Chu Ho Tso Shih Yen Pao Kao (17), 9 (1973). (Japanese). "Influence of Manufacturing Factors upon Properties of Particleboard."

Chapput, A., B. Roussel, and G. Fleury, J. Raman Spectrosc. **1**, 507 (1973). "Raman Spectroscopic Study in the Vapor Phase of Three Deuterated Species of Formaldehyde."

Chattaway, F. D., Proc. Roy. Soc. (London) **A137**, 481 (1932). "Condensation of a Bromal with Urea."

Cherubim, M., Holz als Roh- und Werkstoff **34**, 449 (1976). "Hydrolysis and Formaldehyde Release from Particle Boards."

Cherry, O. A., USP 1,790,461, Jan 27, 1931. "Condensation Product of Urea with Formaldehyde."

Chiavarini, M., R. Bigatto, and N. Conti, Angew. Makromol. Chem. **70**, 49 (1978). "Synthesis of Urea-Formaldehyde Resins: NMR Studies on Reaction Mechanisms."

Chiavarini, M., N. Del Fanti, and R. Bigatto, Angew. Makromol. Chem. **46**, 151 (1975). "Compositive Characterization of Urea-Formaldehyde Adhesives by NMR Spectroscopy."

Chow, S., and G. E. Troughton, For. Prod. J. **25**(8), 54 (1975). "Development of an X-ray, Potassium Bromide Embedding Method for Determination of Resin Cure."

Chow, S., and P. R. Steiner, Holzforschung **29**(1), 4 (1975). "Catalytic Exothermic Reactions of Urea-Formaldehyde Resin."

Chow, S., and P. R. Steiner, Forest Prod. J. **24**(5), 35 (1974). "Resistance to Cyclic Exposure at Extreme Low Temperature of Urea Resin Bonds."

Chow, S., J. Appl. Polym. Sci. **18**(9), 2785 (1974). "Morphologic Accessibility of Wood Adhesives."

Chow, S., Holzforschung **27**(2), 64 (1973). "Softening Temperatures and Durability of Wood Adhesives."

Chrastil, J., and J. T. Wilson, Anal. Biochem. **63**, 202 (1975). "Sensitive Colorimetric Method for Formaldehyde."

Christensen, H. E., E. J. Fairchild, B. S. Carroll, and R. J. Lewis, Sr., U.S. Dept. HEW, Public Health Service, Center for Disease Control, NIOSH, Rockville, Md., 1976. "Registry of Toxic Effects of Chemical Substances."

Christensen, R. L., and P. Robitschek, USP 3,826,770, Júly 30, 1974. "Stabilized Mixture of Urea Formaldehyde and Methylolureas."

Christensen, R. L., and P. Robitschek, Can. P 1,029,915, April 25, 1978. "Stabilized Mixture of Urea Formaldehyde and Methylol-ureas."

Christensen, R. L., Forest Prod. J. **22**(4), 17 (1972). "Test for Measuring Formaldehyde Emission from Formaldehyde Resin Bonded Particleboards and Plywood."

CIBA-GEIGY, Fr. Demande 2,182,168, Jan 14, 1974. "Apparatus for Solidifying Liquid, Non-metallic Substances, Especially Polymerizable or Crystallizable Substances."

CIBA, BP 1,187,352, Apr 8, 1970. "New, Sulphur-Containing Derivatives of Aminotriazine-Formaldehyde Condensation Products, Process for their Preparation, and their Use as Plasticizing Additives for Quick-Curing and Storage-Stable Aminoplast Moulding Compositions."

ISBN 0-201-04558-3

CIBA, BP 342,767, Nov 7, 1928. "Synthetic Resins."

CIBA, BP 342,730, Nov 6, 1928. "Synthetic Resins."

CIBA, BP 342,723, Nov 2, 1928. "Synthetic Resins."

CIBA, BP 281,717, Dec 4, 1925. "Urea-Aldehyde Condensation Products."

CIBA, BP 260,288, Oct 22, 1925. "Urea-Aldehyde Condensation Products."

CIBA, BP 253,094, June 5, 1925. "Condensation Products of Formaldehyde and Urea."

CIBA, BP 249,101, Mar 10, 1925. "Condensation Products of Formaldehyde and Urea."

CIBA, FP 644,648, Nov 28, 1927. "Porous Material."

CIBA, GP 732,498, Feb 4, 1943. "Condensation Products."

CIBA, GP 562,943, Sept 20, 1929. "Artificial Masses."

CIBA, GP 523,181, Dec 12, 1926. "Urea-Formaldehyde Condensation Products."

CIBA, GP 499,710, Mar 14, 1925. "Urea-Formaldehyde Condensation Products."

CIBA, GP 499,589, Jan 20, 1925. "Soluble Condensation Products and Formaldehyde."

CIBA, Swiss P 154,831, Apr 23, 1931. "Artificial Masses."

CIBA, Swiss P 154,520, Nov 12, 1930. "Artificial Masses."

CIBA, Swiss P 114,289, Mar 16, 1926. "Verfahren zur Herstellung von Kondensationsprodukten aus Harnstoff bezw. seinen Derivaten und Formaldehyd."

CIBA Ltd. Cabot Corp., FP 2,068,286, Sept 24, 1971. "Continuous Manufacture of Crosslinked Aminoplast Resins."

Clad, W., and E. H. Pommer, Holz als Roh- und Werkstoff **36**, 383 (1978). "Querschnitt und Umriss, Aktuelle Klebstoffprobleme."

Clad, W., and E. H. Pommer, Mitt. Deut. Ges. Holzforsch. No. 57, 153 (1971). "Importance of Synthetic Resin Content for the Stability of Chipboards."

Clad, W., Holz als Roh- und Werkstoff **18**, 991 (1960). "Die Beurteilung von UF-Harzen und Ihre Prufung."

Clarke, B. J., R. W. Kershaw, and F. J. Lubbock, Australian P 441,798, Nov 8, 1973. "Aminoplast Resin Granules."

Clermont, L. P., and M. N. Carroll, Forest Prod. J. **26**(8), 35 (1976). "Improved Chromotropic Acid Determination of Formaldehyde Evolution from Wood Composites."

Cleveland, J. J., USP 3,943,994, Mar 16, 1976. "Ceramic Cellular Structure Having High Cell Density."

Columbus, P. S., and C. R. Erikson, USP 3,692,185, Sept 19, 1972. "Filter Paper."

Contardi, A., and B. Ciocca, Rend. ist. lombardo sci. **69**, 1057 (1936). "The Reaction Between Formaldehyde and the Polyalcohols."

Conte, C., and G. Devitofrancesco, Ann. Ist. Super. Sanita 9 (Spec. Number 1), 383 (1973). "Choice of Parameters for the Estimation of Hazard Industry Work Environments."

Cowan, W. D., J. Hoggan, and J. E. Smith, Tech. Q., Master Brew. Assoc. Am. **12**(1), 15 (1975). "Introduction of Respiratory-Deficient Mutants in Brewery Yeast."

Coyle, R. P., USP 3,931,072, Jan 6, 1976. "Particleboard, Hardboard, and Plywood Produced in Combination with a Lignin Sulfonate-Phenol Formaldehyde Adhesive System."

ISBN 0-201-04558-3

Crocker, E. C., USP 2,327,871, Aug 24, 1944. "Lining Cedar Chests with Urea-Aldehyde Condensation Products."

Cropp, J. A. D., E. L. D'Ouville, and H. C. Messman, BP 1,378,938, Mar 6, 1972. "Fertilizer Composition."

Curs, A., and H. Wolf, USP 2,076,295, Apr 6, 1933. "Cellular Heat- and Sound-Insulating Material."

Da Cunha, A., and A. Da Silva, FP 2,154,914, June 22, 1973. "Varnish for Floor Coverings."

Dahlberg, H. S., Can. P 991,780, June 22, 1976. "Flame Retardant Treatment of Fibers."

D'Alelio, G. F., USP 2,317,181, Apr 20, 1949. "Urea-Formaldehyde-Triethanolamine Hydrochloride Condensation Products."

Dalhamn, T., and A. Rosengren, Arch. Otolaryngol. **93**(5), 496 (1971). "Effect of Different Aldehydes on Tracheal Mucosa."

Dankelman, W., and J. M. H. Daemen, Analyt. Chem. **48**, 401 (1976). "Gas Chromatographic and Nuclear Magnetic Resonance Determination of Linear Formaldehyde."

Dankelman, W., M. H. Daemen, A. J. J. de Breet, *et al.* Angew. Makromol Chem. **54**(1), 187 (1976). "Modern Methods for the Analysis of Urea Formaldehyde Resins."

Dankelman, W., and J. De Wit, Holz als Roh- und Werkstoff **34**(4), 131 (1976). "Modern Methods for Analysis of Urea-Formaldehyde Resins."

Danner, B., H. Gerber, and H. Pummer, Swiss P 601,360, July 14, 1978. "Poly-(amidamine) Compounds."

Dashkovskaya, Z. F., T. G. Zakharov, T. I. Kravchenko, and K. I. Stankevich, USSR P 480,555, Aug 15, 1975. "Diminishing the Evolution of Formaldehyde from Products Containing Wood Particles and Carbamide Resins."

Date, M., and S. Fukuoka, Kokai 73:56,996, Aug 10, 1973. "Fire-Resistant Resin Treatment of Cellulose Fiber Blend Fabrics."

Davidovits, J., S. African P 73:08,363, Aug 8, 1974. "Sintered Composite Panels."

Davidson, R. W., and C. Schuerch, USP 3,642,042, Feb 15, 1972. "Forming Wood after Treatment with Sulfur Dioxide and Impregnation with Ammonia."

Davies, C. K., G. A. Varga, and R. S. Hinkson, Managing Livest. Wastes, Proc. 3rd Int. Symp., p. 227, Am. Soc. Agric. Eng., St. Joseph, Mich., (1975). "Conversion of Animal Wastes to Feed Supplements via the Organiform Process."

Davydov, V. V., *et al.*, USSR P 403,811, Oct 26, 1973. "Composition for the Stabilization of Soil."

Dead Sea Bromine Co. Ltd., Neth. Appl. 74:06,625, Nov 26, 1974. "Flame-Retardant Plates of Cellulose Materials."

de Breet, A. J. J., W. Dankelman, W. G. B. Huysmans, and J. de Wit, Angew. Makromol. Chem. **62**, 7 (1977). "^{13}C-NMR Analysis of Formaldehyde Resins."

De Bruyne, N. A., Plastics (London) **9**, 228 (1945). "Synthetic Adhesives."

De Bruyne, N. A., and D. A. Hubbard, BP 549,496, Nov 24, 1942. "Synthetic-Resin Adhesives."

De Bruyne, N. A., and the De Havilland Aircraft Co. Ltd., BP 470,331, Aug 3, 1937. "Coated and Compound Fabrics."

ISBN 0-201-04558-3

de Chesne, E. B., Koll. Beihefte **36**, 25 (1932). "Beitrage zur Kenntnis von Harnstoff-Formaldehyd-Kondensations- und Polymerisationsprodukten."

de Chesne, E. B., Koll. Beihefte **36**, 385 (1932). "Urea-Formaldehyde Condensation and Polymerization Products."

Decorit-Industrie-Beratung, BP 1,233,396, May 26, 1971. "Chipboard or Blockboard Covered with Fiber-Reinforced Synthetic Resin Sheets."

de Jong, J. I., and J. de Jonge, Recueil **72**, 1027 (1953). "The Chemical Composition of Some Condensates of Urea and Formaldehyde."

de Jong, J. I., and J. de Jonge, Recueil **72**, 139 (1953). "Kinetics of the Formation of Methylene Linkages in Solutions of Urea and Formaldehyde."

de Jong, J. I., J. de Jonge, and H. A. K. Eden, Recueil **72**, 88 (1953). "The Formation of Trimethylol Urea."

de Jong, J. I., and J. de Jonge, Recueil **71**, 890 (1952). "The Reaction of Urea with Formaldehyde."

de Jong, J. I., and J. de Jonge, Recueil **71**, 643 (1952). "The Reaction between Urea and Formaldehyde in Concentrated Solutions."

de Jong, J. I., and J. de Jonge, Recueil **71**, 661 (1952). "The Formation and Decomposition of Dimethylolurea."

Delepine, M., Ann. Chim. Phys. **15**, 469 and 570 (1898). "Amines et amides derives des aldehydes."

Demenkova, K. M., O. V. Senkevich, and K. A. Telichko, Stroit. Mater (Russ) (11), 18 (1974). "Improving the Properties of Urea Plastic Foams."

Dement'ev, A. G., and P. I. Seliverstov, Stroit. Mater. (Russ) (6), 34 (1975). "Factors Affecting the Heat Conductivity of Cellular Plastics."

Demko, P. R., F. J. Washabaugh, and R. H. Williams, USP 4,018,959, Apr 19, 1977. "Corrugating Adhesive Compositions Containing Thermoplastic Polymer, Thermosetting Resin, and Starch."

Den'gina, N. A., V. D. Marichev, V. M. Kozin, and K. G. Asnach, Vestn. Khar'kov. Politekh. Inst. (Russ), No. 68, 54 (1972). "Preparation of Bars in Hot Boxes with an MFF-M Binder."

Denisova, S. G., and T. V. Terskova, Listvennitsa 5, 115 (1974). "Studying the Kinetics of the Hardening of the Urea-Formaldehyde Resin KS-68M to Design Conditions for Molding Chip Boards from Larch."

Deppe, H. J., and K. Ernst, "Taschenbuch der Spanplatten Technik." DRW-Verlag, Stuttgart, 1977.

Deriancourt, G., Fr. Demande 2,213,963, Aug 9, 1974. "Mat of Lignocellulosic Fibers."

Detushev, I. F., V sb., Issled. Stroit. Konstruktsii s Primeneniem Polimer. Materialov (Russ), (3), 70 (1976). "Strength of Adhesive Lines of Wood Made from Melt Adhesives and a Urea Adhesive."

Di Dio, P., Kongr. "Chem. Pol'nohospod.", [Pr.], 2nd, 1, B15, 11 pp (Ger) (1972). "Use of Urea-Formaldehyde Resinous Foam for Soil Improvement."

Diem, H., and A. Hilt, "Formaldehyd", in "Ullmanns Encyklopädie der technischen Chemie." Volume 11, Verlag Chemie, Weinheim, 1974, page 687.

Dimter, L., and R. Kaps, E. Ger. P 77,033, Nov 12, 1970. "Adhesives for Making Composite Flat Structures Based on Cellulose Derivatives, Preponderantly Cellulose Triacetate."

ISBN 0-201-04558-3

Ding, J.-Y., F.-H. Tsai, and S.-D. Ueng, T'ai-wan Mu Ts'ai Kung Yeh **32**, 6 (1974). "Painting and Finishing Materials. II. Ultraviolet Spectra of Melamine Formaldehyde Resins and Urea Formaldehyde Resins."

Dirska, B., H. Jurek, J. Dalecki, and T. Manczak, Pol. P 73,219, Nov 15, 1974. "Hardening of Photographic Gelatin."

Dixon, A. E., J. Chem. Soc. **113**, 238 (1918). "Interaction of Formaldehyde and Carbamide."

Dixon, A. E., and J. Taylor, J. Chem. Soc. **109**, 1244 (1916). "The Interaction of Aldehydes and Thiocarbamides in the Presence of Acids."

Dixon International Ltd., Neth. Appl. 75 11,572, Apr 5, 1976. "Sealing Composition."

Dockal, P., Czech. P 169,239, May 15, 1977. "Detoxication of Waste Water from the Production of Urea-Formaldehyde Adhesive."

Dolenko, A. J., Rep. – East. For. Prod. Lab. (Can), OPX143F (1976). "Industrial Evaluation of Fire Retardant Amino Resin Coatings."

Domokos, G., USP 3,784,494, Jan 8, 1974. "Artificial Lumber from Untreated Sawdust, a Powdered Urea-Formaldehyde Resin, and Water."

Drews, M., Arch. Gartenbau **21**(1), 21 (1973). "Effect of Different Organic and Synthetic Substances on the Physical and Chemical Characteristics of Soils and Earths Used for Vegetable Growing Under Glass and Plastic Covers."

Drobyazko, V. N., and S. P. Doroshenko, Liteinoe Proizvod. **3**, 22 (1978). "Improving the Technological Properties of Antiscorching Paints."

Druzhinin, S. A., *et al.,* Probl. Gorn. Teplofiz., Mater. Vses. Nauchno-Tekh. Konf. 1973, 102 (1974). "Use of Polymer Foams to Prevent Soils from Freezing."

Druzhinin, S. A., E. V. Khlystunova, Yu. I. Krotov, and P. I. Kozlov, Plast. Massy (7) 35 (1974). "Properties of Urea-Formaldehyde Resin for the Preparation of Polymer Foam of High Multiplicity."

Dubrovin, V. S., E. I. Dromashko, and A. T. Sememova, Sb. Tr., Vses. Nauch.-Issled. Proekt.-Konstr. Inst. Osush. Mestorozhd. Polez. Iskop., Spets. Gorn. Rab., Rud. Geol. Marksheid. Delu, No. 14, 172 (1970). "Hardening Kinetics of Urea-Formaldehyde Resin."

Duclairoir, C., and J.-C. Brial, J. Appl. Polymer Sci. **20**, 1371 (1976). "Analyse par Resonance Magnetique Nucleaire des Polycondensats Uree-Formol."

Dueva, L. A., Gig. Tr. Prof. Zabol. (4), 20 (1974). "Experimental Reproduction of Immunological Tolerance of Urea-Formaldehyde Resin."

Dumov, S. N., *et al.,* USSR P 415,103, July 15, 1978. "Composition for Manufacturing Foam Plastic."

Duval, M., B. Bloch, and S. Kohn, J. Appl. Polymer Sci. **16**, 1585 (1972). "Analysis of Phenol-Formaldehyde Resols by Gel Permeation Chromatography."

Dyno Industrier A. S., Oslo, Norway, Technical Information Bulletin, number **6**, 1977. "Formaldehyde Emission from Particleboard."

Eades, A. G., BP 1,421,119, Jan 14, 1976. "Urea-Formaldehyde Compositions."

Ebdon, J. R., and P. E. Heaton, Polymer **18**, 971 (1977). "Characterization of Urea-Formaldehyde Adducts and Resins by Carbon-13 NMR Spectroscopy."

Eder, H. M., Ger. Offen. 2,716,375, Oct 19, 1978. "Wood Material Molding."

ISBN 0-201-04558-3

Edler, F., Swed. P 397,834, Nov 21, 1977. "Amino or Phenol Resin-Based Adhesives with Sulfite Liquor Additives."

Eek, L., M. Ciutat, Ion (Madrid) **33**(380), 117 (1973). "Determination of Free Formaldehyde in UF-Resin."

Eguchi, T., T. Taura, and S. Iimura, J. Kokai 73 75,489, Oct 11, 1973. "Catalyst for Removing Carbon Monoxide and Formaldehyde from Air."

Einhorn, A., *et al.,* Ann. **361**, 113 (1930). "*N*-Methylol Compounds of Acid Amides (II)."

Einhorn, A., and A. Hamburger, Ber. **41**, 24 (1908). "Die Methylolverbindungen des Harnstoffs."

Eisele, W., H. Petersen, J. Mayer, and O. Wittmann, GP 2,334,380, Sept 5, 1974. "Wood Cement."

El'bert, A. A., and Z. V. Tsareva, Khim. Mekh. Pererab. Drev. Drev. Otkhodov (Russ) 3, 96 (1977). "Study of the Setting of a Urea-Formaldehyde Resin by Gel Permeation Chromatography."

El'bert, A. A., and N. S. Time, Tekhnol. Drevesn. Plit i Plastikov (Russ) (3), 127 (1976). "Production of Dry-Molded Wood Fiberboard Using Binders Based on Urea-Formaldehyde Oligomers."

El'bert, A. A., *et al.* USSR P 518,363, June 25, 1976. "Wood Mass for Producing Board, Including Ground Wood Particles, a Urea-Formaldehyde Resin Binder, Modified by Compositions Based on Lignosulfonates."

El'bert, A. A., I. A. Gamova, and E. V. Viktorova, Derevoobrab. Prom-st. (Russ) (1), 4 (1975). "Effect of the Conditions for Pressing of Particle Boards on the Interaction of Resin and Wood."

El'bert, A. A., N. S. Time, and V. V. Vasil'ev, Derevoobrab. Prom. (Russ) (12), 4 (1973). "Modification of a Urea-Formaldehyde Resin with a Polyvinyl Acetate Emulsion."

El'bert, A. A., B. P. Erykhov, V. G. Liburkin, and V. I. Romitsyn, Derevoobrab. Prom. (Russ) (3), 5 (1973). "Hardening of Urea-Formaldehyde Resins."

El'bert, A. A., and N. S. Time, Derevoobrab. Prom. (Russ) **20**(9), 5 (1971). "Electric Determination of the Degree of Hardening of Urea-Formaldehyde Resins."

Elkins, H. B., and L. D. Pagnotto, in *Treatise on Analytical Chemistry,* Part III, Section B, Interscience, New York, 1971. "Industrial Health Hazards."

Ellis, C., "Chemistry of Synthetic Resins." Pergamon Press, 1935.

Ellis, C., USP 1,952,060, Mar 27, 1934. "Synthetic Resinous Products."

El-Sayad, I. A., Gig. Tr. Prof. Zabol. **16**(8), 46 (1972). "Effect Produced by Formaldehyde-Containing Resins of the Skin."

Endo, A., and T. Okumichi, Japan P 71 28,785, Aug 20, 1971. "Modified Urea Resin Adhesives for Corrugated Board Paper."

Energy Resources Co., Inc., Cambridge, Mass., Report, 489 pp., Nov. 1975. "A Review of Concentration Techniques for Trace Chemicals in the Environment."

Enkvist, T. U., Ger. Offen. 2,334,540, Jan 24, 1974. "Adhesives from Sulfate or Soda Black Liquor,"

Epstein, E., W. Rees, and H. Mailbach, Arch. Dermatol. **98**(1), 18 (1968). "Recent Experience with Routine Patch Test Screening."

Erhardt, K., H. Petersen, *et al.,* Ger. Offen. 2,243,857, Apr 4, 1974. "Aminoplast Adhesives for Wood."

ISBN 0-201-04558-3

Erhardt, K., *et al.,* Ger. Offen. 2,207,921, Aug 30, 1973. "Urea-Formaldehyde Resins."

Erhardt, K., *et al.,* GP 2,243,857, Sept 7, 1972. "Aminoplast Adhesives for Wood."

Erhardt, K., *et al.,* GP 2,207,931, Feb. 19, 1972. "Urea-Formaldehyde Resins."

Erins, P., and V. Cinite, Khim. Drev. (Russ) **9**, 29 (1971). "Action of Aqueous Ammonia Solutions on Birch Wood."

Ernst, K., H. Lobenhoffer, and A. Oberlein, Ger. Offen. 1,920,283, Jan 13, 1972. "Fire-Resistant Lightweight Building Material."

Erykhov, B. P., V. G. Liburkin, V. I. Romitsyn, and A. A. El'bert, Izv. Vyssh. Ucheb. Zaved., Les. Zh. **16**(1), 133 (1973). "Dynamical Method for the Study of Resin Polycondensation Processes."

Eschalier, X., Rev. gen mat. color. **12**, 249 (1908). "Strengthening Artifical Silk."

Établissement Lambiottes Frères and J. Lichtenberger, FP 722,289, Nov 22, 1930. "Synthetic Resins."

Fabriques de Produits de Chimie Organique de Laire, J. Malet, and R. Armenault, BP 340,114, Dec 6, 1928. "Urea-Formaldehyde Condensation Products."

Fabriques de Produits de Chimie Organique de Laire, J. Malet, and R. Armenault, FP 679,321, Dec 6, 1928. "Synthetic Resins."

Fahey, D. J., and D. S. Pierce, Forest Prod. J. **21**(11), 30 (1971). "Resistance of Resin-Impregnated Paper Overlays to Accelerated Weathering."

Fahrni, F., Holz als Roh- und Werkstoff **15**, 24 (1957). "Die Entwicklung der Holzspanplatte in dokumentarischer Sicht und der verfahrensmässige Beitrag von Novopan."

Fahrni, F., USP 2,686,143, Aug 10, 1954. "Process for Manufacturing a Composite Wooden Board."

Fahrni, F., FP 926,024, Sept 19, 1947. "Procédé pour la fabrication des plaques presses et formées de plusieurs couches de particules de bois et de liant et plaque pressé obtenue au moyen de ce procédé."

Fahrni, F., FP 899,487, May 31, 1945. "Procédé de transmission de liants liquides a une masse de particules de bois et similaire pur la fabrication de panneaux our d'autres objets moulés par compression."

Fahrni, F., Swiss P 232,065, Aug 1, 1944. "Verfahren zur Herstellung von Pressholzkunstplatten."

Fahrni, F., Holz als Roh- und Werkstoff **6**, 277 (1943). "Die Holzspanplatte."

Fahrni, F., FP 881,781, May 7, 1943. "Procédé pour la fabrication des plaques artificielles en bois comprimé."

Fahrni Institute AG, Zurich: Formaldehydabgabe: "Auszug aus der Japanese Industrial Standard, Particleboards, JIS A-5908–1977 betreffend Prüf- und Gütevorshriften fur Formaldehydabgabe."

Fahrni, P., FAO, World Consultation on Wood Based Panels; New Delhi, Feb 1974; Published by the Food & Agricultural Organization, UNESCO, Forest Industries and Trade Division; Rome, Italy, 1976. "Particle Board Plant Case Study; Small-Scale Wood-Based Panel Plants."

Faichney, G., Aust. J. Agric. Res. **26**(2), 319 (1975). "Effect of Formaldehyde Treatment of a Concentrate Diet on the Passage of Solute and Particle Markers Through the Gastronintestinal Tract of Sheep."

ISBN 0-201-04558-3

Feldman, Y. G., and T. I. Bonashevskaya, Gig. Sanit. **36**(5), 6 (1971). "Action of Low Concentrations of Formaldehyde on the Body."

Ferber, K. D., Farbe Lack **79**, 968 (1973). "Formaldehyde Separation from Acid-Hardened Products (e.g., Parquetry Sealers). Possibilities of Determination and Odor Reduction."

Feriday, J. E., B. J. Kendall-Smith, and M. L. Bradley, Ger. Offen. 2,322,617, Nov 22, 1973. "Aminoplast Molding Materials."

Feriday, J. E., and B. J. Kendall-Smith, BP 1,483,458, Aug 17, 1977. "Improvements in the Production of Injection or Extrusion Molded Articles."

Fiberglass Ltd., FP 2,053,840, May 21, 1971. "Thermal Insulators Comprising Resin-Bound Mineral Fibers."

Fiberglass Ltd., BP 1,293,744, Oct 25, 1972. "Fiberglass Resin Lignin Coating."

Fiehn, T. L., Zellstoff Papier **25**(6), 166 (1976). "Practical Aspects of the Brightness Stability of Bleached Paper Pulps."

Fife, R. L., USP 3,758,377, Sept 11, 1973. "Paper Sheet Treated with an Aqueous Mixture of Lignosulfonate and Isoprene."

Flick, K., Zentralbl. Arbeitsmed. Arbeitsschutz **25**(9), 257 (1975). "Determination of Formaldehyde Concentration at the Factory in the Production of Chipboards."

Foris, P. L., R. W. Brown, and P. S. Phillips, Jr., Ger. Offen. 2,529,427, Jan 29, 1976. "Small Polymer Capsules."

Frankel, L. S., K. L. Wallisch, *et al.,* Anal. Chem. **44**, 2401 (1972). "Selective Retention by Porous Polymer Adsorbents. Application to Formaldehyde Determination."

Freeman, H. G., G. F. Baxter, and G. T. Tiedeman, USP 3,786,025, Jan 15, 1974. "Rapid Curing Resin Compositions Comprising a Phenol-Aldehyde Condensation Polymer Modified with an Aminonaphthalene."

Freeman, H. G., and W. C. Grendon, Forest Prod. J. **21**(9), 54 (1971). "Formaldehyde Detection and Control in the Wood Industry."

Frisch, K., and J. Saunders, "Plastic Foams", Marcel Dekker Inc., New York, 1973.

Froede, O., and H. Witt, in "Holzwerkstoffe." In "Ullmanns Encyklopädie der technischen Chemie." Volume 12, page 709, Verlag Chemie, Weinheim, 1974.

Frolova, M. K., A. D. Laz'ko, and V. M. Litvinenko, USSR P 462,846, Mar 5, 1975. "Composition for the Production of a Cellular Plastic."

Fujita, K., and K. Ueda, J. Kokai 74:129,753, Dec 12, 1974. "Particle Boards Containing Methylated Urea-Formaledhyde Resins."

Fujitani, J., M. Ohtake, and K. Yamasawa, J. Kokai 74:130,933, Dec 16, 1974. "Stable Emulsion Adhesives for Plywood."

Fukuda, T., J. Kokai 76:41,031, Apr 6, 1976. "Aminoplast Adhesives for Plywood Having Low Levels of Free Formaldehyde."

Fuwa, K., Japan P 93,961, Dec 17, 1931. "Reflector for Ultraviolet Ray."

Gablin, K. A., and L. J. Hansen, USP 4,056,362, Nov 1, 1977. "System for Disposing of Radioactive Waste."

Galil, F., M. Ibrahim, and S. Shakra, Egypt. J. Chem. **15**(5), 501, (1972). "Determination of Formaldehyde Yield of Cellulose Textiles Treated with Urea-Formaldehyde."

ISBN 0-201-04558-3

Gamova, I. A., T. S. Koromyslova, and L. N. Natkina, Izv. Vyssh, Uchebn. Zaved., Lesn. Zh. **17**(4), 113 (1974). "Mechanism of the Formation of Polymer-Wood Material."

Gams, A., and G. Widmer, USP 1,831,706, Nov 10, 1932. "Condensation Products from Aldehydes and Carbamides."

Gams, A., and G. Widmer, USP 1,718,901, June 25, 1929. "Condensation Products of Urea and Formaldehyde."

Gams, A., and G. Widmer, USP 1,676,543, July 10, 1928. "Condensation Product of Urea with Formaldehyde."

Gams, A., and G. Widmer, USP 1,674,199, June 19, 1928. "Condensation of Urea and Formaldehyde under Pressure."

Gams, A., and G. Widmer, Can. P 281,327, June 26, 1928. "Condensation Product of Urea and Formaldehyde."

Gams, A., and G. Widmer, Can. P 281,326, June 26, 1928. "Condensation Products of Urea and Formaldehyde."

Gams, A., and G. Widmer, Can. P 281,325, June 26, 1928. "Condensation Product of Urea and Formaldehyde."

Gams, A., and G. Widmer, Can. P 274,266, Sept 27, 1927. "Urea and Formaldehyde Product."

Gaponenko, I. M., Org. Reagenty Anal. Khim., Korroz. Met., Uchebn. Eksp. (Russ) 71 (1973). "Simple Analysis and Identification of Major Plastics."

Gavel, J., I. Nadudvari, and I. Szaller, Hung Teljes (Pat) 13,033, Mar 28, 1977. "Aqueous Lacquer Compositions for Treatment of Hydrophobic Paper Surfaces."

Gavirati, U. M., Pitture Vernici (Ital) **53**(3), 91 (1977). "Melamine Resin for the Hydrosoluble Catalytic System in Wood."

Gavrilov, V. S., *et al.*, Inform. Soobshch. Gos. Nauch.-Issled. Proekt. Inst. Azotn. Prom. Prod. Org. Sin., No 2 (Pt 1) 72 (1969). "Preparation of Urea-Formaldehyde Compounds with Increased Anion-Exchange Properties for Saline Soils and Solonchaks."

Geissler, T., and P. Starke, Wiss. Z. Humboldt-Univ. Berlin. Math.-Naturwiss. Reihe **21**(3), 234 (1972). "Use of Urea-Formaldehyde Resin Foam in Substrate Management and in the Amelioration of Ground Under Glass and Plastics."

Geissman, T. A., Ch. 3 in *Organic Reactions,* Vol. 2, R. Adams *et al.*, eds., John Wiley & Sons, New York, 1944. "The Cannizzaro Reaction."

Gel'tishcheva, E. A., V. Khrustaleva, and S. K. Osokina, Gig. Vop. Proizvod. Primen. Polim. Mater. (Russ), 208 (1979). "Toxicity of Polymer Building Materials Employed in Children's Institutions."

Gerlachov, T., Drev. Vysk. (Slo.) **16**(3), 187 (1971). "Properties of Particle Material with 15% Content of Cresol Resol."

Getchell, N. F., USP 3,944,388, Mar 16, 1976. "Method for Improving Abrasion and Wear Resistance of Edge Portions of Durable-Press Garments."

Geyer, B., and I. Villwock, Arch. Gartenbau **21**(8), 665 (1973). "Effect of Urea Formaldehyde Foam Used as a Soil Conditioner."

Geyer, B., and I. Villwock, Arch. Gartenbau **21**(3), 223 (1973). "Decomposition of Urea-Formaldehyde Resin-Foam Used as a Soil Conditioner."

Geyer, B., Wiss. Z. Humboldt-Berlin, Math.-Naturwiss. Reihe **21**(3), 242 (1972). "Compatibility with Plants and Decomposition of Urea-Formaldehyde Resin Foam."

ISBN 0-201-04558-3

Gfeller, B., Holz als Roh- und Werkstoff **31**, 131 (1973). "Herstellung und Eigenschaften den Holzspan-Polyurethan-Hartschaumstoff-Mischplatten."

Gibson, D. V., and B. Leary, USP 4,064,090, Dec 20, 1977. "Aqueous Coating Composition of Epoxy-Amine Adduct and an Acid with Crosslinker."

Gilfrich, H. P., and H. Wallhaeusser, Kunststoffe **62**(8), 519 (1972). "Thermal Behavior of Duroplast Molded Materials."

Gillespie, R. H., and B. H. River, Forest Prod. J. **25**(7), 26 (1975). "Durability of Adhesives in Plywood. Dry-Heat Effects by Rate-Process Analysis."

Ginzel, W., Holz als Roh- und Werkstoff **31**(1), 18 (1973). "Hydrolysis in Urea-Resin-Bonded Particle Board."

Ginzel, W., Holz als Roh- und Werkstoff **29**(8), 301 (1971). "Chemical-Hydrolytic Degradation of Urea-Formaldehyde Resin in Particle Boards After Three Years of Weathering."

Girsewald, C. von, and H. Siegens Ber. **47**, 2464 (1914). "Hydrogen Peroxide. II. Tetramethylene diperoxide-dicarbamide."

Glubish, P. A., USSR P 584,062, Dec 5, 1977. "Soil-Repellent Finishing of Textile Materials Made from Cellulose and Polyester Fibers."

Glukhikh, L. S., O. V. Gulyaeva, and N. M. Fokina, Tr. Ural. Lesotekh. Inst. No. 26, 231 (1972). "Internal Stresses in Adhesive Bonds During Veneering."

Glukhikh, L. S., O. V. Gulyaeva, and V. M. Fokina, Izv. Vyssh. Ucheb. Zaved., Les. Zh. **15**(2), 71 (1972). "Effect of the Type and Properties of Adhesives on the Magnitude of Internal Stresses in Glue Seams During Veneering."

Gnatowski, M., Polimery (Warsaw) **23**(2), 55 (1978). "Aging of Urea-Formaldehyde Resins Used for Soil Stabilization. Part I."

Gnatowski, M., Polimery (Warsaw) **22**(7), 234 (1977). "Modified Urea-Formaldehyde Resins for Injection Strengthening and Sealing Soil."

Gnatowski, M., and W. Szlezyngier, Przem. Chem. **55**(7), 343 (1976). "Urea-Formaldehyde Resins for Strengthening and Consolidating Soils by Injection."

Goertz, H. M., USP 4,025,329, May 24, 1977. "Particulate Urea-Formaldehyde Fertilizer Composition."

Goldschmidt, C., Ber. **29**, 2438 (1896). "Über die Einwirkung von Formaldehyd auf Harnstoff."

Goldschmidt, H., and O. Neuss, USP 1,844,570, Feb 9, 1932. "Artificial Resin."

Goldschmidt, S., and R. Mayrhofer, BP 316,144, July 23, 1928. "Urea-Formaldehyde Condensation Products."

Goldschmidt, S., and R. Mayrhofer, GP 561,050, Mar 3, 1928. "Urea-Formaldehyde Condensation Products."

Goldschmidt, S., and R. Mayrhofer, GP 551,422, Sept 18, 1927. "Urea-Formaldehyde Condensation Products."

Goldschmidt, H., and O. Neuss, BP 208,761, Sept 20, 1922. "Condensation Products from Formaldehyde and Urea, etc."

Goldschmidt, H., and O. Neuss, BP 187,605, Oct 17, 1922. "Condensation Products from Formaldehyde and Urea."

Golubovskaya, E. K., V. K. Golyshev, and V. V. Sverkanova, Sb. Tr., Leningrad. Inzh.-Stroit. Inst. **69**, 129 (1971). "Effect of Pollutant Concentration in Waste Water of a Veneer-Furniture Plant on the Oxygen Consumption by Active Sludge."

ISBN 0-201-04558-3

Golyshev, V. K., Sb. Tr., Leningrad. Inzh.-Stroit. Inst. **69**, 127 (1971). "Biochemical Method of Removing Resins from Waste Water in Veneer and Furniture Production."

Gordon, A. F., USP 3,962,166, June 8, 1976. "Stable Urea-Formaldehyde Compositions."

Gordon, M. S., and M. L. Morris, Adv. Mol. Relax. Interaction Processes **13**, 95 (1978). "Localized Orbital Studies of Hydrogen Bonding. IV. Salicylaldehyde. Salicylaldimine, H_2CO-H_2O, and H_2CNH-H_2O."

Gorenko, V. G., and T. A. Zdorovetskaya, USSR P 576,154, Oct 15, 1977. "Composition for Heat-Insulating Articles."

Goren'kov, M. P., Izv. Vyssh. Ucheb. Zaved., Les. Zh. **15**(5), 104 (1972). "Transmission and Absorption of Ultraviolet Rays by Solid and Liquid Coatings."

Gosink, T. A., Env. Sci. Tech. **9**(7), 630 (1975). "GC In Environmental Analysis."

Grabowsky, O., and J. Mayer, Ger. Offen. 2,454,769, May 26, 1976. "Reactive Urea-Formaldehyde Condensation Products."

Grabowsky, O., O. Wittmann, and D. Duermeyer, GP 2,435,793, July 25, 1974. "Aminoplast Resin."

Gracheva, M. P., Yu. P. Tikhomirov, V. N. Kalent'ev, and D. N. Okorokov, Gig. Sanit. (3), 102 (1973). "Burning of Solid Industrial Wastes as a Means of Their Disposal."

Graham, J. L., U.S. Environ. Prot. Agency, Off. Res. Dev., [Rep.] EPA-2-73-195, 57 pp (1973). "Aerobic Secondary Treatment of Plywood Glue Wastes."

Greenberg, W. H., USP 3,734,985, May 22, 1973. "Glass Fiber-Reinforced Cellular Thermoplastics."

Grigor'eva, M. N., and N. G. Shplct, Tr. Leningr. Sanit.-Gig. Med. Inst. **105**, 177 (1974). "Use of a Urea Plastic Foam Sealant in Enclosing Structures."

Grigor'eva, M. N., and N. G. Shplet, Gig. Sanit. (7), 106 (1973). "Hygienic Characteristics of Urea Plastic Foams."

Gruben, S., and S. A. E. Persson, Swed. P 336,470, July 5, 1971. "Preparing Urea Resin Foams Intended as Absorbents for Oil."

Gruntfest, I. J., and D. D. Gagliardi, Textile Res. J. **18**, 643 (1948). "The Modification of Cellulose by Reaction with Formaldehyde."

Guha, S. R. D., M. M. Singh, and S. P. Singh, Indian For. **104**(1), 51 (1978). "High Brightness Pulps as Filler for the Production of Urea Formaldehyde and Melamine Formaldehyde Molding Powder."

Gupta, R. C., B. R. S. Chauhan *et al.*, J. Timber Dev. Assoc. India **20**(4), 9 (1974). "Studies on Glued Laminated Constructions—Preliminary Studies on the Effect of Preservatives on Bonding of *Pinus Roxburghii* (chir.)."

Gusev, V. A., Derevoobrab. Prom. (4), 23 (1974). "Preparation and Use of the Resin-Impregnated Decorative Paper."

Gutheil, H., "Supply and Demand for UF-Resins." 12th Proc. Washington State University Particleboard Symposium, Pullman, Washington, T. Maloney, editor, 1978.

Gvozdev, D. V., *et al.*, Tr. Mosk. Khim.-Tekhnol. Inst. **86**, 85 (1975). "Regulation of the Properties of Urea-Formaldehyde Polymers."

ISBN 0-201-04558-3

Halama, D., and B. Hanckova, Czech. P 157,574, Apr 15, 1975. "Suppressing Fungal Contamination in the Microbial Surface Production of Citric Acid."

Halligan, A. F., and A. P. Schneiwind, Forest Prod. J. **22**(4), 41 (1972). "Effect of Moisture on Physical and Creep Properties of Particleboard."

Ham, G. E., USP 3,080,343, Mar 5, 1963. "Diamine-Urea-Thiourea Copolymers for Fibers."

Hamada, R., Japan P 72:47,293, Nov 29, 1972. "Urea Resin Adhesive Agent of High Adhesive Power and High Water Resistance."

Hanetho, P., 12th Int. Particleboard Symp., Washington State Univ., Pullman, WA, Apr 5, 1978. "Formaldehyde Emission from Particleboard and Other Building Materials; A Study from the Scandinavian Countries."

Hani, Y., *Wood-Bonding and Adhesives,* Morikita Publisher, Tokyo, 1965.

Hanna, G. F., and J. Richardson, Am. Soc. Heat. Refrig. Air-Cond. Eng. Symp.: "Odors and Odorants: The Engineer's View.", Chicago, Jan 1969, ASHRAE, New York, 1969. "A Solution for Industry Odor Problems."

Harkin, J. M., J. R. Obst, and W. F. Lehmann, Forest Prod. J., **24**(1), 27 (1974). "Visual Method for Measuring Formaldehyde Release from Resin-Bonded Boards."

Hart, W. F., USP 3,957,731, May 18, 1976. "Storage-Stable Butylated Urea-Formaldehyde Crosslinking Agent and Process for the Manufacture Thereof."

Hasegawa, M., T. Sakurai, S. Tazawa, and K. Takeuchi, J. Kokai 74:99,191, Sept 19, 1974. "Odorless Urea-Formaldehyde Primary Condensates."

Hata, Y., and M. Kamiyama, Japan P 74:04,542, Feb 1, 1974. "Adhesive Agent for Plywood."

Havinga, R., and P. D. Swaters, Ger. Offen. 2,317,846, Oct 21, 1973. "Acetophenone Photocatalysts for Crosslinking of Polymer Coatings."

Hayasahi, K., S. Munari, C. Rossi, and C. Uliana, Chim. Ind. (Milan) **56**(4), 264 (1974). "Polycondensation Induced by Ionizing Radiation in Urea-Formaldehyde System."

Hayashi, S., and H. Yamazaki, J. Kokai 78:41,393, Apr 14, 1978. "Adhesives for Wood."

Hayashi, S., and H. Yamazaki, J. Kokai 78:41,392, Apr 14, 1978. "Aminoplast Adhesives for Wood."

Hayashi, S., and H. Yamazaki, J. Kokai 78:41,354, Apr 14, 1978. "Adhesives for Wood."

Hayashi, S., and H. Yamazaki, J. Kokai 78:33,253, Mar 29, 1978. "Storage-Stable Amino Resins for Adhesives."

Hayashi, T., and K. Tsuji, J. Kokai 74:99,321, Sept 19, 1974. "Lightweight Gypsum Boards."

Heap, S. A., *et al.,* in *Chemical Aftertreatment of Textiles,* H. Mark *et al.,* eds., Interscience, New York, 1971.

Health Effects Research Lab, Res. Triangle Park, N.C., EPA/600/1-76/023. U.S. Dept. HEW, Environmental Protection Agency, 1976. "Scientific and Technical Data Base for Criteria and Hazardous Pollutants–1975 ERC/RTP Review."

Hebeish, A., and K. Schliefer, Text. Res. J. **46**(6), 465 (1976). "Free Formaldehyde in Fabrics Treated with *N*-methylol Finishing Agents."

Heetman, J. G. A., Swiss P 526,481, Sept 29, 1972. "Fire-Resistant Thermal Insulators."

ISBN 0-201-04558-3

Heetman, J. G. A., Ger. Offen. 2,227,679, Jan 10, 1974. "Perlite Lightweight Panels."

Heiduschka, A., and H. Zirkel, Arch. Pharm. **254**, 456 (1916). "Action of Formaldehyde on Lactose, Maltose, and Sucrose."

Heimburg, R. W., AIChE Symp. Ser. **68**(122), 21 (1972). "Environmental Effects of the Incineration of Plastics."

Hemmelmayr, F. v., Monatsh. **12**, 89 (1891). "Über die Methylenderivate des Harnstoffes und Thioharnstoffes."

Henbest, R. G., and K. McGregor, Ger. Offen. 2,754,525, June 15, 1978. "Material Based on Formaldehyde Resin in Fiber Form."

Hensley, W. L., J. Paint. Technol. **44**(572), 87 (1972). "Fast Curing Methylated Urea-Formaldehyde Resins in Surface Coatings."

Hentschel, K. D., and A. Zott, Wiss. Z. Humboldt-Univ., Berlin, Math.-Naturwiss. Reihe **21**(3), 264 (1972). "Problems of Disinfecting the Soil When Using Ameliorants."

Hermann, D. T., USP 4,039,496, Aug 2, 1977. "Low Formaldehyde Fully Etherified Methylolated Melamine with Urea-Formaldehyde Glyoxal as Textile Resin."

Herzberg, G., *Molecular Spectra and Molecular Structure; III. Electronic Spectra and Electronic Structure of Polyatomic Molecules,* Van Nostrand, New York, 1966.

Higashimura, E., M. Hasegawa, T. Sakurai, and S. Tazawa, J. Kokai 75:09,693, Jan 31, 1975. "Hardening Agents for Urea Resins."

Higuchi, K., J. Kokai 75:34,332, Apr 2, 1975. "Odor-Free Adhesives for Plywoods."

Higuchi, K., and K. Morita, J. Kokai 75:34,330, Apr 2, 1975. "Adhesives for Plywood for Improving Initial Adhesion."

Higuchi, K., J. Kokai 73:60,737, Aug 25, 1973. "Odorless Urea Resin Adhesives."

Higuchi, K., J. Kokai 73:79,241, Oct 24, 1973. "Odorless Urea-Formaldehyde Resin Adhesives."

Hinkel, L. E., and D. H. Hey, Rec. Trav. Chim. **48**, 1280 (1929). "The Condensation of Benzaldehyde and Ethyl Acetoacetate with Urea and Thiourea."

Hinterwaldner, R., GP 1,745,562, Jan 3, 1974. "Multicomponent Polymer Foam."

Hirata, S., and N. Minemura, Rinsan Shikenjo Geppo (8), 5 (1974). "Use of Starch Waste as a Filler of Adhesive."

Hirotsugu, O., and T. Goto, Shimane Daigaku Hogakubu Kenkyu Hokoku (5), 61 (1971). "Wood Gluing. VIII. Effects of Wood Extractives on the Gelation Time of Urea-Formaldehyde Resin Adhesives."

Hiyoshi, K., H. Fukui, K. Yamamoto, and Y. Fujimoto, Japan. P 74:33,682, Sept 9, 1974. "Modified Pulp."

Hoedt, H., FP 2,065,238, Aug 27, 1971. "Manufacture of Artificial Stones and Plates from Minerals and Hardenable Organic Binders."

Hoener, H., E. H. Fietz, and P. Luethge, Ger. Offen. 2,242,548, Mar 7, 1974. "Fire-Resistant Building Materials."

Hoener, H., and H. Hoener, Neth. Appl. 73 11,929, Mar 4, 1974. "Sheets or Structures Containing Cork Particles and Aminoplast Binders."

Hojo, Z., J. Kokai 74:75,709, July 22, 1974. "Removal of Formaldehyde Odor from Plywood."

ISBN 0-201-04558-3

Holmes, C. A., Advan. Fire Retardants 2(Pt. 1), 24 (1973). "Evaluation of Fire-Retardant Treatments for Wood Shingles."

Holtschmidt, U., A. Laqua, and B. Petrik, Ger. Offen. 2,448,472, Apr 22, 1976. "Solutions of Hardenable Urea-Formaldehyde Resin."

Holmgren, T. A. F., Swed. P 53,833, Feb 21, 1923. "Formaldehyde."

Honma, S., and M. Kitayama, Hakkaidoritsu Eisei Kenkyusho Ho (24), 128 (1974). "Formaldehyde Contents in Clothing."

Hope, P., B. P. Stark, and S. A. Zahir, Brit. Polym. J. **5**(5), 363 (1973). "Studies of Urea-Formaldehyde Resins Using Gel Permeation Chromatography."

Horowitz, F., and J. B. Dede, Jr., USP 3,658,622, Apr 25, 1972. "Making Wood Laminates by Bonding with Modified Urea-Formaldehyde Adhesives."

Houwink, R., and G. Salomon, eds., *Adhesion and Adhesives,* Vol. 1, p. 187, Elsevier Publishing Co., Amsterdam, 1965. "Thermosetting Resin Adhesives."

Hse, C.-Y., Mokuzai Gakkaishi **20**(10), 483 (1974). "Characteristics of Urea-Formaldehyde Resins as Related to Glue Bond Quality of Southern Pine Particleboard."

Hse, C.-Y., Mokuzai Gakkaishi **20**(10), 491 (1974). "Reaction pH of Urea-Formaldehyde Resins as Related to Strength Properties of Southern Pine Particleboard."

Hse, C.-Y., Mokuzai Gakkaishi **20**(11), 538 (1974). "Effects of Reaction Catalysts of Urea-Formaldehyde Resins on the Strength Properties of Southern Pine Particleboard."

Hsiao, S.-H., and J. E. Villaume, *A Literature Review – Problem Definition Studies on Selected Toxic Chemicals, Vol. 6: Occupational Health and Safety and Environmental Aspects of Urea-Formaldehyde Resins,* Franklin Institute Research Labs, Philadelphia, 1978.

Huang, D. K., and J. Holden, USP 4,035,328, July 12, 1977. "Continuous Process for Making Urea Formaldehyde Pigment."

Hubbard, D. A., Ger. Offen. 2,726,617, Jan 5, 1978. "Aqueous Formaldehyde-Synthetic Resin Solutions."

Hubbard, D. A., Ger. Offen. 2,532,753, Feb 12, 1976. "Mixed Adhesive and Its Use."

Hubbard, D. A., and K. Atkinson, Neth. Appl. 74 14,340, May 7, 1975. "Urea-Formaldehyde Resin Molding."

Hughes, E. E., and S. H. Lias, Anal. Chem. **32**, 707 (1960). "Detection and Estimation of Low Concentrations of Aldehyde in Air."

Hunt, C. M., and D. M. Burch, U.S. Department of Commerce, Nat'l. Bur. of Standards, Washington, D.C., 1975. "Air Infiltration Measurements in a Four-Bedroom Townhouse Using Sulfur Hexafluoride as a Tracer Gas."

Hurnik, H., Rev. Plast. Mod. **26**(207), 426 (1973). "Blowing Agents for Plastic Foams."

Hurst, H., BP 1,412,559, Nov 5, 1975. "Urea-Formaldehyde Foams for Thermal Insulation."

Hutschneker, K., in SAH Bull. **5**, 6 (1977). "Bindemittel für Feuchtigkeitsbeständige Spanplatten."

Ichimura, Y., T. Aoyagi, and Y. Takayama, J. Kokai 74:112,993, Oct 28, 1974. "Unsaturated Polyester Compositions for Sheet and Bulk Moldings."

ISBN 0-201-04558-3

I. G. Farbenindustrie, FP 848,604, Nov 3, 1939. "Condensation Products."

I. G. Farbenindustrie, FP 847,953, Oct 19, 1939. "Condensation Products."

I. G. Farbenindustrie, BP 507,175, June 12, 1939. "Urea-Aldehyde Condensation Products."

I. G. Farbenindustrie, BP 460,275, Jan 25, 1937. "Urea-Formaldehyde Resins."

I. G. Farbenindustrie, BP 459,788, Jan 11, 1937. "Urea-Aldehyde Condensation Products."

I. G. Farbenindustrie, FP 769,588, Aug 28, 1934. "Joining Surfaces."

I. G. Farbenindustrie, FP 721,828, Aug 21, 1931. "Dimethylolurea."

I. G. Farbenindustrie, FP 697,874, June 25, 1930. "Adhesive."

I. G. Farbenindustrie, FP 552,264, Mar 28, 1930. "Urea-Formaldehyde and Like Condensation Products."

I. G. Farbenindustrie, FP 38,157, Mar 20, 1930. "Synthetic Resins."

I. G. Farbenindustrie, GP 540,071, Dec 7, 1920. "Synthetic Resins."

I. G. Farbenindustrie, GP 528,582, Oct 8, 1929. "Formaldehyde and Urea Condensation Products."

I. G. Farbenindustrie, FP 675,398, May 18, 1929. "Artificial Masses."

I. G. Farbenindustrie, GP 519,892, Jan 16, 1929. "Urea-Formaldehyde Condensation Products."

I. G. Farbenindustrie, GP 536,733, June 26, 1928. "Condensation Products."

I. G. Farbenindustrie, BP 314,908, Jan 5, 1928. "Urea- and Thiourea-Formaldehyde Condensation Products."

I. G. Farbenindustrie, FP 641,770, Oct 3, 1927. "Condensation Products of Urea."

I. G. Farbenindustrie, BP 288,346, Feb. 12, 1926. "Wood Adhesive."

I. G. Farbenindustrie, BP 288,346, Oct 1, 1926. "Urea-Formaldehyde Condensation Products."

I. G. Farbenindustrie, FP 641,420, Sept 11, 1926. "Urea Condensation Products."

I. G. Farbenindustrie, GP 537,611, Jan 29, 1926. "Urea Condensation Products."

I. G. Farbenindustrie, GP 511,979, Jan 13, 1926. "Condensation Products."

I. G. Farbenindustrie, BP 260,253, Oct 21, 1925. "Urea-Formaldehyde Condensation Products."

I. G. Farbenindustrie, GP 535,852, Sept 12, 1925. "Urea-Formaldehyde Condensation Products."

I. G. Farbenindustrie, GP 493,988, Aug 19, 1924. "Condensation Products."

I. G. Farbenindustrie, GP 535,851, June 3, 1922. "Condensation Products."

Iguchi, K., Japan. P 73:29,612, Sept 12, 1973. "Hard-to-Flame Foam Body."

Ikeda, M., N. Sawada, K. Fukunishi, and S. Tochikawa, J. Kokai 74:18,926, Feb 19, 1974. "Deodorized Adhesives."

Ikeda, M., N. Sawada, K. Fukunishi, and S. Tochikawa, J. Kokai 74:18,925, Feb 19, 1974. "Deodorized Urea-Formaldehyde Adhesives."

Immarino, N. J., USP 3,736,275, May 29, 1973. "Particle Production Board Using Crosslinked Xanthomonas Colloid in the Glue Mixture."

Imoto, M., *Formaldehyde,* Asahnla Shoten Publisher, Tokyo, 1965.

Imperial Chemical Industries, Ltd., Fr. Demande 2,240,254, Mar 7, 1975. "Molded Articles of Expanded Urea-Formaldehyde Resin."

ISBN 0-201-04558-3

Imperial Chemical Industries Ltd., Belg. P 818,549, Feb 6, 1975. "Molded Resin Products."

Imura, S., and N. Minemura, Rinsan Shikenjo Geppo, (305), 1 (1977). "Prevention of Formaldehyde Liberation from Plywood and Particleboard by Secondary Treatment."

Imura, S., M. Sato, H. Nakamura, and I. Abe., Rinsan Shikenjo Geppo (3), 9 (1974). "Effects of Bark Powder on the Adhesive Strength of Urea-Formaldehyde Resin."

Inoue, K., T. Ohta, N. Araki, and T. Seto, Japan. P 75:28,482, Sept 16, 1975. "Fire-Resistant Plywood."

Inoue, S., Zairyo **23**(250), 569 (1974). "Effect of Water on the Strength Development of Cold Setting Epoxy Resins."

Inui, K., Y. Kamitaki, and T. Iizuka, J. Kokai 74:44,060, Apr 25, 1974. "Plywood with Low Levels of Formaldehyde Odor."

Inverarity, G., Ger. Offen. 2,318,802, Oct 18, 1973. "Crosslinking Agent for Formaldehyde-Urea Copolymer Foam."

Ionescu Muscel, I., I. Casandra, and D. Petra., Ind. Usoara: Text., Tricotaje, Confectii Text., **28**(8), 358 (1977). "Certain Measures Initiated in Japan Regarding the Use of Urea-Formaldehyde in Textile Finishing Processes."

Irish, D. E., Ch. 7 in *Raman Spectroscopy; Theory and Practice,* H. A. Szymanski, ed., Plenum Press, New York, 1967. "Raman Spectroscopy of Complex Ions in Solution."

Ishida, M., and Y. Oka, J. Kokai 75:142,703, Nov 17, 1975. "Fire-Resistant Decorative Plywood."

Ishida, S., Y. Kura, J. Masumi, and K. Kaneko, Kanazawa Daigaku Kogakuba Kiyo **7**(1), 1 (1973). "Molecular Weight Estimation of Cocondensation Resin of Phenol-Urea-Formaldehyde System."

Ishin, N. D., I. D. Borisyuk, A. G. Yakhno, and V. I. Papeshina, USSR P 332,064, Mar 14, 1972. "Priming Solution for Finishing Wooden and Other Plates."

Isono, S., and T. Yoshinaga, Japan. P 74:04,318, Jan 31, 1974. "Flame-Retardant Particle Boards."

"Isovolta" Oesterreichische Isolierstoffwerke, Austrian P 324,923, Sept 25, 1975. "Lightweight Construction Material."

Ito, Y., S. Suzuki, and S. Takashima, Japan. P 76:06,193, Feb 26, 1976. "Foam-Backed Decorative Panels."

Ivancheva, E. G., and N. I. Zbykovskaya, Lakokrasoch. Mater. Ikh Primen. (2), 49 (1974). "Determination of the Degree of Aging of Coatings Produced from Urea Resins."

Ivanov, N. A., and B. I. Artamonov, Derevoobrab. Prom. (1), 4 (1973). "Quick-Hardening Urea Adhesives for Veneering in Single-Span Presses."

Ivanov, S., K. Slivkov, and G. Andreev, Nauchni Tr.-Plovdivski Univ. **13**(3), 383 (1975). "Study of the Production and Interpretation of a Modified Urea-Formaldehyde Resin Meant to be Used as a Binder for Sandpaper Production. IV. Study of Modifier Behavior Using IR Spectrophotometry."

ISBN 0-201-04558-3

Ivanov, S., and K. Slivkov, Nauchni Tr. – Plovdivski Univ. **13**(3), 347 (1975). "Study of the Production and Interpretation of the Properties of a Modified Urea-Formaldehyde Resin Meant to be Used as a Binder for Sandpaper Production. I. Statement of Purpose and Study of the Effect of Modifiers on the Physicomechanical Properties of the Initial Resin."

Ivanova, T. P., N. K. Statsek, A. I. Buslenko *et al.*, Farmakol. Toksikol. (Kiev) (8), 161 (1973). "Sensitizing Action of Formaldehyde."

Iwada, R., J. Kokai 73:56,731, Aut 9, 1973. "Urea Resin Adhesives."

Iwaki, T., M. Tahara, and Y. Taniguchi, J. Kokai 73:12,386, Feb 16, 1973. "Finely Powdered Cured Amino Resins."

Iwasa, Y., J. Kokai 75:43,182, Apr 18, 1975. "Fire-Resistant Wood Chip-Mineral Composite Boards."

Iwasa, Y., J. Kokai 75:38,778, Apr 10, 1975. "Fire-Proof Panels."

Iwasaki, H., and S. Kobori, Japan. P 72:32,422, Aug 19, 1972. "Application of Surface Coatings to Give a Vein-Like Finish."

Iwata, M., T. Ichinose, and T. Ohnuki, J. Kokai 75:109,933, Aug 29, 1975. "Extenders for Amino Resin Adhesives."

Izumi, H., and T. Koyama, J. Kokai 78:17,690, Feb 17, 1978. "Urea-Melamine Resins."

Izumi, H., and T. Kimura, J. Kokai 78:16,755, Feb 16, 1978. "Modified Melamine-Formaldehyde Copolymer Compositions for Decorative Sheets."

Jaccard, R., USP 4,051,208, Sept 27, 1977. "Conversion of Liquid Phases of Nonmetallic Substances, Especially Polymerizable or Crystallizable Substances, into Solid Phases."

Jaccard, R., P. Hostettler, and R. Siegrist, Ger. Offen. 2,131,717, Jan 20, 1972. "Apparatus for Converting Liquid Phases to Solid Phases."

Jaccard, R., and J. A. Eklund, Ger. Offen. 2,046,496, Mar 23, 1972. "Continuous Manufacture of Crosslinked Amino Resin Polymers."

Jaccard, R., and J. A. Eklund, Swiss P 542,893, Nov 30, 1973. "Continuous Manufacture of Crosslinked Amino Resin Polymers."

Jacenkow, B., and M. Gnatowski, Pol. P 92,502, Nov 30, 1977. "Urea Resin for Soil Strengthening and Packing."

Jackson, L. P., Jr., USP 4,055,974, Nov 1, 1977. "Disintegratable Fertilizer Tablet."

Jain, N. C., R. C. Gupta, and B. R. S. Chauhan, Holzforsch. Holzverwert. **24**(1), 9 (1972). "Gluing of *Pinus Roxburghii.* II. Effect of Additives."

Jain, N. C., R. C. Gupta, and B. R. S. Chauhan, Holzforsch. Holzverwert. **23**(4), 74 (1971). "Gluing of *Pinus Roxburghii.* I. Effect of Extractives."

Jakimavicius, K., Derevoobrab. Prom. (4), 25 (1974). "Printing of Wood Imitation Designs on Paper at Furniture Enterprises."

Jamin, A., P. Monfort, and N. Peschong, Ger. Offen. 2,537,859, Mar 11, 1976. "Flame-Resisting Composition for Polyurethane Foams."

ISBN 0-201-04558-3

Jenks, T. E., "Urea Resins in Hardboard Manufacture." Background Paper, FAO World Consultations on Wood Based Panels, New Delhi, 1974.

Jenks, T. E., and E. R. Bittner, FP 2,161,187, Aug 10, 1973. "Urethane-Coated Synthetic Panels."

John, H., USP 1,355,834, Oct 19, 1921. "Formaldehyde Condensation Products with Carbamide or its Derivatives."

John, H., GP 78,251, Sept 10, 1919. "Verfahren zur Herstellung von Kondensationsprodukten aus Formaldehyd und Harnstoff bzw. Thioharnstoff oder anderen Harnstoffderivaten."

Johns, W. E., Ph.D. Thesis, Paper No. 8372, Agricultural Exp. Station, Univ. Minnesota, St. Paul, 1972. "A Thermodynamic Comparison of Adsorption of Binary Liquid and Vapor on Northern White Cedar Sapwood."

Johnston, T. H., R. V. Marcote, and R. Chand, Final Rep. No. 171, Dynasciences Corp., Instr. Systems Div. Whittaker Corp., Los Angeles, CA, Aug 1971. "Development of a Formaldehyde Analyzer for Motor Vehicle Exhaust Emissions."

Juenger, H., and F. Weissenfels, USP 3,830,894, Aug 20, 1974. "Filled Phenol Resin Foam Materials."

Jumneanbhandhu, N., and F. N. Tamolang, Araneta Res. J. **20**(4), 258 (1973). "Technical Feasibility of Para-Rubber (Hevea Brasiliensis Muell.-Arg.) for Particleboard Manufacture."

Juneja, S. C., and L. R. Richardson, Forest Prod. J. **24**(5), 19 (1974). "Versatile Fire Retardants from Amino-Resins."

Jung, J., GP 1,081,482, 1959. "Slow Release Fertilizer."

Junkermann, H., and G. Pohl, Ger. Offen. 2,218,582, Oct 31, 1973. "Storage-Stable Solid Formaldehyde Concentration."

Just Bravo, M. A., and C. Just Bravo, Span. P 417,401, Feb 16, 1976. "Foam-Forming Plastic Resins."

Kachmar, E. G., and V. A. Khrustaleva, Gig. Sanit. **34**(9), 87 (1969). "Use of Hard Granulated Adsorbents for the Taking of Atmospheric Samples."

Kadowaki, H., Bull. Chem. Soc., Jpn. **11**, 248 (1936). "New Compounds of Urea-Formaldehyde Condensation Products."

Kadowaki, H., and Y. Hashimoto, Rept. Imp. Ind. Research Inst. Osaku (Japan) **7**(6), 2 (1926). "The Condensation Product of Formaldehyde and Urea."

Kakabadze, T. G., Subtrop. Kul't, (3-4), 49 (1976). "Effect of Urea-Formaldehyde Fertilizer on the Qualitative Indexes of Tea Raw Material."

Kamachi, T., J. Kokai 77:95,745, Aug 11, 1977. "Removal of Free Formaldehyde from Amino Resins for Coating Compositions."

Kamenskii, I. V., and Z. K. Zinovich, Tr. Nausk. Khim.-Tekhnol. Inst. No. 70, 164 (1972). "Synthesis and Study of Fusible Furfural-Urea Oligomers."

Kaneda, H., and T. Maku, Mokuzai Gakkaishi, **22**(3), 173 (1976). "Studies on the Weatherability of Composite Wood. V. Weatherability and Durability of Adhesive Joints of Plywood under Loaded Conditions."

Kaniszewski, S., Rocz. Nauk. Roln., Ser. A. **100**(2), 17 (1974). "Effect of Foams of Urea-Formaldehyde Synthetic Substances on Improvement of Physical Properties and Fertility of a Sandy Soil."

ISBN 0-201-04558-3

Kara, V. V., R. G. Il'yushenko, and Yu. N. Tsedrik, Ugol'Ukr. (1), 11 (1978). "Reinforcing Compositions for Strengthening the Roof Rocks at the Working Face of the Mine."

Karnemaat, J. N., USP 3,655,395, Apr 11, 1972. "Treating Odoriferous Industrial and Municipal Waste Materials by Adding Formaldehyde, Nitric Acid, and Urea."

Karnemaat, J. N., Fr. Demande 2,074,766, Nov 12, 1971. "Granular Fertilizer with Slow Nitrogen Release."

Karnemaat, J. N., Ger. Offen. 2,002,873, July 29, 1971. "Granular Relatively Dust-Free Fertilizer Composition with Slow Nitrogen Release."

Kartashova, S. A., and S. P. Khainer, Tekhnol Svoistva Fosfatnykh Mater., p. 95 (1974). "Extra-Light Perlitic Components for Insulation of Buildings."

Käsbauer, F. K., D. Merkel, and O. Wittmann, Z. Anal. Chem. **281**, 17 (1976). "Bestimmung von freiem Formaldehyd und *N*-Methylol-Formaldehyd in Harnstoff-Formaldehyd-Kondensaten."

Katayama, M., W. Ogami, and M. Takaka, Japan. P 75:34,055, Nov 5, 1975. "Powdered Orange Peel for Plywood."

Kato, C., K. Kamei, K. Matsushima, and M. Hirakawa, J. Kokai 75:05,471, Jan 21, 1975. "Fire-Resistant Lightweight Composite Having Patterns."

Kato, C., J. Kokai 74:131,244, Dec 15, 1974. "Fire-Resistant Composite Boards."

Kato, H., Japan. P 73:17,755, May 31, 1973. "Removing Formaldehyde from Amino Resins."

Katsyuba, V. I., and O. D. Likholetov, Tr. NII betona i zhelezo-betona. Gosstroi SSR (11), 154 (1974). "Chemically Stable Urea Resin-Based Mastics and Concretes."

Katuscak, S., K. Horsky, and L. Zatopkova, Drev. Vysk. **17**(2), 111 (1972). "Swelling of Wood in Organic Liquids Used for Production of Wood-Plastic Combinations."

Kaufmann, F., and H. G. Kaufmann, Arch. Gartenbau **21**(2), 125 (1973). "Use of Synthetic Industrial Products as Soil Conditioners in Asparagus *(Asparagus Officinalis)* Growing."

Kawahara, N., and K. Takashima, J. Kokai 73:72,309, Sept 29, 1973. "Odorless Plywood."

Kawahara, N., and K. Takashima, J. Kokai 75:43,181, Apr 18, 1975. "Removal of Free Formaldehyde from Particle Boards."

Kawahara, T., and Y. Kawamura, Niigata Norin Kenkyu (24), 61 (1972). "Adhesive Properties of Urea Resin Adhesives. II. Adhesive Properties of Urea Resins of Various Composition."

Kawakami, K., H. Imagome, S. Hayashi, and T. Hosokawa, Japan. P 73:09,935, Mar 29, 1973. "Flame Resisting Polymer Compositions."

Kawamura, S., H. Semoto, and M. Fukushima, J. Kokai 75:35,230, Apr 3, 1975. "Urea-Formaldehyde Resin Adhesives for Plywood."

Kawashima, M., J. Kokai 74:71,118, July 10, 1974. "Plywood Boards Free of Formaldehyde Odor."

Kay, R. W., Brit. Foundryman **67**, Pt. 1, 1 (1974). "Survey into the Fumes Evolved from Foundry Sand Binders Based on Synthetic Resins."

Keller, A. V., and J. Taylor, FP 699,968, Aug 2, 1930. "Adhesives."

ISBN 0-201-04558-3

Kelly, M. W., Proc. 4th WSU Particleboard Symp., Wash. State Univ. Pullman, Wash., 1970. "Formaldehyde Odor and Release in Particleboard."

Kelly, M. W., USDA FS General Technical Report FPL-10, Forest Products Laboratory, Madison, Wisc., 1977. "Critical Literature Review of Relationships Between Processing Parameters and Physical Properties of Particleboard."

Kharchenko, T. F., and A. M. Shevchenko, Gig. Primen. Polim. Mater. Stroit., Mater. Vses. Soveshch., 1st 1972, p. 97 (1973). "Hygienic Problems in Studying and Regulating Synthetic Adhesive Materials."

Khlystunova, E. V., N. M. Tinyakova, and S. P. Andreyuk, Sint., Anal. Strukt. Org. Soedin. **7**, 86 (1976). "Apparatus for Studying Surfactants."

Khoshkhoo, H., and E. R. Nixon, Spectrochim. Acta, **A29**, 603 (1973). "Infrared and Raman Spectra of Formaldehyde in Argon and Nitrogen Matrices."

Khoshkhoo, H., Diss. Abstr. Int. B., **33**(7), 3023 (1972). "Infrared and Laser Raman Spectroscopy of Crystalline and Matrix Isolated Formaldehyde."

Khrulev, V. M., and G. I. Shusterzon, Izv. Vyssh. Ucheb. Zaved., Stroit. Arkhitekt. **15**(2), 82 (1972). "Water Resistance of Wood Joints."

Kikuchi, K., K. Miyake, I. Takatsuji, and S. Ikeda, J. Kokai 76:67,334, June 10, 1976. "Water-Resistant Adhesive Compositions for Corrugated Paperboard."

Kikuchi, K., *et al.,* J. Kokai 76:67,335, June 10, 1976. "Waterproofing of Adhesives for Corrugated Paperboards."

Kinastowski, S., and A. Cendrowska, Pol. P 91,790, Oct 15, 1977. "Chemocuring of Urea-Formaldehyde Resins."

Kinastowski, S., Z. Morze, and A. Cendrowska, Pol. P 85,670, Oct 15, 1976. "Hardening of Urea-Formaldehyde Binders."

Kinastowski, S., and A. Cendrowska, Pol. P 85,450, Sept 15, 1976. "Hardening of Urea-Formaldehyde Binders."

Kirkeby, W., and A. Pran, Norw. P 124,479, Apr 24, 1972. "Film-forming Suspension of Urea-Formaldehyde Particles in a Dilute Solution of Urea-Formaldehyde."

Kirsch, K., *et al.,* Ger. Offen. 2,410,605, Sept 26, 1974. "Building Boards."

Kishi, A., J. Kokai 75:45,066, Apr 22, 1975. "Reinforcement of Particle Boards."

Kitakado, Y., J. Kokai 74:80,210, Aug 2, 1974. "Odorless Plywoods."

Kitakado, Y., J. Kokai 74:66,808, June 28, 1974. "Odorless Plywood."

Kitakado, Y., J. Kokai 73:72,308, Sept 29, 1973. "Formaldehyde Odor-Free Plywood."

Kitakado, Y., J. Kokai 73:58,109, Aug 15, 1973. "Formaldehyde-Odorless Plywood."

Kitamura, K., *et al.,* J. Kokai 77:74,015, June 21, 1977. "Fibers from Alcohol-Modified Urea Resins."

Kitayama, M., M. Nakamura, and K. Yokoyama, J. Kokai 74:36,732, Apr 5, 1974. "Adhesive Coating of Steel Sheets."

Kitchens, J. F., *et al.,* EPA/560/2-76/009, Atlantic Richfield Corp., Alexandria, Va., Aug 1976. "Investigations of Selected Potential Environmental Contaminants: Formaldehyde."

Kitsuda, Y., and K. Ueda, Japan. P 74:40,617, Nov 2, 1974. "Amino Resin Molding Material."

ISBN 0-201-04558-3

Kitta, Y., Japan Kokai 75:63,097, May 29, 1975. "Early-Stage Condensation of Urea-Formaldehyde."

Kitta, Y., K. Ueda, and H. Izumi, J. Kokai 74:63,784, June 20, 1974. "Alkyletherified Amino Resins."

Kitta, Y., J. Kokai 73:31,290, Apr 24, 1973. "Amorphous Precondensate of Urea Resin."

Klebeko, H., Pol. P 74,012, Feb 15, 1975. "Thermosetting Amino-Formaldehyde Resins."

Knabe, G., H. Koehler, and H. Thielsch, E. Ger. P 127,967, Oct 26, 1977. "Subbing Layer for Cellulose Acetate Photographic Film Supports."

Knat'ko, V. M., V. V. Begletsov, V. E. Parshakov, and E. M. Blyakhman, USSR P 337,358, May 5, 1972. "Asphalt Emulsion."

Knilling, W. v., USP 1,920,451, Aug 1, 1933. "Condensation Products of Urea and Formaldehyde."

Kobayagawa, T., and Y. Nakajima, J. Kokai 74:72,179, July 12, 1974. "Purification of Formaldehyde-Containing Waste Gas."

Kobayashi, K., J. Kokai 74:71,103, Nov 14, 1972. "Mildew-Resistant Decorative Plywood Boards with Low Levels of Formaldehyde Odor."

Koberg, H., Ger. Offen. 2,164,336, June 28, 1973. "Fertilizer Tablets for Forest Plants."

Kobori, K., Japan P 75:33,124, Oct 28, 1975. "Nonflammable Sizing Composition."

Koch, H. G., B. E. H. Sundstedt, and J. O. Osterberg, Ger. Offen. 2,213,269, Oct 5, 1972. "Resin-Hardener Compositions for Joining Wood or Timber Parts."

Kodama, T., *et al.,* Aichi-Ken Kogyo Shidosho Hokoku, **10**, 6 (1974). "Heat Degradation in Wood Adhesives."

Kodama, T., Setchaku **19**(2), 51 (1975). "Improvement of Thermal Stability of Urea Resin Adhesives."

Kodo, K., M. Maeda, and M. Yamao, J. Kokai 78:42,178, Sept 28, 1976. "Removal of Formaldehyde from Wastewater."

Kojima, S., and T. Oba, Bunseki Kagaku **24**(5), 294 (1975). "Analysis of Harmful Chemicals in Household Necessities. I. Determination of Free Formaldehyde in Underwear."

Kolaric, J., Tekstil **23**(4), 243; (5), 333 and 348 (1974). "Crosslinking of Cellulose Fibers with Urea-Formaldehyde Resins by Radiation."

Koleshnya, A. D., N. A. Mikhailov, L. N. Stadnik, and N. I. Popova, Izv. Vyssh. Uchebn. Zaved., Lesn. Zh. **21**(2), 109 (1978). "Study of Wood Gluing Using Adhesives Based on Urea-Formaldehyde Resins Modified by Isocyanates."

Kolesnikov, P. A., Radiobiologiya **14**(5), 778 (1974). "Action of Formaldehyde on the Course and Outcome of Acute Radiation Sickness."

Kollmann, F., "Technologie des Holzes und der Holzwerkstoffe. Springer, New York, 1975.

Kollmann, F., W. Clad, and D. Wittmann, Holz als Roh- und Werkstoff **22**, 325 (1964). "Strength of Wood-Glue Bond with UF-Resin."

Kolotilo, D. M., Liteinoe Proizvod. (3), 27 (1976). "Gassing and Coke Formation of Organic Components of the Mold During Pouring."

ISBN 0-201-04558-3

Konishi, M., M. Fujimoto, and G. Araki, J. Kokai 75:22,493, Mar 10, 1975. "Fireproofing Plywood."

Korf, H. L., USP 4,107,105, Aug 15, 1978. "Cellular Urea-Formaldehyde Resins."

Koromyslova, T. S., *et al.*, USSR P 338,415, May 15, 1972. "Molding Compositions Prepared in Wood-Plastics Production."

Koromyslova, T. S., I. A. Gamova, L. N. Natkina, and N. Ya. Solechnik, Izv. Vyssh. Ucheb. Zaved., Les. Zh. **14**(3), 98 (1971). "Particle Boards Made from Sawdust."

Kotani, Y., and M. Yamakita, J. Kokai 77:128,894, Oct 28, 1977. "Formalin Absorbent."

Kotani, Y., J. Kokai 74:112,952, Oct 28, 1974. "Odor-Free Amino Resin Compositions."

Kotani, Y., and K. Kageyama, J. Kokai 74:35,511, Apr 2, 1974. "Plywood Free of Formaldehyde Odor."

Kotani, Y., K. Takeo, and R. Kobayashi, J. Kokai 73:60,140, Aug 23, 1973. "Amino Resin Compositions."

Koval'chuk, L. M., and I. P. Preobrazhenskaya, Derevoobrab. Prom. (7), 6 (1973). "Standardization of Methods for Monitoring Adhesive Joints."

Koyama, T., S. Yamato, and M. Matsumura, J. Kokai 76:88,595, Aug 3, 1976. "Urea Resin Compositions."

Kozin, V. M., *et al.*, USSR P 403,654, Oct 26, 1973. "Raw Mixture for Production of Heat-Insulating Sheets."

Kozlov, M. K., Yu. V. Surotkin, V. D. Isaenko, and V. P. Loginov, USSR P 317,804, Oct 19, 1971. "Plugging Mixture."

Kozlova, G. I., L. V. Dubikovskaya, D. P. Shvets, and N. I. Sereda, Poluch., Svoistva Primen. Fenoplastov Ionoobmen. Smol., p. 73 (1976). "Chemistry of the Reaction of Phenol, Urea and Formaldehyde."

Krach, H., and B. Gos, Zesz. Nauk. Akad. Roln. Warszawie, Technol. Drewna **5**, 49 (1974). "New Adhesive for Joining Wooden Elements."

Krach, H., and B. Gos., Zesz. Nauk. Akad. Roln. Warszawie, Technol. Drewna **4**, 37 (1973). "Use of Furfural for Producing Urea-Furfural Resin Used for Wood Bonding."

Kramer, F., USP 3,800,696, Sept 3, 1974. "Display Device for Plants."

Kramer, F., USP 3,729,320, Apr 24, 1973. "Animal Feed from Chicken Manure."

Kramer, H., GP 48,035, July 9, 1889. "Verfahren zur Herstellung eines Stoffes aus Hobelspänen und Leinwand."

Kratochvil, I., Prac. Lek. **23**(10), 374 (1971). "Health Hazards Caused by Formaldehyde Liberated from Anticrease Textile Finish."

Kravchenko, T. I., K. I. Stankevich, E. F. Malygina, and T. G. Zakharova, Gig. Sanit. (5), 19 (1974). "Migration of Formaldehyde from Wood Sheets."

Kreibich, R. T., and A. G. Andersen, Forest Prod. J., **24**(1), 36 (1974). "Measuring Durability of Adhesive Resins on Paper Substrates."

Krikunov, N. I., T. V. Kozhukhova, E. K. Bovkun, and M. I. Bukovskii, Metody Anal. Kontrolya Proizvod. Khim. Promsti **2**(1976), 32 (1976). "Modified Method of Determining Formaldehyde in the Air Using Phenyl-Hydrazine."

Kropfhammer, G., Ger. Offen. 2,224,363, Nov 29, 1973. "Building Materials from Garbage."

ISBN 0-201-04558-3

Krueger, A., GP 1,719,151, Aug 21, 1975. "Use of Solvent-Free Compositions Based on Aminoplasts as Adhesives."

Kruglikov, A. A., M. A. Nikolavea, S. B. Vilker, and V. V. Detkov, USP 3,689,463, Sept 5, 1972. "Producing Liquid Water-Soluble Urea-Formaldehyde Resins Employing an Aryl- or Alkylsulfonic Acid."

Kubitzky, C., Ger. Offen. 2,206,696, Aug 16, 1973. "Adhesive Admixtures."

Kubota, M., S. Hirata, and H. Takahashi, Rinsan Shikenjo Geppo **303**, 5 (1977). "Liberation Mechanism of Formaldehyde from Plywood Glued with Urea-Formaldehyde Resin Adhesives. (I). Effects of Free Formaldehyde Emitted During the Manufacture of Plywood."

Kucera, L. J., USP 3,668,064, June 6, 1972. "Hot-Pressed Composition Board."

Kuehn, H. G., Ger. Offen. 2,261,896, July 5, 1973. "Glass Fiber-Polymer Composites."

Kuemmel, R., R. Hohn, and L. Kuegler, Chem.-Tech. Umsch. **10**(2), 13 (1978). "The Urea Phenomenon."

Kulle, T. J., and G. P. Cooper, Arch. Environ. Health **30**(5), 237 (1975). "Effects of Formaldehyde and Ozone on the Trigeminal Nasal Sensory System."

Kumlin, K., and R. Simonson, Angew. Makromol. Chem. **68**(1), 175 (1978). "Urea-Formaldehyde Resin. 1. Separation of Low Molecular Weight Components in Urea-Formaldehyde Resins by Means of Liquid Chromatography."

Kunii, Y., and Y. Nishiyama, J. Kokai 75:24,413, Mar 15, 1975. "Plywood."

Kurashige, H., K. Fujii, and H. Ogawara, Japan. P 76:17,598, June 3, 1976. "Board Composed Mainly of Rock Wool and Slug Fiber."

Kurashige, H., and K. Fujii, USP 3,804,706, Apr 16, 1974. "Inorganic Fiber Board with Binder of Thermosetting Resin and Thermoplastic Vinyl Resin."

Kurata, D., and S. Takahashi, Nippon Sanshigaku Zasshi **43**(3), 245 (1974). "Assay of Formaldehyde Gas Concentration in the Silkworm Rearing Room Sprayed with Formalin Solution."

Kurene, A. F., Derevoobrab. Prom. (7), 24 (1972). "Staining of Furniture Panels."

Kurshev, I., *et al.*, God. Nauchnoizsled. Inst. Khim. Prom-st **12**, 295 (1973). "Production of Urea-Formaldehyde Fertilizers."

Kusushita, T., and K. Miwa, J. Kokai 75:76,148, June 21, 1975. "Urea Resin Adhesive Compositions for Woods."

Kuzmenski, N. M., *et al.*, Vrach. Del (6), 131 (1975). "Sensitizing Effect of Formaldehyde During Production of Plastics."

Kuznetsova, A. E., *et al.*, USSR P 590,318, Jan 30, 1978. "Stable Urea-Formaldehyde Resin Suspensions."

Kuznetsova, A. E., *et al.*, USSR P 333,154, Mar 21, 1972. "Complex Urea-Formaldehyde Fertilizers."

Kyle, B. G., and N. D. Eckhoff, EPA/650/2-74/084. Kansas State Univ., Manhattan, Kan., Sept. 1974. "Odor Removal from Air by Adsorption on Charcoal."

Labsky, O., V. Dvoranova, and M. Plskova, Drevo **33**(4), 115 (1978). "Rapid Method for Determining the Free Formaldehyde in Particleboard."

Labzova, L. V., *et al.*, Plast. Massy (5), 23 (1975). "Purification of Gaseous Emissions During Manufacture of Chipboard."

ISBN 0-201-04558-3

Laer, M. v., Bull. Soc. Chim. Belg. **28**, 381 (1919). "The Products of Condensation of Carbamide and Formaldehyde."

Lafarge, S. A., and Rougier et Fils, S. A., Neth. Appl. 76 00680, July 30, 1976. "Plastic-Gypsum Plate for Use in Construction."

Lahmann, E., and K. Jander, Gesundh. Ing. **89**(1), 18 (1968). "Formaldehyde Determination in Urban Atmosphere."

Lajoie, J. L., USP 4,064,307, Dec 20, 1977. "Molding and Coating Compositions."

Lakshminarayanan, T. R., and M. P. Gupta, Res. Ind. **19**(1), 4 (1974). "Shellac-Amino Resin Blends as Molded Insulators."

Lambru, A., A. Druta, M. Gavan, and A. Stancu, Rom P 56,115, Nov 28, 1973. "Adhesive for Hot Joinings of Plywood."

Lambuth, A. L., and W. M. Hearon, USP 3,909,470, Sept 30, 1975. "Thermosetting Adhesive Compositions Extended with Cellulosic Paper Mill Sludges."

Lander, H. L., Ger. Offen. 2,105,253, Oct 14, 1971. "Zirconium- or Hafnium-Containing Polymers."

Landqvist, N., Acta Chem. Scand. **11**, 776 (1957). "On the Reaction between Urea and Formaldehyde in Neutral and Alkaline Solutions. VI. A Spectrophotometrical Method for Quantitative Determination of Urea and Methylol Ureas of the Reaction Mixture."

Lapitskii, V. A., T. I. Pilipenko, L. K. Popov, and M. B. Ushakova, USSR P 476,296, July 5, 1975. "Epoxy Composition."

Laqua, A., and U. Holtschmidt, Ger. Offen. 2,620,478, Nov 24, 1977. "Solutions of Curable Urea-Formaldehyde Resins."

La Ronite, FP 720,772, Nov 4, 1930. "Synthetic Resins."

Laukhin, Yu. A., N. T. Fateev, and V. T. Osaulenko, Gorn. Zh. **150**(8), 44 (1974). "Strengthening Friable Rocks with Urea Resins in a Mining Operation."

Lauter, F., USP 1,672,848, June 5, 1928. "Resinous Reaction Product of Urea and Formaldehyde."

Lauter, F., USP 1,671,596, May 29, 1928. "Resinous Reaction Product of Urea and Formaldehyde."

Lee, C. A., USP 3,686,021, Aug 22, 1972. "Backing Sheet Material for Laminated Products."

Lee, D. R., J. Am. Assoc. For Contamination Control **4** (Dec), 18,19,21 (1965). "Using Impregnated Activated Carbon."

Lee, W. Y., Anal. Chem. **44**(7), 1284 (1972). "Thin-Layer and Paper Chromatography Analysis of the Reaction Products of Urea and Formaldehyde."

Lehmann, W. F., R. L. Geimer, and F. V. Hefty, U.S. Forest Serv., Res. Pap., FPL No. 208, (1973). "Factors Affecting Particleboard Pressing Time Interaction with Catalyst Systems."

Lehmann, W. F., and F. V. Hefty, U.S. Forest Serv., Res. Pap., FPL No. 207, (1973). "Resin Efficiency and Dimensional Stability of Flakeboards."

Lehn, J.-M., G. Wipff, and H.-B. Bürgi, Helv. Chim. Acta **57**, 493 (1974). "Stereoelectronic Properties of Tetrahedral Species Derived from Carbonyl Groups. *Ab initio* Study of the Hydroxy-methanes."

Lehnert, E. J., and O. H. Mattsson, Ger. Offen., Sept 23, 1976. "Finishing the Edges of Particle Boards."

Lein, F., Holz als Roh- und Werkstoff **36**, 379 (1978). "Tendenzen der MDF-Plattenerzeugung."

ISBN 0-201-04558-3

Leneuf, A., Pap. Carton Cellul. **23**(12), 46 (1974). "Analysis of Water Resistant Paper and Water Resistant Products."

Lenke, G. M., and K. B. McReynolds, USP 3,997,612, Dec 14, 1976. "Polythiaformal Compositions."

Lenz, J., and F. Brunnmüller, Ger. Offen. 2,324,440, Dec 5, 1974. "Formaldehyde-Urea Resins."

Lenz, J., G. Roessler, H. Scheuermann, and H. Henkel, GP 1,570,240, Aug 9, 1973. "Impregnating Resins."

Levadie, B., and S. M. Macaskill, Anal. Chem. **48**(1), 76 (1976). "Analysis of Organic Solvents Taken on Charcoal Tube Samplers by a Simplified Technique."

Levi, B. I., J. Chem. Ind. (Moscow), No. 9, 38 (1934). "The Synthesis of Urea from Ammonia and Carbon Dioxide."

Lewin, M., E. Krakauer, and P. Smith, Israeli P 30,472, Jan 27, 1972. "Bonded Wooden Articles."

Liebig, J., Ann. der Pharmacie **14**, 10 (1835). "Über die Producte der Oxydation des Alkohols."

Liebman, S. A., D. H. Ahlstrom, and P. R. Griffiths, Appl. Spectrosc. **30**(3), 355 (1976). "On-Line Fourier-Transform Infrared Analysis of Pryolysis and Combustion Products."

Lielpeteris, U., and I. Ziedins, Khim. Drev. **9**, 167 (1971). "Weight Losses by Ammonia-Treated Wood During Heat Treatment."

Lindberg, J. J., B. Hortling, and J. Turunen, Ger. Offen. 2,507,586, Aug 28, 1975. "Polymer Prepared from Lignin."

Loehmer, W., S. African P 77 01,123, Nov 30, 1977. "Coated Abrasive Comprising a Layer of Spheric Abrasive Bodies."

Lohenhoffer, H. P., and K. W. Ernst, USP 3,839,523, Oct 1, 1974. "Chipboards from Chip Mixtures Using Material in Powder or Fiber Form."

Lomtadze, Z. K., Subtrop. Kul't. (5-6), 182 (1976). "Effectiveness of Urea-Formaldehyde Fertilizer."

Loos, W., Z. Werkstofftech. **6**(9), 314 (1975). "Corrosion Caused by Thermosetting Materials. I. Content of Corrosive Components in Thermosetting Materials. Corrosive Effects of Ammonia."

Lubis, B., *et al.,* USSR P 596,561, Mar 5, 1978. "Eliminating the Caking of Granulated Nitrogen-Phosphorus-Potassium Fertilizer."

Ludlam, P. R., Analyst (London) **98**(1163), 116 (1973). "Thin-Layer Chromatography of Simple Urea-Formaldehyde-Methanol Reaction Products. II. Quantitative Aspects."

Ludwig, C. H., and A. W. Stout, USP 3,658,638, Apr 25, 1972. "Plywood-Making Process and Product Wherein the Adhesive Comprises a Lignosulfonate-Phenol-Formaldehyde Reaction Product."

Ludwig, O. R., GP 733,710, Mar 4, 1943. "Urea-Formaldehyde Condensation Products."

Luedy, E., Monatsh. **10**, 295 (1885). "Über einige aldehydische Condensationsproducte des Harnstoffes und den Nachweis des letzteren."

Lühr, H. P., "Anforderungen an Schaumkunststoffe im Hochbau," Chapter 1, in "Kunststoffe, Forschungsbericht, Vol 4, pg. 9-20, C. Hauser, publishers, Munich, 1979, H. Baumann, editor.

ISBN 0-201-04558-3

Lutoshkina, G. T., Lakokrasoch. Mater. Ikh Primen. (6), 35 (1973). "Effect of the Type of Wood Material and Application Rate of Varnish on the Brightness and Roughness of the Coating."

Lützow, D., and G. Reuss, "Harnstoff", in "Ullmanns Encyklopadie der technischen Chemie." Volume 12, p. 497. Verlag Chemie, Weinheim, 1974.

Lvovsky, C., and J. Cihelka, Czech. P 144,662, July 15, 1972. "Electrostatic Metal Plating of Electrically Nonconductive Products."

Maeda, T., J. Kokai 74:128,943, Dec 10, 1974. "Stabilized Phenolic Resin Adhesives."

Maksimov, Yu. I., Sb. Tr. Vses. Nauch.-Issled. Inst. Nov. Stroit. Mater., No. 30, 202 (1971). "Synthetic Adhesives for Bonding Decorative Paper Sheet Laminates to Various Materials."

Malakhova, Z. H. V., Melior. Oroshaemykh Zasolennykh Pochv, Dokl. Vses. Soveshch. **2**, Chast 1, 98 (1971). "Effect of Polymers on the Physicochemical Properties of Soil and the Rice Harvest."

Maloney, T. M., *Modern Particleboard & Dry-Process Fiberboard Manufacturing,* Miller Freeman, San Francisco, CA, 1977.

Malygina, E. F., Gig. Sanit. (8), 95 (1974). "Effect of Surfacing and Additional Heat Treatment on the Emission of Harmful Substances from Particle Board Containing Urea Resin Binders."

Marcote, R. V., R. Chand, and T. H. Johnston, Anal. Instrum. (11), 31 (1973). "Development and Evaluation of a Portable Aldehydes Analyzer."

Mara, A. A., *Kolthoff's Treatise on Analytical Chemistry,* Part III, Vol. 4, Section D, Interscience, New York, 1977. "Adhesive Properties."

Marra, A. A., USP 3,671,377, June 20, 1972. "Rigid Composite Products."

Marsh, J. T., "Selfsmoothing Fabrics." Chapman and Hall, London, 1962.

Martirosyan, G. A., *et al.,* USSR P 626,753, Oct 5, 1978. "Fodder Additive."

Martorano, R., W. H. Brendley, Jr., and T. H. Haag, USP 4,020,219, Apr 26, 1977. "Metal Decorating Method Involving Mechanical Shaping."

Marutzky, R., and E. Roffael, Holzforschung **31**, 8 (1977). "Über die Abspaltung von Formaldehyd bei der thermischen Behandlung von Holzspänen. Teil 1: Modell-Versuche."

Masek, V., Gesundh.-Ing. **92**(8), 245 (1971). "Determination of Acrolein and Formaldehyde in Air."

Maslinkovskii, A. S., *et al.,* USSR P 554,233, Apr 15, 1977. "Solution for Finishing Internal Surfaces of Buildings."

Masuda, S., J. Fujitani, M. Ohtake, and K. Yamasawa, J. Kokai 74:130,934, Dec 16, 1974. "Aminoplast-Based Adhesives Containing Saponified Acrylamide-Vinyl Acetate Copolymers for Plywood."

Materiaux Reunis, Fr. Demande 2,219,134, Sept 20, 1974. "Fireproofing Panels."

Matsumoto, T., Ringyo Shikenjo Kenkyu Hokoku **291**, 87 (1977). "Two Parameters Characterizing the Composition of Urea-Formaldehyde Reaction Products as Adhesives."

Matsumoto, T., Ringyo Shikenjo Kenkyu Hokoku **262**, 41 (1974). "Concentration of Formaldehyde Released from Plywood in an Environmental Test Room."

ISBN 0-201-04558-3

Matsumoto, T., Bull. Govt. For. Expt. Sta. (Tokyo) (212), 179 (1968). "Free Formaldehyde Determination in Urea Resin Adhesives by the Ammonium Chloride Method."

Matsumura, Y., and H. Arito, Ind. Health **13**(4), 135 (1975). "Toxic Volatile Components of Organic Soil Consolidating Agents."

Mattsson, O., B. Sundin, B. Åström, P. Hanetho, and E. Ormstad, Proc. Symp. Hållet hos KemaNobel 1 Nacka 1977-11-15, I arrangemang av Casco, Dyno och Svenska Spånskiveföreningen, Stockholm, 1978-02-15. "Formaldehydavigivning från Byggnadsmaterial."

Mattsson, O., and N. H. Wallin, Cellulosa (Stockholm) (1) 104 (1976). "Formaldehyde-Based Hard Plastics-Key Components in the Wood Products Industry."

Marutzky, R., and E. Roffael, Holz-Zentralblatt **103**, 424 (1977). "Über den Stickstoffgehalt in Holz, Rinden und Biomasse."

Mayer, J., and C. Schmidt-Hellerau, Ger. Offen. 2,020,481, Nov 11, 1971. "Phenol-Urea-Melamine Resin Adhesives for Wood."

Mazur, S., and B. Kolarz, Pr. Nauk. Inst. Inz. Ladowej Politch. Wroclaw. No. 2, 15 (1970). "Use of Urea Resin in Building Construction."

McMinimy, L. O., USP 3,674,596, July 4, 1972. "Laminating Untreated Wood to Wood Treated with Fire-Retardant Material."

Mehlhorn, L., E. Roffael, and H. Miertzsch, Holz-Zentralblatt **104**, 345 (1978). "Erfahrungen mit den vom FIHH-Karlsruhe vorgeschlagenen Prüfmethoden zur Bestimmung des Formaldehyds."

Menashi, J., *et al.*, J. Colloid Interface Sci. **66**, 220 (1978). "The Effect of the Morphology of Urea-Formaldehyde Pigments in Paper Filler Applications."

Merkcl, D., *et al.*, GP 2,550,739, Nov 12, 1975. "Binder Composition for Wood Adhesion."

Meyer, B., and W. E. Johns, Holzforsch. **32**, 102 (1978). "Sulfur in Wood Bonding Systems."

Meyer, B., W. Johns, and J. K. Woo, Forest Prod. Journal, in press, "Formaldehyde Release from Sulfur Modified UF-Resin Systems."

Meyer, B., "Formaldehyde Release from UF-Systems," Proc. 13th Int. Particleboard Symposium, Washington State University, T. Maloney, ed., Pullman, 1979.

Meyer, B., and B. Mulliken, "Wood-Sulfur Bonding," Adv. Chem. Series, **165**, 225 (1978).

Meyer, W. T., Plast. Bldg. Constr., Realities Challenges, Nat. Tech. Conf., Soc. Plast. Eng., p. 317 (1972). "Urea-Formaldehyde Foam Insulation for the Construction Industry."

Michal, J., and J. Mitera, Nehorlavost Plast. Hmot, Dreva Text., p. 118 (1974). "Products of Combustion of Plastics."

Midwest Research Institute, FP 2,082,270, Jan 14, 1972. "Fire-Resistant Polyurethane Foam."

Mielke, R., E. Ger. P 103,883, Feb 12, 1974. "Fiber Insulation Composition."

Mikhailova, A. A., *et al.*, Gig. Sanit. (7), 86 (1974). "Improvement in the Sanitary-Hygienic Properties of Wood Filled Urea-Formaldehyde Panels."

Miller, C. H., and J. A. Mayrl, USP 3,878,278, Apr 15, 1975. "Lightweight Reinforced Structural Material."

ISBN 0-201-04558-3

Minemura, N., *et al.*, Rinsan Shikenjo Geppo (284), 5 (1975). "Effects of Reinforcing Materials on Properties of Glue and Plywood."

Minemura, N., and S. Imura, Rinsan Shikenjo Geppo, **298**, 1 (1976). "Starch Waste as an Extender of Urea-Formaldehyde Resin Adhesive."

Minemura, N., and H. Takahashi, Rinsan Shikenjo Geppo (12), 1 (1973). "Analysis of Dilute Formaldehyde Solutions by Gas Chromatography."

Minemura, N., S. Hirata, and H. Takahashi, Rinsan Shikenjo Geppo (1), 13 (1975). "Effects of Chemicals on Free Formaldehyde Emission from Treated Plywood."

Minemura, N., S. Imura, S. Hirata, and H. Takahashi, Rept. Hokkaido Forest Prod. Res. Inst. No. 65: 53 (1976). "Studies on Less Formaldehyde Liberation from Type II Plywood Glued with Urea Formaldehyde Adhesive."

Minnesota Mining and Mfg. Co., Ger. Offen. 2,425,887, Dec 19, 1974. "Encapsulated Abrasives."

Minnesota Mining and Mfg. Co., BP 419,812, Nov 9, 1934. "Flexible Abrasive Articles."

Minoji, A., T. Koma, and O. Onoura, J. Kokai 76:20,225, Feb 18, 1976. "Formaldehyde Odor-Free Coating Compositions."

Minoji, A., T. Koma, and O. Onoura, J. Kokai 74:38,925, Apr 11, 1974. "Decorative Board Top Coating Composition."

Minoji, A., T. Koma, and O. Onodera, J. Kokai 74:38,927, Apr 11, 1974. "Decorative Boards Top Coating Composition."

Mints, B. Ya., *et al.*, USSR P 550,153, Mar 15, 1977. "Composition for Impregnation of Fibrous Materials for Preparation of Hats."

Mironov, O. G., I. E. Putlyaev, V. I. Katsyuba, and O. D. Likholetov, Plast. Massy (11), 24 (1973). "Highly Filled Urea-Formaldehyde Compositions for Agricultural Structures."

Mir-Puig, P., Chem. Ind. (Duesseldorf) **27**(9), 534 (1975). "Production of Urea-Formaldehyde Adhesives."

Mita, A., and H. Hirai, J. Kokai 75:25,802, Mar 18, 1975. "High-Yield Fibrous Materials."

Mitrofanov, E. G., Tr., Leningrad. Lesotekh. Akad. No. 151, 55 (1972). "Hardness and Wear of Three-Ply Composite Slabs."

Mitsua, I., *Urea-Melamine Resins,* Tokyo, 1965.

Miwa, T., and H. Tomokawa, J. Kokai 74:124,207, Mar 30, 1973. "Removal of Formaldehyde from Plywood."

Miwa, T., and H. Tomokawa, J. Kokai 74:124,203, Nov 28, 1974. "Removal of Formaldehyde from Plywood."

Miwa, T., and I. Ochiai, J. Kokai 74:25,108, Mar 6, 1974. "Plywood Free of Formaldehyde Odor."

Miyahara, S., and N. Kuzuba, J. Kokai 73:66,152, Sept 11, 1973. "Fluorescent Coating Compositions."

Mochalova, Yu. Z., I. V. Gorbunova, and E. M. Moskalenko, Izv. Vyssh. Ucheb. Zaved., Khim. Khim. Tekhnol. **16**(7), 1088 (1973). "Hardening Rate of Urea-Formaldehyde Resin Binder M_2 Studied by a Fractional Factorial Analysis."

Mohl, H. R., Holz als Roh- und Werkstoff **36**, 69 (1978). "Saug- und Spaltmethode zur Bestimmung der Formaldehydabgabe von Holzwerkstoffen und Leimen sowie zur allgemeinen Luftanalyse."

ISBN 0-201-04558-3

Mohl, H. R., Holz-Zentralbl. **101**(67), 869, 871 (1975). "A Nondestructive Rapid Test for Determining Formaldehyde Liberation from Particleboards without Overlays."

Mohl, H. R., Technische Mitteilung Nr. 3/75, Forschungsinstitut für Holzwerkstoffe und Holzleime, Karlsruhe-Durlach, 1975. "Zerstörungsfreie Schnellprüfmethode zur Bestimmung der Formaldehydabspaltung aus unbeschichteten Spanplatten."

Mølhave, L., Holzforsch. Holzverwert. **29**(4), 73 (1977). "The Relation Between the Free Formaldehyde Concentration in Stored (Urea-Formaldehyde) Chipboard, and the Formaldehyde Concentration in the Air."

Monsanto Co., BP 1,486,342, Sept 21, 1977. "Storage Stable Urea-Formaldehyde Resin Sirup and Process Therefor."

Monsanto Corp., Lab. Test Method 130,135. Monsanto Corp., Eugene, Oregon, 1964-65. "Determination of Formaldehyde Evolution Rate from Cured Amino-Plast Resin(s): 130–films; 135–in wood particle products."

Monti, L., Gazz. Chim. Ital. **60**, 39 (1930). "Some Condensation Products of Methylolbenzamide."

Moore, W. P., Jr., USP 4,097,419, June 27, 1978. "Stabilized High Strength Urea-Aldehyde Polymers."

Moore, W. P., Jr., USP 4,092,277, May 30, 1978. "Chemically Stable Urea-Formaldehyde Foams."

Moore, W. P., J. E. Sansing, and H. D. Williamson, USP 3,970,625, July 20, 1976. "Urea-Formaldehyde Concentrates."

Moralt, A., BP 1,289,465, Sept 20, 1972. "Molded Wooden Bodies Having Fire-Resistant Properties."

Mori, K., S. Ikebe, Y. Takemoto, and K. Hatanaka, Japan. P 73:21,346, June 28, 1973. "Method of Making Particle Board."

Mori, S., M. Saito, and M. Kubota, Japan. P 73:32,418, Oct 5, 1973. "Filler for Adhesive from Bark."

Morimoto, T., K. Takeyama, and F. Konishi, J. Appl. Polym. Sci. **20**, 1967 (1976). "Composition of Gaseous Combustion Products of Polymers."

Morin, N. C., and H. Kubinski, Ecotoxicol. Environ. Saf. **2**(2), 133 (1978). "Potential Toxicity of Materials Used for Home Insulation."

Moroff, H., and H. Neumann, Ger. Offen. 2,135,072, Feb 1, 1973. "Highly Lustrous Decorative Paper."

Morozova, S. S., and O. P. Demidova, USSR P 559,000, May 25, 1977. "Composition for Saturation of Textured Papers."

Moschandreas, M., "Indoor Air Pollution levels as a function of ventilation rates." GEOMET Report for U.S. Environmental Protection Agency and U.S. Department of Housing and Urban Development, 1976.

Moslemi, A. A., *Particleboard,* Vol. 1 & 2, Southern Illinois Univ. Press, Carbondale, Ill., 1974.

Mostovoy, S., and E. J. Ripling, Appl. Polym. Symp. No. 19, 395 (1972). "Effect of Temperature on the Fracture Toughness and Stress Corrosion Cracking of Adhesives."

Mueller, F. X., and J. A. Miller, Am Lab. **6**(5), 49, 52, 54, 56, 58 (1974). "Determination of Organic Vapors in Industrial Atmospheres."

ISBN 0-201-04558-3

Mukudai, J., H. Kadita, and S. Yata, Nippon Secchaku Kyokai Shi **9**(2), 58 (1973). "Development of the Internal Stress in the Adhesive Layer of Woodworking Adhesives During Cure."

Murav'ev, V. S., and L. I. Sakovich, Sb. Tr. Mosk. Tekhnol. Inst. No. 20, 93 (1971). "Numerical Determination of the Life of Urea Adhesives."

Muresan, L., Gh. Iliescu and Gh. Iliescu, Celul. Hirtie **21**(1), 59 (1972). "Use of Urelit HM in the Manufacture of Kraft Sack Paper."

Myers, G. E., Progress Report, Forest Research Laboratory, U.S. Department of Agriculture, Madison, Wisconsin; FS-FPL-3204; May 1978, Apr 1977, and Sept 1978. "Improved Performance of UF-Resin Bands."

Myers, R. R., "Treatise on Coatings." Marcel Dekker, New York, 1968, J. S. Long, editor.

Nagao, A., J. Kokai 72:43,128, Dec 18, 1972. "Urea Resins Blended with 3,9-bis[2-(2,4-Diamino-*s*-Triazin-6-yl)Ethyl]-2,4,8,10-Tetra-oxaspiro[5.5] Undecane Resin Composition for Wood Adhesives."

Nagao, A., J. Kokai 72:22,441, Oct 7, 1972. "Urea-CTU Guanamine Copolymer Resin Adhesives."

Naidenov, M. N., P. P. Gutsal, B. M. Shemerdyak, and M. S. Mel'nik, USSR P 576,327, Oct 15, 1977. "Composition for Producing Urea-Formaldehyde Resin Foam."

Nakane, B., USP 3,730,825, May 1, 1973. "Decorative Sheet Material Having Plural Adhesion Layers."

Nakanishi, T., and T. Ueta, Japan. P 71:31,015. "Amino Resin Molding Materials."

Nakajima, H., and M. Kubo, Japan P 74:15,071, Apr 12, 1974. "Polyurethane Foamed Resin."

Nakajima, M., and N. Sakato, J. Kokai 74:94,744, Sept 9, 1974. "Additives for Amino Resin Adhesives."

Nakajima, T., and T. Shyaku, USP 3,690,976, Sept 12, 1972. "Overlaid Decorative Wood Panels."

Narasaki, H., and K. Umezawa, Kobunshi Kagaku **29**(6), 438 (1972). "Application of Low-Temperature Ashing to Polymer Analysis."

National Academy of Sciences, Wash., D.C. (1975). "Principles for Evaluating Chemicals in the Environment."

National Research Council, Wash., D.C., Oct (1975). "Vapor-Phase Organic Pollutants—Volatile Hydrocarbons and Oxidation Products."

Nefedov, Y. G., *et al.*, Kosm. Biol. Med. **3**(5), 71 (1969). "Contaminants in the Air Exhaled by Man."

Nelander, B., J. Mol. Structure **50**, 223 (1978). "Complex Formation Between Hydrogen Sulfide and Formaldehyde in Solid Argon and Solid Nitrogen."

Nelyubin, I. A., I. I. Leonovich, and B. Vrublevskii, USSR P 554,344, Apr 15, 1977. "Composition for Building of Road Bases."

Nestler, F. H. M., *The Formaldehyde Problem in Wood-Based Products—An Annotated Bibliography,* U.S. Dept. of Agriculture, Forest Service, Forest Products Lab., Madison, Wisc., USDA Forest Service General Technical Report FPL-8, 1977.

ISBN 0-201-04558-3

Neumann, R., E. Ger. P 96,965, May 30, 1972. "Hydrophilic Open-Pore Urea-Formaldehyde Foams."

Neumann, R., B. Mueller, M. Spichala, and H. Brandner, Ger. Offen. 2,130,577, Dec 30, 1971. "Modified Aminoplast Resins, Especially Urea-Formaldehyde Resins."

Neuss, O., GP 604,194, Oct 16, 1934. "Urea-Formaldehyde Condensation Products."

Neuss, O., BP 379,154, Aug 25, 1932. "Synthetic Resins."

Neuss, O., BP 372,847, May 19, 1932. "Urea-Formaldehyde Condensation Products."

Neuss, O., BP 202,651, Aug 17, 1923. "Condensation Products from Urea and Formaldehyde."

Neusser, H., and W. Schall, Holzforsch. Holzverwert. **24**(5), 108 (1972). "Promising Possibilities of Plywood Improvements. I. Gluing with Aminoplasts."

Neusser, H., and M. Zentner, Holzforsch. Holzverwert. **5**, 1 (1968). "Über die Ursachen und die Beseitigung des Formaldehydgeruches von holzhaltigen Baustoffen, insbesondere von Spanplatten."

Nevin, J. V., USP 2,247,764, July 1, 1941. "Adhesive Suitable for use in Patching Veneers."

Niedermaier, T., and H. Pruen, Landwirt. Forsch., Sonderh. **28**(1), 175 (1973). "Nitrogen Activity Index of Hygromull."

Nishino, S., and H. Shikata, J. Kokai 73:92,541, Nov 30, 1973. "Deodorant for Formaldehyde of Resin Products."

Niwa, T., and I. Ochiai, J. Kokai 74:25,018, Mar 6, 1974. "Amino-Alkyd Resin Coating Compositions with Low Formaldehyde Liberation."

Nizhni Tagil Plastics Plant, J. Kokai 77:155,697, Dec 24, 1977. "Modified Formaldehyde-Urea Resins."

Nobell, A., USP 3,759,687, Sept 18, 1973. "Urea-Formaldehyde Fertilizer."

Noda, H., S. Kobayashi, and M. Akasaka, J. Kokai 78:37,767, Apr 7, 1978. "Continuous Foam Sheets of Urea-Formaldehyde Copolymers."

Nogi, T., *et al.*, J. Kokai 75:148,624, Nov 28, 1975. "Fire-Resistant Aminoplast Fibers."

Nogi, T., *et al.*, J. Kokai 75:126,792, Oct 6, 1975. "Amino Resin Solutions."

Nomura, M., and N. Kobayshi, Japan P 74:0,461, Jan 30, 1974. "Adhesion of Wood Materials."

Novak, V. A., Yu. S. Murashov, V. D. Valgin, and V. V. Baranov, Fr. Demande 2,246,588, May 2, 1975. "Phenol-Aldehyde Polymer Foams."

Nowak, D., *et al.*, Pol. P 80,377, Feb 20, 1976. "Stabilized Anionic Adhesives."

Nowak, D., Z. Hehn, M. Maczynska, and M. Jaglarz, Pol. P 83,871, Jan 30, 1976. "Water-Insoluble Mixed Ethers of Urea-Formaldehyde Condensate."

Obetko, D., and J. Melzer, Bull. Vysk. Ustavu Pap. Celul. **12**(2), 35 pp (1972). "Starch Sizes for the Production of Corrugated Board."

Oblivin, A. N., and G. N. Afanas'ev, Izv. Vyssh. Uchebn. Zaved., Lesn. Zh. **18**(2), 104 (1975). "Determination of Formaldehyde and Methanol Evolved in the Polycondensation of Resins under Chip Board Hot Forming Conditions."

ISBN 0-201-04558-3

Oblivin, A. N., and G. N. Afanas'ev, Nauch. Tr. Mosk. Lesotekhn. In-T (64), 24 (1974). "Determination of the Quantity and Composition of Volatile Products from Polycondensation of Urea-Formaldehyde Resins During Pressing of Particle Board Tiles."

Obraztsov, A. E., *et al.*, Zh. Prikl. Khim. (Leningrad) **42**, 2393 (1969). "Liquid-Vapor Phase Equilibrium of a Formaldehyde-Water System."

Oda, N., *et al.*, USP 3,892,709, July 1, 1975. "Aldehyde-Polyamine or Aldehyde-Polyamine-Aromatic Compound Condensation Resin Adsorbent for Heavy Metals and Heavy Metallic Compounds."

O'Donnell, J. M., Ger. Offen. 2,523,483, Dec 4, 1975. "Apparatus and Method for the Treatment of Clarifier Sludge."

Oesterreichische Hiag-Werke, FP 2,712,630, Nov 2, 1973. "Wood Particle Boards."

Oesterreichische Hiag-Werke, FP 2,161,199, Aug 10, 1973. "Particle Board."

Oetgen, W. R., and J. Cooley, USP 3,681,274, Aug 1, 1972. "Ammoniacal Fluoride Catalysts for Formaldehyde Resin Systems."

Ogden, D. H., Ger. Offen. 2,416,270, Oct 24, 1974. "Lightweight Building Materials."

Ogiwara, T., Plast. Age **17**(12), 97 (1971). "Properties and Applications of Plastic Foams. 12. Urea Foams."

Ohe, T., and K. Ikeda, J. Kokai 74:05,891, Jan 19, 1974. "Formaldehyde Scavengers."

Ohhara, O., and T. Maeno, J. Kokai 74:131,243, Dec 16, 1974. "Formaldehyde Capture Agents for Formaldehyde Thermosetting Resin Adhesives."

Ohhara, O., and Y. Matsutomi, J. Kokai 74:92,154, Sept 3, 1974. "Adhesives."

Ohhashi, K., and S. Ishikawa, J. Kokai 74:119,948, Nov 15, 1974. "Odorless Phenolic Resin Powders."

Ohmura, M., Japan. P 75:06,527, Mar 14, 1975. "Decorative Plywood."

Ohno, K., and R. Furudera, Japan P 74:43,126, Nov 19, 1974. "Deodorizing Formaldehyde Odor from Resin-Treated Lumber."

Ohtomi, M., T. Ohno, and S. Akabori, Kogyo Yosui (202), 21 (1975). "Determination of Formaldehyde in Water with 3-Methyl-2-Benzothiazolone Hydrazone."

Ohtsuka, M., and S. Uchihara, Japan. P 76:03,355, Feb 2, 1976. "Interior Building Material."

Oka, Y., K. Fujita, T. Murakami, and K. Kobayashi, J. Kokai 75:13,512, Feb 13, 1975. "Decorative Plywood."

Okabe, J., and T. Nakamoto, Japan. P 74:15,242, Apr 13, 1974. "Water-Insoluble Lignin Products."

Okada, K., K. Yasuda, and H. Minamida, J. Kokai 74:131,259, Dec 16, 1974. "Fire-Resistant Boards."

Okamura, S., K. Hayashi, and I. Kaetsu, Japan P 73:14,793, May 10, 1973. "Curing Amino-Type Low Molecular Weight Condensates by Radioactive Rays."

Okresny Stavebno-Montazny Podnik Poprad Podnikove Riaditelsto, J. Kokai 75:75,257, June 20, 1975. "Fireproofed Cellular Urea-Formaldehyde Resin Compositions."

Olefjord, I., Scand. Corros. Congr., Proc., 6th **11**, 1 (1971). "Application of ESCA (Electron Spectroscopy for Chemical Analysis) in Corrosion Research."

ISBN 0-201-04558-3

Olszowski, T., *et al.,* Ger. Offen. 2,400,908, Dec 5, 1974. "Formaldehyde-Urea Copolymer Binder for Mold Sand."

Oohara, O., and T. Maeno, J. Kokai 75:104,293, Aug 18, 1975. "Formaldehyde Resins for Adhesives."

Orth, G. O., Jr., S. African P 72:04,876, Jan 22, 1973. "Hydrophobic, Oleophilic Cellulosic Fibers."

Oshima, K., Y. Ito, and S. Suzuki, USP 3,764,428, Oct 9, 1973. "Foaming Fibrous Fleece."

Osipova, N. A., A. V. Sandalov, V. I. Samoilov, and V. F. Tsibizov, USSR P 557,859, May 15, 1977. "Binder for Cast Products."

Ostapovich, I. K., Gig. Sanit. (2), 9 (175). "Features of the Sensitizing Action of Sulfur Dioxide and Formaldehyde in Various Regimens of their Inhalation."

Otsuka, E., and T. Takahashi, Kogyo Kagaku Zasshi, **63**, 1218 (1960). Equilibrium of the $Na_2CO_3-NH_3-CO_2-H_2O$ System."

Otsuka, E., S. Yoshimura, M. Yakabe, and S. Inoue, Kogyo Kagaku Zasshi **63**, 1214 (1960). "Equilibrium of Ammonia-Carbon Dioxide-Water System."

Otsuka, E., S. Yoshimura, and M. Yakabe, Kogyo Kagaku Zasshi, **63**, 1121 (1960). "Separation of the Mixed Gas of NH_3 and CO_2 by Using the Aqueous Solutions of Ammonium Salts."

Pace, L. S., Ger. Offen. 2,516,327, Nov 6, 1975. "Hardener System for Aminoplast Resin Adhesives."

Pahla, P., USSR P 329,897, Feb 24, 1972. "Skis."

Pal, R. N., and S. S. Negi, Indian J. Anim. Sci. **47**(12), 782 (1977). "Utilization of Alkali-Sprayed Paddy Straw Supplemented with Urea or Sparingly Soluble Urea-Formaldehyde Complex by Cattle."

Pal, Mrs. K., Termeszet Vilaga **102**(11), 488 (1971). "Fire Hazard of Plastics."

Panov, V. V., Izv. Vyssh. Ucheb. Zaved., Les. Zh. **16**(2), 122 (1973). "Urea Resins Studied by an Ultrasonic Method."

Parnell, W. E., USP 3,666,597, May 30, 1972. "Catalyzing Adhesive Cure."

Paronyan, R. V., and V. K. Sarkisyan, Zh. Anal. Khim. **29**(3), 604 (1974). "Determination of Formaldehyde by Gas-Liquid Chromatography in a Mixture of Primary Aldehydes, Alcohols and Acids."

Partridge, L. J., S. African P 70 04,991, Jan 21, 1972. "Urea Condensation Products."

Pashkov, D. N., Plast. Massy (2), 61 (1975). "Urea-Formaldehyde Resin-Based Filled Polymeric Compositions."

Pashkov, N., Neft. Khoz **50**(1), 60 (1972). "Curing Urea-Formaldehyde-Furfural Resin Solutions with a Concentrate of Low-Molecular-Weight Acids."

Pastyr, J., and L. Kuniak, Czech. P 170,983, Jan 15, 1978. "Modification of Cellulose Sheets."

Paszner, L., H. T. Cho, T. Kaneko, and M. M. Micko, For. Prod. J. **24**(10), 39 (1974). "One-Step Method of Making Veneer Polymer Composite Overlaid Plywood by Gamma Radiation."

Paterson, A. R., and A. C. Reimschuessel, USP 3,790,417, Feb 5, 1974. "Fiberboard Having Improved Dimensional Stability."

ISBN 0-201-04558-3

Patterson, R. M., M. I. Bornstein, and E. Garshick, Vol. VIII, GCA-TR-75-32-G(8), GCA Technology Div., GCA Corp., Bedford, Mass., 1976. "Assessment of Formaldehyde as a Potential Air Pollution Problem."

Paul, R. L., and P. J. Hendra, Miner. Sci. Eng. **8**(3), 171 (1976). "Laser Raman Spectroscopy in Surface Chemical Problems."

Paulitsch, M., Holz als Roh- und Werkstoff **30**(11), 437 (1972). "pH-Value of Aqueous Extractions from Urea Resin-Glued Particleboard."

Pavlenko, S. M., and V. A. Guseva, Itogi Nauki Tekh., Farmakol., Khimioter. Sredstva, Toksikol., Probl. Toksikol. **5**, 110 (1973). "Dynamics of the Development of Adaptive Reactions under the Long Term Effect of Industrial Poisons Entering an Organism by Different Means."

Pearson, D., *The Chemical Analysis of Foods,* Churchill Livingston, New York, 1976.

Pellizzari, E. D., J. E. Bunch, B. H. Carpenter, and E. Sawicki, Environ. Sci. Technol. **9**(6), 552 and 556 (1975). "Collection and Analysis of Trace Organic Vapor Pollutants in Ambient Atmospheres. Thermal Desorption of Organic Vapors from Sorbent Media."

Penczek, P., and K. Stepien, Pol. P 67,486, Mar 20, 1973. "Phenol-Melamine-Urea-Formaldehyde Resins."

Perevertov, A. S., A. G. Romashova, O. P. Blinkova, and M. G. Dranovskii, Stabil. Polim. Mater. Izdelii Iz Nikh, No. 2, 84 (1971), "Physical Properties of Thermosetting Plastics (Upon Storage)."

Peters, T. E., USP 3,649,397, Mar 14, 1972. "Manufacture of Products from Comminuted Wood."

Petersen, H., Kunststoff-Jahrbuch 10 Folge, Wilhelm Pansegrau Verlag, Berlin (1958). "Grundzüge der Aminoplastchemie."

Petersen, H., Holz als Roh- und Werkstoff **36**, 397 (1978). "Application of Statistical Test Planning in Particleboard Manufacture with Urea-Formaldehyde Binders. (3) Planning and Evaluation of 2^n Test Factors."

Petersen, H., Holz als Roh- und Werkstoff **35**, 369 (1977). "Application of Statistical Test Planning in Particleboard Manufacture with Urea-Formaldehyde Binders. (2) Studies of the Influencing Factors Resin Catalyst Rate, Pressing Temperature, Pressing Period, and Moisture of Glued Chips during Particleboard Production."

Petersen, H. Holz als Roh- und Werkstoff **34**, 365 (1976). "Application of Statistical Test Planning in Particleboard Manufacture with Urea-Formaldehyde Binders. (1) Studies of Binder Application, Catalyst Amount, Pressing Temperature, and Pressing Period as Factors Influencing Particleboard Properties."

Petersen, H., J. Melzer, R. Hummerich, and M. Rohmann, Ger. Offen. 2,547,966, May 5, 1977. "Water-Insoluble Urea-Formaldehyde Condensate."

Petersen, H., W. Reuther, W. Eisele, and O. Wittmann, Holz Roh-Werkst **32**, 402 (1974). "Formaldehyde Liberation During Particleboard Production with Urea-Formaldehyde Adhesives. III. Effect of Hardener Type, Amount of Hardener, and Formaldehyde-Binding Agents."

Petersen, H., P. Scharwaechter, and J. Lenz, Ger. Offen. 2,241,715, Mar 7, 1974. "Fire- and Shrinkage-Resistant Aminoplast Foams."

ISBN 0-201-04558-3

Petersen, H., W. Reuther, W. Eisele, and O. Wittmann, Holz Roh-Werkst. **31**, 463 (1973). "Formaldehyde Liberation During Particleboard Production with Urea-Formaldehyde Adhesives. II. Effect of Resin Quantity, Pressing Time, and Temperature."

Petersen, H., W. Reuther, W. Eisele, and O. Wittmann, Holz Roh-Werkst. **30**, 429 (1972). "Formaldehyde Liberation During the Manufacture of Particleboard with Urea-Formaldehyde Adhesives."

Petersen, H., Angew. Chem. **76**, 909 (1964). "New Findings in Aminoplast Chemistry."

Petersen, S., A. Mitrowsky, and A. Dorlars, "Preparation and Reactions of Monomeric Carbonic Acid Derivatives." p. 460 in "Houben-Weyl, Mthoden der Organischen Chemie, Vol. 8, "Oxygen Compunds (III)," 4th Edition, Thieme Verlag, Stuttgart, 1952, E. Müller, editor.

Petropoulos, J. C., I. H. Updegraff, and I. L. Williams in *Applied Polymer Science,* J. K. Craver and R. W. Tess, eds., Organic Coatings and Plastics Chemistry Div., Am. Chem. Soc., 1975. "Amino Resins for Surface Coatings."

Petrusenko, P. I., and O. V. Petrova, Khim. Ind. (Sofia) **44**(4), 180 (1972). "Use of Higher Alcohols in the Production of Urea-Formaldehyde Resins."

Pfohl, F., Swiss P 193,139, Jan 12, 1937. "Platte fur Mobel und Bautischler und Verfahren zur Herstellung."

Philpotts, A. R., E. O. Evans, and N. Sheppard, Trans. Faraday Soc. **51**, 1051 (1955). "The Infrared Spectra and Structure of Some Polymers of Formaldehyde."

Piechota, H., and H. Roehr, *Integral Foam Materials. Principles, Manufacture, and Applications of Integral and Structural Foam Materials,* Hanser, Munich, Ger. 1975.

Pizzi, A., Adhes. Age., **20**(12), 27 (1977). "Hot-Setting Tannin-Urea-Formaldehyde Exterior Wood Adhesives."

Pizzi, A., J. Appl. Polym. Sci. **22**(8), 2397 (1978). "Tannin-Formaldehyde Exterior Wood Adhesives through Flavonoid B-Ring Crosslinking."

Plath, L., Staerke **24**(9), 306 (1972). "Starch Products in Gluing of Wood Materials by Synthetic Resin Adhesives."

Plath, L., Adhäsion (9), 229 (1971). "Microscopic Studies of Painted or Coated Wood Surfaces."

Plath, L., Holz als Roh- und Werkstoff **24**, 312 (1966). "Determination of Formaldehyde Liberation from Particle Boards by Microdiffusion."

Plomley, K. F., United Nations Industrial Development Organization ID/WG.248/6, Aug 24, 1977. "The Formulation and Industrial Application of Naturally Occurring Polyphenol (Tannin) Adhesives in the Wood Based Panel Industry."

Pocker, Y., and J. E. Meany, Biochem. **4**, 2535 (1965). "The Catalytic Versatility of Erythrocyte Carbonic Anhydrase. I. Kinetic Studies of the Enzyme-Catalyzed Hydration of Acetaldehyde."

Pocker, Y., and J. E. Meany, J. Am. Chem. Soc. **87**, 1809 (1965). "The Catalytic Versatility of Carbonic Anhydrase from Erythrocytes. The Enzyme-Catalyzed Hydration of Acetaldehyde."

ISBN 0-201-04558-3

Polak, A. F., and E. I. Mulyukov, Sb. Tr. Nauch.-Issled. Inst. Prom. Stroit. Nüpromstroi (10), 212 (1971). "Supramolecular Structure of Urea Resins."

Polinek, V., Czech. P 170,877, Jan 15, 1978. "Treatment of Industrial Waste Waters from the Production of Urea-Formaldehyde Resins."

Pollak, F., USP 1,950,746, Mar 13, 1934. "Resin-like Condensation Products."

Pollak, F., BP 375,609, June 30, 1932. "Synthetic Resins."

Pollak, F., Kunstharzfabrik, BP 339,601, May 11, 1929. "Urea-Formaldehyde Condensation Products."

Pollak, F., Kunstharzfabrik, Austrian P 111,534, July 15, 1928. "Urea-Aldehyde Condensation Products."

Pollak, F., Kunstharzfabrik, FP 657,794, June 14, 1928. "Artificial Resins."

Pollak, F., Kunstharzfabrik, BP 291,366, May 31, 1927. "Condensation Products of Aldehydes with Urea and Thiourea."

Pollak, F., Kunstharzfabrik, BP 301,798, Dec 5, 1927. "Urea-Formaldehyde Condensation Products."

Pollak, F., Kunstharzfabrik, FP 637,318, May 14, 1927. "Artificial Glass."

Pollak, F., Kunstharzfabrik, GP 484,972, Aug 18, 1922. "Artificial Masses."

Pollak, F., BP 248,729, Mar 3, 1927. "Urea-Aldehyde Condensation Products."

Pollak, F., Austrian P 109,532, Nov 15, 1926. "Shaped Articles from Urea-Aldehyde Condensation Products."

Pollak, F., BP 261,409, Nov 14, 1925. "Urea-Aldehyde Condensation Products."

Pollak, F., and K. Ripper, Chem. Zeit. **48**, 569 (1924). "Über Pollopas, ein neues synthetisches Kolloid."

Pollak, F., BP 238,904, Aug 25, 1924. "Urea-Aldehyde Condensation Products."

Pollak, F., BP 206,512, July 23, 1923. "Condensation Products of Formaldehyde with Urea, Thiourea, etc."

Pollak, F., BP 201,906, July 23, 1923. "Condensation Products from Formaldehyde and Urea, Thiourea, etc."

Pollak, F., BP 213,567, Mar 31, 1923. "Condensation Products from Formaldehyde and Urea."

Pollak, F., BP 193,420, Feb 17, 1923. "Condensation Products of Formaldehyde and Urea, etc."

Pollak, F., BP 181,014, May 30, 1922. "Condensation Products."

Pollak, F., BP 171,096, Nov 1, 1921. "Condensation Products from Formaldehyde, Urea, etc."

Pollak, F., GP 499,792, Oct 29, 1921. "Condensation Products of Urea and Formaldehyde."

Pollopas, Ltd., FP 718,093, June 3, 1931. "Synthetic Resins."

Pollopas, Ltd., FP 680,110, Aug 8, 1929. "Condensation Products."

Pollopas, Ltd., FP 674,999, May 13, 1929. "Condensation Products."

Pollopas, Ltd., BP 301,626, Nov 11, 1927. "Urea-Formaldehyde Condensation Products."

Pollopas, Ltd., E. C. C. Baly, and E. J. Baly, BP 299,487, June 24, 1927. "Dimethylolurea."

Potekhina, E., *et al.*, Derevoobrab. Prom-St. (4), 11 (1975). "Quick-Hardening Urea-Formaldehyde Resin (UKS-72)."

ISBN 0-201-04558-3

Potutkin, G. F., Tr. Arkhangel'sk. Lesotekh. Inst. (32), 91 (1972). "Infrared Spectra of the Product of the Condensation of Dioxane Lignin with the Urea Resin M-60."

Potutkin, G. F., L. V. Shiryaeva, and G. L. Dranishinikov, Tr. Arkhangel'sk. Lesoekh. Inst. (32), 85 (1972). "Reaction of Dioxane Lignin with Urea Resins in 30:1 and 1:1 Ratios."

Pratt, B. C., USP 2,349,756, May 23, 1944. "Formaldehyde Resin Modified with a Polyisocyanate or Polyisothiocyanate."

Preobrazhenskaya, I. P., Tr. Mezhvuz. Konf. Primen. Plastmass Stroit., 3rd 1970, p. 156 (1972). "Strength of Bonding with a Urea Adhesive Modified with a Poly-Vinyl Acetate Emulsion."

Prilukov, A. D., and V. N. Il'inskaya, Deposited Doc. VINITI 3831-76 (1976). "Mix Formulation and Technology for the Preparation of Polymer Binders at the Site of Production of Loess-Urea Cellular Plastics and Study of Methods for Rendering Them Water-Repellent."

Pritchard, G. W., and L. S. M. Fisscher, Austr. P 481,197, Feb 23, 1977. "Improved UF-Condensation Process."

Przybylak, A., J. Bakiewicz, and U. Frackowiak, Pr. Kom. Technol. Drewna, Paznan. Tow. Przyj. Nauk **3**(3), 19 (1971). "Properties of a 70% Odorless Urea-Formaldehyde Resin Used for Cold Gluing."

Pshenitsyna, V. P., N. N. Molotkova, M. P. Noskova, and B. Ya. Aksel'rod, Zh. Prikl. Spektrosk. **29**(3), 486 (1978).

Pungs, W., and K. Eisenmann, USP 1,967,685, July 24, 1934. "Urea-Aldehyde Condensation Products."

Purcell, R. F., J. H. Hunsucker, and D. K. Sausaman, USP 3,761,339, Sept 25, 1973. "Laminating a Structurally Weak Member to a Structurally Strong Member."

Pusch, G., Ger. Offen. 2,639,754, Mar 9, 1978. "Condensation Products of Urea, Formaldehyde, and CH-Acidic Aldehydes."

Pushkina, I. K., Gig. Vop. Proizvod. Primen. Polim. Mater. p. 38 (1969). "Industrial Hygiene in the Production of Woodchip Particleboards."

Putseiko, L. K., M. N. Pershin, and A. P. Platonov, USSR P 361,251, Dec 7, 1972. "Compositon for Strengthening Soils."

Putylyaev, I. E., O. A. Mamykina, N. Z. Mukhibova, and T. V. Rusakova, Mastiki, Polimerbetony Polimersilik., p. 8 (1975). "Modification of Mastics Based on Furfural-Acetone Resins."

Putlyaev, I. E., A. F. Tikhomirova, Zh. M. Suleimanova, M. A. Khor'kova, Tr. Nauchno-Issled. Inst. Betona Zhelezobetona **11**, 163 (1974). "Chemical Resistance of Epoxide-Amine-Aldehyde Composition."

Putlyaev, I. E., *et al.*, Stroit. Mater (4), 20 (1973). "EIS-1 Resin-Based Materials."

Quade, F., USP 1,062,501, May 20.

Radian Corp., Gov. Rep. Announcement Index (U.S.) **78**(22), 130 (1978). "HAOS Aldehydes Monitoring Program."

Raizman, M. Sh., and Ya. Sh. Shkol'nik, USSR P 327,224, Jan 26, 1972. "Adhesive."

ISBN 0-201-04558-3

Raknes, E., Norsk. Skogind. **30**(6), 168 and 175 (1976). "Long Term Durability of Glues for Load-Bearing Timber Constructions: Results after 10 Years of Exposure."

Raknes, E., Nor. Skogind. **25**(11), 325 (1971). "Long-term Durability of Structural Adhesives for Wood."

Rasp, H., Z. Pflanzenernaehr. Bodenk. **133**(1-2), 111 (1972). "Influence of Soil Conditioner on Structure and Yields on Horticultural Soils and Substrates."

Rathi, A. K. A., and M. Chanda, J. Applied Polym. Sci. **18**, 1541 (1974). "Kinetics of Resinification of Furfuryl Alcohol in Aqueous Solution."

Rayner, C. A. A., Ch. 4 in *Adhesion and Adhesives*, Vol. 1, R. Houwink and G. Salomon, eds., Elsevier Publishing Co., Amsterdam, 1965. "Synthetic Organic Adhesives."

Rechner, L., Fr. Demande 2,306,225, Oct 29, 1976. "Urea-Formaldehyde-Phenol Copolymer Foam with Weak Catalyst."

Redfarn, C. A., Brit. Plastics **5**, 238 (1933). "Theoretical Aspects of Urea-Formaldehyde Resins."

Reeves, B. H., USP 1,982,539, Nov 27, 1935. "Insulation of Electric Conductors."

Rehn, T., and S. Petterson, Omgivningshygieniska avdelningen, statens naturvårdsverk 1977-03021, 1977. "Lukttröskelbestämning av Formaldehyd."

Renner, A., USP 3,931,063, Oct 6, 1976. "Porous Solids Consisting of Crosslinked Urea-Formaldehyde Polycondensation Products."

Renner, A., Ger. Offen. 2,521,689, Dec 18, 1975. "Urea-Formaldehyde Condensation Polymers Containing Sulfo Groups."

Renner, A., S. African P 70 05,676, May 6, 1971. "Manufacture of Highly Dispersed Solids Consisting of Crosslinked Urea-Formaldehyde Polycondensation Products."

Reuss, G., E. Luecke, E. Renauer, and A. Kerbeck, Ger. Offen. 2,403,443, July 31, 1975. "Cationic Condensation Products."

Reuther, W., H. Petersen, J. Lenz, J. Mayer, O. Wittmann, and M. Graser, GP 2,110,264, Mar 4, 1971. "Hardening Accelerator for Phenolic Adhesives."

Reuther, W., H. Petersen, F. Becke, and O. Wittmann, GP 2,058,405, Nov 27, 1970. "Crosslinking Catalysts for Phenol-Formaldehyde Resin Adhesives."

Richard, B., and A. Gourdenne, FATIPEC Congr. **13**, 550 (1976). "Physiocochemical Characterization of Urea-Formaldehyde and Thiourea-Formaldehyde Precondensates."

Richard, B., and A. Gourdenne, C. R. Hebd. Seances Acad. Sci., Ser. C **281**(2), 815 (1975). "Nuclear Magnetic Resonance Analysis of a Urea-Formaldehyde Resin."

Riesmeier, W., Ger. Offen. 2,111,651, Sept 21, 1972. "Wood Gluing."

Rikunov, E. G., I. P. Yashina, and A. S. Roman'ko, Mekh. tekhnologiya drevesiny (3), 31 (1974). "Hardening of Urea-Formaldehyde Resins *in Vacuo*."

Ripper, K., USP Reissue 19,463, Feb 12, 1935. "Urea-Aldehyde Condensation Products."

Ripper, K., USP 1,967,261, July 24, 1934. "Condensation Product of Urea and Formaldehyde."

Ripper, K., USP 1,833,868, Nov 24, 1932. "Resinous Condensation Products."

ISBN 0-201-04558-3

Ripper, K., USP 1,762,456, June 10, 1930. "Condensation Products from Urea and Formaldehyde."

Ripper, K., Can. P 284,790, Nov 13, 1928. "Hydrophobe Resins."

Ripper, K., USP 1,625,283, Apr 19, 1928. "Homogeneous Glass-Like Product from Condensation of Formaldehyde and Urea."

Ripper, K., USP 1,687,312, Oct 9, 1928. "Urea-Formaldehyde Condensation Products."

Ripper, K., BP 323,047, Sept 21, 1927. "Plastic Condensation Products."

Ripper, K., BP 287,568, Mar 25, 1927. "Condensation Products of Thiourea, etc., and Aldehydes."

Ripper, K., USP 1,460,606, July 3, 1923. "Condensation Product of Urea with Formaldehyde."

Robitschek, P., and R. L. Christensen, Forest Prod. J. **26**(12), 43 (1976). "Degradation Phenomena in Urea Formaldehyde Resin-Bonded Particle Board."

Roff, W. J., J. Textile Inst. Trans. **50**,T353 (1959). "A Study of the Sorption of Formaldehyde from Aqueous Solution by Cellulose."

Roff, W. J., J. Textile Institute Trans. **49**,T646 (1958). "A Review of the Literature Relating to the Interaction of Cellulose and Formaldehyde."

Roffael, E., Proc. 12. Wash. State Univ. Int. Particleboard Symp. of 5. Apr 1978, p. 233 (1978). "Progress in the Elimination of Formaldehyde Liberation from Particleboards."

Roffael, E., and G. Kossatz, Deutsches Architektenblatt **10**(1), 1101 (1978). "Particleboard Floor Underlayment Causes Formaldehyde Nuisance."

Roffael, E., and A. Schneider, Holz als Roh- und Werkstoff **36**, 393 (1978). "Absorption by Particleboard, Part 1: Influence of Resin Type and Quantity on Water Uptake."

Roffael, E., Adhasion **22**(6), 180 (1978). "Effect of Storage on Formaldehyde Release by Particleboard."

Roffael, E., D. Greubel, and L. Mehlhorn, Holz-Zentralblatt **104**(1), 24 (1978). "Formaldehyde Determination by the Perforator Method and the WKI Method."

Roffael, E., Adhäsion **22**(2), 54 (1978). "Differential Formaldehyde Measurement in Particleboard."

Roffael, E., and L. Mehlhorn, Holz- Kunststoffverarbeitung **12**(1), 770 (1977). "Methods for Measuring Formaldehyde Release."

Roffael, E., Holz- Zentralblatt **103**(27), 400 (1977). "Comparison of Several Methods for Determination of Formaldehyde Release."

Roffael, E., Holz als Roh- und Werkstoff **34**, 385 (1976). "The Effect of the Formaldehyde Content of UF-Resins on the Formaldehyde Release from the Finished Product."

Roffael, E., Adhäsion **20**, 297 (1976). "Über die Reaktivität von wässrigen Rindenextrakten gegenüber Formaldehyd."

Roffael, E., Holzforsch. **30**, 9 (1976). "Über die Bedeutung von Rindeninhaltsstoffen für ihre Verwertung in Spanplatten."

Roffael, E. Holz- Zentralblatt **101**(111), 1403 (1975). "Messung der Formaldehydabgabe in Spanplatten."

ISBN 0-201-04558-3

Roffael, E., W. Rauch, and C. v. Bismarck, Holz als Roh- und Werkstoff **33**(7), 271 (1975). "Formaldehydabgabe und Festigkeitsausbildung bei der Verleimung von Eichenspänen mit Harnstoffformaldehydharzen."

Roffael, E., and W. Rauch, Holz- Zentralblatt **150**, 1461, 1462 (1974). "Sulfitablauge als Spanplattenbindemittel."

Roffael, E., and W. Rauch, Holzforsch. **27**, 214 (1973). "Über die Herstellung von Holzspanplatten auf Basis von Sulfitablauge."

Roffael, E., and W. Rauch, Holzforschung **27**, 178 (1973). "Verwendung von Sulfitablauge in Kombination mit alkalischen Phenolharzen bei der Spanplatten-Verleimung."

Roffael, E., Holz-Zentralblatt **99**, 845 (1973). "Zur Verminderung der Formaldehydabgabe von Spanplatten."

Romanenkov, I. G., and V. A. Ivanov, Mekh. Polim. (5), 936 (1972). "Strength of Honeycomb Plastic in Planar Stress State."

Roos, O., GP 767,897, June 10, 1954. "Verfahren zur Herstellung von Pressplatten insbesondere Sperrholzmittellagen aus zerkleinerten Holzabfällen."

Rosenkranz, H. S., Bull. Environ. Contam. Toxicol. **8**(4), 242 (1972). "Formaldehyde as a Possible Carcinogen."

Rosenthal, D., USP 3,651,182, Mar 21, 1972. "Making Multicellular Thermoset Foamed Plastic Beads."

Rosner, H., Ger. Offen. 2,063,815, July 6, 1972. "Poorly Flammable Coatings."

Rossiter, E. C., USP 2,097,895, Nov 2, 1944. "Resinous Reaction Products Suitable for Molding and Coating Compositions."

Rossiter, E. C., Can. P 274,738, Oct 18, 1927. "Artificial Resin."

Rossiter, E. C., BP 248,477, 1924, BP 258,950, 1925, BP 266,028, 1925. "Beetle."

Rossiter, W. J., Jr., R. G. Mathey, D. B. Burch, and E. T. Pierce, Nat. Bur. of Standards Tech. Note (U.S.) **946**, 83p (1977). "Urea-Formaldehyde Based Foam Insulations: An Assessment of Their Properties and Performance."

Rostenberg, A., Jr., and M. B. Sulzberger, Arch. Dermatol. **35**, 433 (1937). "Some Results of Patch Tests."

Rothera, W. S., S. Blythen, and H. R. Gillespie, GP 567,271, Feb 3, 1925. "Condensation Products from Urea, etc., and Aldehydes."

Rothera, W. S., S. Blythen, and H. R. Gillespie, GP 568,629, Nov 20, 1921. "Urea-Formaldehyde Condensation Products."

Rothera, W. S., S. Blythen, and H. R. Gillespie, GP 563,037, June 2, 1921. "Urea-Formaldehyde Condensation Products."

Rougier, M. F. A., and J. C. J. Pages, Fr. Addn. 95,499, Jan 15, 1971. "Particle or Fiber Panels."

Roy, D. C., *et al.*, Indian Plywood Ind. Res. Inst. J. **7**(1), 5 (1977). "Gelatinized Starch Blended UF (Urea-Formaldehyde) Resin Adhesives for Plywood."

Roy, D. C., and S. Chowdhury, Indian Plywood Ind. Res. Inst. J. **5**(2), 64 (1975). "Identification of Glue in Plywood Glue Line – Part II."

Rudnev, A. V., E. P. Kalyazin, K. S. Kalugin, and G. V. Kovalev, Zh. Fiz. Khim. **51**(10), 2603 (1977). "Formaldehyde. II. Kinetics of the Desolvation of Formaldehyde in Aqueous and Methanol Solutions."

ISBN 0-201-04558-3

Saarbergwerke, A. G., Fr. Demande 2,270,221, May 8, 1974. "Urea-Formaldehyde-Based Fertilizer."

Sakaguchi, H., Mokuzai Kogyo **26**(291), 16 (1971). "Glueability of Shina (Tillia Species) Wood. II. Effect of Water-Extractives on the Glue Bond Quality."

Sakarada, S., Y. Miyazaki, and T. Nakai, J. Kokai 73:51,032 (1973). "Adhesives."

Sakovich, L. I., and V. S. Murav'ev, Sb. Tr. Mosk. Tekhnol. Inst. (21), 113 (1971). "Characteristics in the Gelation of Urea Resins."

Sakuno, T., T. Goto, and R. Katsube, Shimane Daigaku Nogakubu Kenkyu Hokoku (5), 66 (1971). "Wood Gluing. IX. Influence of Surface Aging Prior to Gluing on Wettability and Gluability of Wood."

Sander, B., E. Bonitz, H. Berbner, and M. Hoffmann, Ger. Offen. 2,352,190, Apr 30, 1975. "Water Suspension of Polyolefin Short Fibers."

Sandler, S. R., and W. Karo, *Polymer Synthesis,* Vol. II, Ch. 1, Acad. Press, New York, 1977. "Urea, Melamine, Benzoguanamine-Aldehyde Resins (Amino Resins or Aminoplasts)."

Sano, S., and N. Miura, J. Kokai 76:89,537, Aug 5, 1976. "Phenolic or Aminoplast Adhesive Compositions."

Sanotskii, I. V., *et al.,* Gig. Tr. Prof. Zabol. (1), 25 (1976). "Study on the Effect of Pregnancy on the Sensitivity of Animals to Chemical Agents."

Sarig, Y., R. W. Little, and L. J. Segerlind, J. Appl. Polym. Sci. **22**(2), 419 (1978). "Mechanical Properties of Urea-Formaldehyde Foam."

Sarkisyan, R. R., Tr. Inst. – Nauchno-Issled. Inst. Stroit. Arkhit. (yerevan) **25**, 103 (1977). "Poly(vinyl Acetate) Cement Concretes Modified with Urea-Formaldehyde Resin."

Sarrut, J., and G. Mallet, Fr. Demande 2,203,847, May 17, 1974. "Rapid Hardeners for Aminoplast Resins."

Sasaki, T., J. Kokai 74:125,504, Dec 2, 1974. "Formalin Odor Free Plywoods."

Sato, T., *et al.,* USP 3,962,491, June 8, 1976. "Resin-Coated Sand Particles for Use in Shell Molds."

Sato, T., *et al.,* Fr. Demande 2,161,936, Aug 17, 1973. "Granular Phenol/Formaldehyde Resinous Composition."

Sausman, D. K., USP 3,893,963, July 8, 1975. "Urea-Formaldehyde Composition."

Sautin, A. I., Tr. Nauch. Konf., Nauch.-Issled. Inst. Vod. Transp. (2), 102 (1972). "Hygienic Evaluation of Footwear Made From Synthetic Materials."

Sawicki, E., and C. R. Sawicki, *Aldehydes–Photometric Analysis,* Vol. 5, "Formaldehyde Precursors," Academic Press, New York, 1978.

Sawicki, E., and C. R. Sawicki, *Aldehydes– Photometric Analysis,* Vols. 1 and 2, Academic Press Inc., New York, 1975.

Sax, N. I., *Dangerous Properties of Industrial Materials,* 4th Ed., Van Nostrand Reinhold, New York, 1975.

Scheiber, J., *Chemie und Technologie der künstlichen Harze,* Wissenschaftliche Verlagsges. m.b.H., Stuttgart, 1943.

Scheibler, H., Z. Ang. Chem. **53**, 303 (1940). "Über Harnstoff-Formaldehyd-Harze und deren Mischpolymerisate mit Acrylsäure."

ISBN 0-201-04558-3

Scheibler, H., F. Trostler, and E. Scholz, Angew. Chem. **41**, 1305 (1928). "Über hochmolekulare Kondensationsprodukte des Harnstoffs mit Formaldehyd."

Scheuermann, H., and E. Späth, GP 734,094, Mar 11, 1943. "Insulation."

Scheuermann, H., and J. Lenz, GP 729,029, Nov 12, 1942. "Condensation Products of Urea and (or) Thiourea with Formaldehyde."

Schliefer, K., F. Nassar, and M. Kamel, Textilveredlung **8**(4), 207 (1973). "Thin-Layer Chromatography of Crosslinking Agents and Quantitative Analytical Determination of Free and Covalently Bonded Formaldehyde."

Schmidt-Hellerau, C., and O. Grabowsky, Ger. Offen. 2,354,928, May 15, 1975. "Improving the Cold Adhesion of Wood Glues."

Schmidt-Hellerau, K. C., Ger. Offen. 2,043,440, Mar 9, 1972. "Adhesives for Wood."

Schneider, A., Holz als Roh- und Werkstoff **31**(11), 425 (1973). "Sorption Behavior of Phenolic-Bonded and Urea-Bonded Particle Boards."

Schneider, H., Ger. Offen. 2,422,238, Nov 27, 1975. "Urea-Formaldehyde Fertilizer."

Schneider, W. G., and H. J. Bernstein, Trans. Faraday Soc. **52**, 13 (1956). "Molecular Association and Infrared Spectrum of Solid Formaldehyde and Acetaldehyde."

Schoellhorn, W. D., Ger. Offen. 2,436,465, Feb 12, 1976. "Urea-Formaldehyde Resin Foams."

Schorning, P., and G. Stegmann, Holz als Roh- und Werkstoff **30**, 329 (1972). "New Wood-to-Wood Bonds in Chip Materials by Chemical Polyfunctional Systems. II. Hot Pressing with Wooden Boards and Particles in a Weak Acid Bond System."

Schorning, P., and G. Stegmann, Ver. Foerder. Deut. Brandschutzes Z. **21**(4), 133 (1972). "Test of the Burning Properties of Wood Construction Material by a New Fully Automatic, Electronically Recording Small Fire Test Apparatus (VER Fire Test Apparatus). 2."

Schorr, W. F., E. Keran, and E. Plotka, Arch. Dermatol. **110**(1), 73 (1974). "Formaldehyde Allergy."

Schorr, W. F., Am. Perfum. Cosmt. **85**(3), 39 and 45 (1970). "Allergic Skin Reactions from Cosmetic Preservatives."

Schulz, H., and H. Wallhäusser, Ger. Offen. 2,238,742, Feb 7, 1974. "Degradable Aminoplast Moldings."

Schulze, W., and H. Baumann, Zentralblatt für Chirurgie **85**, 210 (1960). "Wundverbände unter Einstaz von Aracidpuder und Trockenschaumplatten (Methylol-Harnstoffpräparate)."

Schwab, J., F. Boehm, and G. Ruthner, Ger. Offen. 2,219,033, Nov 2, 1972. "Foams Containing Material of Active Surface."

Schwager, P., Ger. Offen. 2,209,109, Sept 7, 1972. "Cardboard for Gun Training Targets."

Schwarz, F. E., Proc. 5th Particleboard Symp., Wash. State Univ., Pullman, Wa., 1971. "Advances in Detecting Formaldehyde Release."

Seberini, M., Czech. P 141,245, May 15, 1971. "Loose Construction Material."

Secrest, P. J., and S. E. Heckman, J. Paint Technol. **43**(562), 81 (1971). "Determination of Nonvolatile Contents of Amino Polymers."

ISBN 0-201-04558-3

Seeholzer, J., H. Michaud, J. Mangs, and W. Poschinger, Ger. Offen. 2,260,186, June 12, 1974. "Thermosetting Plastic Coatings for Wood Fiberboards."

Segond, M., Rev. gen. mat. plastiques **12**, 230 (1936). "A Stable and Transparent Urea Resin."

Seifert, J., Holz als Roh- und Werkstoff **30**(9), 332 (1972). "Sorption and Swelling of Wood and Wood-Base Materials. 3. Volume Contraction between Wood and Water."

Seifert, J., Holz als Roh- und Werkstoff **30**(3), 99 (1972). "Sorption and Swelling of Wood and Wood-Base Materials. I. Influences on the Sorption of Wood Base Materials."

Seldon, M. M., and L. J. Garbini, Ger. Offen. 2,157,573, June 8, 1972. "Preventing the Penetration of Gases into Steel Objects by Coating with Gold and Tantalum."

Senn, S., Swiss P 516,992, Nov 24, 1969. "Making Single or Multilayer Plates by Hot-Pressing Wood Chips Treated with a Thermosetting Binder."

Sergeeva, M. P., Nauch. Tr., Leningrad. Lesotekh. Akad. (151), 48 (1972). "Determination of the Hardening Time of Urea Resins *in Vacuo.*"

Serova, M. N., A. P. Zhukovskii, and S. P. Girs, Deposited Doc. VINITI 1099-77, 8 pp (1977). "Spectroscopic Study of Intermolecular Interactions in a Methaldehyde-Water System."

Sgibnev, A. K., Gig. Tr. Prof. Zabol. **12**(7), 20 (1968). "Effect of Low Formaldehyde Fume Concentrations on Humans."

Shank, H. C., Jr., Ger. Offen. 2,154,620, May 10, 1973. "Polymer Sheathings for Glass Bottles."

Sharphouse, J. H., and D. Mukherjee, Indian P 120,113, June 19, 1971. "Urea-Formaldehyde Resins."

Shastri, S., Ger. Offen. 2,041,566, Feb 24, 1972. "Sodium-Dithionite-Containing Bleaching Agents for Wood Pulp, Stabilized Against Self-Ignition."

Shchedro, D. A., Derevoobrab. Prom-st. (6), 4 (1978). "Use of Low Toxic Carbamide Resin KS-M 0.3P for Production of Particle Boards."

Shearer, H. N., USP 3,832,201, Aug 27, 1974. "Asphalt Paving Compositions."

Shelton, T. S., USP 3,951,595, Apr 20, 1976. "Vapor-Phase Process for Cellulose Fabrics."

Sheridan, J. L., USP 3,677,808, July 18, 1972. "Manufacturing Moisture-Resistant Consolidated Articles by Use of Steam-Atomized Wax."

Shevlyakov, L. V., Uch. Zap. Petrozavodsk. Gos. Univ. **21**(4), 223 (1974). "Mechanism of Allergic Dermatitis Caused by Urea-Formaldehyde Resins."

Shigeta, T., and M. Kitamura, J. Kokai 75:05,421, Jan 21, 1975. "Gypsum-Sulfur Composite Materials."

Shimokai, K., *et al.*, Japan. P 74:27,672, July 19, 1974. "Fiberglass-Reinforced Resin Sheets."

Shiota, T., and T. Sato, J. Kokai 75:119,892, Sept 19, 1975. "Opened-Cell Porous Materials for Microfilters."

Shiota, T., T. Sato, J. Kuroda, and S. Okamura, J. Kokai 75:112,430, Sept 3, 1975. "Urea Resin Adhesives Containing Microorganisms as Filler."

Shiota, T., T. Hino, T. Tanno, and S. Okamura, J. Kokai 74:48,782, May 11, 1974. "Odor-Free Urea-Formaldehyde Resins."

ISBN 0-201-04558-3

Shirai, K., and H. Okamura, J. Kokai 74:109,277, Oct 17, 1974. "Adsorbent for Gaseous Pollutants."

Shirokorodyuk, V. K., *et al.,* USSR P 615,049, July 15, 1978. "Binder for Mineral Wool Articles."

Shishkov, I. T., Nauch. Tr., Vissh Lesotekh. Inst., Sofia, Ser. Mekh. Tekhnol. Durv. **18**, 37 (1971). "Duration of Adhesion of Beech Plywood with a Urea-Formaldehyde Adhesive."

Shofuda, H., K. Okura, and K. Ueda, J. Kokai 74:128,031, Dec 7, 1974. "Adhesives for Woods."

Sholokhova, A. B., A. N. Pukhovitskaya, S. I. Tarasova, and A. S. Freidin, Tr. TSNII Stroit. Konstruktsii (53), 13 (1975). "Study of Properties of Wood Joints Using Modified Urea Adhesives."

Shumilina, A. V., Gig. Tr. Prof. Zabol. (12), 18 (1975). "Menstrual and Child-Bearing Functions of Female Workers Occupationally Exposed to the Effects of Formaldehyde."

Shutov, G. M., A. F. Shut'ko, V. M. Khrulev, and G. P. Khanenya, USSR P 431,008, June 5, 1974.

Shvartsman, G., R. Temkina, and M. Svitkin, Derevoobrab. Prom. **21**(5), 3 (1972). "Reduction of Formaldehyde Evolution from Particle Boards."

Shwartsman, G. M., *et al.,* Derevoobrab. Prom-st. (3), 12 (1975). "Intensification of the Pressing Process of Nontoxic Wood-Chip Tiles on Resin KS-68M."

Siegler, M., H. Lehnert, M. Graser, and R. Goettsche, Ger. Offen. 2,401,554, July 24, 1975. "Improvement of the Cold Adhesive Strength of Wood Cement."

Siegler, M., H. Lehnert, M. Graser, and R. Goettsche, Ger. Offen. 2,351,981, Apr 24, 1975. "Composition for Improving the Cold Adhesiveness of Wood Glue."

Siegmann, A., N. Tirosh, M. Narkis, and Y. Rechav, J. Appl. Polym. Sci. **22**(4), 1015 (1978). "Controlled Release of Propham from Plastic Granules."

Silbernagel, H., and K. Kratzl, Holzforsch. Holzverwert. **5**, 13 (1968). "Zur quantitativen Bestimmung der Formaldehydabgabe von Spanplatten."

"Silur" Tech. & Chem. Produkts, GP 569,342, Apr 6, 1927. "Urea-Formaldehyde Condensation Products."

Simeonov, I., V. Cheshkov, and G. Zakhariev, Fr. Demande 2,329,688, May 27, 1977. "Hardening of Urea-Formaldehyde Resin."

Sklyarenko, B. S., A. F. Natalich, B. P. Zolotarev, and N. A. Osnach, USSR P 456,730, Jan 15, 1975. "Multilayer Reinforced Boards Produced from Wood Wastes."

Slonim, I. Ya., *et al.,* Vysokomol. Soedin., Ser. A **20**(10), 2286 (1978). "Structural Change of Urea-Formaldehyde Resins in Synthesis and Curing."

Slonim, I. Ya., *et al.,* Vysokomol. Soedin., Ser. A **20**(7), 1477 (1978). "Study of Equilibriums in the Urea-Formaldehyde System by Carbon-13 NMR."

Slonim, I. Ya., *et al.,* Vysokomol. Soedin., Ser. A. **20**(6), 1418 (1978). "Proton NMR Structural Study of Urea-Formaldehyde Resins."

Slonim, I. Ya., *et al.,* Polymer Sci. (USSR) **19**, 920 (1977). "Determination of the Structure of Urea-Formaldehyde Resins of Cyclochain Structure by the Carbon-13 NMR Method."

Smidth, L., USP Reissue 18,943, Sept 12, 1933. "Condensation Product of Formaldehyde and Urea."

ISBN 0-201-04558-3

Smidth, L., USP 1,893,911, Jan 10, 1933. "Urea-Formaldehyde Condensation Product."

Smidth, L., USP 1,704,347, Mar 5, 1929. "Condensation Products of Formaldehyde and Urea."

Smidth, L., BP 294,253, July 21, 1927. "Urea-Formaldehyde Condensation Products."

Smirnov, B. A., USSR P 487,860, Oct 15, 1975. "Gypsum-Polymer-Concrete Mixture."

Smirnova, E. A., P. G. Sekachev, and D. M. Flyate, USSR P 509,615, Apr 5, 1976. "Modified Urea-Formaldehyde Resin."

Smith, A. R., British Plastics, Aug, p. 402 (1962). "Ten Years' Progress in Amino Resins."

Smith, J. J., and B. Meyer, J. Chem. Phys. **50**, 456 (1969). "Fluorescence and Induced Phosphorescence of Formaldehyde in Solid Low-Temperature Solutions."

Smith, R. G., *et al.,* Health Lab. Sci. **9**(1), 75 (1972). "Tentative Method of Analysis for Low-Molecular-Weight Aliphatic Aldehydes in the Atmosphere."

Smith, T. F., and A. Timmons, Ger. Offen. 2,631,038, Jan 27, 1977. "Fuel Composition."

Snedecor, G. C., and W. G. Cochran, *Statistical Methods,* Iowa State University Press, Columbus, Ohio, 1967.

Snuparek, J., and J. Cerny, *Aminoplasts,* Prague, 1963.

Snyder, A. D., F. N. Hodgson, M. A. Kemmer, and J. R. McKendree, MRC-DA-567, EPA/600/2-76/201, Monsanto Research Corp., Dayton, Ohio, 1976. "Utility of Solid Sorbents for Sampling Organic Emissions."

Soc. Nobel francaise, FP 804,714, Oct 31, 1936. "Solvents for Synthetic Resins."

Soc. Nobel francaise, FP 736,744, Sept 8, 1931. "Synthetic Resins."

Sokolova, V. A., A. A. Sorokina, D. P. Tsar'kov, and V. V. Lyubimov, USSR P 580,939, Nov 25, 1977. "Mixture for Casting Molds and Cores."

Solvay et Cie, Belg. P 848,093, Mar 1, 1977. "Synthetic Paste for Manufacture of an Electrical Insulating Paper."

Soto Urbina, F. G., and H. C. v Bismarck, Forschungsber. Landes Nordrhein-Westfalen **2429**, 44 pp (1974). "Burned Condition of Unprotected Particle Boards Depending on Board Type, Leading to an Optimal use of Fireproofing Agents."

Sparkes, A. J., Adhäsion (8), 282 (1972). "Synthetic Adhesives for Furniture Manufacture."

Speiser, P. M., H. P. Merkle, and L. Schibler, Ger. Offen. 2,233,428, Jan 25, 1973. "Spray Condensation Process for Encapsulating Finely Divided Substances in a Liquid Medium."

Spence, R., and R. Wild, J. Chem. Soc. **1935**, 506 (1935). "Thermodynamic Data for the System Formaldehyde-Polyoxymethylene."

Speyer, F. B., USA P 4,022,727, May 10, 1977. "Thermoplastic Resinoid Hot Melt Compositions."

Spirin, Yu. L., V. N. Vel'sovskii, and V. V. Slesarev, USSR P 321,505, Nov 19, 1971. "Binder for Producing Articles from Silicate Fibers."

ISBN 0-201-04558-3

Splawa-Neyman, S., Pr. Inst. Technol. Drewna **18**(3), 69 (1971). "Resistance of Lignite Particleboard to Biological Corrosion."

Stahly, E. E., R. D. Johnson, and R. G. Rice, J. Elastoplast. **4**(Jan), 60 (1972). "Phosphonitrilic Additives in Preparation of Fire-Resistant Resins."

Stakavich, A. J., Am. Soc. Heat. Refrig. Air-Cond. Eng. Symp.: "Odors and Odorants: The Engineer's View," Chic., Ill., Jan 27, 1969, ASHRAE, New York, 1969. "The Capacity of Activated Charcoal under Dynamic Conditions for Selected Atmospheric Contaminants in the Low Parts-per-Million Range."

Stamm, A. J., *Wood and Cellulose Science,* Ronald Press Co., New York, 1964.

Stamm, A. J., Tappi, **42**(1), 39 (1959). "Dimensional Stabilization of Wood by Thermal Reactions and Formaldehyde Cross-Linking."

Stankevich, K. I., V. A. Tsendrovskaya, E. F. Malygina, and I. S. Reisig, Gig. Sanit. **50**(9), 101 (1972). "Comparative Hygienic Assessment of Glass-Reinforced Plastics Used in Building."

Starzynska, K., Pol. P 74,999, Oct 10, 1975. "Modified Urea Resins."

Starzynska, K., Polimery (Warsaw) **19**(7), 314 (1974). "Urea-Formaldehyde Resins Modified with Furfuryl Alcohol as Foundry Binders."

Starzynska, K., Przem. Chem. **52**(3), 192 (1973). "Urea Resins of Low Content of Free Formaldehyde."

Starzynska, K., H. Stasiak, and J. Dworakowski, Pol. P 68,454, Aug 6, 1973. "Urea-Formaldehyde Resins with Small Content of Free Formaldehyde."

Stashevski, A. M., and H. J. Deppe, Holz als Roh- und Werkstoff **31**(11), 417 (1973). "Application of Tannin Resins as Adhesives for Wood Particle Board."

Staudinger, H., Ber. **59**, 3022 (1926). "Die Chemie."

Steele, R., Textile Res. J. **25**, 545 (1955). "The Reaction of Formaldehyde with Cellulose."

Steiner, P. R., and S. Chow, Wood Fiber **6**(1), 57 (1975). "Comparison of Modifiers for Durability Improvement of Urea-Formaldehyde Resin."

Steiner, P. R., and S. Chow, For. Prod. J. **25**(8), 26 (1975). "Low-Temperature Durability of Common Wood Adhesives."

Steiner, P. R., and S. Chow, U.S. NTIS, PB Rep. PB-249983, from NTIS, from Gov. Rep. Announce. Index (U.S.) **76**(10), 133 (1976). "Factors Influencing Western Hemlock Bark Extracts Use as Adhesives."

Steiner, P. R., Wood Science **7**, 99 (1974). "Study of Urea-Formaldehyde Resin Cure by Bromination and X-Ray Spectrometry."

Steiner, P. R., and S. Chow, Wood & Fiber **6**, 57 (1974). "Comparison of Modifiers for Durability Improvement of Urea-Formaldehyde Resin."

Steiner, P. R., Forest Prod. J. **32**(12), 32 (1973). "Durability of Urea-Formaldehyde Adhesives. Effects of Molar Ratio, Second Urea, and Filler."

Stene, R. M., Ger. Offen. 2,350,820, Apr 18, 1974. "Coating Plastics with a Thin Metal Layer."

Stenzel, J., and I. Gros, Rom. P 60,253, Mar 30, 1976. "Urea-Formaldehyde-Furfuryl Resins."

Steppes, F. E. K., USP 1,658,359, Feb 7, 1928. "Condensation Product of Urea and Formaldehyde."

Stockel, R. F., S. J. O'Brien, and R. G. Weyker, BP 1,325,806, Aug 8, 1973. "Flame-Retardant Finish for Cellulosic Textiles."

ISBN 0-201-04558-3

Stofft, E., I. Nitsche, and A. Mayet, Zentr. Bakteriol. Parasitenk. Abt. I, Orig. **155**, 131 (1971). "Formaldehyde in Dissecting Rooms."

Stofko, J., and E. Zavarin, S. African P 75 00,495, Nov 25, 1975. "Bonding of Solid Lignocellulosic Materials."

Stout, R. M. T., and D. E. Rogerson, USP 3,965,056, June 22, 1976. "Adhesive Extended Composition."

Strelkov, V. P., V. Yu. Miretskii, and V. V. Fefilov, Izv. Vyssh. Uchebn. Zaved., Lesn. Zh. **19**(2), 82 (1976). "Use of Urea-Formaldehyde Resin in Fiber Board Production."

Stubblefield, D. J., J. P. R. Falconer, and T. B. Moore, SPI (Soc. Plast. Ind.) Int. Cell. Plast. Conf., 3rd, p. 320 (1972). "Innovative Urethane Foam Composites for Housing."

Suematsu, A., and H. Yamada, J. Kokai 75:24,404, Mar 15, 1975. "Odorless Plywood."

Suen, T. J., in *Polymer Processes,* C. E. Schildknecht, ed., Interscience, N.Y., 1956. "Condensation with Formaldehyde."

Suetaka, W., Junkatsu **21**(4), 272 (1976). "Infrared Spectroscopy for Investigating Thin Films on Metal Surfaces."

Sukovatov, V. I., T. Galitsina, A. I. Isaev, and V. V. Kharchenko, Tr. Nauchno-Issled. Inst. Stroit. Ugol. Gornorud. Predpr. No. 5, 311 (1968). "Evolution of Phenol and Formaldehyde from Articles Made of Molded Wood Chips."

Sundie, R. D., and W. L. Pennie, USP 3,979,492, Sept 7, 1976. "Fiberboard."

Sundie, R. D., and W. L. Pennie, Ger. Offen. 2,261,800, July 5, 1973. "Hardenable Amino Resins."

Sundin, B., Proc. 12th Particleboard Symp. Wash. State Univ., Pullman, Wash., Apr 5, 1978. "Formaldehyde Emission from Particleboard and Other Building Materials; A Study from the Scandinavian Countries."

Sunshine, N. B., Flame Retardancy Polym. Mater **2**, 201 (1973). "Flame Retardancy of Phenolic Resins and Urea- and Melamine-Formaldehyde Resins."

Supkis, S. J., USP 4,038,046, July 26, 1977. "Coated Abrasive Bonded with Urea-Formaldehyde-Phenolic Resin Blends."

Susicky, K., M. Simek, and O. Krumphanzl, Czech. P 157,191, Feb 15, 1975. "Laminated Fiber Boards."

Suwalska, W., and W. Russek, Pr. Inst. Wlok., Lodz **22**, 217 (1972). "Chromatographic Analysis of Reactive Resins on Fabrics."

Suzuki, J., M. Odagiri, K. Suzuki, and K. Inoue, J. Kokai 74:30,504, Mar 19, 1974. "Wood Adhesives."

Suzuki, S., Japan. P 73:44,341, Sept 2, 1970. "Cellular Urea Resin."

Suzuki, T., K. Awano, and M. Endo, J. Kokai 73:14,747, Feb 24, 1973. "Urea Resin Moldings with Low Soluble Formaldehyde Content."

Suzuki, Y., and S. Sugano, J. Kokai 77:36,899, Mar 22, 1977. "Powdery Fire-Extinguishing Agent."

Svrdiik, R., Medzinar Konf. Org. Povlakoch [Pr], 12th, p. 93 (1973). "Qualitative and Quantitative Analysis of Urea-Formaldehyde and Melamine-Formaldehyde Resins in Combination with Alkyl Resins Using IR Spectrophotometry."

Swire, J. R., BP 1,401,960, Aug 6, 1975. "Manufacture of Fire-Resistant Pulp Moldings."

ISBN 0-201-04558-3

Szemiotowicz, J., Pol. P 78,613, Aug 30, 1975. "Urea-Mealmine-Formaldehyde Resin."

Szlezyngier, W., and A. Zmihorska, Polimery **18**(12), 639 (1973). "Lowering the Free Formaldehyde Content in Urea-Formaldehyde Adhesive Resins."

Szlezyngier, W., S. Wilaszek, A. Maczuzak, and E. Szewczyk, Pol. P 74,060, July 31, 1975. "Urea-Formaldehyde Resins for Adhesives."

Tajima, S., and K. Kobayashi, Japan. P 71:28,783, Aug 20, 1971. "Cyclic Urea-Formaldehyde Primary Condensation Product."

Takahashi, H., and A. Nagao, Japan. P 75:13,305, May 19, 1975. "Urea Resin Adhesive Composition."

Takahashi, I., *et al.*, Japan. P 73:16,066, May 19, 1973. "Polymethylenepolyalkylidene Polyurea."

Takahashi, M., Japan. P 77:20,516, June 3, 1977, "Odorless Aqueous Resin Solutions with Improved Storage Stability."

Takahashi, M., and S. Oshika, J. Kokai 73:55,223, Aug 3, 1973. "Plywood Adhesive."

Takahashi, M., Japan. P 73:17,877, June 1, 1973. "Urea-Formaldehyde Resin Solution."

Takahashi, T., M. Kitazawa, and T. Anazawa, Rinsan Shikenjo Geppo **283**, 9 (1975). "Formaldehyde Liberation from Particle Boards."

Takahashi, T., M. Kitazawa, and T. Anazawa, Rinsan Shikenjo Geppo (8), 9 (1975). "Effects of Manufacturing Conditions on Formaldehyde Liberation from Particleboards."

Takashima, H., T. Hishita, and T. Miyazaki, Japan. P 71:09,210, Mar 9, 1971. "Granular Fertilizers."

Takashima, S., Plast. Age **17**(11), 97 (1971). "Properties and Applications of Plastics Foams. 11. Phenolic Foams."

Takeshita, S., Japan. P 73:19,865, June 16, 1973. "Polyamide Fiber-Reinforced Plastic."

Takeyama, K., T. Morimoto, and F. Konishi, Natl. Tech. Rep. (Matushita Electr. Ind. Co. Osaka) **20**(5), 562 (1974). "Composition of Gaseous Combustion Products of Plastics and Wood."

Takhirov, M. T., Gig. Sanit. (5), 100 (1974). "Combined Action of Six Air Pollutants on the Human Body."

Tamura, Y., J. Kokai 74:130,935, Dec 16, 1974. "Adhesives for Plywood Applied Through Curtain-Coating Apparatus."

Tanaka, A., N. Kawahara, K. Ishii, and K. Takashima, J. Kokai 73:58,032, Aug 15, 1973. "Adhesives for Plywood."

Tanaka, M., Y. Uemura, and Y. Nonaka, Japan P 71:28,784, Aug 20, 1971. "Treatment of Free Formaldehyde with Ethylenediamine in the Preparation of Urea-Formaldehyde Resin."

Tange, Y., and H. Matsuda, J. Kokai 74:261,072, Sept 3, 1974. "Urea-Formaldehyde Resin Adhesives."

Tange, Y., and H. Matsuda, J. Kokai 74:92,160, Sept 3, 1974. "Urea-Formaldehyde Resin Adhesives."

ISBN 0-201-04558-3

Tanioka, H., J. Kokai 74:110,585, Oct 21, 1974. "One-Liquid Waterproof Lining Agents for Roofs."

Tanner, D. J., S. African P 72:00,485, Aug 24, 1972. "Firelighter Compositions."

Tapia, O., A. Nogales, and P. Campano, Chem. Phys. Letters **24**, 401 (1974). "Hydrogen Bond: Second Order Effects on Potentials Calculated by CNDO/2 Method."

Tarasenko, L. D., *et al.*, USSR P 471,348, May 25, 1975. "Raw Material Mixture for the Production of Heat-Insulating Materials."

Tashiro, F., T. Tanno, and T. Sando, J. Kokai 74:101,418, Sept 25, 1974. "Phenolic Resin-Neoprene Rubber Blend Adhesives Free of Formaldehyde Odor."

Taylor, J., and A. V. Keller, BP 331,428, Aug 29, 1929. "Adhesive or Size."

Taylor, L. J., USP 4,018,727, Apr 19, 1977. "Chemically Convertible Tape with *in situ* Synthesized Polymer."

Taylor, L. J., USP 3,901,752, Aug 26, 1975. "Laminating Process Using Mixtures of Pyrolyzable and Polymerizable Binders."

Technical Assn. of the Pulp and Paper Industry, TAPPI Official Standard T600os-76. TAPPI, Atlanta, Ga., 1976. "Analysis of Formaldehyde in Aqueous Solutions and of Free Formaldehyde in Resins."

Temkina, R. Z., G. G. Yudina, and M. M. Svitkina, Novoe v teckhn. i tekhnol. proiz-va fanery, drevesnostruzhech. plit i drevesnosloist. plastikov, p. 64 (1974). "Effect of Impurities Contained in Urea on Properties of Adhesive Urea Resins."

Temkina, R. Z., M. M. Svitkina, and G. G. Yudina, Nov. Proizvod. Fanery, Drevesnostruzh. Plit Drevesnosloistykh Plast. p. 74 (1973). "Urea-Formaldehyde Resin KS 68M."

Temkina, R. Z., *et al.*, USSR P 358,195, Nov 3, 1972. "Chip Boards."

Tendrovskaya, V. A., Gig. Sanit. (12), 77 (1973). "Use of the Mathematical Planning of an Experiment During Sanitary-Chemical Studies."

Teodorescu, L., E. Petre, and R. Horea, Rom. P 57,015, Mar 25, 1975. "Urea-Formaldehyde Resin with High Water Resistance and Low Formaldehyde Release During Setting."

Terporten, H., and W. Schaefer, Ger. Offen. 2,401,440, July 24, 1975. "Adhesive Coating Film."

Teukros, Co., Ger. Offen. 2,745,951, Apr 13, 1978. "Reactive Catalyst for the Polycondensation of Urea-Formaldehyde Resins."

Thinius, K., Plaste Kautsch. **24**(7), 479 (1977). "Studies on the History of Plastics. XIV. 50th Anniversary of Amino Plastics. Reaction between Formaldehyde and Urea or Analogs in their Development for Use as Organic Materials up to 1950."

Thomas, S., Arch. Acker-Pflanzenbau Bodenk. **17**(7-8), 623 (1973). "Combined Effects of the Application of Urea-Formaldehyde Resin Foam, Mineral Fertilization, and Sprinkler Irrigation on the Growth of Grasses in Sandy Soil."

Throne, J. L., and R. G. Griskey, Mod. Plast. **49**(11), 96, 98 (1972). "Heating Values and Thermochemical Properties of Plastics."

Tiedeman, G. T., Ger. Offen. 2,225,478, Jan 11, 1973. "Acyl Hydrazide-Modified Polycondensate Resin Materials as Adhesives."

Time, N. S., A. A. El'bert, N. Ya. Solechnik, and L. Yu. Sokolova, USSR P 368,067, Jan 26, 1973. "Particle Boards."

Tobol'skii, G. F., Fr. Demande 2,243,917, Apr 11, 1975. "Porous Structures."

ISBN 0-201-04558-3

Tobol'skii, G. F., and L. M. Vasilenko, Svoistva Pererab. Shlakov Stroit. Mater. Izdeliya, p. 218 (1971). "Slag Cotton Acoustic Tiles."

Tokuoka, M., J. Agr. Chem. Soc. Japan **10**, 1333 (1934). "Synthesis of Urea from Carbon Dioxide and Ammonia. I."

Toledo Scale Manufacturing Co., FP 674,589, May 4, 1929. "Synthetic Resins."

Toledo Synthetic Products, Inc., FP 715,153, Apr 13, 1931. "Plastic Compositions."

Toledo Synthetic Products, Inc., FP 711,395, Feb 17, 1931. "Synthetic Resins."

Toledo Synthetic Products, Inc., FP 711,333, Feb 16, 1931. "Synthetic Resins."

Tollens, B., Ber. **29**, 2751 (1896). "Über den Methylen-Harnstoff."

Tollens, B., Ber. **17**, 653 (1885). "Über einige Derivate des Formaldehyds."

Tomita, B., and S. Hatono, J. Polym. Sci. **16**, 2509 (1978). "Urea-Formaldehyde Resins. III. Constitutional Characterization by ^{13}C Fourier-Transform NMR Spectroscopy."

Tomita, B., J. Polym. Sci. **15**, 2347 (1977). "Melamine-Formaldehyde Resins: Molecular Species Distributions of Methylolmelamines and Some Kinetics of Methylolation."

Tomita, B., and Y. Hirose, Mokuzai Gakkaishi **22**(1), 59 (1976). "Urea-Formaldehyde Resins. II. Formation of Methylol-Type Hemiformals in Reaction of Formaldehyde with Urea in Deuterium Oxide."

Tomita, B., and Y. Hirose, J. Poly. Sci. **14**, 387 (1976). "Urea-Formaldehyde Resins: NMR Study on Base-Catalyzed Reaction of Formaldehyde with Urea in Deuterium Oxide."

Touati, R., Fr. Demande 1,242,344, Mar 28, 1975. "Synthetic Stone."

Toyota, H., and Y. Hosota, Japan. P 71:18,570, May 24, 1971. "Slowly Active Nitrogen Fertilizers."

Traganos, F., *et al.*, J. Histochem. Cytochem. **23**(6), 431 (1975). "Denaturation of Deoxyribonucleic Acid *in situ* Effect of Formaldehyde."

Trautvetter, R., and J. Froehlich. E. Ger. P 79,962, Feb 12, 1971. "Light Construction Materials."

Trinkler, H., Med. Lab. **21**(12), 283 (1968). "Working with Formaldehyde."

Trocino, F. S., USP 3,616,201, Oct 26, 1971. "Douglas Fir Bark Extender for Thermosetting Resins."

Troughton, G. E., Wood Sci. **1**(3), 172 (1969). "Accelerated Aging of Glue-Wood Bonds."

Troughton, G. E., and S. Chow, Holzforsch. **29**, 216 (1975). "Effect of Fortifier Addition on the Curing Reactions of Urea-Formaldehyde Adhesives."

Troughton, G. E., Forest Prod. Lab., Information Rep. VP-X-26, Vancouver, B.C., 1967. "Kinetic Evidence for Covalent Bonding Between Wood and Formaldehyde Glues."

Tsai, C.-M., Kuo Li Tai-wan Ta Hsueh Nung Hsueh Yuan Shih Yen Lin, Yen Chiu Pao Kao **116**, 343 (1975). "Effect of Extractive Removal on the Adhesiveness of some Southeastern Asian Hardwoods."

Tsubota, Y., and K. Seki, J. Kokai 74:130,977, Dec 16, 1974. "Thermal Insulator Boards from Bark Chips and Cellular Polystyrene."

ISBN 0-201-04558-3

Tsuge, M., T. Miyabayashi, and S. Tanaka, Bunseki Kagaku **23**(10), 1146 (1974). "Gel Chromatography of Amino Resins. II. Gel Chromatography of Urea-Formaldehyde Resins."

Tsukamoto, A., J. Kokai 73:18,338, Mar 8, 1973. "Impact-Resistant Amino Resins."

Tsuruta, K., and Y. Nishida, Japan. P 75:06,526, Mar 14, 1975. "Plywood with Foamed Glue."

Tsutsumoto, T., *et al.,* Ringyo Shikenjo Kenkyu Hokoku, **269**, 95 pp. (1974). "Properties of Tropical Woods. 20. Utilization of the Nine Species from New Guinea and Other Areas."

Tsvetkov, V. E., N. I. Moskvitin, and V. I. Azarov, USSR P 411,108, Jan 15, 1974. "Binder."

Tsvetkov, V. E., *et al.,* Plast. Massy (9), 7 (1972). "Modification of Urea Resins."

Tsyvin, M. M., G. V. Balakhonkina, I. N. Gur'evskaya, and A. A. Kozhevnikova, Nauch. Tr. Tsent. Nauch.-Issled. Inst. Mekh. Obrab. Drev., No. 27, 46 (1972). "Compatibility of Some Types of Wood with Phenolic and Urea Adhesives."

Tunik, S. P., O. A. Lavrova, and T. M. Lesteva, Deposited Doc., VINITI 2171-77, 14 pp (1977). "Liquid-Vapor Phase Equilibriums in Formaldehyde-Alcohols-Water Systems. II. Procedure for Processing Data on Liquid-Vapor Equilibrium in a Methanol-Formaldehyde-Water System Complicated by a Chemical Reaction."

Tunik, S. P., T. M. Lesteva, and V. I. Chernaya, Zh. Fiz. Khim. **51**, 1268 (1977). "Phase Equilibria in the Water-Alcohols-Formaldehyde Systems. I. The Liquid Vapor Equilibria in the Systems Water-C_6 and C_7 Alcohols."

Tymochowicz, S., Nukleonika **23**(8), 821 (1978). "Selectivity of Ion-Exchange Reactions on Stabilized Bentonite. II."

Uchida, T., *et al.,* J. Kokai 78:71,677, June 26, 1978. "Treatment of Wastewaters Containing Formaldehyde and Poly(Vinyl Alcohol)."

Uchida, T., A. Kitamura, T. Miyagi, and Y. Yano, Japan. P 78:34,184, Sept 19, 1978, "Treatment of Wastewaters Containing Phenol and Formaldehyde."

Uchimi, S., Japan. P 71:39, 479, Nov 20, 1971. "Particle Boards."

Ueda, K., and K. Fujita, J. Kokai 75:16,753, Feb 21, 1975. "Particle Boards."

Ueda, K., J. Kokai 74:73,450, July 16, 1974. "Amino Resin Molding Compositions."

Ueda, K., Y. Kitsuda, and A. Tsukamoto, J. Kokai 74:73,447, July 16, 1974. "Urea Resin Molding Compositions."

Ueda, K., J. Kokai 73:58,048, Aug 15, 1973. "Impact-Resistant Amino Resin Molding Materials."

Ueda, M., and T. Saeki, J. Kokai 76:28,148, Mar 9, 1976. "Fireproofed Particle Boards."

Ueda, M., and T. Horike, J. Kokai 75:25,716, Mar 18, 1975. "Fire-Resistant Plywood."

Ulbricht, J., Ger. Offen. 2,639,470, Mar 9, 1978. "Insulating Molding."

ISBN 0-201-04558-3

Updegraff, I. H., and T. J. Suen, in *Polymerization Processes,* C. E. Schildknecht, ed., Wiley-Interscience, New York, 1975. "Condensation with Formaldehyde."

U.S. Dept. of Health, Education and Welfare, National Institute for Occupational Safety and Health, Public Health Service Center for Disease Control, Rockville, Md., Report, 1975. "Working With Formaldehyde."

U.S. Dept. of Health, Education and Welfare, National Institute for Occupational Safety and Health, Fed. Regist. **40** (121, June 23), 26445 (1975). "Part II. Toxic Substances List."

U.S. Dept. of Labor, Occupational Safety and Health Administration, Fed. Regist. **36** (105, May 29), 10503 (1971). "Subpart G–Occupational Health and Environmental Control. 1910.93 Air Contaminants (Gases, Vapors, Fumes, Dust and Mists)."

Usmanov, S., V. A. Budkov, B. M. Beglov, and O. D. Rudnik, Deposited Publ. VINITI 6434-73, 16 pp (1973). "Effect of Additions of Urea-Formaldehyde Fertilizers on the Properties of Urea."

Ustinovakaya, I. A., *et al.,* Izv. Sib. Otd. Akad. Nauk SSR, Ser. Khim. Nauk (6), 131 (1971). "Analysis of Some Oxygen-Containing Substances by Gas Chromatography on Polysorb with a Liquid Phase."

Utsumi, S., J. Kokai 75:21,072, Mar 6, 1975. "Water-Resistant Particle Boards."

Vad, J., *et al.,* Ger. Offen. 2,361,085, June 27, 1974. "Soil Conditioner."

Vale, C. P., and W. G. K. Taylor, *Aminoplastics,* London Iliffe Books Ltd., London, 1964.

Valgin, V. D., and V. A. Novak, Plast. Massy (10), 44 (1974). "Phenol-Formaldehyde Cellular Plastics of Vilares Type."

Valgin, V. D., and N. I. Borodkina, Plast. Massy (10), 46 (1974). "Urea-Formaldehyde Cellular Plastic MFP-2."

Valgin, V. D., V. A. Novak, and Yu. S. Murashov, USSR P 328,152, Feb 2, 1972. "Foam Plastics."

Van Atta, F. A., Chem. Technol. **6**(1), 28 (1976). "Developing Occupational Health Standards."

Vargiu, S., G. Mazzoleni and U. Nistri, Ger. Offen. 2,263,125, July 19, 1973. "Formaldehyde-Urea Resins."

Vargiu, S., G. Mazzoleni, U. Nistri., Ger. Offen. 2,262,197, July 19, 1973. "Formaldehyde-Urea Resins."

Varlacheva, L. A., Nauch. Tr., Tsent. Nauch.-Issled. Inst. Mekh. Obrab. Drev., No. 27, 8 (1972). "Atmospheric and High-Speed Testing of Coatings on Wood."

Vassiliades, A. E., USP 3,993,831, Nov 23, 1976. "Microcapsules and Transfer Sheet Record Material Coated Therewith."

Vassiliades, A. E., USP 3,886,084, May 27, 1975. "Microencapsulation System."

Verbestel, J. B., USP 3,699,202, Oct 17, 1972. "Manufacture of Pressed Boards."

Vertruete, W., M. Claes, R. De Backere, and J. P. Voets, Pflanzenern. Bodenkunde **135**, 258 (1974). "Effect of Crotodur, Isodur and Nitroform on Some Soil Microbiological Populations."

Vinogradov, G. I., I. A. Chernichenko, and E. M. Makarenko, Gig. Sanit. (8), 10 (1974). "Allergenic Activity of Motor Traffic Exhaust Gas."

ISBN 0-201-04558-3

Vinogradov, V. M., Termoplasty Konstr. Naznacheniya, p. 83 (1975). "Residual Stresses in Articles Made From Thermplastics."

Virpsha, Z., and Ya. Bzhezinskii, *Aminoplasts,* Khimiya Publishers, Moscow, 1973.

Vintila, E., and I. Alexandru, Rom. P 52,770, Mar 24, 1971. "Fungicidal, Insecticidal, and Fire-Proofing Composition for Wood."

Volkova, Z. A., and E. A. Sidorova, Gig. Tr. Prof. Zabol. **95**(5), 44 (1971). "Formaldehyde Content in the Blood of Persons Working in Contact with Urea-Formaldehyde Resins."

Vologodskii, A., and M. D. Frank-Kamenetskii, J. Ther. Biol. **55**(1), 153 (1975). "Theoretical Study of DNA Unwinding Under the Action of Formaldehyde."

Volotskoi, D. V., Izv. Vyssh. Ucheb. Zaved., Stroit. Arkhitekt. **17**(4), 147 (1974). "Stabilization of Roadbeds by Chemical Processes."

Waddleton, N., BP 1,428,334, Mar 17, 1976. "A Process for Making Insulating Materials of Low Combustibility on the Basis of Foam-Like Urea Formaldehyde Resins."

Wagner, R. E., and E. Pacsu, Textile Res. J. **22**, 12 (1952). "Cellulose Studies. Part XVI. The Reaction of Cellulose with Formaldehyde."

Walker, J. F., in *Kirk-Othmer Encyclopedia of Chemical Technology,* 2nd Ed., V. 10, p. 77, Interscience Publ., John Wiley and Sons, New York, 1966. "Formaldehyde."

Walker, J. F., 3rd Ed., Am. Chem. Soc. Monogr. Ser. **159** (1964). "Formaldehyde."

Walker, J. F., USP 2,321,958, June 15, 1958. "CH_2O Addition Products Suitable for Use in Aqueous Solutions."

Walter, G., and H. Lutwak, Koll. Beihefte **40**, 158 (1934). "Synthetic Resins, Part 9: Phys-Chem Investigations of the UF-Condensation."

Walter, G., and M. Gewing, Koll. Beih. **34**, 163 (1932). "Synthetic Resins, Part 2: Theoretical Considerations, and Quantitative Study of UF-Condensation."

Walter, G., and K. Oesterreich, Koll. Beih. **34**, 115 (1932). "Structure of Synthetic Resins, Part 1: Condensation Products from Thiourea, Formaldehyde and Copper Chloride."

Walter, G., USP 1,863,426, June 14, 1932. "Transparent Products from Methylol Derivatives of Urea."

Walter, G., BP 375,843, June 27, 1932. "Artificial Resins."

Walter, G., Kolloid Zeitschr. **58**, 229 (1931). "Das Harnstoff-Formaldehyd-Kolloid. Ein Beitrag zur Konstitution der Künstlichen Harze."

Walter, G., and M. Gewing, Kolloid Beih. **34**, 163 (1931). "The Constitution of Artificial Resins. II. Theoretical Considerations and Quantitative Investigations of the Urea-Formaldehyde Condensation."

Walter, G., and K. Oesterreich, Kolloid Beih. **34**, 115 (1931). "Constitution of Artificial Resins. I. Condensation Products from Thiourea, Formaldehyde and Cuprous Chloride."

Walter, G., Austrian P 123,847, Mar 15, 1931. "Synthetic Resins."

Walter, G., Austrian P 130,030, July 15, 1930. "Methylol Derivatives of Urea, etc."

Walter, G., Austrian P 121,999, Nov 15, 1930. "Synthetic Resins."

ISBN 0-201-04558-3

Walter, G., BP 291,712, Nov 29, 1926. "Methylol Compounds of Organic Acid Amides."

Walter, G., BP 284,272, Nov 28, 1925. "Methylol Ureas."

Walter, G., BP 262,148, Nov 28, 1925. "Condensation Products of Methylolureas, etc."

Warneck, P., W. Klippel, and G. K. Moortgat, Ber. Bunsenges. Phys. Chem. **82**, 1136 (1978). "Formaldehyd in troposphärischer Reinluft."

Wayne, L. G., R. J. Bryan, and K. Ziedman, U.S. Dept. HEW, Public Health Service, Center for Disease Control, NIOSH, Div. Biomed. Behv. Sci., Final Rep. on Contr. No. CDC 99-74-21, Cincinnati, 1976. "Irritant Effects of Industrial Chemicals: Formaldehyde."

Webb, R. J., BP 1,380,442, Jan 15, 1975. "Shaped Heat-Insulating Refractory Compositions."

Weeraratne, W. G., *et al.*, J. Inst. Rubber Ind. **8**(2), 74 (1974). "Use of Natural Rubber Latex-Resin Blends as an Adhesive for Plywood."

Wegler, R., in *Methoden der Organischen Chemie* (Houben-Weyl), Vol. 14, Pt. 2, E. Müller, Ed., Georg Thieme Verlag, Stuttgart, 1963. "Polyadditions- und Polykondensationsprodukte von Carbonyl- und Thiocarbonyl-verbindungen mit Harnstoffen, Melaminen, Urethanen, Carbonsäureamiden, Dicyandiamid, Guanidin, Sulfurylamid, Sulfonsäureamiden, Nitrilen, Ammoniak, aliphatischen Aminen und Phosphin."

Weill und Reineke Co., Ger. Offen. 2,162,223, June 28, 1973. "Adhering a Sealing Sheet to a Supporting Surface."

Weizman, J., Polim. Vehomarim Plast. **4**(1), 12 (1974). "Foamed Plastics in Construction."

Wellington, Ch., and B. Tollens, Ber. **18**, 3298 (1885). "Uber einige Derivate des Formaldehydes."

Wellons, J. D., and V. T. Stannett, Forest Prod. J. **21**(10), 95 (1971). "Radiation Can Catalyze Urea-Formaldehyde Resins."

Wenzel, H. F. J., *The Chemical Technology of Wood,* Trans. F. E. & D. A. Brauns, Academic Press, New York, 1970.

Wenzel, W., and D. Dieterich, Ger. Offen. 2,637,690, Feb 23, 1978. "Aqueous Dispersions or Solutions of Polyurethanes."

West, H. J., USP 2,327,984, Aug 31, 1944. "Low-Viscosity Urea-CH_2O-Alcohol Resin Solutions."

White, J. T., and D. L. Gumprecht, USP 3,827,995, Sept 6, 1974. "Copolymer Blends."

Whiteside, I. R., S. African P 77 00,345, Oct 25, 1977. "Urea-Formaldehyde Resin Compositions."

Widmann, M., USP 3,979,341, Sept 7, 1976. "Urea Formaldehyde Foam."

Widmann, M., Can. P 1,013,499, July 5, 1977. "Urea Formaldehyde Foam."

Widmer, G., and H. P. Wohnsiedler, in *Encyclopedia of Polymer Science Technology,* Vol. 2, H. F. Mark *et al.*, eds., Interscience, New York, 1965. "Amino Resins."

Wilkinson, J. F., Agron. J., **69**(4), 657 (1977). "Effect of IBDU and UF Rate, Date, and Frequency of Application on Merion Kentucky Bluegrass."

ISBN 0-201-04558-3

Willert, J., *et al.*, Pol. P 73,289, Nov 15, 1974. "Amine-Formaldehyde Resins."

Wilton, J., USP 3,770,466, Nov 6, 1973. "Heat-Insulating Shaped Compositions."

Winkler, H. D., and K. Welzer, Wasser, Luft Betr. **6**(7), 213 (1972). "Formaldehyde Emissions from Pressboard Production."

Wirpsza, Z., Pol. P 72,885, Dec 20, 1974. "Amino Resins with Low Content of Free Formaldehyde."

Wirpsza, Z., Polimery **18**(4), 192 (1973). "Acidic Catalysts for Curing Amino Plastics. IV. New Types of Latent Catalysts."

Wirpsza, Z., Pr. Nauk. Inst. Technol. Org. Tworzyw Sztucznych Politech. Wroclaw, No. 6, 46 pp (1972). "Use of Urea for Synthesizing Plastics."

Wirpsza, Z., Polimery **17**(9), 465 (1972). "Amino Resins for Chemically Hardening Varnishes. III. Twice Etherified Resins."

Wirpsza, Z., *et al.*, Pol. P 66,858, Dec 20, 1972. "Hardening of Aminoplastics."

Wirpsza, Z., and H. Klebeko, Pol. P 66,154, July 31, 1972. "Urea Resins."

Wismer, M., K. F. Schimmel, and R. G. Temple, Ger. Offen. 2,415,762, Oct 24, 1974. "Pearls of Urea – or Phenol-Formaldehyde Resins and Their Use as Delustering Compositions."

Wittmann, O., Holz als Roh- und Werkstoff **34**, 427 (1976). "The Gluing of Wood with Isocyanate."

Wittmann, O., Holz als Roh- und Werkstoff **31**, 419 (1973). "Alkali in Phenolic Resin-Bonded Particle Board."

Wittmann, O., Holz als Roh- und Werkstoff **29**, 259 (1971). "Manufacture of Nonhydroscopic Particle Boards."

Wnuk, M., I. Szozda, and W. Czechowski, Pr. Inst. Technol. Drewna **20**(2), 139 (1973). "Shortening of the Pressing Cycle of Particle Boards."

Wnuk, M., Pr. Inst. Technol. Drewna **19**(1-2), 311 (1972). "Possibility of Pressing Time Reduction in Manufacture of Flat-Pressed Particleboard through the Application of Gas Shock at 160°."

Wöhler, F., Pogg. Annalen der Physik **12**, 253 (1828). "Über künstliche Bildung des Harnstoffs."

Wojciak, S., Ger. Offen. 2,629,889, Jan 20, 1977. "Microencapsulation of Metallocenes."

Wood, A. D., "Plywoods of the World." W. & A. K. Johnston, London, 1963.

Wu, S.-C., and C.-M. Yeh, T'ai-wan Mu Ts'ai Kung Yes **33**, 1 (1974). "Effect of Adhesives on Thickness Swelling and Water Absorption of Particleboards."

Wulkan, E. K. H., "UF-Foam as Core Insulation in two-layer Wall Construction," in "Kunststoff Fortschrittsberichte, 1979, C. Hanser, Vienna, 1979; H. Baumann, editor.

Yakovlev, D. A., *et al.*, USSR P 481,584, Aug 25, 1975. "Heat Insulating Mass."

Yamagishi, Y., N. Kawai, and S. Ono, Mokuzai Kogyo **27**(309), 588 (1972). "Gluability of Tropical Woods. II. Gelation Time of the Urea-Formaldehyde Resin Adhesive."

Yamamoto, N., J. Kokai 75:42,011, Apr 16, 1975. "Decorative Boards."

ISBN 0-201-04558-3

Yamamoto, Y., K. Higuchi, and K. Okada, J. Kokai 75:76,147, June 21, 1975. "Water-Soluble Adhesives for Woods."

Yamamoto, Y., K. Higuchi, and K. Okada, J. Kokai 75:76,146, June 21, 1975. "Water-Soluble Adhesvies for Woods."

Yamamoto, Y., K. Higuchi, and K. Okada, J. Kokai 75:76,145, June 21, 1975. "Water-Soluble Adhesives for Woods."

Yamamoto, Y., and K. Higuchi, J. Kokai 75:04,133, Jan 17, 1975. "Adhesives for Laminated Wood Panels."

Yamashita, H., W. Yasutake, and Y. Hata, Japan. P 74:44,936, Nov 30, 1974. "Amino Resin Composition."

Yamashita, K., and T. Yamashita, Japan. P 75:25,491, Aug 23, 1975. "Adhesive Composition of Water-Soluble Resin."

Yoshida, A., K. Asahara, H. Imai, and T. Isa, J. Kokai 75:19,909, Mar 3, 1975. "Fire-Resistant Plywood."

Yoshihara, T., *et al.*, Tokyo Toritsu Eisei Kenkyusho Kenkyu Nempo (25), 103 (1974). "Testing Method for Household Articles Containing Poisonous Chemical Substances. II. Determination of Formaldehyde in Commercial Underwear for Ladies and Infants."

Yoshihiro, K., and T. Kawahara, Niigata Norin Kenkyu (24), 55 (1972). "Adhesive Properties of Urea Resin Adhesives. I. Reaction Heat of the Urea Resin Adhesive in the Curing Process."

Yoshimitsu, T., Y. Kawashima, and H. Katsube, J. Kokai 73:103,705, Dec 26, 1973. "Deodorizing Plywood."

Yoshina T., J. Kokai 73:93,580, Dec 4, 1973. "Adsorbent for Removing Formaldehyde from Air."

Yudina, G. G., and M. M. Svitkina, Derevoobrab. Prom-st. (12), 6 (1975). "Effect of Urea Quality on Properties of Urea-Formaldehyde Resins."

Yudina, G. G., *et al.*, USSR P 485,887, Sept 30, 1975. "Nontoxic Wood-Particle Boards."

Yui, H., S. Moriwaki, Y. Ichikawa, and S. Ohi, J. Kokai 76:82,334, July 19, 1976. "Self-Extinguishing Thermoplastic Resin Compositions with Improved Processability."

Yukin, N. A., and M. T. Tsupikov, USSR P 627,109, Oct 5, 1978. "Slow-Acting Nitrogen-Containing Fertilizer."

Zabrodkin, A. G., A. E. Anokhin, and T. V. Zakharova, USSR P 496,291, Dec 25, 1975. "Amino-Formaldehyde Resin Adhesive."

Zapletal, J., and L. Sedlak, Ger. Offen. 2,341,553, Mar 20, 1975. "Urea-Formaldehyde Foam as Insulation Material with Decreased Combustibility."

Zapletal, J., and L. Sedlak, Czech. P 162,810, Mar 15, 1976. "Workup of Foamed Urea-Formaldehyde Resins to Insulants with Reduced Flammability."

Zastanchenko, M. A., and N. A. Osnach, USSR P 395,280, Aug 28, 1973. "Particle Boards."

Zastanchenko, M. A., and N. A. Osnach, USSR P 390,969, July 25, 1973. "Particle Boards."

Zastanchenko, M. A., and N. A. Osnach, USSR P 387,847, June 22, 1973. "Particle Board."

ISBN 0-201-04558-3

Zeigerson, E., and M. R. Bloch, Ger. Offen. 2,247,036, Apr 5, 1973. "Brominated, Sulfonated Lignin for Flameproofing Combustible Materials."

Zenkteler, M., and D. Srama, Rocz. Akad. Roln. Poznaniu **99**, 101 (1978). "Study of the Bonding of Particleboard with Urea Resin Without Hardener."

Zenkteler, M., and D. Majsjej-Kubik, Rocz. Akad. Roln. Poznaniu **99**, 93 (1978). "Susceptibility of Some Glue Joints to Shearing and Splitting."

Zigeuner, G., and W. Rauter, Monatsh. **96**, 1950 (1965). "Über Heterocyclen, 6. Mitt: Cyclische Harnstoff-Aldehyd-Kondensate."

Zigeuner, G., and M. zur Hausen, Monatsh. **92**, 278 (1961). "Über Heterocyclen, 4. Mitt.: Über beta-Ureidoaldehydacetale."

Zigeuner, G., and W. Nischk, Monatsh. **92**, 79 (1961). "Über Heterocyclen, 3. Mitt.: Zur Reaktion von Homologen des Acetaldehyds mit Carbamiden."

Zigeuner, G., M. Wilhelmi, and B. Bonath, Monatsh. **92**, 42 (1961). "Über Heterocyclen, 2. Mitt.: Über das 2-Oxo-1,4-dimethyl-6-(3-methylureido)-hexahydropyrimidin und das 2,7-Dioxo-3,4,5,8-tetramethyl-dekahydro-pyrimido[4,5-d]-pyrimidin."

Zigeuner, G., E. A. Gardziella, and G. Bach, Monatsh. **92**, 31 (1961). "Über Heterocyclen, 1. Mitt.: Zur Reaktion von Aldehyden mit Carbamid."

Zigeuner, G., and W. Hoselmann, Monatsh. **87**, 406 (1956). "Studien auf dem Gebiet der Harnstoff-Formaldehyd-Kondensation. XV. Mitt.: Über Hydroxybenzylurone."

Zigeuner, G., R. Pitter, and H. Rauch. Monatsh. **86**, 173 (1955). "Studien auf dem Gebiete der Harnstoff-Formaldehyd-Kondensation. XII. Mitteilung: Zur Kenntnis der Oxybenzylcarbamide."

Zigeuner, G., R. Pitter, H. Berger, and H. Rauch, Monatsh. **86**, 165 (1955). "Studien auf dem Gebiete der Harnstoff-Formaldehyd-Kondensation. XI. Mitt.: Über Methylenharnstoffe."

Zigeuner, G., and R. Pitter, Monatsh. **86**, 57 (1955). "Studien auf dem Gebiete der Harnstoff-Formaldehyd-Kondensation. X. Mitt.: Über das Auftreten von Dimethylenätherbrücken in Harnstoff-Formaldehyd-Kondensaten."

Zigeuner, G., Monatsh. **82**, 175 (1951). "Studien auf dem Gebiete der Harnstoff-Formaldehyd-Kondensationsprodukte. I. kurze Mitt.: Über die Einwirkung von 2,4-Dimethyl-phenol auf Harnstoff-Formaldehyd-Kondensate."

Zinin, A. V., and V. S. Murav'ev, Plast. Massy (11), 77 (1975). "Rheological Characteristics of Thermosetting Adhesives and Molded Materials."

Zisman, D., D. Vaskevich, A. Norkute, and P. Konovalov, Zh. Prikl. Khim. **44**(12), 2685 (1971). "Polymerization of Formaldehyde in Media of Different Polarity."

Zlodovskii, L. I., *et al.*, USSR P 108,597, Oct 25, 1977. "Wool-like Viscose Fibers."

Znachko-Yavorskii, I. L., Cem.-Wapno-Gips **28**(12), 379 (1973). "Effective Building Material."

ISBN 0-201-04558-3

Indexes

Author Index

This Index lists all authors quoted in the text. When a name is followed by another name, rather than a page number, the person is co-author of a paper or patent listed in the Bibliography.

The spelling of names follows that used in the original reference or source. In the case of Japanese, Russian, and other foreign-alphabet names, the reader should check under all possible forms of spelling because such authors might appear in more than one entry.

Several older patents do not list initials of authors. Anonymous patents are listed under the company name. Company names are integrated with author names throughout this Index.

Abe, I., 111, 242, *see also* Imura
Adamca, M., 108, 110
Afanas'ev, N. V., 109, *see also* Oblivin
Agron, J., 210, *see also* Wilkinson
Ahlstrom, D. H., 89, *see also* Liebman
Ahmad, S., 202
Aignesberger, A., 106, 120
Akabane, J., 262
Akabori, S., *see* Ohtomi
Akasaka, M., *see* Noda
Akhmedzhanov, K. A., 258
Akino, H., *see* Ando, S.
Akutin, M. S., 105, 203, *see also* Afanas'ev
Aksel'rod, B. Ya., *see* Pshenitsyna
Aleksandrovich, Kh. M., 214
Alekseev, A. V., 107
Alexandru, I., *see* Vintila
Allers, 6
Allgemeine Elektrizitaets-Ges., 13
Allied Chemical Co., 186, 276
Allyn, C. L., 101
Almassy, G., 208
Altshuller, 255
Amand, 121
American Conference of Government Industrial Hygienists, ACGIH, 129
American Cyanamid Co., 10, 15, 18, 202, 276
American Industrial Hygiene Assoc., AIHA, 129
American Reinforced Paper, 18
American Society for Testing Materials, ASTM, 157
American Society of Heating, Refrigerating, and Air-Conditioning Engineers, ASHRAE, 148
Anzawa, T., *see* Takahashi
Andersen, A. G., *see* Kreibich
Anderson, E., *see* Bertsch
Andersen, I., 22, 68, 111, 148, 182, 236, 237, 238, 240, 252, 253, 254, 258, 261, 262, 291
Anderson, M. P., *see* Burnett
Ando, S., 109
Andreyuk, S. P., *see* Khlystunova
Adnrezen, L. M., 114
An Foras Taluntais, 210
Anokhin, A. E., 105, 109, 174, 175, 178, 262, *see also* Belova, Zabrofkin
Ansreev, G., *see* Ivanov
Antlfinger, G. J., 203
Antonia, K. A., 164
Aono, I., 121
Aoyagi, T., *see* Ichimura
Araki, N., *see* Inoue
Arbuzov, V. V., 173
Argirova, M., *see* Basmadzhieva
Argo, W. B., 210
Aries, R., 106
Arima, Y., 203
Arito, H., *see* Matsumura
Artamonov, B. I., *see* Ivanov

Asahara, K., *see* Yoshida
Ashall, R. J., 188, 205
ASHRAE, *see* American Society of Heating, Refrigerating and Air-Conditioning Engineers
Asnach, K. G., *see* Den'gina
ASTM, *see* American Society for Testing Materials
Åström, B., 240, *see also* Mattsson
Atkinson, K., 110, 248, *see also* Hubbard
Atsuke, K., 17, 203
Auer, L., 14
Averell, P. R., 132
Awano, K., *see* Suzuki
Azarov, V. I., 111, *see also* Tsvetkov

Baccareda, 98
Bach, G., *see* Zigeuner
Backeland, 72
Baekland, 6
Baer, R. L., 52, 256
Bagley, J. M., 114
Bakelite Corp., 98
Bakiewicz, J., *see* Przybylak
Balakhonkina, G. V., *see* Tsyvin
Ball, G. L., *see* Boettner
Ball, G. R., 174
Bally, E. C. C., 12
Bally, E. J., 12
Baranov, V. V., *see* Novak
Baranowski, P., 89
Barbera, A., 210
Barghoorn, A. W., 234, 235, 241, 242, 286
Bartashevich, A. A., 111
Barthelemy, H., 14
Barzynski, H., 204
Basaroff, A., 5, 99
BASF, 6, 16, 18, 19, 98, 106, 276
Basmadzhieva, K., 291
Basterfield, S., 14
Batalina, E. S., *see* Belova
Batty, J. W., 199
Batyuk, V. Ts., 207, 208
Bauer, W., 19, 186, 193
Baumann, H., 19, 79, 107, 191, 194, 195, 196, 197, 198, 199, 207, 208, 209, 217, *see also* Schulze
Baxter, G. F., 121
Bayersdorf, F., 105, *see also* Braum
Baymiller, J. W., 177
Beck, K. M., 114
Becke, F., *see* Reuther
Becker, R. S., *see* Bercovici
Begletsov, V. V., *see* Knat'ko
Beglov, B. M., *see* Usmanov
Behar, R., 106
Bekkert, L. G., *see* Akutin
Belman, 128
Belova, T. A., 204
Belyi, V. A., 175
Benteli, A., 17
Berbner, H., *see* Sander
Bercovici, T., 23
Bergin, E. G., 111
Bergsund, K., 109
Bernstein, H. J., 30, *see also* Schneider
Bertagnini, 52
Bertsch, W., 146
Bieber, 26
Bigatto, R., *see* Chiavarini
Billingsfors Bruks AB, 109
Binder, K., 203
Biondi, E., *see* Baer
Bitterli, W., 129
Bittner, N., *see* Jenks
Black, W. R., 107, 109
Blackadder, 55
Blaga, A., 196
Blais, J. F., 107
Blake, D., 189
Blandin, H. M. F. F., 205
Blandin, P. P. P. C., *see* Blandin
Blank, W. J., 110, 125
Blinkova, O. P., *see* Perevertov
Bloch, B., *see* Duval
Bloch, M. R., *see* Zeigerson
Blommers, E. A., 106, 121, 248
Blumer, 6
Blyakhman, E. M., *see* Knat'ko
Blythen, S., 7, 11, *see also* Rothera
Boehm, F., 188, *see also* Schwab
Boehm, W., 101, 107, 109, 286
Boettner, E. A., 206, *see also* Ball
Bögemann, M., 76, 77, 78
Bogosavlejevic, M., 114
Bokareva, E. Z., 212
Bohlen, J. C., 111
Bokov, A. N., 257, *see also* Anokhin
Bonashevakaya, T. I., *see* Feldman
Bonath, B., *see* Zigeuner
Bonastie, 6
Bondi, E., 106
Bonitz, E., *see* Sander
Bonner, W. H., 105
Bonnington, M. G., 212
Borden (UK) Ltd., 172, 212
Borden Products, 190, 276
Borisyuk, I. D., 203, *see also* Ishin
Bornstein, M. I., *see* Paterson
Borodkina, N. I., 188, 189, *see also* Valgin
Borozdin, V. S., 189, *see also* Krikunov
Bradley, M. L., *see* Feriday
Brancato, J. J., 109
Brandani, V., 38

Brandenburger, 6
Brandeis, J., 202
Brandner, H., *see* Neumann
Brendley, W. H., *see* Martorano
Breysse, P., 148, 258, 262
Brial, J. C., 106, *see also* Duclairoir
British Cyanides Co., 10, 15
British Industrial Plastics Ltd., 106, 186, 190, 197
British Power Boat Co., 161
Brooks, S. H. W., 174, 175
Brose, H., 114
Brown, R. W., *see* Foris
Brown-Boveri Co., 189, 197
Brunner, K., 149, 150, 152, 154
Brunnmüller, F., 101, 107, 109, 188, 207, 248, 267, *see also* Lenz
Bryan, R. J., 148, 258, 262, *see also* Wayne
Budkov, V. A., *see* Usmanov
Bukovskii, M. I., *see* Krikunov
Bunch, J. E., *see* Pelli
Burch, D. B., *see* Rossiter
Burch, D. M., *see* Hunt
Bürgi, H. -B., *see* Lehn
Burkova, T., *see* Basmadzhieva
Burmester, A., 178, 286
Burnett, R. D., 146
Buschfeld, A., 109
Buslenko, A. I., *see* Ivanova
Butlerov, 5, 6, 26, 47, 48
Bzhezinskii, Ya., *see* Virpsha

Cailey, G., 161
Cairns, 51
Campano, P., *see* Tapia
Canepa, D., *see* Barbera
Carbone, R. D., 148, 258, 262
Carpenter, B. H., *see* Pellizzari
Carroll, B. S., *see* Christensen
Carroll, M. N., *see* Clermont
Carson, F., 164
Casandra, I., *see* Ionescu Muscel
Casco, 276
Casebier, R. L., 121
Catoni, M., 109
Celanese, 276
Cendrowska, A., *see* Kinastowski
Central Laboratory in Sofia, 106
Cerny, J., *see* Snuparek
Chalmin, 235, 236
Champion Paper Co. Ltd., 204
Chand, R., *see* Johnston, Marcote
Chanda, M., *see* Rathi
Chang, T., 111
Chapput, A., 25
Chattaway, F. D., 16
Chauhan, B. R. S., *see* Gupta, Jain
Chernaya, V. I., *see* Tunik
Chernichenko, I. A., *see* Vinogradov
Cherubim, M., 240
Cherry, O. A., 15
Cheshkov, V., *see* Simeonov
Chiavarini, M., 39, 73, 79, 89, 92, 105, 134, 135, 136, 137
Cho, H. T., *see* Paszner
Chow, S., 108, 109, 110, 112, 113, 114, 115, 116, 117, 176, *see also* Steiner, Troughton
Chowdhury, S., *see* Roy
Chrastill, J., 129
Christensen, H. E., 266
Christensen, R., 106
CIBA, 4, 11, 12, 13, 14, 15, 18, 101, 110, 185, 202
CIBA-GEIGY, 162, 193, 212, 213
Cinite, V., *see* Erins
Ciocca, B., *see* Contardi
Ciutat, M., *see* Eek
Ckhelka, J., *see* Lvovsky
Clad, W., 111, 122, 157, 161, 178, 291, *see also* Kollman
Claes, M., 210
Clarke, B. J., 106
Claus, 65
Clermont, L. P., 127
Cleveland, J. J., 190, 197
Cochran, W. G., *see* Snedecor
Columbus, P. S., 204
Contardi, A., 58
Conte, C., 262
Conti, N., *see* Chiavarini
Cooley, J., *see* Oergen
Cooper, C. H., 17
Copper, G. P., *see* Kulle
Cowan, W. D., 263
Coyle, R. P., 247
Craver, J. K., *see* Polymer Science
Crist, 216
Crocker, E. C., 131, 233, 260
Cropp, J. A. D., 210
Crowe, 92
Curs, A., 4, 16, 19, 185, 196
Czechowski, W., *see* Wnuk

Da Cunha, A., 203
Daemen, J. M. H., 130, *see also* Dankelman
Dahl, G., *see* Boehme
Dahlberg, H. S., 202
Dalecki, J., *see* Dirska
Dalham, T., 262
Dammer, S., 214, *see also* Boehme
Dankelman, W., 27, 38, 39, 40, 43, 44, 72, 73, 79, 89, 130, 134, 135, *see also* de Breet

Dashkovskaya, Z. F., 246
Da Silva, A., *see* Da Cunha
Date, M., 202
Davidovits, J., 176
Davidson, R. W., 107, 245
Davis, 60, 210
Davydov, V. V., 208
Dead Sea Bromine Co. Ltd., 206
De Backere, R., *see* Claes
de Breet, A. J. J., 27, 66, 67, 79, 135
De Bruyne, N. A., 18, 157, 202
de Chesne, E. B., 15, 90
Dede, J. B. Jr., *see* Horowitz
Degussa Co., 98
de Havilland Aircraft Co., 202
Del Fanti, N., *see* Chiavarini
de Jong, J. I., 4, 88, 100, 103, 131, 133
de Jonge, J., 4, 88, 92, 100
Delepine, M., 48, 52
Demenkova, K. M., 189
Dement-ev, A. G., 194, 217
Demidova, O. P., *see* Morozova
Demko, P. R., 106
Den'gina, N. A., 211
Denisova, S. G., 109, 114
Deppe, H. J., 110, 122, 160, 166, 167, 178, 219, 235, 287, 288, *see also* Stashevski
Deriancourt, G., 178
Detkov, V. V., *see* Kruglikov
Detushev, I. F., 178
Devitfrancesco, G., 262, *see also* Conte
de Wit, J., 72, *see also* de Breet, Dankelman
Diamond, P., *see* Burnett
Dickson, 49
Di Dio, P., 198
Diem, H., 267
Dieterich, D., *see* Wenzel
Dillan, 59
Dimter, L., 173
Ding, J.-Y., 88
Dirska, B., 204
Ditmer, 289
Dixon, A. E., 7, 64
Dixon International Ltd., 206
Dockal, P., 211
Doebereiner, 5
Döhler, F., 198
Dolenko, A. J., 206
Domokos, G., 176
Doroshenko, S. P., *see* Drobyazko
D'Ouville, E. L., *see* Cropp
Dranovskii, M. G., *see* Perevertov, Potutkin
Drechsel, H., *see* Boehm
Drews, M., 208
Drobyazko, V. N., 206
Dromashko, E. I., *see* Dubrovin
Druta, A., *see* Lambru
Druzhinin, S. A., 189, 199
Dubikovskaya, L. V., *see* Kozlova
Dubrovin, V. S., 114
Dubrovskii, S. L., *see* Akhmedzhanov
Duclairoir, C., 73, 79, 89, 135, *see also* Brial
Ducorit-Consultants, 174
Duermeyer, D., *see* Grabowsky
Dueva, L. A., 263
Dumov, S. N., 190
Dunn, 131, 132
DuPont Corp., 98, 122
Dusenbury, 59
Duval, M., 134
Dvoranova, V., *see* Labsky

Eades, A. G., 110
Ebdon, J. R., 73, 79, 138
Eckhoff, N. D., *see* Kyle
Eder, H. M., 178
Edler, F., 247
Edwards, J. W., *see* Argo
E. G. Budd Manufacturing Co., 12
Egorenkov, N. I., *see* Belyi
Egorova, T. N., *see* Belova
Einhorn, A., 6, 15, 63, 64
Eisele, W., 107, 108, 109, 155, 248, *see also* Petersen
Eisenmann, K., 14, 17, *see also* Pungs
Eklund, J. A., 101, *see also* Jaccard
El'bert, A. A., 89, 106, 108, 111, 114, 121, *see also* Erykhov, Time
Ellis, C., 4, 5, 11, 15, 17, 18, 19, 52, 70, 79, 105, 112, 121, 211, 226, 229, 248, 255
Ellis-Foster Co., 17
El-Sayad, I., 263
Endo, A., 204, *see also* Suzuki
Energy Resources Co., 146
Enkvist, T. U., 246
Erhardt, K., 106, 107, 109
Erikson, C. R., *see* Columbus
Erins, P., 111
Ernst, K., 245, *see also* Deppe, Lohenhoffer
Erykhov, B. P., 114, *see also* El'bert
Eschalier, X., 59, 199
Etablissements Lambiottes Freres, 14
Evans, E. A., *see* Philpotts
European Federation of Associations of Particleboard Manufacturers, FESYP, 146, 147, 150, 152, 153, 154, 155

Fahey, D. J., 204
Fahrni, F., 4, 110, 159, 165, 166, 170, 171, 285, 288
Faichney, G., 263
Fairchild, E. J., *see* Christensen
Falconer, J. P. R., *see* Stubblefield
Fateev, N. T., *see* Laukhin
Fefilov, V. V., *see* Strelkov

Feinauer, A., 202, *see also* Brandeis
Feldman, Y. G., 67, 147, 262
Fenenko, L. M., *see* Batyuk
Ferber, K. D., 246
Feriday, J. E., 211, 212, 215
Fiberglass Ltd., 121, 187, 203
Fiehn, T. L., 204
Fietz, E. H., *see* Hoener
Fife, R. L., 204
Filippov, G. S., *see* Borozdin
Finenbeiner, 125
Finklea, 254
Fischer, 254
Fishbein, L., 254
Fleury, G., *see* Chapput
Flick, K., 148
Flory, K., 202, *see also* Brandeis
Flueck, K., 257
Flyate, D. M., *see* Smirnova
Fokina, V. M., *see* Glukhikh
Foris, P. L., 110
Foscolo, P. U., *see* Brandani
Frackowiak, U., *see* Przybylak
Frank-Kamenetskii, M. D., *see* Vologodskii
Frankel, L. S., 146
Freeman, H. G., 106
Freidin, A. S., *see* Sholokhova
Froehlich, P., *see* Boehme, Trautvetter
Frolova, M. K., 189
Frydrych, J., *see* Boehm
Fujii, K., 148, 237, 239, 240, *see also* Kurashige
Fujimoto, M., *see* Konishi
Fujimoto, Y., *see* Hiyoshi
Fujita, K., 176, *see also* Oka, Ueda
Fujitani, J., 176
Fukuda, T., 247, *see also* Masuda
Fukui, H., *see* Hiyoshi
Fukuoka, S., 202, *see also* Date
Fukushima, M., *see* Kawamura
Furudera, R., *see* Ohno

Gablin, K. A., 211
Gagliardi, D. D., 59, 199, *see also* Gruntfest
Galeotti, 67
Galil, F., 199, 203
Galitsina, T., *see* Sukovatov
Gamova, I. A., 111, *see also* El'bert, Koromyslova
Gams, A., 11, 12, 13, 14, 15, 16, 19, 66, 185
Gaponenko, I. M., 108
Garbini, L. J., *see* Seldon
Gardziella, E. A., *see* Zigeuner
Garshick, E., *see* Paterson
Gavan, M., *see* Lambru
Gavel, J., 204
Gavirati, U. M., 106
Gebtishchevra, 256
Geimer, R. L., *see* Lahmann
Geisse, 62
Geissler, T., 198
Geissman, T. A., 36, 47
Gel'tishcheva, E. A., 257
Georgia Pacific, 276
Gerber, H., *see* Danner
Gerlachov, T., 111
Gerwing, M., 16, *see also* Walter
Getchell, N. F., 202
Geyer, B., 198
Gfeller, B., 287
Giacomo, G. Di., *see* Brandani
Gibson, D. V., 214
Gillespie, H. R., 7, 11, 109, *see also* Rothera
Girs, S. P., *see* Serova
Gilfrich, H. P., 211
Ginzel, W., 111
Gipstein, 51
Glubish, P. A., 203
Glukhikh, L. S., 111
Gnatowski, M., 210, *see also* Jacenkos
Godin, V., *see* Bergin
Goertz, H. M., 210
Goettsche, R., *see* Siegler
Goldman, 146
Goldschmidt, C., 4, 6
Goldschmidt, H., 10, 15, 64
Goldschmidt, S., 4, 13
Golubovskaya, E. K., 210
Golyshev, V. K., 210, 257, *see also* Golubovskaya
Gorbunova, I. V., *see* Mochalova
Gordon, A. F., 24, 110, 204
Gorenko, V. G., 190
Gos, B., *see* Krach
Gosink, T. A., 146
Goto, T., *see* Hirotsugu, Sakuno
Gourdenne, A., *see* Richard
Grabowsky, O., 109, 188, 248, *see also* Brunnmüller, Schmidt-Hellerau
Gracheva, M. P., 211
Grad, 131, 132
Graham, J. L., 210
Grantham, F., *see* Ball
Graser, M., *see* Siegler
Grassie, 229, 284
Greenberg, W. H., 211
Grendon, W. C., *see* Freeman
Greubel, D., *see* Roffael
Griessbach, R., 14
Griffiths, P. R., *see* Liebman
Grigoreva, N. N., 196
Griskey, R. G., *see* Throne
Gros, I., *see* Stenzel
Gruben, S., 197
Gruntfest, I. J., 59, 60, 149

Guha, S. R. D.
Gulyaeva, O. V., *see* Glukhikh
Gumprecht, D. L., *see* White
Gupta, R. C., 181, 212, *see also* Jain, Lakshminarayanan
Gur'evskaya, I. N., *see* Tsyvin
Guseva, V. A., *see* Pavlenko
Gutheil, H., 268, 269, 270
Gutsal, P. P., *see* Naidenov
Gvozdev, D. V., 109

Haag, T. H., *see* Martorano
Halama, D., 211
Halligan, A. F., 111
Hall, 27
Ham, G. E., 105
Hamada, R., 105
Hamburger, A., 6, 15, 64, *see also* Einhorn
Hanckova, B., *see* Halama
Hanetho, P., 236, 240, *see also* Mattsson
Hanif, 202
Hansen, L. J., *see* Gablin
Harmon, 52
Hasegawa, M., 246
Hashimoto, Y., *see* Kadowaki
Hata, Y., 109, *see also* Yamashita
Hatanaka, K., *see* Mori
Hatono, S., *see* Tomita
Haui, 242
Havinga, R., 111
Hayasahi, K., *see* Kawakami, Okamura
Hayashi, S., 106, 109
Hayashi, T., 205
Heap, S. A., 202
Hearon, W. M., *see* Lambuth
Heaton, P. E., *see* Ebdon
Hebeish, A., 199
Heckman, S. E., *see* Secrest
Heetman, J. G. A., 205
Hefty, F. V., *see* Lehmann
Hehn, Z., *see* Nowak
Heiduschka, A., 58
Heming, 50
Hemmelmayr, F. v., 6
Henck, C., 12, 14
Hendra, P. J., *see* Paul
Henkel, H., *see* Lenz
Henry, 61
Hensley, W. L., 105
Hentisch, W., 18
Hentschel, K. D., 207, *see also* Brose
Hepp, 68
Herbert, 106
Hermann, D. T., 109, 203, *see also* Brancato
Heubel Co., 18
Hey, D. H., 14, *see also* Hinkel
Hiag Co., 176
Higashimura, E., 110, 112, 114
Higuchi, K., 105, 110, 245, 247, *see also* Yamamoto
Hills, 52
Hilt, A., *see* Diem
Himmelheber, 166
Hinkel, L. E., 14
Hinkson, R. S., *see* Davies
Hino, T., *see* Shiota
Hinshelwood, 55
Hinterwaldner, R., 189
Hirai, H., *see* Mita
Hirakawa, M., *see* Kato
Hirata, S., 109, 242, *see also* Kubota, Minemura
Hirose, Y., *see* Tomita
Hirotsugu, O., 111
Hishita, T., *see* Takashima
Hiyoshi, K., 204
Ho, 92
Hodgson, F. N., *see* Snyder
Hoedt, H., 205
Hoelzer, 7
Hoffman, R. K., *see* Braswell
Hoffmann, M., *see* Sander
Hoggan, J., *see* Cowan
Hohn, R., *see* Kuemmel
Holden, J., *see* Huang
Holmes, C. A., 206
Holmgren, T. A. F., *see* Laqua
Holtschmidt, U., 110
Holzer, G., *see* Bertsch
Holzer, 6
Homann, 51
Hope, P., 89, 134
Horea, R., *see* Teodorescu
Horike, T., *see* Ueda
Horima, 252
Horsky, K., *see* Katuscak
Hoselmann, W., *see* Zigeuner
Hosokawa, T., *see* Kawakami
Hosota, Y., *see* Toyota
Hostettler, P., 110, *see also* Jaccard
Houwink, R., 97, 157
Hse, C.-Y., 111
Hsiao, S.-H.
Hubbard, D. A., 107, 110
Huebs, 99
Hughes, E. E., 146
Humik, H., 188
Hummerich, R., *see* Petersen
Hunsucker, J. H., *see* Purcell
Hurst, H., 189
Hutschneker, K., 122, 178
Huysmans, W. G. B., *see* de Breet
Hybernia Mining Co., 99

Ibrahim, M., *see* Galil
Ichikawa, M., 203, *see also* Arima, Yui

Ichinose, T., *see* Iwata
ICI Corp., 98, 212
Idamura, 212
I. G. Farbenindustrie, 4, 10, 12, 13, 15, 16, 17, 19, 185, 276
Ikebe, S., *see* Mori
Ikeda, M., 246, *see also* Kikuchi, Ohe
Iizuka, T., *see* Inui
Iliceto, 26
Il'inskaya, V. N., *see* Prilukov
Il'yushenko, R. G., *see* Kara
Imagone, H., *see* Kawakami
Imai, H., *see* Yoshida
Immarino, N. J., 176
Imoto, M., 23, 92
Imura, S., 111, 242, *see also* Eguchi, Minemura
Inoue, K., 178, 246, *see also* Suzuki
Inoue, S., 114, *see also* Otsuka
Inverarity, G., 106
Isaenko, V. D., *see* Kozlov
Ishida, M., 178
Ishida, S., 114
Ishii, K., *see* Tanaka
Isaev, A. I., *see* Sukovatov
Ishikawa, S., *see* Ohhashi
Ishin, N. D., 181, *see also* Borisyuk
Isono, S., 176
"Isovolta" Co., 206
Ito, Y., 92, 188, 190, 197, *see also* Oshima
Ivancheva, E. G., 109
Ivanov, N. A., 111
Ivanov, S., 89, 203
Ivanov, V. A., *see* Romanenkov
Ivanova, T. P., 263
Iwaki, T., 108
Iwasa, Y., 178
Iwasaki, H., 286
Iwata, M., 108, 110
Izumi, H., 106, *see also* Kitta

Jaccard, R., 101
Jacenkow, B., 210
Jackson, L. P., Jr., 210
Jaglarz, M., *see* Nowak
Jain, N. C., 111
Jakimavicius, K., 181
Jamil, N. A., *see* Ahmad, S.
Jamin, A., 190
Jander, K., *see* Lahmann
Jenks, T. E., 205
John, H., 4, 6, 7, 8, 10, 13, 16, 122
Johns, W. E., 111, 118, 119, 149, 152, 178, 243, 247, *see also* Meyer
Johnston, T. H., *see* Marcote
Johnson, R. D., *see* Stahly
Jones-Sefton, 58
Jorgensen, 60
Juenger, H., 189
Jumneanbhandhu, N., 176
Juneja, S. C., 206
Junkermann, H., 106
Jung, J., 18
Junger, 217
Jurek, H., *see* Dirska
Just Bravo, M. A., 190
Justice, G. H., 186, 187

Kabayashi, R., *see* Kotani
Kadita, H., *see* Mukudai
Kadowaki, H., 12, 19, 20, 22, 64, 90, 91, 92, 96
Kaesbauer, F. K., 103, 133
Kaetsu, I., *see* Okamura
Kageyama, K., *see* Kotani
Kakabadze, T. G., 210
Kalugin, K. S., *see* Rudnev
Kalyazin, E. P., *see* Rudnev
Kamachi, T., 247
Kamei, K., *see* Kato
Kamel, M., *see* Schliefer
Kamenskii, I. V., 105
Kamitaki, Y., *see* Inui
Kamiyama, M., *see* Hata
Kaneko, K., *see* Ishida
Kaneko, T., *see* Paszner
Kaniszewski, S., 198
Kanof, N. B., *see* Baer
Kaps, R., *see* Dimter
Kara, V. V., 206
Karahasanovich, 236
Karlent'ev, V. N., *see* Gracheva
Karnemaat, J. N., 207, 210
Karo, W., *see* Sandler
Kartashova, S. A., 188
Katayama, M., 178
Kato, C., 176, 178
Kato, H., 246
Katsube, R., *see* Sakuno
Katsube, H., *see* Yoshimitsu
Katsyuba, V. I., 205, *see also* Mironov
Katuscak, S., 111
Kaufmann, F., 208
Kawahara, N., 245, 246, 247, *see also* Tanaka, Yoshihiro
Kawai, N., *see* Yamagishi
Kawamura, S., 110, *see also* Kawahara
Kawashima, M., 246, *see also* Yoshimitsu
Kay, R. W., 207
Keller, A. V., 14, *see also* Taylor
Kelley, F. L., 186, 236
Kemmer, M. A., *see* Snyder
Kendall-Smith, B. J., *see* Feriday
Keran, E., *see* Schorr
Kerbeck, A., *see* Reuss
Kerber, M. L., *see* Akutin

Kershaw, R. W., *see* Clark
Khainer, S. P., 188, *see also* Kartashova
Khanenya, G. P., *see* Shutov
Kharchenko, T. F., 263, *see also* Sukovatov
Khlystunova, E. V., *see* Druzhinin
Khomyakov, A. M., *see* Andrezen
Khor'kova, M. A., *see* Putlyaev
Khoshkhoo, H., 30
Khrulev, V. M., 111, *see also* Bezverkhaya, Shutoy
Khrustaleva, V., 257, *see also* Gel'tischeva
Khrustaleva, V. A., *see* Kachmar
Khviler, V. P., *see* Akutin
Khylstunova, E. V., 189
Kikuchi, K., 204
Kilpelainen, 121
Kimura, T., *see* Izumi
Kinastowski, S., 106, 114
King, J., *see* Bercovici
Kirkeby, W., 289
Kirsch, K., 176
Kishi, A., 178
Kitakado, Y., 246
Kitamma, 247
Kitamura, K., *see* Shigeta, Uchida
Kitayama, M., 203, *see also* Honma
Kitazawa, M., *see* Takahashi
Kitchens, J. F., 263
Kitsuda, Y., 106
Kitta, Y., 106
Klatte, 6
Klauditz, 166
Klebeko, H., 109, *see also* Wirpsza
Klemm, H., 164
Klimashevich, N. M., *see* Belyi
Klingenberg, 14
Klippel, W., *see* Warneck
Knabe, G., 204
Knapp, 194, 217
Knat'ko, V. M., 205
Knilling, W. v., 17
Knopf, 198
Kobayashi, K., 176, *see also* Oka, Tajima
Kobayashi, S., *see* Noda
Kobayshi, N., *see* Nomura
Koberg, H., 208
Kobori, K., 206
Kobori, S., *see* Iwasaki
Kobuta, 233, 234, 237, 242
Koch, H. G., 107
Kodama, T., 106, 109
Kodo, K., 211
Koehler, H., 18, *see also* Knabe
Kohlrausch, 109
Kohn, S., *see* Duval
Kojima, S., 252
Kolaric, J., 109
Kolarz, B., *see* Mazur
Kolbe, H., 5
Koleshnya, A. D., 122
Kolesnikov, P. A., 291
Kolino, 51
Kolochinskaya, I. B., *see* Aleksandrovich
Kolotilo, D. M., 207
Koma, T., *see* Minoji
Konishi, M., 206, *see also* Morimoto, Takeyama
Konovalov, P., *see* Zisman
Konrad, H., *see* Boehme
Kopitovic, 236
Koromyslova, T. S., 111, 211, *see also* Gamova
Kossatz, G., *see* Roffael
Kostin, N. N., *see* Anokhin
Kotani, Y., 105, 106, 246
Koval'chuk, L. M., 111
Kovalev, G. V., *see* Rudnev
Koyama, T., 110, *see also* Izumi
Kozheznikova, A. A., *see* Tsyvin
Kozhukhova, T. V., *see* Krikunov
Kozin, V. M., 188, *see also* Den'gina
Kozlov, M. K., 105, *see also* Druzhinin
Kozlova, G. I., 139
Krach, H., 108, 109
Kraineva, L. V., *see* Bokareva
Krakauer, E., *see* Lewin
Kramer, F., 208, 210
Kramer, H., 164
Kratochvil, I., 291
Kratzl, K., *see* Silbernagel
Kravchenko, T. I., 236
Kreibich, R. T., 109
Kreibaum, 166
Kreidl, I., 16, 118
Krikunov, N. I., 146
Kropfhammer, G., 211
Krotov, Yu. I., *see* Druzhinin
Krueger, A., 110
Krumphanzl, O., *see* Susicky
Kubinski, H., 246, *see also* Morin
Kubitzky, C., 107
Kubo, M., 189, *see also* Nakajima
Kubota, M., *see* Mori
Kucera, L. J., 174
Kuchenbuck, J., 14
Kuegler, L., *see* Kuemmel
Kuehn, H. G., 205
Kuemmel, R.
Kugriklov, 105
Kulle, T. J., 262
Kumlin, K., 79, 92, 95, 105, 134
Kuniak, L., *see* Pastyr
Kunstharzfabrik Fritz Pollak, *see* Pollak
Kura, Y., *see* Ishida
Kurashige, H., 176, 178
Kurene, A. F., 181

Kuroda, J., *see* Shiota
Kurshev, I., 208
Kurtenacker, 125
Kusushita, T., 110
Kuzmenki, N. M., 291
Kuznetsova, A. E., 207, 212
Kuzuba, N., *see* Miyahara
Kyle, B. G., 146

Laer, M. V., 7
Lafarge, S. A., 206
Laire, Malet, and Armenault Co., 13
Lajoie, J. L., 106
Lakshminarayanan, T. R., 211
Lal, 51
Lambru, A., 108
Lambuth, A. L., 110
Lander, H. L., 105
Landqvist, N., 133
Lane Co., 181
Lapitskii, V. A., 106
Laqua, A., 106
Larcher, A., *see* Bondi
Latinerer, 194, 217
Laukhin, Yu. A., 205
Lauch, 58
Lauter, F., 13, 53
Lavrova, O. A., *see* Tunik
Laz'ko, A. D., *see* Frolova
Leary, B., *see* Gibson
Lederer, 71
Lee, C. A., 286
Lee, D. R., 146
Lee, W. Y., 89
Lehman, W. F., 108, 111, 146, *see also* Harkin
Lehn, J.-M., 24, 26
Lehnert, E. J., 286
Lehnert, H., 178, *see also* Siegler
Lein, F., 178
Lemme, 125
Lenke, G. M., 106
Leneuf, A., 205
Lenz, J., 18, 106, 109, 176, 188, 189, 246, *see also* Brunnmüller, Petersen, Scheuermann
Leonovich, I. I., *see* Nelyubin
Lesteva, T. M., *see* Tunik
Le Terre, 52
Levadie, B., 146
Levi, B. I., 17
Lewin, M., 174
Lewis, R. J., Sr., *see* Christensen
Lias, S. H., *see* Hughes
Liburkin, V. G., *see* El'bert, Erykhov
Lichtenberger, J., *see* Etablissements Lambiottes Freres
Liebig, J., 5, 66
Liebman, S. A., 89
Lielpeteris, U., 111
Likholetov, O. D., *see* Katsyuba, Mironov
Lindberg, J. J., 106
Lineken, 60
Lipkin, G., *see* Baer
Little, A. P., *see* Argo
Little, R. W., *see* Sarig
Litvinenko, V. M., *see* Frolova
Lobenhoffer, H., *see* Ernst
Lobering, 42, 43
Loehmer, W., 203
Loew, 47
Loginov, V. P., *see* Kozlov
Lohenhoffer, H. Ph., 177
Lomtadze, Z. K., 210
Lubbock, F. J., *see* Clark
Lubis, B., 210
Ludlam, P. R., 89, 133
Ludwig, C. H., 245
Ludwig, O. R., 18
Luecke, E., *see* Reuss
Lüdersdorf, 6
Luedy, E., 6, 7
Luethge, P., *see* Hoener
Lühr, H. P., 217
Lukesch, K., *see* Binder
Lumb, W. M., *see* Argo
Lundqvist, G. R., 88, 258, *see also* Andersen
Lupa, D., *see* Branowski
Luther, M., 12, 14
Lutoshkina, G. T., 111
Lützow, D., 99
Lutwak, H., *see* Walter
Lvovsky, C., 214
Lyubimov, V. V., *see* Kokolova

Macaskill, S. M., *see* Levadie
Maczuzak, A., *see* Szlezyngier
Maczynska, M., *see* Nowak
Maeda, M., *see* Kodo
Maeda, T., 109
Maeno, T., *see* Ohhara
Mailbach, H., *see* Epstein
Majsjej-Kubik, D., *see* Zenkteler
Makarenko, E. M., *see* Vinogradov
Maksimov, Yu. I., 181
Maku, T., *see* Kanida
Malafeeva, Z. V., *see* Bulygin
Malakhova, V. Zh., 207, *see also* Batyuk
Maller, G., *see* Sarrut
Maloney, T. M., 110, 166, 168, 170, 178, 219, 286
Malygina, E. F., 237, *see also* Kravchenko, Stankevich
Mamykina, O. A., *see* Putylyaev
Manasse, 71
Manczak, T., *see* Dirska

Mangs, J., *see* Seeholzer
Mannick, 62
Marcote, R. V., 149, *see also* Johnston
Marichev, V. D., *see* Den'gina
Markevich, P. N., *see* Bartashevich
Markin, A. D., *see* Aleksandrovich
Marks, M., 17
Marra, A. A., 157, 158, 159, 175, 229
Marsh, 199, 200
Martirosyan, G. A., 210
Martynov, K. Ya., *see* Bezverkhaya
Masek, V., 146
Maslinkovskii, A. S., 286
Massa, U., *see* Catoni
Masuda, S., 109
Masumi, J., *see* Ishida
Materiaux Reunis Co., 205
Mathey, R. G. Jr., *see* Rossiter
Matsuda, H., *see* Tange
Matsumoto, T., 110, 126, 151, 155, 237, 242
Matsumura, M., *see* Koyama
Matsumura, Y., 210
Matsunaga, 242, *see also* Mitsua
Matsuoka, K., 18, 203
Matsushima, K., *see* Kato
Matsutomi, Y., *see* Ohhara
Mattsson, O., 110, 233, 240
Mattsson, O. H., *see* Lehnert
Mavoks, 118
Mayer, J., 107, 248, *see* Brunnmüller, Eisele, Grabowsky
Mayet, A., *see* Stofft
Mayrhofer, R., 13, *see also* Goldschmidt
Mayrl, J. A., *see* Miller
Mazur, S., 205
Mazurova, L. A., *see* Andrezen
Mazzoleni, G., *see* Vargiu
McAffee, R. C., *see* Anderson
McGrego, K., *see* Henbest
McKendree, J. R., *see* Snyder
McMinimy, L. O., 175
McReynolds, K. B., *see* Lenke
Meany, J. E., *see* Pocker
Mecke, 73
Mehlhorn, L., 150, *see also* Roffael
Melia, 30
Mel'nik, M. S., *see* Naidenov
Melzer, J., *see* Obetko, Petersen
Menashi, J., 205
Merkel, D., 110
Merkle, H. P., *see* Speiser
Messman, H. C., *see* Cropp
Meyer, B., 24, 25, 30, 31, 39, 41, 51, 52, 53, 55, 56, 104, 118, 119, 125, 141, 142, 143, 149, 150, 152, 159, 178, 194, 210, 243, 247, 249, 254, 256, 289, *see also* Smith
Meyer, W. T., 196
Meyer, L. S., 186, 198
Michael, J., 257
Michaud, H., *see* Seeholzer
Michel, M. R. A., *see* Blandin
Micko, M. M., *see* Paszner
Midwest Research Institute, 188
Miertzsch, H., *see* Mehlhorn
Mikhailov, N. A., *see* Koleshnya
Mikhailova, A. A., 237
Mildenstein, 236
Miller, C. H., 206
Miller, J. A., *see* Mueller
Minamida, H., *see* Okada
Minemura, N., 110, 146, 242-244, *see also* Hirata, Imura
Minnesota Mining and Mfg. Co., 17, 203
Minoji, A., 177, 247, 286
Mints, B. Ya., 202
Miretskii, V. Yu., *see* Strelkov
Mironov, O. G., 205
Mita, A., 121
Mitera, J., 257, *see also* Michal
Mitrofanov, E. G., 111
Mitsua, I., 242
Miura, N., *see* Sano
Miura, S., *see* Ando
Miwa, K., *see* Kusushita
Miwa, T., 246
Miyabayashi, T., *see* Tsuge
Miyake, K., *see* Kikuchi
Miyazaki, T., *see* Takashima
Miyazaki, Y., *see* Sakarada
Mochalova, Yu, Z., 114
Mohl, H. R., 155, 241
Møhlhave, L., 236, 240, 258, *see also* Andersen
Molotkova, N. N., *see* Pshenitsyna
Monfort, P., *see* Jamin
Monsanto Co., 98, 106, 203
Monti, L., 15
Moore, T. B., *see* Stubblefield
Moore, W. P., 106, 114, 119, 210, 248
Moortgat, G. K., *see* Warneck
Moralt, A., 174
Mori, S., 176
Morimoto, T., 206, *see also* Takeyama
Morin, N. C., 291
Morita, K., *see* Higuchi
Morizaki, S., *see* Yui
Moroff, H., 176, 286
Morozova, S. S., 204
Morschauser, C., 257, 276
Morze, Z., *see* Kinastowski
Moschandreas, 148
Moskalenko, E. M., *see* Mochalova
Moskvitin, N. I., *see* Tsvetkov
Moslerin, A. A., 219

Mostovoy, S., 107
Moult, R. H., *see* Blommers
Mueller, B., *see* Neumann
Mukherjee, D., *see* Sharphouse
Mukhibova, N. Z., *see* Putylyaev
Mukudai, J., 108
Mulyukov, E. I., *see* Polak
Muminov, A. M., *see* Akhmedzhanov
Munari, S., *see* Hayasahi
Murakami, T., *see* Oka
Murashov, Yu, S., 188, *see also* Novak, Valgin
Murav'ev, V. S., 107, *see also* Sakovich, Zinin
Muresan, L., 204
Muyagi, T., *see* Uchida
Myers, G. E., 139, 142, 143, 144, 182, 232, 240
Myers, R. R., 203

Nadudvari, I., *see* Gavel
Nagao, A., 105, *see also* Takahashi
Naidenov, M. N., 190, *see also* Sakarada
Nakajima, H., 189
Nakajima, M., 109
Nakajima, T., 204
Nakajima, Y., *see* Kobayagawa
Nakamoto, T., *see* Okabe
Nakamura, H., *see* Imura
Nakamura, M., *see* Kitayama
Nakane, B., 176
Nakanishi, T., 105
Narasaki, H., 111
Narkis, M., *see* Siegmann
Nash, 128
Nassar, F., *see* Schliefer
Natalich, A. F., *see* Sklyarenko
National Institute for Occupational Safety and Health (NIOSH), 145, 146, 148, 149
National Particleboard Association, 174, 175
Natkina, L. N., *see* Gamova, Koromyslova
Nefedov, Y. G., 257
Negi, S. S., *see* Pal
Nelyubin, I. A., 206
Nestler, F. H. M., 149, 182, 236, 267
Neumann, H., *see* Moroff
Neumann, R., 187, 194, 217, 245
Neuss, O., 10, 15, 16, 17, 257, *see also* Goldschmidt
Neusser, H., 107, 111, 182, 234, 235, 236, 256
Nevin, J. V., 18
Niedermaier, T., 210
Nielsen, 26
Nischk, W., *see* Zigeuner
Nishida, Y., *see* Tsuruta
Nishino, S., 245
Nishiyama, Y., *see* Kunii
Nistri, U., *see* Vargiu
Nitsche, I., *see* Stofft
Nixon, E. R., 25, 30, *see also* Khoshkhoo
Nizhni, 247
Nobell, A., 210
Noda, H., 190
Nogales, A., *see* Tapia
Nogi, T., 110, 202
Noguchi, 236
Nomura, M., 109
Nonaka, Y., *see* Tanaka
Norkute, A., *see* Zisman
Noskova, M. P., *see* Pshenitsyna
Novak, V. A., 188, 189, *see also* Valgin
Nowak, D., 121
Novopan-Keller Co., 205

Oba, T., *see* Kojima
Oberlein, A., *see* Ernst
Obetko, D., 204
Oblivin, A. N., 111, *see also* Azarov
O'Brien, S. J., 202, *see also* Stockel
Obst, J. R., *see* Harkin
Ochiai, I., *see* Miwa
Oda, N., 110
Odagiri, M., *see* Suzuki
O'Donnell, J. M., 211
Oesterreich, K., 16, *see also* Walter
Oetgen, W. R. 114
Ogawar, H., *see* Kurashige
Ogden, D. H., 205
Ogiwara, T., 194
Ohhara, O., 106, 110, 246
Ohhashi, K., 246
Ohi, S., *see* Yui
Ohmura, M., 178
Ohno, K., 246
Ohno, T., *see* Ohtomi
Ohnuki, T., *see* Iwata
Ohta, T., *see* Inoue
Ohtake, M., *see* Fujitani, Masuda
Ohtomi, M., 129
Oka, Y., 178, *see also* Ishida
Okabe, J., 178
Okada, K., 206, *see also* Yamamoto
Okamura, H., *see* Shirai
Okamura, S., 108, 189, 198, *see also* Shiota
Okotokov, D. N., *see* Gracheva
Okrensny Stavebno, 189
Okumichi, T., *see* Endo
Okura, K., *see* Shofuda
Olszowski, T., 106
O'Neill, 248
Ono, S., *see* Yamagishi
Onoura, O., *see* Minoji
Orlova, L. F., *see* Andrezen
Ormstad, E., *see* Mattsson
Orth, G. O., Jr., 210

Osaulenko, V. T., *see* Laukhin
Oshika, S., *see* Takahashi
Oshima, K., 188, 197
Osipova, N. A., 106
Osnach, N. A., *see* Sklyarenko, Zastanchenko
Osokrisa, S. K., 257, *see also* Gel'tishcheva
Ostapovich, I. K., 291
Osterberg, J. O., *see* Koch
Othmer, K., 97
Otsuka, E., 99, 206
Ozlorskaya, 130

Pace, L. S., 110
Pacific Resin, 276
Pacsu, E., 59, 199, *see also* Wagner
Pages, J. C. J., *see* Rougier
Pagnotto, L. D., *see* Elhins
Pahla, P., 175
Pal, R. N., 210
Pal, Mrs. K., 210
Panov, V. V., 114
Papeshina, V. I., *see* Borisyuk, Ishin
Parnell, W. E., 107
Paronyan, R. V., 146
Parshakov, V. E., *see* Knat'ko
Pashkov, D. N., 109
Pashkov, N., 205
Pastyr, J., 181
Paszner, L., 111, 181
Paterson, A. R., 177
Patrick, 52
Paul, R. L., 143
Pavlenko, S. M., 262
Pearson, D., 267
Pellegrini, P., *see* Catoni
Pellizzari, E. D., 146
Pelshenke, H., *see* Brose
Penczek, P., 106
Pennis, W. L., *see* Sundie
Perevertov, A. S., 107
Perrier, A., *see* Behar
Pershin, M. N., *see* Putseiko
Persson, S. A. E., *see* Gruben
Peschong, N., *see* Jamin
Peters, T. E., 175
Petersen, H., 75, 79, 81, 82, 83, 84, 86, 87, 95, 96, 107, 115, 155, 178, 183, 184, 189, 204, 236, 248, *see* Eisele, Erhardt, Reuther
Petersen, S., *see* Bögemann
Peterson, S., 236, *see also* Rehn
Petra, D., *see* Ionescu Muscel
Petre, E., *see* Teodorescu
Petrik, B., *see* Holtschmidt
Petrova, O. V., *see* Petrusenko
Petrusenko, P. I., 105

Pfenning-Schumacher Industries, 118
Pfohl, F., 164
Phillips, P. S., Jr., *see* Foris
Philpotts, A. R., 30
Piechota, H., 197
Pierce, D. S., *see* Fahey
Pierce, E. T., 30, *see also* Rossiter
Pilipenko, T. I., *see* Lapitskii
Piret, 27
Piria, 71
Pitter, R., *see* Zigeuner
Pizzi, A., 106
Plath, L., 107, 111, 150, 166, 235, 236
Platonov, A. P., *see* Putseiko
Platzer, 274
Pleskachevskii, M. Yu., *see* Belyi
Plomley, K. F., 178
Plotka, E., *see* Schorr
Plskova, M., *see* Labsky
Pohl, G., *see* Junkermann
Polinek, V., 211
Pollak, F., 4, 7, 10, 11, 12, 13, 14, 16, 17, 65, 114, 118, 185, 215, 229, 255
Polukhin, V. V., *see* Akutin
Pommer, E. H., *see* Clad
Popov, L. K., *see* Lapitskii
Popova, N. I., *see* Koleshnya
Popovic, M., *see* Bogosavljevic
Poschinger, W., *see* Seeholzer
Potekhina, E., 109
Potutkin, G. F., 89, 107
Poulds, 200
Powell, E. C., 14, *see also* Basterfield
Pratesi, 28
Prahl, 53
Pratt, B. C., 122
Preobrazhenskaya, I. P., 107, *see also* Koval'chuk
Prilukov, A. D., 106
Pritchard, 106
Proust, 4, 5
Pruen, H., *see* Niedermaier
Przybylak, A., 245
Pshenitsyna, V. P., 89
Pukhovitskaya, A. N., *see* Sholokhova
Pummer, H., *see* Danner
Pungs, W., 14, 17
Purcell, R. F., 105, 176
Pusch, G., 203
Pushkina, I. K., 257
Putlyaev, I. E., 105, 203, *see also* Mironov
Putseiko, L. K., 207

Quade, R., 58

Rachmar, 146
Rackwitz, 166

Raddin, H. A., *see* Brooks
Raizman, M. Sh., 107
Raknes, E., 111, 178
Ramanauskiene, V., *see* Burakinskaya
Raschig, 6, 52, 53
Rauch, W., *see* Roffael
Rauter, W., *see* Zigeuner
Rayner, C. A. A., 163
Rechav, Y., *see* Siegmann
Rechner, L., 189
Redfarn, C. A., 17
Redmond, 49
Rees, W., *see* Epstein
Reeves, B. H., 205
Rehn, T., 236, 251, 252
Reichhold Chemicals, 276
Reimschuessel, A. C., *see* Paterson
Reinhart, R. F., *see* Antlfinger
Reisig, I. S., *see* Stankevich
Renner, A., 101, 110, 248
Reuss, G., 110
Reuther, W., 114, 155, *see also* Petersen
Reutskii, V. A., *see* Akutin
Rice, R. G., *see* Stahly
Richard, B., 89, 110
Richardson, J., *see* Hanna
Richardson, L. R., *see* Juneja
Riesmeier, W., 107
Rikunov, E. F., 109
Ripling, E. J., *see* Mostovoy
Ripper, K., 7, 10, 12, 13, 14, 16, 17, 52, 121, 125, 185, *see also* Pollak
River, B. H., *see* Gillespie
Robitschek, P., 178, *see also* Christensen
Rodgers, 145
Roehr, H., 197, *see also* Piechota
Roessler, G., *see* Lenz
Roff, W. J., 59, 60, 61, 199, 201, 202
Roffael, E., 109, 150, 154, 155, 178, 182, 236, 241, 286, *see also* Marutzky, Mehlhorn
Rogerson, D. E., *see* Stout
Rohm and Haas Co., 18
Rohmann, M., *see* Petersen
Roman'ko, A. S., *see* Rikunov
Romashova, A. G., *see* Perevertov
Romitsyn, V. I., *see* El'bert, Erykhov
Ron-Werkst, H., *see* Paulitsch
Roos, O., 166
Rosenbauer, H. G., *see* Aignesberger
Rosenberg, P., 58
Rosengren, A., *see* Dalham
Rosenkranz, H. S., 254, 263
Rosenthal, D., 187
Rossi, C., *see* Hayasahi
Rossiter, E. C., 4, 10, 13, 17, 18, 226
Rossiter, W. J., Jr., 194, 196, 217
Rothera, W. S., 7, 11
Rouelle, G. F., 4, 72
Roussel, B., *see* Chapput
Roy, D. C., 121
Rudnev, A. V., 26, 40, 44
Rudnik, O. D., *see* Usmanov
Runauer, E., *see* Reuss
Rusakova, T. V., *see* Putylyaev
Russek, W., *see* Suwalska
Ruthner, G., 188, *see also* Schwab

Saarbergwerke, A. G., 210
Sachs, 198
Saeki, T., *see* Ueda
Saito, M., *see* Mori
Sakaguchi, H., 111
Sakarada, S., 121
Sakato, N., *see* Nakajima
Sakovich, L. I., 114, *see also* Murav'ev
Sakuno, T., 111
Sakurai, T., *see* Hasegawa, Higashimura
Salomon, G., *see* Houwink
Samoilov, V. I., *see* Osipova
Sandalov, A. V., *see* Osipova
Sander, B., 3, 204
Sandler, S. R., 79, 107
Sando, T., *see* Tashiro
Sanin, I. K., *see* Afanas'ev
Sano, S., 110
Sansing, J. E., *see* Moore
Sarig, Y., 194
Sarkisyan, R. R., 206
Sarkisyan, V. K., *see* Paronyan
Sarrut, J., 114
Sasaki, T., 246
Sato, M., *see* Imura
Sato, T., 106, 189, 198, 207, *see also* Shiota
Sausman, D. K., 106, *see also* Purcell
Sautin, A. I., 214
Sawicki, C. R., 124, 127, 129, *see also* Sawicki
Sawicki, E., 124, 129, *see also* Pellizzari
Sax, N. I., 266
Schall, W., *see* Neisser
Schaefer, W., *see* Terporten
Scharwaechter, P., 189, *see also* Petersen
Schatz, F. H., *see* Brunnmüller
Schedro, 178
Scheiber, J., 19, 70, 79, 88, 105, 115, 248
Scheibler, H., 13, 14, 16, 18
Scheuermann, H., 18, 19, 96, 186, *see also* Lenz
Schibler, L., *see* Speiser
Schimmel, K. F., *see* Wismer
Schiff, 127
Schliefer, K., *see* Hebeish

Schmidt, G., 197
Schmidt, M., 51, 73
Schmidt-Hellerau, C., 105, *see also* Mayer
Schmidt-Hellerau, K. C., 247
Schmiking, A., 14
Schmitzler, 166
Schneider, A., 108, *see also* Roffael
Schneider, H., 210
Schneider, W. G., 30
Schneiwind, A. P., 111, *see also* Halligan
Schoellhorn, W. D., 190
Schoenthaler, 79
Scholz, E., *see* Scheibler
Schorning, P., 117
Schorr, W. F., 26, 263
Schradere, 73
Schuerch, C., *see* Davidson
Schulz, H., 210
Schulze, W., 197
Schurath, V., *see* Becker
Schwab, J., 188
Schwager, P., 204
Schwenkler, 58
Scott-Paine, H., 161
Sears, K. D., *see* Casebier
Seberini, M., 174
Sedlak, L., *see* Zapletal
Seeholzer, J., 203
Segerlind, L. J., *see* Sarig
Segond, M., 18
Seifert, J., 111
Sekachev, P. G., *see* Smirnova
Seki, K., *see* Tsubota
Seldon, M. M., 214
Seliverstov, P. I., *see* Dement ev
Sell, 183
Sememova, A. T., *see* Dubrovin
Semenov, V. I., *see* Azarov
Semoto, H., *see* Kawamura
Sempevil Co., 188
Senkevich, O. V., *see* Demenkova
Senn, S., 175
Sereda, N. I., *see* Kozlova
Sergeeva, M. P., 107
Seto, T., *see* Inoue
Sevliakov, 263
Sgibnev, A. K., 262
Shank, H. C., Jr., 211
Shakra, S., *see* Galil
Shamshura, I., *see* Bokov
Sharkovskii, V. A., *see* Akutin
Sharphouse, J. H., 105
Shastri, S., 245
Shearer, H. N., 205
Shelton, T. S., 203
Shemerdyak, B. M., *see* Naidenov
Shen, 12
Sheppard, N., *see* Philpotts
Sheridan, J. L., 175
Shevchenko, A. M., *see* Kharchenko
Shigeta, T., 205
Shikata, H., *see* Nishino
Shimokai, K., 205
Shiota, T., 110, 189, 198, 246
Shirokorodyuk, V. K., 206
Shiryaeva, L. V., *see* Potutkin
Shiskov, I. T., 111
Shkol'nik, Ya. Sh., *see* Raizman
Shofuda, H., 106
Sholokhova, A. B., 111
Shplet, N. G., *see* Grigor'eva
Shriner, 53
Shumilina, A. V., 263
Shusterzon, G. I., *see* Khrulev
Shut'ko, A. F., *see* Shutov
Shutov, G. M., 177
Shvartsman, G., 245
Shvets, D. P., *see* Kozlova
Shyaku, T., *see* Nakajima
Sidorova, E. A., *see* Volkova
Siegens, H., *see* Girsewald
Siegler, M., 110
Siegmann, A., 210
Siegrist, R., 101, *see also* Jaccard
"Silur" Tech. & Chem. Produkts, 12
Simek, M., *see* Susicky
Simeonov, I., 114
Simonson, R., 95, 134, *see also* Kumlin
Singh, M. M., *see* Guha
Singh, S. P., *see* Guha
Sire, J. M., *see* Bitterli
Sklyarenko, B. S., 178
Slesarev, V. V., *see* Spirin
Slivkov, K., *see* Ivanov
Slonim, I. Ya., 69, 79, 89, 93, 95, 135, 203, 204
Smidth, L., 12, 14, 17
Smirnov, B. A., 205
Smirnova, E. A., 110
Smith, J. E., *see* Cowan
Smith, P., *see* Lewin
Smith, R. G., 146
Smith, T. F., 203
Smythe, 92
Snyder, A. D., 146
Soboll, R., *see* Binder
Sofue, 92
Société Nobel francaise, 15, 17
Sokolova, L. Yu., 212, *see also* Time
Solechnik, N. Ya., *see* Koromyslova, Time
Solvay et Cie, 205
Sorensen, 67
Sorokina, A. A., *see* Kokolova
Soto Urbina, F. G., 111

Sparkes, A. J., 107
Spãthe, E., 186, *see also* Scheuermann
Speiser, P. M., 204
Spence, R., 23
Speyer, F. B., 214
Spichala, M., *see* Neumann
Spiers, 68
Spiner, D. R., *see* Braswell
Spinner, 73
Spirin, Yu. L., 205
Splawa-Neyman, S., 111
Srama, D., *see* Zenkteler
Stacke, P., 198
Stadnik, L. N., *see* Koleshnya
Stakavich, A. J., 146
Stamm, A. J., 61
Stancu, A., *see* Lambru
Stankevich, K. I., *see* Kravchenko
Stannett, V. T., *see* Wellons
Stark, B. P., *see* Hope
Starke, P., *see* Geissler
Starzynska, K., 121, 246
Stashevski, A. M., 111
Statsek, N. K., *see* Ivanova
Staudinger, H., 12, 15, 29, 30, 57, 70, 90
Steeger, 166
Steele, R., 60, 199, 200
Stegmann, G., 166, *see also* Schorning
Steiger, 248
Steiner, P. R., 109, 111, 112, 113, 114, 115, 116, 117, 166, 176, *see also* Chow
Stene, R. M., 202
Stenzel, J., 106
Stepien, K., *see* Penczek
Steppes, F. E. K., 13
Stockel, R. F., 202
Stofft, E., 254, 257
Stofko, J., 178
Stolger, 235, 236
Stout, A. W., *see* Ludwig
Stout, R. M. T., 245
Strelkov, V. P., 111
Stubblefield, D. J., 196
Suematsu, A., 247
Suen, T. J., *see* Updegraff
Suetaka, W., 89
Sugano, S., *see* Suzuki
Sukovatov, V. I., 236
Suleimanova, Zh. M., *see* Putlyaev
Sulzberger, M. B., *see* Rostenberg
Sundie, R. D., 106, 178
Sundin, B., 236, 240, 258, 259, 261, *see also* Mattsson
Sundstedt, B. E. H., *see* Koch
Supkis, S. J., 203
Surotkin, Yu. V., *see* Kozlov
Susicky, K., 178
Suwalska, W., 114
Suzuki, J., 73, 106
Suzuki, K., *see* Suzuki
Suzuki, S., 188, 197, *see also* Ito, Oshima
Suzuki, T., 246
Suzuki, Y., 206
Sverkanova, V. V., *see* Golubovskaya
Svitkin, M., *see* Shvartsman
Svitkina, M. M., *see* Temkina, Yudina
Svrdiik, R., 89
Swarfsman, 111
Swaters, P. D., *see* Havinga
Swigart, M. L., *see* Burnett
Swire, J. R., 206
Synthetic Organic Chemicals Manufacturers Association, 148
Synthetic Plastics Co., 17
Syrov, N. V., *see* Belyi
Szaller, I., *see* Gavel
Szemiotowicz, J., 105
Szewczyk, E., *see* Szlezyngier
Szlezyngier, W., 121, 246, 247, *see also* Gnatowski
Szozda, I., *see* Wnuk

Tahara, M., *see* Iwaki
Tajima, S., 105
Takahashi, H., 242, *see also* Kubota, Minemura
Takahashi, I., 210
Takahashi, M., 106, 108, 110, 247
Takahashi, S., *see* Kurata
Takahashi, T., *see* Otsuka
Takaka, M., *see* Katayama
Takashima, H., 207
Takashima, K., *see* Kawahara, Tanaka
Takashima, S., 194, *see also* Ito
Takatsuji, I., *see* Kikuchi
Takayama, Y., *see* Ichimura
Takenoto, Y., *see* Mori
Takeo, K., *see* Kotani
Takeshita, S., 106
Takeuchi, K., *see* Hasegawa
Takeyama, K., 257
Takhirov, M. T., 262
Tamolang, F. N., *see* Jumneanbhandhu
Tamura, Y., 177
Tanaka, A., 105, 108
Tanaka, M., 245
Tanaka, S., *see* Tsuge
Tange, Y., 246
Taniguchi, Y., *see* Iwaki
Tanioka, H., 109
Tanno, T., *see* Shiota, Tashiro
Tapia, O., 24, 26
TAPPI, 125
Tarakhtunov, *see* Akutin

Tarasenko, L. D., 189
Tarasova, S. I., *see* Sholokhova
Tartar, 125
Tashiro, F., 246
Taskina, T. A., *see* Andrezen
Tatarczyk, T., *see* Becker
Taubkina, R. S., *see* Burakinskaya
Taura, T., *see* Eguchi
Taylor, J., 7, 14, *see also* Dixon, Keller
Taylor, L. J., 121, 204
Taylor, W. G. K., *see* Vale
Tazawa, S., *see* Hasegawa, Higashimura
Telichko, K. A., *see* Demenkova
Temkina, R., *see* Shvartsman
Temkina, R. Z., 108, 114, 175
Temple, R. G., *see* Wismer
Tendrovskaya, V. A., 237, 291
Teodorescu, L., 247
Terporten, H., 110
Terskova, T. V., *see* Denisova
Teukros Co., 114
Teveb, 111
Thiel, 248
Thielsch, H., *see* Knabe
Thinius, K., 15
Thomas, S., 198
Tiedeman, G. T., 106, *see also* Freeman
Tikhomirov, Yu. P., *see* Gracheva
Tikhomirova, A. F., *see* Putlyaev
Time, N. S., 176, *see also* El'bert
Timmons, A., *see* Smith
Tinyakova, N. M., *see* Khlystunova
Tirosh, N., *see* Siegmann
Tobol'skii, G. F., 189, 205
Tokeyama, K., *see* Morimoto
Tokuoka, M., 17
Toledo Scale Manufacturing Co., 14
Toledo Synthetic Products, 16, 211
Tollens, B., 4, 6, 58, 88, *see also* Wellington
Tomita, B., 39, 73, 79, 89, 91, 92, 105, 135, 138, 139, 140
Tomokawa, H., *see* Miwa
Tootal Broadhurst Lee Co. Ltd., 17, 200
Torfit Co., 164
Towati, R., 206
Toyota, H., 207
Traganos, F., 263
Trautvetter, R., 174
Trinkler, H., 263
Trocino, F. S., 121
Trostler, F., *see* Scheibler
Troughton, G. E., 112, *see also* Chow
Trubitskaya, G. P., *see* Anokhin
Truempler, 26, 38
Tsai, F. H., 111, *see also* Ding
Tsareva, Z. V., *see* El'bert
Tsar'kov, D. P., *see* Kokolova
Tsedrik, Yu. N., *see* Kara
Tsendrovskaya, V. A., *see* Stankevich
Tsibizov, V. F., *see* Osipova
Tsubota, Y., 177
Tsuge, M., 89, 246
Tsuji, K., *see* Hayashi
Tsukamoto, A., 106
Tsupikov, M. T., *see* Yukin
Tsuruta, K., 178
Tsutsumoto, T., 111
Tsvetkov, V. E., 105, 109
Tsyvin, M. M., 111
Turunen, J., *see* Lindberg

Uchida, T., 211
Uchihara, S., *see* Ohtsuka
Uchimi, S., 174
Ueda, K., 106, 178, 211, *see also* Fujita, Kitsuda, Kitta, Shofuda
Ueda, M., 178
Uemura, Y., *see* Tanaka
Ueng, S. D., *see* Ding
Ugine-Kuhlmann Co., 15
Ulbricht, J., 212
Uliana, C., *see* Hayasahi
Ullmann, 97, 230
Umezawa, K., *see* Narasaki
UNESCO, 178
Unyte Co., 17
Updegraff, I. H., 270, 272
Updegraff, I. L., 79, 100, 107, *see also* Petropoulos
U.S. Dept. of Housing and Urban Development (HUD), 148
U.S. Environmental Protection Agency (EPA), 148
Ushakova, M. B., *see* Lapitskii
Usmanov, S., 208
Ustinovakaya, I. A., 146
Utsumi, S., 178

Vad., J., 210
Valgin, V. D., 188, *see also* Novak
Van Atta, F. A., 266
Van Slyhe, 254
Varga, G. A., *see* Davies
Vargiu, S., 101, 106, 108
Varlacheva, L. A., 245
Vaskevich, D., *see* Zisman
Vassil'ev, V. V., *see* El'bert
Vassiliades, A. E., 204
v Bismarck, H. C., *see* Soto Urbina
Vel'sovakii, V. N., *see* Spirin
Verbestel, J. B., 150, 184, 235, 236, 243, 245, 261, 286
Vierling, 14
Viktorova, E. V., *see* El'bert
Vilker, S. B., *see* Kruglikov

Villaume, J. E., *see* Hsiao
Villwock, I., 198, *see also* Geyer
Vinogradov, G. I., 291
Vinogradov, V. M., 212
Vintila, E., 174
Voets, J. P., *see* Claes
Volkova, Z. A., 291
Vologodskii, A., 263
Volotskoi, D. V., 205
Vrublevskii, B., *see* Nelyubin

Waddleton, N., 190
Wagner, R. E., 59, 199
Walker, J. F., 6, 22, 23, 26, 30, 32, 33, 34, 37, 38, 45, 50, 51, 57, 65, 118, 124, 130, 131, 149
Wallhaeusser, H., *see* Gilfrich, Schulz
Wallin, N. H., *see* Mattsson
Wallisch, K. L., *see* Frankel
Walter, G., 11, 12, 15, 16, 17, 65
Warneck, P., 22, 46, 255
Washabaugh, F. J., *see* Demko
Wayne, L. G., 148, 258, 262
Webb, R. J., 178
Weeraratne, W. G., 108
Wegler, R., 79, 100
Weidenbusch, 5
Weise, B., *see* Boettner
Weissenfels, F., 189, 217, *see also* Juenger
Weizman, J., 196
Wellons, J. D., 114
Wells, R. L., 186
Welzer, K., *see* Winkler
Weyker, R. G., *see* Stickel
White, J. T., 181
Whiteside, I. R., 106
Widmann, M., 189, 190
Widmer, G., 11, 12, 13, 14, 15, 18, 19, 66, 79, 120, 185, 199, 271, 273, *see also* Gams
Wilaszek, S., *see* Szlezyngier
Wild, R., 23, *see also* Spence
Wilhelmi, M., *see* Zigeuner
Wilkie, 178
Wilkinson, J., 210
Willegger, 248
Willert, J., 109
Williams, I. L., *see* Petropoulos
Williams, R. H., *see* Demko
Williamson, H. D., *see* Moore
Willie, W. E., *see* Burmester
Wilton, J., 176
Winkler, H. D., 236
Wipff, G., *see* Lehm
Wirpsza, Z., 105, 108, 114, 246
Wismer, M., 214
Wittmann, D., *see* Kollman
Wittmann, O., 107, 111, 155, 235, 236, 248, *see also* Eisele, Grabowsky, Reuther
Wnuk, M., 111
Wohl, 51, 52
Wöhler, F., 4, 5, 6, 72
Wohnsiedler, H. P., *see* Widmer
Wojciak, S., 204
Wolf, H., 16, 185, *see also* Curs
Woo, 59, 199
Wood, A. D., 200, 221
Woodward, J., 198
Wu, S.-C., 111
Wulkan, E. K. H., 217, 248, 249

Yagoda, 146
Yakabe, M., *see* Otsuka
Yakhno, A. G., *see* Borisyuk, Ishin
Yakovlev, D. A., 190
Yamada, H., *see* Suematsu
Yamagishi, Y., 111
Yamaguchi, 73, 74
Yamakita, M., *see* Kotani
Yamamoto, K., *see* Hiyoshi
Yamamoto, N., 178
Yamamoto, Y., 106, 110
Yamao, M., *see* Kodo
Yamasawa, K., *see* Fujitani, Masuda
Yamashita, H., 109
Yamashita, T., *see* Yamashita
Yamato, S., *see* Koyama
Yamazaki, H., *see* Hayashi
Yano, Y., *see* Uchida
Yarlykov, B. V., *see* Afanas'ev
Yashina, I. P., *see* Rikunov
Yasuda, K., *see* Okaba
Yasutake, W., *see* Yamashita
Yata, S., *see* Mukudai
Yeh, C.-M., *see* Wu
Yokoyama, K., *see* Kitayama
Yoshida, A., 178
Yoshihara, T., 252
Yoshihiro, K., 107, 114
Yoshimitsu, T., 245
Yoshimura, S., *see* Otsuka
Yoshinaga, T., *see* Aono, Isono
Yudina, G. C., 99, 178
Yudina, G. G., *see* Temkina
Yui, H., 106
Yukin, N. A., 210

Zabrodkin, A. G., 110, *see also* Anokhin
Zahir, S. A., *see* Hope
Zakhariev, G., *see* Simeonov
Zakharov, T. G., *see* Dashkovskaya
Zakharova, T. G., *see* Kravchenko
Zakharova, T. V., *see* Zabrodkin
Zapletal, J., 190
Zasshi, K. K., *see* Otsuka

Zastanchenko, M. A., 176
Zatopkova, L., *see* Katuscak
Zavarin, E., *see* Stofko
Zbykovskaya, N. I., *see* Ivancheva
Zdorovetskaya, T. A., *see* Gorenko
Zeigerson, E., 246
Zenkteler, M., 178
Zentner, M., 182, 234, 235, 236, 256, 257, *see also* Neusser
Zhukovskii, A. P., *see* Serova
Ziedins, I., *see* Lielpeteris
Ziedman, K., *see* Wayne
Zigeuner, G., 86, 88, 89, 91, 92, 96
Zinovich, Z. K., *see* Komenskii
Zirkel, H., *see* Heiduschka
Zirnin, 109
Zlodovskii, L. I., 203
Zmihorska, A., *see* Szlezyngier
Znachko-Yavorskii, I. L., 205
Zolotarev, B. P., *see* Sklyarenko
Zott, A., *see* Hentschel
zur Hausen, M., *see* Zigeuner

Patent Index

This Index follows the *Chemical Abstracts* procedures. Countries are listed alphabetically. Whenever possible, the U.S. Patent number, rather than the patent of the country of origin, is listed. The correlation between patents issued in different countries can be established with the help of the *Chemical Abstracts Patent Concordance,* which is published yearly. U.S. Patents can be ordered from Box 9, U.S. Patent and Trademark Office, Washington, D.C. 20231.

Throughout the entire world, patent law is currently in flux. Japan has two systems: the Japanese patent system (Japan), and the Japanese Kokai (Kokai). German patents (GP) are now first published as Offenlegung (Ger. Offen.), and only later converted to GP. The same system is followed in several other countries.

The date given in this Index is the year during which the patent was granted. In all Japanese entries, the year in which the Patent or Kokai was issued is included as an integral part of the Patent or Kokai number: the first two digits represent the last two digits of the year of assignment.

Brit. P. stands for British Patent.

Belg. P. stands for Belgian Patent.

Neth. P. stands for Netherlands Patent,

etc.

Patents issued in the Democratic Republic of Germany are listed as Ger. P. (E.) according to the *Chemical Abstracts* convention.

When a patent listing is followed by name and year, the reference will be found in the Bibliography.

Australian P. 481,197, *see* Pritchard, G., 106
Australian P. 441,798, *see* Clarke, B. J., 106
Austrian P. 324,923, *see* "Isovolta" Oesterreichische Isolierstoffwerke, 206
Austrian P. 130,030, *see* Walter, G., (1930), 15
Austrian P. 123,847, *see* Walter, G., (1931)
Austrian P. 121,999, *see* Walter, G., (1930), 15
Austrian P. 111,534, *see* Kunstharzfabrik, Fritz Pollak, (1928), 13
Austrian P. 109,532, *see* Pollak, F., 1926, 12
Austrian P. 78,251; 7

Belg. P. 848,093, *see* Solvay et Cie, 205
Belg. P. 818,549, *see* Imperial Chemical Industries, Ltd., (1975), 212
Brit. P. 1,499,285, *see* Central Lab. Phys. Chem., 106
Brit. P. 1,486,342, *see* Monsanto Co., 106
Brit. P. 1,483,458, *see* Feriday, J. E., (1977), 212

Brit. P. 1,483,150, *see* An Foras Taluntais
Brit. P. 1,428,334, *see* Waddleton, N., 190
Brit. P. 1,421,994, *see* Borden (UK) Ltd., 212
Brit. P. 1,421,119, *see* Eades, A. G., 110
Brit. P. 1,412,559, *see* Hurst, H., 181
Brit. P. 1,401,960, *see* Swire, J. R., 206
Brit. P. 1,380,919, *see* Bagley, J. M., 114
Brit. P. 1,380,442, *see* Webb, R. J., 178
Brit. P. 1,378,938, *see* Curs, A.
Brit. P. 1,333,470, *see* Fiberglass Ltd., 121, 203
Brit. P. 1,325,806, *see* Stockel, R. F., 202
Brit. P. 1,316,911, *see* Ashall, R. J., 73, 121, 188, 203
Brit. P. 1,293,744, *see* Fiberglass Ltd., (1972), 121, 203
Brit. P. 1,289,465, *see* Moralt, A., 175
Brit. P. 1,273,152, *see* Ashall, R. J., 68, 205
Brit. P. 1,270,408, *see* Champion Paper Co. Ltd., 204
Brit. P. 1,233,396, *see* Decorit-Industrie Beratung, 174
Brit. P. 1,187,352, *see* CIBA, (1970)
Brit. P. 549,496, *see* De Bruyne, N. A., (1942), 18
Brit. P. 545,409, *see* Aero Research Ltd., 18
Brit. P. 536,493, *see* De Bruyne, 18
Brit. P. 507,175, *see* I. G. Farbenindustrie, (1939), 18
Brit. P. 499,818, *see* Tootal Broadhurst Lee, 17
Brit. P. 470,331, *see* De Bruyne, N. A., 202
Brit. P. 460,275, *see* I. G. Farbenindustrie, (1937), 18
Brit. P. 459,788, *see* I. G. Farbenindustrie, (1937), 17, 18
Brit. P. 455,008, *see* Hentisch, W., (1936), 18
Brit. P. 449,243, *see* Tootal Broadhurst Lee, 17
Brit. P. 419,812, *see* Minnesota Mining and Mfg. Co., 17
Brit. P. 379,154, *see* Neuss, O., (1932), 16
Brit. P. 378,938, *see* Cropp, 210
Brit. P. 375,843, *see* Walter, G., (1932)
Brit. P. 375,609, *see* Pollak, F., (1932)
Brit. P. 372,847, *see* Neuss, O., (1932), 16
Brit. P. 366,065, *see* Pfenning-Schumacher Industrie, 118
Brit. P. 366,009, *see* Pfenning-Schumacher Industrie, 118
Brit. P. 359,498, *see* British Cyanides Co.
Brit. P. 359,163, *see* I. G. Farbenindustrie, (1930), 18
Brit. P. 342,767, *see* CIBA, (1928), 13
Brit. P. 342,730, *see* CIBA, (1928), 13
Brit. P. 342,723, *see* CIBA, (1928), 13
Brit. P. 340,114, *see* Fabriques de Produits de Chimie Organique de Laire, (1928), 13
Brit. P. 339,601, *see* Pollak, F., (1929), 14
Brit. P. 333,763, *see* Keller, A. V., 14
Brit. P. 331,428, *see* Taylor, J., 14
Brit. P. 325,843, *see* Kreidl, I., 16
Brit. P. 323,047, *see* Ripper, K., (1927), 12
Brit. P. 319,251, *see* I. G. Farbenindustrie, (1928), 17
Brit. P. 316,194, *see* Allgemeine Elektrizitaets-Ges., 13
Brit. P. 316,144, *see* Goldschmidt, S., (1928), 13
Brit. P. 314,908, *see* I. G. Farbenindustrie, (1928), 14
Brit. P. 313,455, *see* Pfenning-Schumacher Ind., 118

Brit. P. 301,798, *see* Pollak, F., (1927), 12
Brit. P. 301,696, *see* I. G. Farbenindustrie, (1926), 17
Brit. P. 301,626, *see* Pollopas, Ltd., (1927), 12
Brit. P. 299,487, *see* Pollopas, Ltd., (1927), 12
Brit. P. 294,253, *see* Smidth, L., (1927), 12
Brit. P. 291,712, *see* Walter, G., 12
Brit. P. 291,366, *see* Pollak, F., (1927), 12
Brit. P. 291,473, *see* Poulds, 200
Brit. P. 288,346, *see* I. G. Farbenindustrie, (1926), 12
Brit. P. 287,568, *see* Ripper, K., (1927), 12
Brit. P. 284,272, *see* Walter, G., (1925), 11
Brit. P. 281,717, *see* CIBA, (1925), 12
Brit. P. 266,028, *see* Rossiter, E. C., (1925), 10
Brit. P. 262,818, *see* I. G. Farbenindustrie, 17
Brit. P. 262,148, *see* Walter, G., (1925)
Brit. P. 261,409, *see* Pollak, F., (1925), 11
Brit. P. 261,029, *see* I. G. Farbenindustrie, 17
Brit. P. 260,288, *see* CIBA, (1925)
Brit. P. 260,253, *see* I. G. Farbenindustrie, (1925), 12
Brit. P. 260,235, *see* I. G. Farbenindustrie, (1925), 17
Brit. P. 258,950, *see* Rossiter, E. C., (1925), 10
Brit. P. 253,094, *see* CIBA, (1925), 12
Brit. P. 249,101, *see* CIBA, (1925), 12
Brit. P. 248,729, *see* Pollak, F., (1927), 12
Brit. P. 248,477, *see* Rossiter, E. C., 10
Brit. P. 248,476, *see* Rossiter, E. C., (1924)
Brit. P. 246,886, *see* Rossiter, E. C., 17
Brit. P. 246,126, *see* Gams, A., (1925), 11
Brit. P. 238,904, *see* Pollak, F., (1924), 10
Brit. P. 213,567, *see* Pollak, F., (1923), 10
Brit. P. 208,761, *see* Goldschmidt, H., (1922), 10
Brit. P. 206,512, *see* Pollak, F., (1923), 10
Brit. P. 202,651, *see* Neuss, O., (1923), 10
Brit. P. 201,906, *see* Pollak, F., (1923), 10
Brit. P. 193,429, *see* Pollak, F., (1923)
Brit. P. 193,420, *see* Pollak, F., (1923), 17
Brit. P. 187,605, *see* Goldschmidt, H., (1922), 10
Brit. P. 181,014, *see* Pollak, F., (1922), 17
Brit. P. 171,096, *see* Pollak, F., (1921), 17

Can. P. 1,029,915, *see* Christensen, R. L., (1978), 106
Can. P. 1,013,499, *see* Widmann, M., (1977), 189
Can. P. 991,780, *see* Dahlberg, H. S., 202
Can. P. 986,100, *see* Casebier, R. L., 121
Can. P. 351,666, *see* Mavoks, 118
Can. P. 284,790, *see* Ripper, K., (1928), 13
Can. P. 281,327, *see* Gams, A., (1928), 13
Can. P. 281,326, *see* Gams, A., (1928), 13
Can. P. 281,325, *see* Gams, A., (1928), 13
Can. P. 274,266, *see* Gams, A., (1927), 13
Can. P. 274,738, *see* Rossiter, E. C., (1927), 13
Czech. P. 170,983, *see* Pastyr, J., 181

Czech. P. 170,877, *see* Polinek, V., 211
Czech. P. 169,239, *see* Dockal, P., 211
Czech. P. 162,810, *see* Zapletal, J., (1976), 190
Czech. P. 157,574, *see* Halama, D., 211
Czech. P. 157,191, *see* Susicky, K., 178
Czech. P. 144,662, *see* Lvovsky, C., 214
Czech. P. 141,245, *see* Seberini, M., 174

Fr. Addition 95,499, *see* Rougier, 174
Fr. Demande 2,338,956, *see* Brial, J. C., (1977), 106
Fr. Demande 2,329,688, *see* Simeonov, I.
Fr. Demande 2,306,225, *see* Rechner, L., 189
Fr. Demande 2,270,221, *see* Saarbergwerke, 210
Fr. Demande 2,243,917, *see* Tobol'skii, G. F., 189
Fr. Demande 2,240,254, *see* Imperial Chemical Industries, Ltd., (1975), 212
Fr. Demande 2,230,348, *see* Aries, R., 106
Fr. Demande 2,219,134, *see* Materiaux Reunis, 205
Fr. Demande 2,213,963, *see* Deriancourt, G., 178
Fr. Demande 2,203,847, *see* Sarrut, J., 114
Fr. Demande 2,193,699, *see* Brown Boveri, 197
Fr. Demande 2,182,168, *see* CIBA-GEIGY
Fr. Demande 2,163,578, *see* BASF, 106
Fr. Demande 2,161,936, *see* Sato, T., (1973), 106
Fr. Demande 2,081,925, *see* Ball, G. R., 174
Fr. Demande 2,074,766, *see* Karnemaat, J. N., (1971), 207
Fr. Demande 1,242,344, *see* Touati, R., 206
Fr. P. 2,246,588, *see* Novak, V. A., 189
Fr. P. 2,192,552, *see* Blandin, H. M., 205
Fr. P. 2,172,630, *see* Oesterreichische Hiag-Werke, (1973), 176
Fr. P. 2,161,199, *see* Oesterreichische Hiag-Werke, (1973), 176
Fr. P. 2,161,187, *see* Jenks, T. E., 205
Fr. P. 2,154,914, *see* Da Cunha, A., 203
Fr. P. 2,117,512, *see* Becker und Van Huellen Niederrheinische Machinenfabrik K.-G.
Fr. P. 2,102,406, *see* British Industrial Plastics Ltd., (1972), 188
Fr. P. 2,098,692, *see* Black, W. R., (1972), 106
Fr. P. 2,082,270, *see* Midwest Research Institute, 188
Fr. P. 2,068,286, *see* CIBA Ltd. Cabot Corp., 101
Fr. P. 2,065,238, *see* Hoedt, H., 205
Fr. P. 2,053,840, *see* Fiberglass Ltd., (1971), 187
Fr. P. 926,024, *see* Fahrni, F., (1947), 165
Fr. P. 899,487, *see* Fahrni, F., (1945), 165
Fr. P. 881,781, *see* Fahrni, F., (1943), 165
Fr. P. 848,604, *see* I. G. Farbenindustrie, (1939)
Fr. P. 847,953, *see* I. G. Farbenindustrie, (1939), 18
Fr. P. 821,840, *see* American Reinforced Paper, 18
Fr. P. 804,714, *see* Soc. Nobel française, (1936), 17
Fr. P. 769,588, *see* I. G. Farbenindustrie, (1934), 17
Fr. P. 749,177, *see* Antoni, K. A., 164
Fr. P. 736,744, *see* Soc. Nobel francaise, (1931), 15
Fr. P. 729,029, *see* Lenz, J., 18
Fr. P. 722,289, *see* Établissements Lambiottes Frères, 14

Fr. P. 721,828, *see* I. G. Farbenindustrie, (1931), 15
Fr. P. 720,772, *see* La Ronite
Fr. P. 718,093, *see* Pollopas, Ltd., (1931)
Fr. P. 715,153, *see* Toledo Synthetic Products, Inc., (1931), 211
Fr. P. 714,101, *see* British Cyanides Co., 15
Fr. P. 711,395, *see* Toledo Synthetic Products, Inc., (1931), 15
Fr. P. 711,333, *see* Toledo Synthetic Products, Inc., (1931), 15
Fr. P. 699,968, *see* Keller, A. V., 14
Fr. P. 697,874, *see* I. G. Farbenindustrie, (1930), 14
Fr. P. 689,410, *see* Pollak, F., 118
Fr. P. 688,269, *see* Amand, 121
Fr. P. 680,110, *see* Pollopas, Ltd., (1929), 14
Fr. P. 679,321, *see* Fabriques de Produits de Chimie Organique de Laire, (1928), 13
Fr. P. 675,398, *see* I. G. Farbenindustrie, (1929), 14
Fr. P. 674,999, *see* Pollopas, Ltd., (1929), 14
Fr. P. 674,703, *see* Auer, L., 14
Fr. P. 674,589, *see* Toledo Scale Manufacturing Co., 14
Fr. P. 657,794, *see* Pollak, F., (1928), 13
Fr. P. 644,648, *see* CIBA, (1927), 13, 185
Fr. P. 641,770, *see* I. G. Farbenindustrie, (1927), 13
Fr. P. 641,420, *see* I. G. Farbenindustrie, (1926), 12
Fr. P. 637,318, *see* Pollak, F., (1927), 12
Fr. P. 552,264, *see* I. G. Farbenindustrie, (1930)
Fr. P. 38,157, *see* I. G. Farbenindustrie, (1930), 14

Ger. Offen. 2,754,525, *see* Henbest, R. G., 107
Ger. Offen. 2,745,951, *see* Teukros, A.-G., 114
Ger. Offen. 2,726,617, *see* Hubbard, D. A., 107
Ger. Offen. 2,716,375, *see* Eder, H. M., 178
Ger. Offen. 2,702,421, *see* Brial, J. C., (1977)
Ger. Offen. 2,639,754, *see* Pusch, G., 203
Ger. Offen. 2,639,470, *see* Ulbricht, J., 212
Ger. Offen. 2,637,690, *see* Wenzel, W.
Ger. Offen. 2,631,038, *see* Smith, T. F., 203
Ger. Offen. 2,629,889, *see* Wojciak, S.
Ger. Offen. 2,620,478, *see* Laqua, A., 106
Ger. Offen. 2,610,286, *see* Lehnert, E. J., 178, 286
Ger. Offen. 2,547,966, *see* Petersen, H., (1977)
Ger. Offen. 2,546,668, *see* Yul, 106
Ger. Offen. 2,537,859, *see* Jamin, A.
Ger. Offen. 2,532,753, *see* Hubbard, D. A., (1976), 110
Ger. Offen. 2,529,427, *see* Foris, P. L.
Ger. Offen. 2,523,483, *see* O'Donnell, J. M., 211
Ger. Offen. 2,521,689, *see* Renner, A.
Ger. Offen. 2,516,327, *see* Pace, L. S., 110
Ger. Offen. 2,507,586, *see* Lindberg, J. J., 106
Ger. Offen. 2,505,578, *see* Aignesberger, A., 106, 120
Ger. Offen. 2,454,769, *see* Grabowsky, O., (1976), 109
Ger. Offen. 2,448,472, *see* Holtschmidt, U., 110
Ger. Offen. 2,447,941, *see* Novak, V. A., 189
Ger. Offen. 2,436,465, *see* Schoellhorn, W. D., 190
Ger. Offen. 2,435,793, *see* Grabowsky, 109

Ger. Offen. 2,422,238, *see* Schneider, H., 210
Ger. Offen. 2,421,445, *see* Takahashi, 110
Ger. Offen. 2,416,270, *see* Ogden, D. H., 205
Ger. Offen. 2,415,762, *see* Wismer, M., 214
Ger. Offen. 2,413,362, *see* Yamashita, 110
Ger. Offen. 2,410,605, *see* Kirsch, K., 176
Ger. Offen. 2,403,443, *see* Reuss, G., 110
Ger. Offen. 2,402,441, *see* Blake, A. D., 189
Ger. Offen. 2,402,411, *see* Blake, A. D.
Ger. Offen. 2,401,554, *see* Siegler, M., (1975), 110
Ger. Offen. 2,401,440, *see* Terporten, H., 110
Ger. Offen. 2,400,908, *see* Olszowski, T., 106
Ger. Offen. 2,361,085, *see* Vad, J.
Ger. Offen. 2,354,928, *see* Schmidt-Hellerau, C., (1975), 247
Ger. Offen. 2,354,607, *see* Anderson, R. W.
Ger. Offen. 2,354,149, *see* Chand, R.
Ger. Offen. 2,352,190, *see* Sander, B., 204
Ger. Offen. 2,351,981, *see* Siegler, M., (1975), 110
Ger. Offen. 2,350,820, *see* Stene, R. M., 203
Ger. Offen. 2,347,401, *see* Behar, R., 106
Ger. Offen. 2,341,553, *see* Zapletal, J., (1975)
Ger. Offen. 2,337,233, *see* Buschfeld, 109
Ger. Offen. 2,334,540, *see* Enkvist, T. U., 246
Ger. Offen. 2,324,440, *see* Lenz, J., (1974), 109, 246
Ger. Offen. 2,322,617, *see* Feriday, J. E., (1973)
Ger. Offen. 2,318,802, *see* Inverarity, G., 106
Ger. Offen. 2,317,846, *see* Havinga, R., 111
Ger. Offen. 2,263,125, *see* Vargiu, S., (1973)
Ger. Offen. 2,262,197, *see* Vargiu, S., (1973), 108
Ger. Offen. 2,261,896, *see* Kuehn, H. G., 205
Ger. Offen. 2,261,800, *see* Sundie, R. D., (1973), 106
Ger. Offen. 2,260,186, *see* Seeholzer, 203
Ger. Offen. 2,259,680, *see* Brandeis, J., 202
Ger. Offen. 2,251,433, *see* Barzynski, H., 204
Ger. Offen. 2,247,036, *see* Zeigerson, E., 246
Ger. Offen. 2,243,857, *see* Erhardt, K., (1974), 109
Ger. Offen. 2,242,548, *see* Hoener, H., (1974), 205
Ger. Offen. 2,241,996, *see* Brunnmüller, F., (1974)
Ger. Offen. 2,241,715, *see* Petersen, H., (1974), 189
Ger. Offen. 2,238,742, *see* Schulz, H., 210
Ger. Offen. 2,233,428, *see* Speiser, P. M., 204
Ger. Offen. 2,227,679, *see* Heetman, J. G. A., (1974), 205
Ger. Offen. 2,225,478, *see* Tiedeman, G. T., 106
Ger. Offen. 2,224,363, *see* Krapfhammer, G., 211
Ger. Offen. 2,219,033, *see* Schwab, J., 188
Ger. Offen. 2,218,582, *see* Junkermann, H., 106
Ger. Offen. 2,213,269, *see* Koch, H. G., 107
Ger. Offen. 2,209,109, *see* Schwager, P., 204
Ger. Offen. 2,207,921, *see* Erhardt, K., (1973), 106, 107
Ger. Offen. 2,206,696, *see* Kubitzky, C., 107, 246
Ger. Offen. 2,202,202, *see* Boehm, W., (1972), 101
Ger. Offen. 2,164,336, *see* Koberg, H., 208

Ger. Offen. 2,162,223, *see* Weill und Reineke, GmbH
Ger. Offen. 2,161,570, *see* Ashall, R. J., (1972), 205
Ger. Offen. 2,157,573, *see* Seldon, M. M., 214
Ger. Offen. 2,154,630, *see* Shank, H. C., 211
Ger. Offen. 2,154,620, *see* Shank, H. C.
Ger. Offen. 2,135,072, *see* Moroff, H., 176, 286
Ger. Offen. 2,131,717, *see* Jaccard, R., (1972), 101
Ger. Offen. 2,130,577, *see* Neumann, R., 105, 245
Ger. Offen. 2,110,939, *see* Brunnmüller, F., (1972), 188
Ger. Offen. 2,109,754, *see* Brunnmüller, F., (1972), 101
Ger. Offen. 2,105,253, *see* Lander, H. L., 105
Ger. Offen. 2,064,102, *see* Brunnmüller, F., (1972), 188
Ger. Offen. 2,063,815, *see* Rosner, H.
Ger. Offen. 2,055,074, *see* Brunnmüller, F., (1972), 207
Ger. Offen. 2,046,496, *see* Jaccard, R., (1972), 101
Ger. Offen. 2,043,440, *see* Schmidt-Hellerau, C., (1972), 105
Ger. Offen. 2,041,566, *see* Shastri, S., 245
Ger. Offen. 2,020,481, *see* Mayer, J., 107
Ger. Offen. 2,002,873, *see* Karnemaat, J. N., (1971), 207
Ger. Offen. 1,966,796, *see* Brose, H., 114
Ger. Offen. 1,920,283, *see* Ernst, 245

Ger. P. 2,550,739, *see* Merkel, D., 110
Ger. P. 2,435,793, *see* Grabowsky, O., (1974), 109
Ger. P. 2,334,380, *see* Eisele, W.
Ger. P. 2,243,857, *see* Erhardt, K., (1972), 107
Ger. P. 2,215,947, *see* Brunnmüller, F., (1972), 107
Ger. P. 2,207,931, *see* Erhardt, K., (1972)
Ger. P. 2,111,651, *see* Riesmeier, W., 108
Ger. P. 2,110,264, *see* Reuther, W., (1971), 114
Ger. P. 2,058,405, *see* Reuther, W., (1970)
Ger. P. 1,745,562, *see* Hinterwaldner, R., 189
Ger. P. 1,719,151, *see* Krueger, A., 110
Ger. P. 1,645,016, *see* Bondi, E., 106
Ger. P. 1,570,240, *see* Lenz, J., (1973), 106, 176
Ger. P. 1,081,482, *see* Jung, J., 18
Ger. P. 839,944, *see* Döhler, F., 198
Ger. P. 800,704, *see* BASF, 19, 189
Ger. P. 767,897, *see* Roos, O.
Ger. P. 734,094, *see* Scheuermann, H., (1943), 19, 186
Ger. P. 733,710, *see* Ludwig, O. R., 18
Ger. P. 732,498, *see* CIBA, (1943), 18
Ger. P. 729,029, *see* Scheuermann, H., (1942)
Ger. P. 636,658, *see* Curs, A., 16, 19
Ger. P. 604,194, *see* Neuss, O., (1934), 17
Ger. P. 588,879, *see* Pfenning-Schumacher Ind., 118
Ger. P. 569,342, *see* "Silur" Tech. & Chem. Produkts, 12
Ger. P. 568,629, *see* Rothera, W. S., (1921), 7
Ger. P. 567,271, *see* Rothera, W. S., (1925), 11
Ger. P. 563,037, *see* Rothera, W. S., (1921), 7
Ger. P. 562,943, *see* CIBA, (1929), 14
Ger. P. 561,050, *see* Goldschmidt, S., (1928), 13

Ger. P. 552,264, *see* I. G. Farbenindustrie, 14
Ger. P. 551,422, *see* Goldschmidt, S., (1927), 13
Ger. P. 550,647, *see* Vierling, 14
Ger. P. 540,071, *see* I. G. Farbenindustrie, (1929), 14
Ger. P. 537,611, *see* I. G. Farbenindustrie, (1926), 12
Ger. P. 536,733, *see* I. G. Farbenindustrie, (1928), 14
Ger. P. 535,852, *see* I. G. Farbenindustrie, (1925), 12
Ger. P. 535,851, *see* I. G. Farbenindustrie, (1922), 10
Ger. P. 528,582, *see* I. G. Farbenindustrie, (1929), 14
Ger. P. 523,818, *see* CIBA, (1926)
Ger. P. 523,181, *see* CIBA, (1926), 12, 185
Ger. P. 519,892, *see* I. G. Farbenindustrie, (1929), 14
Ger. P. 511,979, *see* I. G. Farbenindustrie, (1926), 12
Ger. P. 510,513, *see* I. G. Farbenindustrie, 12
Ger. P. 499,792, *see* Pollak, F., (1921), 7
Ger. P. 499,710, *see* CIBA, (1925)
Ger. P. 499,589, *see* CIBA, (1925), 12
Ger. P. 493,988, *see* I. G. Farbenindustrie, (1924), 10
Ger. P. 487,871, *see* Gams, A., 11
Ger. P. 484,972, *see* Pollak, F., (1922), 10
Ger. P. 409,847, *see* I. G. Farbenindustrie, 10
Ger. P. 394,488, *see* John, 7
Ger. P. 392,183; 7
Ger. P. 189,036, *see* Rosenberg, P.
Ger. P. 148,345, *see* Kramer, (1889), 164
Ger. P. 97,164; 7
Ger. P. 78,251, *see* John, H., (1919)
Ger. P. 48,035, *see* Kramer, H.

Ger. P. (E.) 127,967, *see* Knabe, G.
Ger. P. (E.) 124,312, *see* Boehme, P., 286
Ger. P. (E.) 103,883, *see* Mielke, R.
Ger. P. (E.) 96,965, *see* Neumann, R., 187
Ger. P. (E.) 88,196, *see* Boehm, W., (1972), 101, 107
Ger. P. (E.) 79,962, *see* Trautvetter, R., 174
Ger. P. (E.) 77,033, *see* Dimter, L., 289

Hung. P. 13,033, *see* Gavel, 204

Indian P. 120,113, *see* Sharphouse, J. H.

Israeli P. 30,472, *see* Lewin, M., 174

Japan Kokai 78 71,677, *see* Uchida, T., 211
Japan Kokai 78 42,178, *see* Kodo, K., 211
Japan Kokai 78 41,393, *see* Hayashi, S., 106
Japan Kokai 78 41,392, *see* Hayashi, S., 106
Japan Kokai 78 41,354, *see* Hayashi, S., 106
Japan Kokai 78 37,767, *see* Noda, H., 190
Japan Kokai 78 33,253, *see* Hayashi, S., 106
Japan Kokai 78 17,690, *see* Izumi, H., 106
Japan Kokai 78 16,755, *see* Izumi, H.

Japan Kokai 77 155,697, *see* Nizhni Tagil Plastics Plant, 247
Japan Kokai 77 128,894, *see* Kotani, Y.
Japan Kokai 77 95,745, *see* Kamachi, T., 247
Japan Kokai 77 74,015, *see* Kitamura, K., 247
Japan Kokai 77 36,899, *see* Suzuki, Y., 206
Japan Kokai 76 89,537, *see* Sano, S., 110
Japan Kokai 76 88,595, *see* Koyama, T., 110
Japan Kokai 76 82,334, *see* Yui, H.
Japan Kokai 76 67,335, *see* Kikuchi, K., 204
Japan Kokai 76 67,334, *see* Kikuchi, K.
Japan Kokai 76 55,499, *see* Arima, Y., 203
Japan Kokai 76 41,031, *see* Fukuda, T., 247
Japan Kokai 76 28,148, *see* Ueda, M., 178
Japan Kokai 76 20,225, *see* Minoji, A., 247
Japan Kokai 76 06,269, *see* British Industrial Plastics Ltd., 190, 197
Japan Kokai 75 148,624, *see* Nogi, T., 202
Japan Kokai 75 142,703, *see* Ishida, M., 178
Japan Kokai 75 126,792, *see* Nogi, T., 110
Japan Kokai 75 119,892, *see* Shiota, T., 189
Japan Kokai 75 112,430, *see* Shiota, T., 110
Japan Kokai 75 109,933, *see* Iwata, M., 110
Japan Kokai 75 104,293, *see* Ohhara, O., 110
Japan Kokai 75 76,147, *see* Yamamoto, Y., 110
Japan Kokai 75 76,146, *see* Yamamoto, Y., 106
Japan Kokai 75 76,145, *see* Yamamoto, Y., 106
Japan Kokai 75 75,257, *see* Okresny Stravebno-Montazny
Japan Kokai 75 63,097, *see* Kitta, Y.
Japan Kokai 75 45,066, *see* Kishi, A., 178
Japan Kokai 75 43,182, *see* Iwasa, Y., 178
Japan Kokai 75 43,181, *see* Kawahara, N., 247
Japan Kokai 75 42,011, *see* Yamamoto, N., 178
Japan Kokai 75 38,778, *see* Iwasa, Y., 178
Japan Kokai 75 35,230, *see* Kawamura, S., 110
Japan Kokai 75 34,332, *see* Higuchi, K., 247
Japan Kokai 75 34,330, *see* Higuchi, K., 110
Japan Kokai 75 25,802, *see* Mita, A., 121
Japan Kokai 75 25,716, *see* Ueda, M., 178
Japan Kokai 75 24,413, *see* Kunii, Y.
Japan Koaki 75 24,404, *see* Suematsu, A., 247
Japan Kokai 75 22,493, *see* Konishi, 206
Japan Kokai 75 21,072, *see* Utsumi, S., 178
Japan Kokai 75 19,909, *see* Yoshida, A., 178
Japan Kokai 75 16,753, *see* Ueda, K., 178
Japan Kokai 75 13,512, *see* Oka, Y., 178
Japan Kokai 75 09,693, *see* Higashimura, E., 110
Japan Kokai 75 05,471, *see* Kato, C., 178
Japan Kokai 75 05,421, *see* Shigeta, T., 206
Japan Kokai 75 04,133, *see* Yamamoto, Y., 106
Japan Kokai 75 257, *see* Okreshy, S., 189
Japan Kokai 75 148, *see* Kusushita, 110
Japan Kokai 74 261, 072, *see* Tange, Y., 246

Japan Kokai 74 131,259, *see* Okada, K., 206
Japan Kokai 74 131,244, *see* Kato, C., 176
Japan Kokai 74 131,243, *see* Ohhara, O., 246
Japan Kokai 74 130,977, *see* Tsubota, Y., 177
Japan Kokai 74 130,935, *see* Tamura, Y., 177
Japan Kokai 74 130,934, *see* Masuda, S., 109
Japan Kokai 74 130,933, *see* Fujitani, J., 176
Japan Kokai 74 129,753, *see* Fujita, K., 176
Japan Kokai 74 128,943, *see* Maeda, T., 109
Japan Kokai 74 128,031, *see* Shofuda, H., 106
Japan Kokai 74 125,504, *see* Sasaki, T., 246
Japan Kokai 74 124,207, *see* Miwa, T., 246
Japan Kokai 74 124,203, *see* Miwa, T., 246
Japan Kokai 74 119,948, *see* Ohhashi, K., 246
Japan Kokai 74 112,993, *see* Ichimura, Y., 212
Japan Kokai 74 112,952, *see* Kotani, Y., 246
Japan Kokai 74 110,585, *see* Tanioka, H., 109
Japan Kokai 74 109,277, *see* Shirai, K.
Japan Kokai 74 101,418, *see* Tashiro, F., 246
Japan Kokai 74 99,775, *see* Anonymous, (1974), 178
Japan Kokai 74 99,321, *see* Hayashi, T., 205
Japan Kokai 74 99,191, *see* Hasegawa, M., 246
Japan Kokai 74 97,744, *see* Nakajima, 109
Japan Kokai 74 94,744, *see* Nakajima, M.
Japan Kokai 74 92,160, *see* Tange, Y., 246
Japan Kokai 74 92,154, *see* Ohhara, O., 106
Japan Kokai 74 80,210, *see* Kitakado, Y., 246
Japan Kokai 74 75,709, *see* Hojo, Z.
Japan Kokai 74 73,450, *see* Ueda, K., 211
Japan Kokai 74 73,447, *see* Ueda, K., 211
Japan Kokai 74 72,179, *see* Kobayagawa, T.
Japan Kokai 74 71,118, *see* Kawashima, M., 246
Japan Kokai 74 71,111, *see* Aono, I., 121
Japan Kokai 74 71,103, *see* Kobayashi, K., 176
Japan Kokai 74 66,808, *see* Kitakado, Y., 246
Japan Kokai 74 63,789, *see* Kitta, Y., 106
Japan Kokai 74 63,784, *see* Kitta, Y.
Japan Kokai 74 48,782, *see* Shiota, T., 246
Japan Kokai 74 44,581, *see* Ando, S., 109
Japan Kokai 74 44,060, *see* Inui, K., 246
Japan Kokai 74 38,927, *see* Minoji, A., 177, 286
Japan Kokai 74 38,925, *see* Minoji, A.
Japan Kokai 74 36,732, *see* Kitayama, M., 203
Japan Kokai 74 35,511, *see* Kotani, Y., 246
Japan Kokai 74 30,504, *see* Suzuki, J., 106
Japan Kokai 74 25,108, *see* Miwa, T., 246
Japan Kokai 74 25,018, *see* Miwa, T.
Japan Kokai 74 18,926, *see* Ikeda, M., 246
Japan Kokai 74 18,925, *see* Ikeda, M., 246
Japan Kokai 74 05,891, *see* Ohe, T.
Japan Kokai 74 04,542, *see* Hata, Y., 109
Japan Kokai 73 103,705, *see* Yoshimitsu, T., 245

Japan Koaki 73 93,580, *see* Yoshina, T.
Japan Kokai 73 92,541, *see* Nishino, S., 245
Japan Kokai 73 79,241, *see* Higuchi, K., 105
Japan Kokai 73 75,489, *see* Eguchi, T.
Japan Kokai 73 72,309, *see* Kawahara, N., 246
Japan Koaki 73 72,308, *see* Kitakado, Y.
Japan Kokai 73 66,152, *see* Miyahara, S.
Japan Kokai 73 60,737, *see* Higuchi, K., 245
Japan Kokai 73 60,140, *see* Kotani, Y., 105, 106
Japan Kokai 73 58,109, *see* Kitakado, 246
Japan Kokai 73 58,048, *see* Ueda, K., 106
Japan Kokai 73 58,032, *see* Tanaka, A., 105, 108
Japan Kokai 73 56,996, *see* Date, M., 202
Japan Kokai 73 56,731, *see* Iwada, R., 108
Japan Kokai 73 55,223, *see* Takahashi, M., 108
Japan Kokai 73 51,032, *see* Sakarada, S., 121
Japan Kokai 73 31,290, *see* Kitta, Y., 106
Japan Kokai 73 18,338, *see* Tsukamoto, A., 106
Japan Kokai 73 14,747, *see* Suzuki, T., 246
Japan Kokai 73 12,386, *see* Iwaki, T., 108
Japan Kokai 72 43,128, *see* Nagao, A., 105
Japan Kokai 72 22,441, *see* Nagao, A., 105
Japan P. 78 34,189, *see* Uchida, T., 211
Japan P. 78 34,184, *see* Uchida, T.
Japan P. 77 20,516, *see* Takahashi, M., 247
Japan P. 76 17,598, *see* Kurashige, H., 178
Japan P. 76 06,193, *see* Ito, Y., 190
Japan P. 76 03,355, *see* Ohtsuka, M., 206
Japan P. 75 34,055, *see* Katayama, M., 178
Japan P. 75 33,124, *see* Kobori, K., 206
Japan P. 75 28,482, *see* Inoue, K., 178
Japan P. 75 25,491, *see* Yamashita, K.
Japan P. 75 13,305, *see* Takahashi, H.
Japan P. 75 06,527, *see* Ohmura, M., 178
Japan P. 75 06,526, *see* Tsuruta, K., 178
Japan P. 74 44,936, *see* Yamashita, H., 109
Japan P. 74 43,126, *see* Ohno, K., 246
Japan P. 74 40,617, *see* Kitsuda, Y., 106
Japan P. 74 33,682, *see* Hiyoshi, K., 204
Japan P. 74 27,672, *see* Shimokai, K., 205
Japan P. 74 15,242, *see* Okabe, J., 178
Japan P. 74 15,071, *see* Nakajima, H., 189
Japan P. 74 04,318, *see* Isono, S., 176
Japan P. 74 04,461, *see* Nomura, M., 109
Japan P. 73 44,341, *see* Suzuki, S., 188
Japan P. 73 29,612, *see* Iguchi, K.
Japan P. 73 32,418, *see* Mori, S., 176
Japan P. 73 21,346, *see* Mori, K., 176
Japan P. 73 19,865, *see* Takeshita, M., 106
Japan P. 73 17,877, *see* Takahashi, M., 106
Japan P. 73 17,755, *see* Kato, H., 246
Japan P. 73 16,066, *see* Takahashi, I., 210

Japan P. 73 14,793, *see* Okamura, S., 107
Japan P. 73 09,935, *see* Kawakami, K.
Japan P. 72 47,293, *see* Hamada, R., 105
Japan P. 72 32,422, *see* Iwasaki, H., 286
Japan P. 71 39,479, *see* Uchimi, S., 174
Japan P. 71 31,015, *see* Nakanishi, T., 105
Japan P. 71 28,785, *see* Endo, A., 204
Japan P. 71 28,784, *see* Tanaka, M., 245
Japan P. 71 28,783, *see* Tajima, S., 105
Japan P. 71 18,570, *see* Toyota, H., 207
Japan P. 71 09,210, *see* Takashima, H., 207
Japan P. 93,961, *see* Fuwa, K., (1931)

Neth. P. 79 24,479, *see* Kirkeby, 289
Neth. Appl. 76 00373, *see* British Industrial Plastics Ltd., 106
Neth. P. 76 068, *see* Lafarge Co., 206
Neth. Appl. 75 11,572, *see* Dixon International Ltd., 206
Neth. Appl. 74 14,340, *see* Hubbard, D. A.
Neth. Appl. 74 06,753, *see* Billingsfors Bruks AB, 109
Neth. Appl. 74 06,625, *see* Dead Sea Bromine Co. Ltd., 206
Neth. Appl. 73 11,929, *see* Hoener, H., 205

Norw. P. 131,891, *see* Bergsund, K., 109
Norw. P. 124,479, *see* Kirkeby, W., (1972)

Pol. P. 92,502, *see* Jacenkow, B., (1977), 210
Pol. P. 91,790, *see* Kinastowski, S., (1977), 106
Pol. P. 85,670, *see* Kinastowski, S., (1976)
Pol. P. 85,450, *see* Kinastowski, S., (1976)
Pol. P. 83,871, *see* Nowak, D., (1976), 106
Pol. P. 80,377, *see* Nowak, D., (1976), 121
Pol. P. 78,613, *see* Szemiotowicz, J., 105
Pol. P. 74,999, *see* Starzynska, K., (1975), 121
Pol. P. 74,060, *see* Szlezyngier, W., 121, 247
Pol. P. 74,012, *see* Klebeko, H.
Pol. P. 73,289, *see* Willert, J., 109
Pol. P. 73,219, *see* Dirska, 204
Pol. P. 72,885, *see* Wirpsza, Z., (1974), 246
Pol. P. 68,454, *see* Starzynska, K., (1973), 246
Pol. P. 67,486, *see* Penczek, P., 106
Pol. P. 66,858, *see* Wirpsza, Z., (1972)
Pol. P. 66,154, *see* Wirpsza, Z., (1972), 105

Rom. P. 60,253, *see* Stenzel, J., 106
Rom. P. 57,015, *see* Teodorescu, L., 247
Rom. P. 56,115, *see* Lambru, 108
Rom. P. 52,770, *see* Vintila, E., 174

S. African P. 77 01,123, *see* Loehmer, W., 203
S. African P. 77 00,345, *see* Whiteside, I. R., 106
S. African P. 76 00,197, *see* Bonnington, M. G., 212
S. African P. 75 00,495, *see* Stofko, J., 178

S. African P. 73 08,363, *see* Davidovits, J., 176
S. African P. 72 04,876, *see* Orth, G. O., 210
S. African P. 72 00,485, *see* Tanner, D. J.
S. African P. 70 05,676, *see* Renner, A., 101
S. African P. 70 05,055, *see* Black, W. R.
S. African P. 70 04,991, *see* Partridge, L.

Span. P. 417,401, *see* Just Bravo, M. A., 190

Swed. P. 397,834, *see* Edler, F., 247
Swed. P. 336,470, *see* Gruben, S., 197
Swed. P. 53,833, *see* Holmgren, T. A. F.

Swiss P. 601,360, *see* Danner, B.
Swiss P. 542,893, *see* Jaccard, J., (1973), 101
Swiss P. 526,481, *see* Heetman, J. G. A., (1972), 205
Swiss P. 516,992, *see* Senn. S., 175
Swiss P. 232,065, *see* Fahrni, F., (1944), 110
Swiss P. 198,723, *see* Torfit Co., 164
Swiss P. 193,139, *see* Pfohl, F., 164
Swiss P. 161,052, *see* Benteli, A., 17
Swiss P. 154,831, *see* CIBA, (1931), 15
Swiss P. 154,520, *see* CIBA, (1930), 14
Swiss P. 126,827, *see* Gams, A., (1926), 185
Swiss P. 114,289, *see* CIBA, (1926), 11

USP 4,107,105, *see* Korf, H. L.
USP 4,097,419, *see* Moore, W. P., (1978), 106, 248
USP 4,092,277, *see* Moore, W. P., (1978)
USP 4,065,421, *see* Allyn, C. L., 101
USP 4,064,307, *see* Lajoie, J. L., 106
USP 4,064,090, *see* Gibson, D. V., 214
USP 4,064,088, *see* Renner, 110, 248
USP 4,056,362, *see* Gablin, K. A., 211
USP 4,055,974, *see* Jackson, L. P., 210
USP 4,051,208, *see* Jaccard, R., (1977), 101
USP 4,051,070, *see* Argo, W. B., 211
USP 4,039,496, *see* Hermann, D. T., 203
USP 4,038,046, *see* Supkis, S. J., 203
USP 4,035,426, *see* Hubbard, 110, 181, 248
USP 4,035,328, *see* Huang, D. K.
USP 4,033,745, *see* Moore, 210
USP 4,032,515, *see* Blommers, E. A., 106, 121, 248
USP 4,025,329, *see* Goertz, H. M., 210
USP 4,022,727, *see* Speyer, F. B., 214
USP 4,021,413, *see* Eisele, 107, 108, 109, 248
USP 4,020,219, *see* Martorano, R., 203
USP 4,018,966, *see* Antlfinger, G. J., 203
USP 4,018,959, *see* Demko, P. R., 106
USP 4,018,727, *see* Taylor, L. J., (1977), 204
USP 4,010,132, *see* Renner, 110
USP 4,001,140, *see* Foris, 110

USP 3,997,612, *see* Lenke, G. M., 106
USP 3,996,190, *see* O'Neill, 248
USP 3,994,850, *see* Willeger, 248
USP 3,993,831, *see* Vassiliades, A. E., (1976), 204
USP 3,986,988, *see* Zapletal, J., 190
USP 3,979,492, *see* Sundie, R. D., (1976), 178
USP 3,979,341, *see* Widmann, M., (1976), 190
USP 3,970,625, *see* Moore, W. P., (1976), 114
USP 3,965,056, *see* Stout, R. M. T.
USP 3,962,491, *see* Sato, T., (1976)
USP 3,962,166, *see* Gordon, A. F., 110, 204
USP 3,957,731, *see* Hart, W. F.
USP 3,957,700, *see* Feriday, 211,215
USP 3,951,595, *see* Shelton, T. S., 203
USP 3,944,388, *see* Getchell, N. F., 203
USP 3,943,994, *see* Cleveland, J. J., 190, 197
USP 3,931,072, *see* Coyle, R. P., 247
USP 3,931,063, *see* Renner, A., 110
USP 3,930,066, *see* Brown-Boveri, 189
USP 3,928,272, *see* Brancato, J. J.
USP 3,909,470, *see* Lambuth, A. L. 110
USP 3,907,740, *see* Blank, W. J., 110
USP 3,905,847, *see* Black, W. R., (1975), 109
USP 3,901,752, *see* Taylor, L. J., (1975), 121
USP 3,893,963, *see* Sausman, D. K., 106
USP 3,892,709, *see* Oda, N., 110
USP 3,886,084, *see* Vassiliades, A. E., (1975), 204
USP 3,878,278, *see* Miller, C. H., 206
USP 3,869,414, *see* Campbell, C. C.
USP 3,862,914, *see* Anderson, 106
USP 3,853,685, *see* Brown-Boveri, 189
USP 3,842,039, *see* Vargin, 101, 108
USP 3,839,523, *see* Lohenhoffer, H. Ph., 177
USP 3,833,696; 198
USP 3,832,201, *see* Shearer, H. N., 205
USP 3,830,894, *see* Juenger, H., 189, 217
USP 3,830,783, *see* Vargin, 106
USP 3,827,995, *see* White, J. T., 181
USP 3,826,770, *see* Christensen, R., (1974)
USP 3,816,376, *see* Brunnmüller, F., (1974), 101, 248
USP 3,816,236, *see* Baymiller, J. W., 177
USP 3,804,706, *see* Kurashige, H., (1974), 176
USP 3,800,696, *see* Kramer, F., (1974), 210
USP 3,790,417, *see* Paterson, A. R., 177
USP 3,786,025, *see* Freeman, H. G., 106
USP 3,784,494, *see* Domokos, G 176
USP 3,770,466, *see* Wilton, J., 176
USP 3,764,428, *see* Oshima, K., 188, 197
USP 3,761,339, *see* Purcell, R. F., 105, 176
USP 3,759,687, *see* Nobell, A., 210
USP 3,758,377, *see* Fife, R. L., 204
USP 3,736,275, *see* Immarino, N. J., 176

USP 3,734,985, *see* Greenberg, W. H., 211
USP 3,730,825, *see* Nakane, B., 176
USP 3,729,320, *see* Kramer, F., (1973), 208
USP 3,725,323, *see* Beck, K. M.
USP 3,697,355, *see* Black, W. R., (1972), 107
USP 3,692,185, *see* Columbus, P. S., 204
USP 3,690,976, *see* Nakajima, T., 204
USP 3,689,463, *see* Kruglikov, A. A., 105
USP 3,686,021, *see* Lee, C. A., 286
USP 3,681,274, *see* Oetgen, W. R.
USP 3,677,808, *see* Sheridan, J. L., 175
USP 3,674,596, *see* McMinimy, L. O., 175
USP 3,671,377, *see* Marra, A. A., 175
USP 3,668,286, *see* Brooks, S. H. W., 174, 175
USP 3,668,064, *see* Kucera, L. J., 174
USP 3,666,597, *see* Parnell, W. E., 107
USP 3,658,638, *see* Ludwig, C. H., 245
USP 3,658,622, *see* Horowitz, F.
USP 3,655,395, *see* Karnemaat, J. N., 210
USP 3,651,182, *see* Rosenthal, D., 187
USP 3,649,397, *see* Peters, T. E., 175
USP 3,645,981, *see* Bonner, W. H., Jr., 105
USP 3,642,042, *see* Davidson, R. W., 107, 245
USP 3,616,201, *see* Trocino, F. S., 121
USP 3,572,275; 197
USP 3,473,252; 196
USP 3,470,115; 186
USP 3,312,639, *see* Justice, 187
USP 3,306,861, *see* Justice, 187
USP 3,231,525, *see* Kelley, F. L., 186
USP 3,172,808, *see* Baumann, 197
USP 3,110,129, *see* Baumann, 198
USP 3,108,990, *see* Baxter, G. F., 121
USP 3,080,343, *see* Ham, G. E., 105
USP 2,860,856, *see* Bauer, W., 186, 193
USP 2,835,330; 196
USP 2,686,143, *see* Fahrni, F., (1954), 110, 165
USP 2,652,377, *see* Allied Chemical Co., 186, 187
USP 2,349,756, *see* Pratt, B. C., 122
USP 2,327,984, *see* West, H. J.
USP 2,327,871, *see* Crocker, E. C., 181, 260
USP 2,321,958, *see* Walker, J. F.
USP 2,317,181, *see* D'Alelio, G. F.
USP 2,273,367, *see* Meyer, L. S., 186
USP 2,247,764, *see* Nevin, J. V., 18
USP 2,121,076, *see* Ellis, 121
USP 2,113,485, *see* Kreidl, 118
USP 2,097,895, *see* Rossiter, E. C., (1944), 18
USP 2,076,295, *see* Plaskon, 186
USP 2,033,411, *see* Carson, F., 164
USP 1,991,765, *see* Marks, M., 17
USP 1,982,539, *see* Reeves, B. H., 205

USP 1,967,685, *see* Pungs, W., 17
USP 1,967,261, *see* Ripper, K., (1934), 17
USP 1,952,060, *see* Ellis, C., 17
USP 1,950,746, *see* Pollak, F., (1934), 17
USP 1,920,451, *see* Knilling, W. v., 17
USP 1,899,109; 7
USP 1,893,911, *see* Smidth, L., (1933), 17
USP 1,863,426, *see* Walter, G., (1932), 16
USP 1,846,853, *see* Ellis, C., 11
USP 1,844,570, *see* Goldschmidt, S., (1932), 15
USP 1,833,868, *see* Ripper, K., (1932), 16
USP 1,831,706, *see* Gams, A., (1932), 15
USP 1,793,666, *see* Baldwin, J. T.
USP 1,791,062, *see* I. G. Farbenindustrie, 12
USP 1,790,461, *see* Cherry, O. A., 15
USP 1,762,456, *see* Ripper, K., (1930), 14, 17
USP 1,734,516; 200
USP 1,718,901, *see* Gams, A., (1929), 14
USP 1,704,347, *see* Smidth, L., (1929), 14, 17
USP 1,691,427, *see* Barthelemy, H., 14
USP 1,687,312, *see* Ripper, K., (1928), 13
USP 1,676,543, *see* Gams, A., (1928), 13
USP 1,674,199, *see* Gams, A., (1928), 11, 13
USP 1,672,848, *see* Lauter, F., (1928), 13
USP 1,671,596, *see* Lauter, F., (1928), 13
USP 1,658,359, *see* Steppes, F. E. K.
USP 1,625,283, *see* Ripper, K., (1928), 13
USP 1,536,882, *see* Ellis, C., 11
USP 1,536,881, *see* Ellis, C., 11
USP 1,482,358, *see* Ellis, C., 11
USP 1,460,606, *see* Ripper, K., (1923), 10
USP 1,355,834, *see* John, H., (1921), 4, 7, 8-9
USP 1,062,501, *see* Quade, F., (1913)
USP Reissue 19,463, *see* Ripper, K., (1935)
USP Reissue 18,943, *see* Smidth, L., (1933)

USSR P. 627,109, *see* Yukin, N. A., 210
USSR P. 615,103, *see* Dumov, S. N., 190
USSR P. 615,049, *see* Shirokorodyuk, V. K., 206
USSR P. 596,561, *see* Lubis, B., 210
USSR P. 590,318, *see* Kuznetsova, A. E., 212
USSR P. 584,062, *see* Glubish, P. A., 203
USSR P. 580,939, *see* Sokolova, V. A., 212
USSR P. 576,327, *see* Naidenov, M. N., 190
USSR P. 576,154, *see* Gorenko, V. G., 190
USSR P. 559,000, *see* Morozova, S. S., 204
USSR P. 557,859, *see* Osipova, 106
USSR P. 554,344, *see* Nelyubin, I. A., 206
USSR P. 554,233, *see* Maslinkovskii, A. S., 286
USSR P. 550,153, *see* Mints, B. Y., 202
USSR P. 518,363, *see* El'bert, A. A., 121
USSR P. 509,615, *see* Smirnova, E. A., 110

USSR P. 496,291, *see* Zabrodkin, A. G., 110
USSR P. 487,860, *see* Smirnov, B. A., 205
USSR P. 485,887, *see* Yudina, G. G., 178
USSR P. 481,584, *see* Yakovlev, D. A., 190
USSR P. 480,555, *see* Dashkovskaya, Z. F., 247
USSR P. 478,724, *see* Anokhin, A. E., (1975), 178
USSR P. 476,296, *see* Lapitskii, V. A., 106
USSR P. 476,294, *see* Bokareva, E. Z., 212
USSR P. 471,348, *see* Tarasenko, L. D., 189
USSR P. 462,846, *see* Frolova, M. K., 189
USSR P. 456,730, *see* Sklyarenko, B. S., 178
USSR P. 448,208, *see* Borodkina, N. I., 189
USSR P. 431,199, *see* Borozdin, V. S., 189
USSR P. 431,008, *see* Shutov, G. M., 177
USSR P. 421,531, *see* Anokhin, A. E., (1974), 178
USSR P. 421,529, *see* Anokhin, A. E., (1974), 178
USSR P. 415,103, *see* Dumov, S. N.
USSR P. 411,108, *see* Tsvetkov, V. E., 109
USSR P. 403,811, *see* Davydov, V. V., 208
USSR P. 403,654, *see* Kozin, V. M., 188
USSR P. 395,280, *see* Zastanchenko, M. A., 176
USSR P. 390,969, *see* Zastanchenko, M. A., 176
USSR P. 387,847, *see* Zastanchenko, M. A., 176
USSR P. 384,843, *see* Akutin, M. S., (1973), 105
USSR P. 368,150, *see* Anokhin, 175
USSR P. 368,067, *see* Time, N. S., 176
USSR P. 364,426, *see* Anokhin, A. E., (1972), 105
USSR P. 362,855, *see* Borisyuk, I. D.
USSR P. 361,251, *see* Putseiko, L. K., 207
USSR P. 358,195, *see* Temkina, R. Z., 175
USSR P. 358,150, *see* Anokhin, A. E., (1972), 175
USSR P. 353,847, *see* Anokhin, A. E., (1972), 174
USSR P. 353,846, *see* Anokhin, A. E., (1972), 174
USSR P. 346,396, *see* Belyi, V. A., 175
USSR P. 338,415, *see* Koromyslova, T. S., 210
USSR P. 337,358, *see* Knat'ko, V. M., 205
USSR P. 333,154, *see* Kuznetsova, A. E., (1972), 207
USSR P. 332,064, *see* Ishin, N. D., 181
USSR P. 329,897, *see* Pahla, P., 175
USSR P. 328,152, *see* Valgin, V. D., 188
USSR P. 327,224, *see* Raizman, M. S., 107
USSR P. 322,348, *see* Arbuzov, V. V., 173
USSR P. 321,505, *see* Spirin, Y. L., 205
USSR P. 317,804, *see* Kozlov, M. K., 105
USSR P. 108,597, *see* Zlodovskii, L. I., 203

Subject Index

Chemicals are listed under the full name. Some important abbreviations are explained in the Glossary.

abatement, formaldehyde, 183-184
  see also reduction
abrasives, UFR, 203
absorption kinetics, formaldehyde, textiles, 201
accelerated curing, 111
accelerated weathering, 287
acetaldehyde, 77, 85
acetaldol, 85
acetals, hydrolysis, 232
acetamide, 64
  for UF, 80
  methylol, 63
  reagents, 80
acetate, formaldehyde reduction, 246
acetonitryl, 68
acetylaceton method, 128, 150
acetyl urea, 65
acid catalysis, UF, 81, 90-92
  salts, urea, 73
acidic cocondensation, 120-121
acidimetric sulfite, 131
acidimetry, formaldehyde, 125
acidity constant, formaldehyde, 26, 36
acoustic tiles, 205
acrolein, 85
acrylamide, UF, 109
acrylate resin, 110
acrylic, 105
acrylonitrile, 68
ACS – *see* American Chemical Society
acute formaldehyde exposure, 290
addition of alcohols, 115
  chemistry, 79-85
  mechanism, UF, 90
  reactions, 48, 100
  stages, 100
additives, foam, 192
addresses, for standards, 222
adducts, urea, 73-74
adenine, 67
adenosine, 67
adherend, 157-159
  surface, 158
adhesives, 107-112, 156-161
  additives, wax, 170
  adherent interface, 158
  bond strength, standards, 218
  cost, particleboard, 167
  dip, 72
  film, definition, 157
  formaldehyde release, mechanism, 233
  formulation, 107-111
  formulation blending, 172
  particleboard, 169
  penetration, 16
  performance, 111
  properties, 157-159
  requirements, plywood, 224
  resins, 107-111
    uses, 270, 285-288
  soaking, 16
  see also glue
adipamide, 65
ærolite, 18
aging, resin, 109
A-grade veneer, 223
agriculture, 207-211
agricultural foams, 207-210
  uses, 279
AIHA – *see* American Industrial Hygiene Association
air concentration, formaldehyde, formula, 148
  formaldehyde and water, 240
  formaldehyde analysis, 127
  indoor, 257-263
  methanol, explosive limit, 98
  quality standards, 291
  sampling, 144-168

standards, for homes, 266
tentative, 259
workplace, 255
airplane assembly line, 162
airplane frames, 18
airplane fuselage, 186
airplane hangars, roofs, 161
airplane plywood, 161
alanine, 67
albumin, 12
alcohol addition, 115
in resins, 115
aldehydes, dismutation, 47
resins, 5, 85
aldol condensation, 47
aldoles, 85
alkali addition, UFR, 107
chlorides, 114
hardening, 114
in phenols, 111
alkaline earth addition, UFR, 107
reactions, UF, 92
salts, addition, UFR, 1, 109
alkyl sulfonic acid, 105
alkylaminomethanols, 61
alkylation, ureido, 79-85
all-wool plaid, 201
alpha-ureido-alkylation, 79-83
alum-citric resin, 109
phosphate addition, 110
to prevent bleeding, 175
aluminum acetate, textiles, 201
chloride hardener, 11, 114
coating, 203
hydroxide, fireproofing
phosphates, 245
sulfate, 120
ambient air, formaldehyde in, 22, 255
sampling, 144-146
standards, 236
ambient formaldehyde standards, 259
American Chemical Society, 124
method, 125
American Industrial Hygiene Association, 25
American Plywood Association, 224-226
American Society of Heating, Refrigerating, and Air-Conditioning Engineers, 148
American Society of Testing Materials
mailing address, 222
(test D 1037-72a), 179
weather test, 287
American Standards Association, 124
method, 124
amide, primary, hydrolysis, 232
secondary, hydrolysis, 232
yields, in UF, 137
amides, 63
reactions with formaldehyde, 63
amidomethylol, hydrolysis, 232
amines, formaldhehyde reductions, 246
amino, acids, 67
compounds, formaldehyde reduction, 247
reagents, 79, 80
yields, UF, 137
aminoacetic acids, 67
aminoform, 48
aminoplasts, 79
formation, 79-85
aminotriazines, 80
aminouracil, 68
ammonia, 4, 5
above soil, 210
addition, 18
consumption, 268-269
effect on cure, 112
formaldehyde reduction, 246
in resin, 107
markets, 268
reactions with UF, 96-97
with wood, 111
uses, statistics, 268
wood treatment, 107
ammonium, effect on formaldehyde release, 243, 244
carbimate, 99
carbinate, 99
chloride, formaldehyde, analysis
chloride, heat release, 115
cyanate, 72
polysulfide, 105
salts, influence, 120
sulfide, 5, 49
sulfide resin, 110
analysis, elemental, 5
analytical chemistry, new methods, 19
revolution, 228
aniline, 120
dimethylol, 62
reaction with formaldehyde, 62-63
resins, 6
anthropogenic formaldehyde, 46
antibacterial foam, 217
antioxidants, 121
antiscorching, 206
APA – *see* American Plywood Assocation
aqueous formaldehyde, 23, 25, 32-43
formaldehyde, composition, 26, 27, 32-43
formaldehyde, density, 32
formaldehyde, partial pressure, 34-35, 37
formaldehyde, sampling, 149

aqueous resin, sampling, 149
artificial silk, 13
  reactions with formaldehyde, 59-61
ASA – *see* American Standards Association
asbestos, addition, 111
  fiber, 175
  in foam, 187
  mats, 189
  UF, 205
ASHRACE – *see* American Society of Heating, Refrigerating, and Air-Conditioning Engineers
asparagine, 67
asparagus, foam, 207
Asplund process, 180
assay, formaldehyde, 125
  formic acid, 131
  methanol, 130
  urea, 124
A-stage, 72
ASTM – *see* American Society of Testing Materials
austenitic steel, 98
autoclaving, of wood, 121
automobile exhaust, formaldehyde in, 127
automobile roofs, 186
auto-redox reactions, formaldehyde, 42
average condensation, UFR, 140
azulene method, 129-130

bacteria resistance, plywood, 224-225
BAM test, weathering, 287
bark chips, in boards, 177
  extenders, 121
  extractives, resin, 109
  powder, filler, 111
  resin, 111, 121, 208
base catalysis, UF, 81, 92-95
batch reactors, 101-102, 103
beams, laminated, 161
beer sludge in resin, 110
beetle, 3, 226
bentonite in boards, 178
benzaldehyde, 117
benzamide, 64
benzene, 68
benzoguanime addition, 108
benzoguanime resin, 108
B-grade Veneer, 223
biological corrosion, 111
bioresistant board, 178
biscarbamidomethyl ether, 88
bis(phenylamino)-methane, 62
bisulfite, adhesive, 110
  air sampling, 146
    resin, 109, 121-122
biuret, 73

blood albumin, 111
blowing agents, foam, 192, 217
board properties, 168
boats, UFR, 18
boiling point, aqueous formaldehyde, 32
  test, plywood, 225
bond distance, formaldehyde, 24
  formation, 157-159
  strength, 110
breath, human, formaldehyde from, 257
bromine, reaction with formaldehyde, 56
B-stage, 72
BSTFA, *N, O*-bistrimethyl-silyl-trifluoroacetamide, 89, 132
buffer, formic acid, 17
  glue, 169
  phosphate, 107
  systems, 10
building board consumption, 282
  fireproofing, foam, 207
  markets, 286
  materials, from waste, 211
built-in hardeners, UFR, 163
bulk adherent, 157
  board, formaldehyde from, 256
  density, 215
  glue line, 158
butyl-phenol resin, 109

cabinet makers, odor, 229
calcium chloride, charring, 190
camphor in foam, 187
canned fish, resin, 110
  sludge in resin, 11
Cannizzaro reaction 42, 47, 95
carbamate, 65
carbamide, 5, 72
carbenium ion, 80, 86, 90
carbinate, ammonium, 99
carbonate, 10
  buffer, in foam, 186
  fire extinguisher, 206
carbonic acids, 86, 95
carbonyl reagents, 79, 85
carborundum, UFR, 203
carboxylic acid, 110
  addition, 106
carcinogen, 254, 266
  formaldehyde, 263
cardboard, boxes, 204
casting, binders, 207
  foundry, 180
  materials, 15
  resin, 211, 214
catalysts, 10
  built-in, 17
  foam, 191

for formaldehyde, 97-99
for urea, 99
methanol oxidation, 98
cationic resins, 110
cattle feed, 208
caul, particleboard, 170-171
CBU, cyclobutylene-urea, 83
CD plugged plywood, 224
CD-X plywood, 224, 226
cedar, odor, reduction, 181, 233
ceiling panels, 212
UF in, 286
cellophane paper, 181
UF bonding, 162
cellular ceramic, 197
celluloid, 7, 9
cellulose, 58-61, 95
acetate, 204
crosslinking, 59
IR, 60
materials, 121
methylol, 61
reactions with formaldehyde, 58, 59-61
resin, 121
swelling, 58
cement, UFR, 206
ceramic binders, 190
cereals, as extenders, 164
C-grade veneer, 223, 224
chain length, polymer, 16
charcoal, 12, 13
CH-bending mode, UFR, 142
CH_2 stretch, UFR, 142
cheese, formaldehyde in, 267
chemical producers, UFR, 276
chemicals, formaldehyde reducing, 245-249
chemistry, UFR, 73-100
chicken manure, 208
chintz-effect, textiles, 201-202
chloral, 85
chloraldehyde, 185
chlorate, 10
chlorella, addition, UFR, 110
chlorine, reaction with formaldehyde, 56
chloromethanol, 61
chromtropic acid, 148, 267
acid method, 127
chronic exposure, 253-255
chronic formaldehyde exposure, 290
cinnamic acid, cure, 114
citrate, 10
citric acid, 60
clathrates, 77
thiourea, 77
urea, 73
clay, filler, 109
CN-mode in UFR, 142
CNDO, molecular calculations, formaldehyde, 26
^{13}C-NMR, 134-140
formaldehyde, 40-41
CO, effect on formaldehyde, 148
CO-stretch in UFR, 142
coal mines, UF foam, in, 19
coated UF, weathering, 287
coatings, 106
compositions, 17
UFR, 203
co-condensations, 118-119
coconut shells, 105
shell flour, 164
coffins, 180
cohesion, 157-158
cold, curing, gamma rays, 181
setting resin, 14
colloidal quartzboard, 205
colored particleboard, 176
tiles, 176
combined formaldehyde, 140
combustibility of foam, 189
combustion of formaldehyde, 47
products, IR, 206
commercial formaldehyde solutions, 45-47
comminuted wood, moldings, 211
urea mix, 175
composite board standards, 220
compressive strength, 159, 217, 220
concentrated UFR, 106
condensation, 100
chemistry, 79-85
delayed, 89
kinetics, 87
mechanism, 90
products, 20-21
products, UF, 79-85
reactions, 48, 64-65, 86-87
water, 15
construction, board, 205
industry, 283
market, 289
materials, 205-206
consumer, complaints, 236, 256
definition, 2
expectations, 227
product safety commission, 291
consumption, resin types, 271
contact allergies, 256
continuous, manufacture, UFR, 108, 109, 248, 285
Conway cell, 147
copper in urea, 124
sulfate in rayon, 60
core resins, 207
cork, bonding with UF, 162

UF board, 205
corrosion, by foam, 189
plywood, 243
protection, 213
urea manufacture, 98, 99
corrugated cardboard, 204
cost breakdown, particleboard, 16
cotton, excess formaldehyde, 59-61
fireproofing, 202
formaldehyde absorption, 60-62
formaldehyde kinetics, 201
cotton underwear, 252-254, 262
crack resistance, 121
craze, 163
creep, 111
creeping, particleboard, 287
creosol, 71
creosol resin, 111
crossed Cannizzaro reaction, 47
crosslinked resin, 106
crosslinking agents, 109
crotonaldehyde, 85
crushing strength, plywood, 223
C-stage, 72
CTB weathering test, 287
cumulative effect, formaldehyde, 254, 263
cups, UFR, 211
cure studies, 112, 114
cure tests, 104
curing, 13
DTA, 113
gamma rays, 181
TGA, 114
viscosity, 108
curing agents, 10
curing of resin, 112-114
curtains, fireproofing, 202
cyanamide, 10, 66, 80, 122
for UF, 80
in resin, 121
cyanogen, history, 4
cyanuric acid, 73
cyclic ethylene urea resins, 93
cyclic UF, 93-94
products, NMR, 93-94
cyclobutylene urea, 83
cyclopentanone, 85
cyclorite, 50
Czechoslovakia, formaldehyde regulations, 256

Dade County, Florida, foam use, 217
dairy wastes in resin, 110
Danish homes, formaldehyde in, 258
decomposition, 23
temperature, foam, 217
UFR, 216
degradable moldings, 210
dehydrogenation, methanol, 96
delamination, 225
delta acid method, 128-129
polyoxymethylene, 57
Denige's method, 127
density, aqueous formaldehyde, 32
bulk, 215
foam, 216
particleboard, 220, 221
UFR, 230
wet, foam, 216
dermatitis, 263
desensitization, 263
desiccator test, German, 150
Japanese, 151
detection limit, formaldehyde, 129
deutero formaldehyde, 55
D-grade veneer, 224
diacetamine, methylene, 63
diacetone alcohols, 85
aldehydes, 85
dialdehydes, 85
diatomaceous earths, board, 178
dibenzyl ether, 70
formal, 57
dichloromethyl sulfide, 51
dicyanamide, 122, 187
for UF, 80
resin, 122
dicyandiamide, 108
dicyandiamide in foam, 190
dicyandiamine resin, 108
dielectric constant, UFR, 215
heating, 110
strength, foam, 197
strength, moldings, 211
diethanolamine, 106
diethyl formal, 57
diglycine, 67
dihydroxy acetone, 47
diisobutyl formal, 57
diisopropyl formal, 57
dimedon method, formaldehyde, 127
dimensional stability, 111
dimethoxymethane, 56
dimethyl acetal, 56
dimethyl formamide, 57
dimethyl sulfoxide resin, 110
urea, 12
dimethylene ether, total, UFR, 140
dimethylol aniline, 62
dimethylol dissociation, 82
dimethylol peroxide, 46
dimethylol urea, 64, 101
dimethylolurea, diethylether, 21
NMR, 135, 138, 140

dimethyloluron, 64, 92
DIN, German DIN system, 179, 217
  mailing address, 222
dioctyl formal, 57
dioxane, 28
dioxolane, 58
dip adhesives, 72
diphenyl-methane-isocyantes, 122
dipropyl formal, 57
dipropylene glycol, 106
  resin, 106, 110
dirigibles, foam in, 196
discomfort, formaldehyde, curves, 253
dishes, UFR, 211
dismutation, formaldehyde, 47
display, plants, foam, 198
dissecting rooms, 256
dissociation energy, formaldehyde, 24
distillation, of formaldehyde, 35
disulfite, 106
  gellation, 104
  resin, 106
disulfur dichloride, 77
dithionite, 55
  formaldehyde, reduction, 245, 247
  gellation, 104
  reaction with formaldehyde, 55
diurea, 118
  addition, 117
  methylene, 86
dixanthylurea, 124
DMOMDU, NMR, 135, 136-141
DMU, dimethylol urea, 92, 88, 101
DMU manufacture, 101
DMU, NMR, 135, 136, 137-140
DNA, formaldehyde effect, 254
dodecylmercaptan, 89
door panels, 174
dose-response, formaldehyde, 234, 252, 253, 262-266
dry-molding, 106
dry-out point, glue, 169
dry-resin, manufacturer, 101
  particleboard, 168
drying effect, textile, 201-202
DTA of resin cure, 112, 114
durability of foam, 217
Dutch Bouwcentrum, 217

ebonite, 7, 9
economy, postindustrial, 227
edges, odor from, 226, 240, 242
  panels, formaldehyde from, 286
Eegriwe method, formaldehyde, 127
egg albumin adhesives, 121
  resin, 121
elasticity of adhesives, 109
electric, conductivity, 108
  foam, 217
  resin, 108
  resistance, 215
electrical insulating paper, 205
electrical insulators, 197, 211, 215, 289
electrolytes, 16
electroplating, 214
elemental sulfur, 50-52
  reaction with formaldehyde, 50, 51
embossed textiles, 202
emotions, in odor, 228, 235
emulsifiers, 120
emulsions, UFR, 107
encapsulated ink, 204-205
  resin, 107, 289
energy conservation, 2
  cost, particleboard, 167
  crisis, 286
  homes, 260
England, formaldehyde regulations, 256
  UFR price trends, 273
environmental compatibility, 18
enzymes, 68
  effect on formaldehyde, 254
  formaldehyde reduction, 246
  urease method, 124
EPA, 148, 266, 291
epichlorohydrin, 105
  resin, 105
epidemiology, formaldehyde, 254
equations, formaldehyde release, 238
equilibrium kinetics, formaldehyde, 40-41
  of aqueous formaldehyde, 38
  urea manufacture, 99
ESCA, 284
ester formation, 115
esterification of formaldehyde, 56
ethers, 57
  formation, 115
ethylamine on rayon, 60
  effect on textiles, 201
ethylene, chlorhydrin, 12
  diamine, 105
    formaldehyde reduction, 245
  glycol, 12, 95
    in UF foam, 186
  urea, 83
  urea, kinetics, 81
ethyl-urea, 6
EU, ethylene, urea, 83
eucalyptus pulp, 212
excess, formaldehyde, 260
  methylol, 260
exhibition halls, roofs, 161
explosive limit, formaldehyde, 24
  methanol, 98

extenders, 110, 111, 164, 169-171
exterior glues, 222-223
  grades, plywood, 225
extruded boards, 178

face density, particleboard, 110
factory built housing standards, 219
Fahrni test, 150
Falima, 241
fatty alcohols, 120
Federal Register (U.S.), 266
Feldman cell, 147
fertilizers, 207-210
  coatings, 230
  formaldehyde in, 127
  slow release, 18
FESYP, European Federation of Particleboard Manufacturers, 222
  iodometric test, 147
  perforator tests, 234
  sampling method, 152
fiberboard, export, 278
  future uses, 275
  import, 278
  standards, 219
  statistics, 279-280
fiberglass, board, 178
  mats, 205
fillers, foam, 217
  moldings, 212
fire resistance, 206, 207, 285, 286
  foam, 191
fire resistant, board, 178
  sealers, 206
fire retardants, board, 170
  UFR, 206-207
fireproof particleboard, 245
fireproofing, 111, 206-207
  UFR materials, 206
Fischer titration, 124
fishing nets, 109
flammability, foam, 194, 217
  resin, 109
flash ignition, foam, 194, 195
flexual strength, 215
  foam, 217
floculating agents, 213, 214
floor finishes, 203
  odor, 233
  underlayment, 177, 276
Floramin, 18
flower exhibits, 210
fluid stage, glue, 169
fluorescence analysis, 285
fluoride, fire retardant, 171
  hardener, 114
fluorosilicate, fireproofing, 174
foams, 110
  additives, 192
  allergies, 197
  applications, future, 289
  charring agents, 189
  chemistry, 185-191
  density, 12, 186, 194
  field apparatus, 193
  formaldehyde, release, mechanism, 233, 249
  free formaldehyde, 187
  gun, 192-193
  inducing agents, 189
  injected into soil, 197
  injecting, fireproofing, 206
  manufacture, 192-193
  odor reduction, 187
  panels, 212
  plant display in, 198
  preparation, 192-193
  properties, 194
  resin, 191-193
  resin sulfite, 121
  use, 217
foamed resin adhesive, 110
foaming, agents, 16, 187-188
  in resin, 107
  machines, portable, 19
foirc tests, 111
foliate board, 176
food, formaldehyde in, 267
footwear, 214
forest areas, largest, 277
formal, 56
Formaldehyde
  abatement, 183-184, 226-249
  absorption, by cellulose, 60
    by wood, 233-234
    human body, 234
    on particleboard, 59
    rate, textiles, 201
  acidity constant, 26
  adducts with hydroxy, 58-60
  analysis, 125-130
    with sulfite, 52
  as carcinogen, 254
  as mutagen, 254
  atmospheric reactions, 46-47
  chemistry, 22, 72
  commercial, 43-45
    properties, 45
  concealed by wood, 233-234
  concentrated solutions, 38-43
  consumption, 9, 268
  cumulative health effect, 263
  determination, comparison, 154
  dilute solutions, 38-39

discomfort curves, 253
discovery, 5
donors, 63
dose response curve, 251-253
effect, on eyes, 262, 266
on health, 262, 266
on nose, 262, 263, 266
emission, from boards, 166
exhalation, 257
free, and in air, 235
and humidity, 237
methods for, 131
perforator values, 235
guidelines, 266
hemiformal, 23
human response, to, 263-265
in air, 235
and water, JIS, 240
effect of SO_2, 238
workplaces, 254-256
in automobile exhaust, 127
in Danish homes, 258
in dissecting rooms, 253
in food, 267
in hospitals, 253
in mortuaries, 253
in water and air, 240
in workplaces, 148
influence of surface, 241
ingestion, 253
inhaled, 253
Institute, 266, 291
latent, 22
lingering on surfaces, 259
manufacture, 97, 100
monomeric, 23-26
NMR, 137, 140
odor, desentization, 251
oligomers, methods for, 130
on cellulose, 58-60
IR, 199, 200
perdeutero, Raman spec., 55
properties, 22-29
public attitude, 255
reaction with acids, 54-55, 56
aliphatic hydroxy, 56-61
amines, 61
ammonia, 48-49, 50, 53
aniline, 62
bases, 50
halogen, 56
heterocyclics, 68-72
hydrogen chloride, 53
hydrogen sulfide, 49-52
in air, 46
mercaptans, 61
ozone, 46
Raman spectra, 53
sulfite, 52-54
sulfur compounds, 50-56
sulfur oxide, 52-56
ultraviolet light, 46
recognition threshold, 251
reduction, bond strength, 242
by additives, 245-249
by ammonia, 242-249
by sulfones, 234
by sulfur compounds, 243-244
by ventilation, 261
chemical, 242-249
half potential, 48
stop-gap, 290
surface treatment, 242
regulations, 256
release, 290
ceilings, 256
coatings, 241
curves, 239
equations, 238
JIS curve, 239
mechanism, 233, 259-260
molded UFR, 257
paper overlay, 241
school furniture, 256
seven rules, 235
soil, 210
standards, 182
sulfur, effect, 247
surface effects, 241
surface treatment, 242
surface types, 241
testing, 236
treatment, 236-249
ventilation, 234
retention, by cellulose, 59-61
by hydroxy, 58
sampling, 144-156
air bag, 155
Mohl method, 154-155
subcutaneous, 253
substitutes, 118
sulfoxylate, 54
synergism, 262
titration, with ammonium chloride, 50
with sulfite, 54
total, methods for, 131
vaporizable, 260-261
toxicity, 250
formals, 57-60
Formanol, 50
formic acid, 12
above soil, 210
methods, 131
formin, 48
formyl radical, 47
foundry casting, 180

fourth resin stage, formaldehyde reduction, 248
free formaldehyde, 10, 103
  ammonia treatment, 245
  and formaldehyde in air, 235
  and humidity, 237
  in foam, 187, 194
  methods for, 131
free methylol, in foam, 187
freeway abutments, 207, 209
freezing point, aqueous formaldehyde, 32
frost protection, 210
fugitive binders, molding, 190
fumaric acids, 60
fumigation, 262, 286
fungi, resistance, UF, 156
furfural, 68
  antiscorching, 206
  resin, 105, 108
furfuryl, addition, 106
  alcohol, 68
  aldehyde, 164
  resin, 106, 109
furniture, 180, 287
  assembly, 175
  coatings, 211
  consumer product, 2
  panels, 111
  panels, odor, 181
F-U feed ratio, 140

gallic acid method, formaldehyde, 130
gamma-polyoxymethylene, 57
gap, in adherent, 159
  filling cement, 164
  joints, 18
gas chromatography of formaldehyde, 130
gas-liquid chromatography, aqueous formaldehyde, 38
gas-solution equilibrium, formaldehyde, 40
gel chromatography, UFR, 89
gelation, 104-105
  influence of glycols, 59
  time, 103, 109
  time, particleboard glue, 169
Geomet study, 257
Germany, air standards, 259
  formaldehyde regulations, 256
  formaldehyde standards, 266, 267
glass, artificial, 10, 11
  bonding, 162
  fiber, bonding, UF, 162
    coating, 203
  in foam, 187
  insulation mats, 189
  UF, intrinsic problems, 12
GLC – *see* gas liquid chromatography
  total formaldehyde, 131, 132

glue, 157-159
  blender, 172
  lines, 159
  particleboard, 169
  spray control, 170
  see also adhesive
gluten, filler, 111
glyceraldehyde, 47
glycerols, 58-59
  reactions with formaldehyde, 58-60
  UFR, 109
glycidyl methacrylate, 105
glycine, 67
glycols, 58-59
  NMR, 39, 40
  resin, 106, 110
glyoxal, 85, 118
glyoxylic acid, 85
GPC, gel permeation chromatography, 89
grabite-UF mix, 205
grades, plywood, 223-226
grapevines, UF foam, 207
greenhouses, 11, 207-208
  effect, foam, 198
growth potential, foams, 289
guanamine, 80, 105
  resin, 105
guanidine, 80, 86
  formaldehyde reduction, 245-249
  UFR, 118
gun training targets, 204
gypsum board, 205, 289
  fireresistant, 288
  as extender, 164
  foam stabilizer, 186
  in boards, 178
  lightweight board, 206

halogens, reactions with formaldehyde, 5
hardboard, consumption, 180
  standards, 220
hardener, 109, 120
  built-in, 106
  formulations, 114
hardening by radiation, 114
hardness, UFR, 215, 230
hardwood consumption, 278
  extractives, 111
health standards, 266-267
heat, durability, adhesive, 226
  of combustion, foam, 194, 217
  of formation, 49
  release, urea addition, 116
  resistance, 105
  vaporization, monomeric formaldehyde, 24
Hehner test, 267
helmitol, 50

hemiacetals, 47, 98
hydrolysis, 234
NMR, 40
hemiformal, 91
NMR, 137, 140
hemp shive board, formaldehyde from 245
HEW, (U.S.), 266
hexamethylene tetramine, 6, 48-50, 106, 174, 267
resin, 106
vapor pressure, 50
hexa-iso-urea, 6
hexamine, 48
hexylsulfamate hardener, 114
high nitrogen fertilizer, 207-208
home, air standards, 266
decorations, 17
insulation, 196-197, 260
homo-saligenin, 71
honeycomb, board, 174
horse feed, 208
hot press, 172-173
hot-setting resins, 159
household utensils, 211
housing, consumer product, 2
prefabricated, 213
standards, 219
starts, 282
HUD, (U.S.), 148, 291
foam approval, 217
human skin, formaldehyde effect, 263
humic acid resin, 110
humidity, and free formaldehyde, 237
hydantoin, 64, 65
hydrazide, resin, 106
hydrazodiurea, 118
hydrochinone adduct, 73
hydrogen, bromide, 12
chloride, 12
cyanide, 50
peroxide, 10
foam, 190
sulfide, 62, 66, 86, 118
reaction with formaldehyde, 50, 51
hydrolysis, by acute exposure, 234
by chronic exposure, 234
surface, 215
UFR, 143-144
hydrolytic stability, 111
hydrophobic, foam, 190
point, 104
resins, 13
hydroponics, 198
hydroxy compounds, reactions with formaldehyde, 56
hydroxyl amines, 120
formaldehyde reduction, 243
hydroxylamine method, 126
hydroxymethylation, 100
hydrolysis, UFR, 143
hygro mulch, 210

ice rinks, roofs, 161
idantrione, 67
identification limit, formaldehyde, 129
imidazolidine, 69, 95
imidazolidone urea, 84
impact, resistant UFR, 106
standards, 218
impinger, sampling, 146, 152
impregnated felt hats, 255
increased adhesion, 109
incubation, curing, UF, 112
indoor air, quality, 255
sampling, 145
standards, 256-259
induction period, 89
industrial, applications, UF, changes in, 16
batch reactor, 102
revolution, 228
waste treatment, 210
infrared bands, UF resin, 142
see also IR
inorganic board, 176-177
in situ analysis, UFR, 136-140
insulating board, rockwool, 176
slag, 176
insulation, foam, 18, 101, 188, 260
markets, 275
paper, electrical, 205
interior decoration, 176, 212
intermediates, MW, 108
in UFR, 13
viscosity, 108
intra-adhesive boundary, 158
iodide, 10
iodine titration, 125
iodometric formaldehyde, 126
ion, exchange resins, formaldehyde reaction, 48
sputtering analysis, 284
Iporka, 19
IR (infrared) analysis, UFR, 139-144
cellulose, formaldehyde, 199, 200
curing, 114
formaldehyde on cellulose, 60
hydrolysis, UFR, 143, 144
IR spectra, 283
formaldehyde, 24
kinetics, 83
monomethylol compounds, 83
UF formation, 82
UFR resin, 142

iron, formaldehyde reduction, 242
oxide, catalyst, 98
in foam, 189
salts in boards, 178
urea, 124
isobutylene, 86
isocyanates, 1, 111, 122, 284-285
adhesive, 178
isomerism, thiourea, 76
isophthalic, UFR, 212
isoprene, 86
iso-urea, 14
chemistry, 74-75

J-acid method, 129
Japan, UFR consumption, 273
Japanese dessicator test, 150-151
odor standards, 182
JIS, Japanese standards, 150, 151, 182, 221, 237, 267
formaldehyde release curves, 239
mailing address, 222
Jong's method, 131

kaolinite, 109
boards, 178
Karatex, 121
Kaurit, 12, 16
ketones, for UF, 84, 85
kick-over, glue, 169
kinetics, condensation, 86
hardening, 109
reaction, 87
UF-reactions, 81
Kjeldahl method, 124
Knapp's test, 267
kraft pulping resin, 120-121
KS-68M, Russian resin, 108

labor, cost, particleboard, 167
intensive plants, 285
laboratory reactor, 102, 103
lactic acid, 60, 121
formaldehyde reduction, 245
laminated, board, 176
foam-board, 187
historic trends, 271
paper, 105
paper resin, 110
rafters, 161
timber, 161
Lanaset, 202
large manufacturers, UF resin, 285
late addition of formaldehyde, 175
of reagents, 115
UF resin, 115
latent hardeners, 112
lauryl triethonalamine sulfate, 189
legislation, odor, 228
licensing, 292
light, as hardener, 114
sensitive pigments, cure, 11, 114
ligning, bonding surface, 159
extenders, 121
UF-chemistry, 89
UFP resin, 121
wastes, UFR, 109
lignocellulosic adhesive, 174
lignosulfonate, 106, 120-121, 203, 245-247
resin, 106, 120, 121
UF gelation, 104
linseed oil, resin, 117
liquid adhesive, tensile test, 218
composition, aqueous formaldehyde, 36
lissapol on rayon, 60
liver enzymes, 290
formaldehyde, 254
Los Angeles, foam use, 217
low, odor board, 178
resins, 107, 110, 243-249
UFR, 108
lumber, end uses, 281
production, 278
Lyofix, 202

"mad hatter," 255
mailing address, standards, 222
maize, extender, 164
MAK, 259
maleic anhydride resin, 110
malonamide, 65
malonic dialdehyde, 85
ester, 74
Mannick reaction, 62, 68, 71
manufacture, resin, 101-106
manufacturing parameters, 183
marble, synthetic, 206
marmitol, 58
Masonite process, 180
mastic, 203, 205
mat, laying, 164
moisture, particleboard, 168
preparation, 170-171
MBTH method, 129
MDF, medium density fiberboard, 17
MDU, NMR, 135, 136-140
meat wrappers, UFR, 229
mechanical strength, foam, 217
medicinal foams, 197
medium density fiber board, 179, 276
meerschaum, 10, 185
melamine, 11, 18, 66-67, 80, 86, 105, 108, 110, 111, 118, 120
melamine, addition, 107

coating, 241
foams, strength, 192
in foam, 187-188
in laminated wood, 161
modified UF, 95
NMR, 67
on textiles, 201-203
resin, 105, 106
with formaldehyde, 66
melting points, polyoxymethylene ethers, 57, 61-62, 86
mercaptans, 65
mercurimetric formaldehyde method, 126
mercury poisoning, felt hats, 255
metabolism, formaldehyde, 257, 262
metal, casting, binders, 207
impurities, 131
oxide extenders, 121
in resin, 121
metallocenes, encapsulated, 204
metal-plastic films, 203
methacrylate resin, 105
methane oxidation, 98, 99
methanedithiol, 51
methanol, assay, 130
methanol, consumption, 268, 269
explosion limit, 98
NMR, 137
oxidation, 46, 96, 98-99
solubility of formaldehyde in, 36, 45
methanolic UFR, 108
metheneamine, 48
methionine, 210
methone method, formaldehyde, 127
methoxy, total, 140
methyl ether, total, 140
ethylene urea, 89
methyl propylene urea, 83
methyl siliconate, 189
methylal, 56
methylamine, 30
methylene, acetoacetate, 105
bis-acetylurea, 20
bis-methylolurea, 20, 21
bis-urea, 65
bridges, 60
diamine, 63
diurea, 20, 64
diurea, NMR, 135, 138, 140
ether, hydrolysis, 23
glycol, 22, 26-28, 48
molfraction in water, 28
sampling, 144
linkage, 60
sulfonic acid, 54
urea, 6
methylmercaptan, 51
methylol, addition to UF, 106
cellulose, 61
measurement, 103
on textiles, 199-201
total in UFR, 140
ureas, manufacture, 11
methylol-methylene-diurea, 120
methylolation, 100
first rate constant, 84
kinetics, 81, 83
methylthioformaldin, 52
MF properties, comparison, 230
MHHT, methyl hexahydrotriazinone, 83
MHI, Manufactured Housing Institute, 290
microelectronics, 2
microencapsulated inks, 110
microfilters, foam, 198
microorganisms in resin, 110
microparticles, UFR, 109
MIK, 259
mildew proofing, textiles, 201
mildew resistant board, 176
milk, formaldehyde in, 68, 267
mill sludges, resin, 110
mineral fillers, 206
mineral oil, additive, 257
mineral wool mats, 205
mining, tunnels, UF foams, 205
walls, foaming, 196
mixed epoxi-UFR, 109
mixed UFP resin, 121
MMU, 64, 88, 101
dissociation kinetics, 82
formation kinetics, 81
in resins, 115
manufacture, 101
monomethylol urea, 92
NMR, 135, 136-140
rate constant, 83
MMUME, NMR, 135, 136-140
MMUT in resins, 115
mobile homes, 1, 228, 256, 260, 282
decking standards, 276
formaldehyde in, 148, 219, 258, 262
modern chemical tools, 283-284
modified, adhesives, 107-111
resins, 95-96, 115-122
UF, 18
UFR, performance, 118
modulus of rupture, MOR, 119
of rupture, molding, 230
MOE, modulus of elasticity, 119
of UF-sulfur resin, 119
Mohl test, 237
moisture content, stand, 218
resistance, UF, 156
uptake, 221

molding, powder, 17, 289
resins, 106, 121
strength, 211
UFR, 211-214
molecular weight distribution, 89
molfraction of methylene glycol, 28
molybdenum catalyst, 98
monoglycerides, 117
monomethylol, kinetics, 81, 83
urea, 64
ethylether, 21
methylether, 21
NMR, 135-138, 140
MOR, modulus of rupture, 118, 220
MPU, methyl propylene urea, 83
MUF-furfural, 109
melamine UFR, 109
resin, 108
mulches, 181
agricultural, 207
municipal waste treatment, 210
mutagen, 254
formaldehyde, 263

naphthalene sulfonate, 110, 185
naphthalene sulfonic acid, 19
naphthyl-amine, 6
National Bureau of Standards (U.S.), 217
Natzsch reaction, 128
neoprene rubber, foam, 189
nerve response, 262
neutron scattering, 284
New York City, foam use, 217
NH-for UF chemistry, 80
NH-stretch, in UFR, 142
ninhydrine, 67
NIOSH, 266
sampling, 146
standards, 253, 291
nitrate, 10
determination, 55
nitric acid, reaction with formaldehyde, 55-56
nitric oxide, formaldehyde method, 128
nitriles, 68
nitrocellulose, 17
nitrophenol, 71
NMR, 284
adhesive resin, 136
analysis, 134, 140
^{13}C, 134-140
formaldehyde, 23
hemiformals, 137, 140
melamine, 67
methanol, 137
paraformaldehyde, 57
PF, 72
proton, 134-140
shifts, UF, 138, 140
solvents, 134-135
UFR, 89, 103
urea, 73
urons, 93-95
NO, effect on formaldehyde in air, 148
non-burning board, 205
UFR, 206-207
non-destructive testing, 28
non-shattering articles, 17
Novolaks, 72, 87
N-P fertilizers, 207-208
NPA, National Particleboard Association, 219
NPK fertilizers, 207-208
N-type plywood, 223
nucleophilic, attack, 47
exchange, UF, 86
nylon, 65, 66
rugs, fireproofing, 202
nynyl acetate, addition, 106

odor, 120, 286
abatement, 226-249
acute, 233
adhesive, 226-249
cedar, reduction, 181
chronic, 233
factors in foam, 194
free adhesives, 110
masking, foam, 190
obnoxious, 182
parameters, 184
problem, 2
reduction, 110, 115
foam, 187
stop gaps, 290
threshold, formaldehyde, 251-254
odorless resin, 105
oil absorbers, foam, 197
oil removal, from sea, foam, 189
olefactory reactions, 262
olefins, 86
oligomers, formaldehyde, 39
oligomers, formaldehyde, determination, 130
optical absorption, UFR, 215
tools, 13
Oracid, 197
ore flotation, 213, 214
organic acid, cure, 112
OSHA standards, 253, 291
outdoor, air, sampling, 144
weathering, 287
overlay, board, 173
lustrous, 176

melamine, 174
plywood, 226
oxalate, 10
oxalic acid, 60
oxamide, 65
oxidative dehydrogenation, 97
oxonium compounds, urea, 74
oxonium ion, 90
ozone, interference, formaldehyde, 128
reaction with formaldehyde, 46

panelboard per unit housing, 283
panels, ceiling, 212
foam, 212
sandwich, 180, 287
paper, 17
bonding, UF, 162
coatings, UFR, 109
cups, 229
food wrapping, 204
insulated electrical, 205
laminates, 256
mill sludges, resin, 110
pressure sensitive, 289
reactions with formaldehyde, 59-61
treated, 204, 205
wet-strength, 18
paraffin addition, 110
para-rosaniline method, 127-128
rubber, boards, 176
paraformaldehyde, 13, 23, 25, 29-31, 45
NMR, 39
oligomers, 57
see also polyoxymethylene glycol
parquets, formaldehyde from, 257
partial, pressure, monomeric formaldehyde, 24
particleboard, 1, 110, 164-168
adhesive, 168
binder, 168
classification, 220
consumption, 276, 278
creeping, 287
factors, 167
fire resistant, 288
foam resin, 166
formaldehyde from, 245-249
formaldehyde release rules, 235
future, 282-283
future uses, 275
high density, 220
leading nations, 278-280
low density, 220
plants, formaldehyde, 257
plant, outline, 171
press, 172-173
price, 167
production, 282
reduced density, 285
roofing standard, 174, 288
standards, 219
statistics, 278
structure, 165
uses, 230
use per home, 283
paste adhesive, 110
patches, plywood, 223
patent, UFR, first, 4, 8, 9
pathogenic, foam, 208
PC, paper chromatography, 89
penetration of adhesive, 159
penetration depth, 111
pentachlorophenol, 224
additive, 170
pentaerythritol, 58-61
in boards, 178
pentamethylene-hexaurea, 20
pentane, as blowing agent, 190
pentanedione method, 128
peptides, 67
per capita UFR consumption, 273
perdeuterated formaldehyde, 55
perforator test, 150-153
test, FESYP, 150
value, free formaldehyde, 235
values for coatings, 241
perfuming, 290
peripheral vision, 262-265
perlite, 206
board, 178, 205
in foam, 190
permanganate, 76
thiourea, 77
peroxide, method, 131
peroxides, 86, 95, 98
petroleum products, 2
PF properties, comparison, 230
resin consumption, 270
pH change, during UFR, 163
control, 10
dependence, addition react, 100
condensation, 100
of methylolation, 84
effect on addition, 100
condensation, 100
degradation, 82
dissociation, 82
hydrolysis, 144
methylolation, 84
resin viscosity, 108
UF kinetics, 81
influence on formaldehyde on cellulose, 58-59
rayon, formaldehyde absorption, 60

resin intermediates, 108
UF on textiles, 201-202
phenanthroline method, 124
phenol, 15, 68, 70-72, 115, 284
cocondensation, 121
copolymer, 105
linkage, 71
reactions with formaldehyde, 70-73
resin, 121
sulfonic acid, 120
UFR, 108, 109
weathering, 287
phenolic, adhesives, 107
resin, 107
resin, formaldehyde reduction, 247
phenols, 111
phenol-urea adduct, 73
phenosulfonic acid, 89
phenylhydrazine, 6
phoron, 85
phosgene, 74
phosphate, buffer, fireproofing, 206
fire proofing, 174
fire retardant, 170
formaldehyde reductions, 243-244
resin, 121
phosphogypsum, 114
phosphoric acid, reaction with formaldehyde, 56
photocuring, 111
photographic gelatin, 204
photographs, finishing, 11
photolysis of formaldehyde, 46
photometric sampling, 147
phthalate resins, 105
phthalic acid, 13
cure, 112
phthalic esters, 78
picoline, 69
alpha, 69
pigments, 17
pine shavings, hardener, 114
pinene in foam, 187
pipeline insulation, foam, 208
plant, conditioners, foam, 207
display, foam, 198
nutrients, 207-209
pots, degradable, 210
plaster board, light, 206
plastic consumption, 273
plastoponics, 198, 207-208, 275, 289
plate glass, 13
plywood, 1, 162-164
adhesive standards, 218
consumption, 278
export, 278
formaldehyde release, 242-245
future, 280-281
glued, standards, 218
grades, 224-226, 281
import, 278
industrial stands, 219
leading producers, 280
low odor, 178
markets, 281
plants, formaldehyde in, 258
resin, 106
species, strength, 222-223
standards, 218-219, 223-226
statistics, 278-280
wood species, 222-223
polarography, formaldehyde, 130
Pollopas, 10, 11, 13
pollution, definition, 256
polyamide fibers, 106
polyamides, 68
polybutadiene resin, 110
polyepoxides, 95
polyester in foam, 192
polyethylene, 105
glycol, charring, 190
in foam, 192
resins, 105
polyformals, 58
hydrolysis, 232
polyhydroxy formals, 58
polyisobutylene, 192
polymer blending, 285
molecular weight, 16
polymethoxy, stretch, IR, 142
polymethylene sulfide, 51
polyoxymethylene, 12
beta, 55
delta, 57
gamma, 57
glycol, 23, 29-31
NMR, 39
see also para-formaldehyde
hemiformal, 23
vapor pressure, 57
polyphenol resins, 121
polypropylene, addition, 106
resin, 106
polystyrene, bonding, UF, 162
foam, 189, 192
polysulfide, formaldehyde reduction, 245
resins, 105
polysulfides, 51, 118
polyvinyl alcohols, 58, 111
resin, 108
porcelain, 10
porous products, 186
post-cure heating, 237
potassium cyanide method, 126

potato, extender, 164
prefabricated foams, 197
prefabricated housing, 161, 213
press corrosion, 122
press cycles, 171-172
price trends, UFR, 272
printing dyes, 11
processed wood products, 278
product standards, 228
profiles, wood, 121
propellers, wood, 161
properties, moldings, 211
propionaldehyde, 85
propylene glycol, UFR, 212
  resin, 110
propylene urea, 83
  dissociation, 82
  kinetics, 82
proteins, 68
  extenders, 121
  formaldehyde reaction, 254
  resins, 121
proton NMR, 134-140
  formaldehyde, 38-40
PU, propylene urea, 83
pulping bisulfite, formaldehyde reduction, 247
pulping wastes, resin, 110
PVC, in boards, 176
PVC, in foam, 192
pyridine, 7, 65
pyridine chloroacetic acid, 162
pyrimidone, 84
pyrolysis, foam, 217
pyrrolidine, 69
pyrrolidone, 236

quality control, 2, 19, 228, 284
quaternary ammonium, 118
quick curing resin, 108, 109
quinoline, 69

radiation, curing, 114
  hardener, 114
radioactive, metering, glue, 171
  waste, sulfite, 211
railway ties, consumption, 278
Raman, analysis UFR, 141-142
  frequencies of UFR, 142
  spectrum, formaldehyde, 24, 25, 284
  spectrum of polymethoxy, 142
  spectrum, UF formation, 82
  spectrum, urea, 73
rare gas matrices, formaldehyde in, 25
rate constants, pH dependence, 284
  UF formation, 82, 140

rayon, excess formaldehyde, 60-61
  formaldehyde, 58-60
    absorption, kinetics, 201
    catalysts, 60
    kinetics, 201
    weight gain, 60
  pretreatment, 60
  reactions with formaldehyde, 59-61
  weight gain with formaldehyde, 60
RDX, 50
reaction control, 19, 103
reagents, for UF, 80, 85
rearrangement in UF, 86
reduction of formaldehyde, 48
refractive index, aqueous formaldehyde, 32
  UFR, 215, 230
refractory insulators, 205
regulations, 291
release agents, for isocyanate, 122
residual tack, glue, 169
resin, additives, formaldehyde reduction, 245
  analysis, 109
  chemistry, 88-97, 283
  cure, 104, 112-114
  degradation, in board, 178
  design, 156
  flow, 169
  formation, principles, 79-85
  formulations, 105-107
  infusible, 72
  intermediates, 108
  manufacture, 97, 101
  performance, 107
  preparation, particle, 167
  shipping, 101
  uses, 270
  viscosity, 108
  wood interface, aging, 231-234
resite, stage C, 72
resitol, B-stage, 72
resole, 71
  stage A, 72
resonance, in thiourea, 76
  in urea, 65
resorcinal, 15
  formaldehyde reductions, 246
resorcinol, 72, 109, 121, 189
  addition, 106
  phenol, 284
  resin, 106
retarding salts, 10
rheology of resin, 108-109
rhodamide, 10, 13
rice, foam on, 208
ricinemic resin, 110

road, access treatment, 209
beds, stabilized, 205, 209
construction, 207-210
Roffael method, formaldehyde, 150
Rongalite, 53
roofing, particleboard, 229
Rosaniline method, 127
rubber, 7, 9
latex, UFR, 108
rugs, fireproofing, 202
Russia, formaldehyde regulations, 256
rye, extender, 164

saligenin, 71
salt, effect on gelation, 104
salycilate, 10
samples, enriching, 146
sampling, aqueous resin, 149
bonded material, 149
sampling methods, 146-148
sanding belts, 203
sandpaper, 203
sand-UF mix, 205
sandwich panels, 180, 212
saporified UFR, 109
sappamine, on rayon, 60
Schellan, 11
Schiff's reagent, 6, 127
schools, odor in, 228
Schryver's test, 267
scorching, anti, furfural, 206
sealing, edge of panels, 286
second urea addition, heat, 116
secondary amines, UFR, 118
selenophene, 69
self-extinguishing, UF, 106
self-ignition, foam, 194, 217
pulp, 245
self-supporting walls, 287
sensitivity of formaldehyde methods, 129
serine, 68
shaped articles, wood, 121
shaving board, standards, 220
shear strength, plywood, 222-223
shelf-life, resin, 109
sheep feed, 208
shims, plywood, 223
ship, insulation, foam, 197
shirt factories, formaldehyde in, 258
Shrewsbury test, 267
shrinkage, 111, 201
foam, 189, 194, 216
silica gel vs. foam, 197
silicate, embrittlement, 204
formaldehyde reduction, 257
sintered boards, 176
sizing, agent, 170
wood, 170
skin allergies, 252, 260, 290
skis, from sawdust, 175
skyscrapers, boards for, 205
slow-release fertilizers, 207-208, 289
sludge, dewatered, 211
paper mill resin, 110
smoke, density, foam, 217
from foam, 194, 217
soaking of glue, 159
soap dispensers, UFR, 17
sodium, chloride, gelation, 104
silicate resin, 110
silicate, UF, 204
sulfite method, 125
softboard, consumption, 278
standards, 219
softening point, UFR, 230
soil, conditioners, 210
disinfection, 208
protection, 198
rehabilitation, 207
soild content, 157
solubility, formaldehyde, 26, 27, 33
formaldehyde, dependence on pH, 42, 43
solvent extraction, 15
sorbitol, 58
sound absorption, 217
by foam, 194-195
particleboard, 221
Southern Building Conference, 217
soybean, formaldehyde reduction, 246
specific gravity, of boards, 220
specific heat, UFR, 230
spectral properties, formaldehyde, 24, 25
spray coating, agriculture, 289
spray-dried resin, 157, 186
resin, consumption, 271
springback, 111
stability of bond-types, 232
staircases, 181
molded, 212
standards, bond, 218
indoor, 257-259
starch, 58-60
addition, UFR, 107
extender, 121
reaction with formaldehyde, 58
resin, 107, 121
starved joints, 163
steel coatings, 203
sthenosage, 59-61, 199
stiffening, textiles, 201-202
stone, synthetic, 206
stop-gap, odor reduction, 290

storable fertilizer, liquid, 210
storable syrup, UFR, 106
storage, life, 107, 108
    temperature for formaldehyde, 45
    vaults, mortuaries, 256
strip mining, foam, 198
stripping textiles, 54
structural effect on CO, 83
structure, hexamethylene tetramine, 49
    urea, 73
styrole, 86
succinamide, 65
suction method, 154-155
    test, 237
sugar, additive, 121
    cane, UF foam, 207
    extender, 121
sugars, 47, 58-60
    reactions with formaldehyde, 58
sulfamate, curing, 114
    formaldehyde reduction, 246
sulfamic resin, 110
sulfamide resin, 120
sulfamoylanthranilic acid, 106
sulfate, 10
sulfide, formaldehyde reduction, 245, 249
sulfides, reaction with formaldehyde, 50
sulfite, 86, 95, 120
    cellulose, 204
    cure, 112
    effect on formaldehyde release, 243-249
    effect on gelation
    hardener, 112
    method, 103, 125, 187
    pulp, 14
    reactions, with formaldehyde, 50, 52-55
    reduction, formaldehyde, 242-248
    resins, 120-122
    testing, 103
    textiles, 202
    titration, 54, 91
    wastes, 120-121
sulfonamide, resin, 106
sulfonamides, 86
sulfonate, formaldehyde reduction, 234, 242
sulfones, 61
sulfonic acid, amides, 80
    resins, 80, 105
sulfoxylate, 54
sulfur, compounds, 95
    corrosion, urea, 99
    dioxide, 12, 16, 107, 112, 121, 290
        cure, 112
        formaldehyde effect, 254
        hardener, 112
        in air, 46
        reaction with formaldehyde, 50, 52-55
        reduction, 243
        resin, 107, 120-121
    effect on formaldehyde in air, 148
    effect on gelation, 104
    formaldehyde reduction, 247
    gellation, 104
    hexafluoride, 217
    hexafluoride, foam, 197
    in resin, 119
    oxyacids, formaldehyde reduction, 247
        resins, 121
    UF resins, 118, 178
sulfuric acid, 16, 185
    fuming, reaction with formaldehyde, 56
sulfurylamide, for UF, 80
sunken vessels, raising with foam, 197
sunlight, 120
surface, contact, UFR, 157
    hydrolysis, 231-234
    weakened, adherent, 159
surfactants, foam, 191
surgical dressing, foam, 197
Sweden, air standards, 259
Swedish building specifications, 217
Swedish homes, formaldehyde in, 259
swelling, 111
swimming pool, roofs, 161
synthesis, UF, 77-96
synthetic, fillers, 223
    marble, 206
    resins, history, 6
    resins, review, 11
    stone, 206

tables, 180
tack, (ASTM D-2979-71), 157
    adhesives, 110
    loss, 169
talcum powder, 197
tall, buildings, board for, 205
    structures, 229
tannic acid, 68
tannin resins, 111
tapioca, extender, 164
TAPPI method, 125
targets, gun, 204
taurine resin, 110
tautomers, 88
    thiourea, 76
    urea, 65
telegraph poles, 159
temperature, upper, for use, 230
tempered board, 180
tensile properties, adhesive, 218
    strength, 159, 220, 230

plywood, 223
termite resistance, UF, 156
testing standards, 222
tetra, aza-tricyclo-decane, 48
hydrofuran, 68
resin, 105
methylene urea, 64
methylol urea, 92
oxane, 29
propylene benzene sulfonic acid, 197
thiane, 51
textiles, 289
crease resistant, 95
discharging, 54
fireproofing agents, 202
freshly treated, 260
impregnation, 109
odor, 120
resins, 109
resin chemistry, 203
stripping, 54
treating resins, uses, 270
treatment, 199-202
yarn, formaldehyde, absorption, 60
thermal conductivity of board, 215, 216, 220
thermal, conductivity, foam, 194
decay, UFR, 229
expansion, UFR, 230
stability, 111
thermogram of cures, 112, 114
thermosetting, foamed beads, 187
resins, 14
THF, 68
thioaldehydes, 5
thioformaldehyde, 118
thional, 51
thiophene, 68
thiosulfate, formaldehyde reductions, 53, 243
UF gelation, 104, 118
thiourea, 13, 65-67, 76-78, 80, 91, 95
for UF, 80
formation, 65, 66
isomers, 76
reactions, 65-67, 76-78
resin, 106
three layer board, 165, 173
threshold standards, formaldehyde, 259
timber, uses, 281
tire cord, 72
Tischenko reaction, 47, 57
TLC, thin layer chromatography, 89
TLV, 259
TMU, trimethylol urea, 92
TMY, 137
toluene, sulfonic acid resin, 110, 112
sulfonamide resin, 109
tomatoes, UF foam, 207-208
tool wear, UFR, 107
torpedo boats, plywood, 161
total, formaldehyde, methods for, 131
methyl ether, UFR, 140
methylol, UFR, 140
vapor pressure, formaldehyde, 33
vaporizable formaldehyde, foam, 187
toughness, UFR, 107
toxicity, burning foam, 194, 217
burning UF, 206-207
of resin, 109
UFR, 109
trace contaminants, 146
tracheal mucus, 262
trans-alkylation, 86, 95
triazine, 18, 117
triazionone, 69, 94
triethanol amine, 70
resin, 109
tri-(hydroxymethyl)phosphine, 109
trimethylene, tetraurea, 20
triamine, 49
trinitramine, 50
urea, 64
trimethylol uron, 92
trioxane, 28-29, 70
trithiane, 51, 118
triureid, 73
tropical woods, 111
tropical worsted, 201
troposphere, formaldehyde in, 22
tropospheric reactions, formaldehyde, 46
Trouton constant, formaldehyde, 24
trusses, laminated, 161
turning point, resin, 104
turpentine, 187
TWA, 259

UF, analysis, 89
bonded products, standards, 218
carbonyl components, 79-83
dissociation kinetics, 82
foam, 207
formation, 137
kinetics, 81
nitrocellulose coating, 203
nucleophile steps, 79-83
paste, aqueous, 14
production, history, 15
properties, comparison, 230
ratio, 13
kinetics, 137-140
yield, 137-140
research, 283
resin, analysis, *in situ,* 136-140

consumption, 270
continuous manufacture, 101
fillers, 10
fire resistance, 229
formation, 79-88
intermediates, 108
manufacture, NMR, 140
producers, 276
sales, 273
uses, 270
wood interface, 157
sulfide resins, 17
sulfur fertilizers, 210
sulfur resin, 178
triethylene glycol, 188
uses, US, 270, 271
uv spectra, 88
weathering, 287
wood properties, 181, 184
UMF dicyanamide board, 176
UMF polymer, 181
underwear, formaldehyde in, 263
upper temperature for use, 230
Uralite, UFR, 205
urban air, formaldehyde in, 22
urea, addition to UFR, 116
adducts, 73-74
analysis, 124
elemental, 5
as fire retardant, 170
assay, 124
chemistry, 72-75
clathrates, 73
consumption, 268
discovery, 4, 5
for UF, 80
formaldehyde reductions; 243
heat of formation, 5
human body, 250
importance in history, 5
influence on viscosity, 117
manufacture, 99-101
monomethylol, 64
oxonium compounds, 74
patents, IG Farben, 5
perdeutero, 74
production, 5, 99-101
properties, 73
reactions, 63-65
salts, 73
second addition, 10
substitution, 117-118
toxicity, 248, 250
treatment, rayon, 60
vibrational spectrum, 74
ureido alkylation, 79-85
ureids, 65-66
urethane, as reagent, 80
bridges, 122
for UF, 80
urethanes, 95
urinary disinfectant, 50, 122
uron resins, 202-204
urons, 69, 95, 289
NMR, 93, 94
urotropine, 48, 50
U.S. air standards, tentative, 259
formaldehyde, 43
method, 125
U. S. Pharmocopeia, 124
U.S.S.R. air standards, 259
UV exposure, UFR, 229
light, influence on formaldehyde, 46
spectrum, formaldehyde, 24

V-20 boards, 221
vacuum, concentration of UFR, 13
curing, 108
drying, 15
gelation, 107
impregnated wood, 177
pressure test, 225
vapor, above soil, 210
barriers, foam, 196
composition, formaldehyde, 36
pressure, aqueous formaldehyde, 27, 31, 37
formula, 34
of polyoxymethylene ethers, 57
paraformaldehyde, 31
polyoxymethylene, 31
saturated aqueous formaldehyde, 34-35
vaporization, of formaldehyde, 44
rate of formaldehyde from solutions, 44
varnishes, 12, 14, 19
veneer, adhesives, 111
consumption, 278
patching, 18
sheet, export, 278
standards, 218
ventilation, 224, 226, 241, 261
vermiculite board, 176, 289
vibration frequencies, CO, 83
formaldehyde, 25
urea, 74
vinyl acetate resin, 106, 109
vinyl methylketone, 85
viscose, reactions with formaldehyde, 59-61
viscosity, aqueous formaldehyde, 32, 103
change, UFR, 117
of resin intermediates, 108
second urea, 117
visual acuity, 262, 266

vulcanization of UFR, 118
vulcanized rubber foam, 189

walnut shell flour, 164
warm sand coating resins, 207
warping temperature, 230
warplanes, 202
waste, liquors, resin, 110
    sludges, resin, 110
    treatment, 210-211
    water, formaldehyde in, 257
water, absorption, boards, 220
    absorption, by foam, 194
    diffusion, 221
    exposure, UFR, 229
    formaldehyde in, 267
    formaldehyde in and air, 240
    in foam, 191-192
    in urea, 124
    insoluble UFR, 106
    toleration, 157
    uptake, UFR moldings, 230
    vapor transmission, foam, 217
waterproof resin, 109
wax, as glue additive, 170
    particleboard, 170
WCAMA weathering, 287
weather resistance, 19, 287
weathering, effects, 231-233
weathering tests, 111
weight loss, during curing, 114
wet density, foam, 194, 216
wet-spinning, 202
wet strength, 120
    cellulose, formaldehyde, 201
wheat starch, extender, 164
WKI method, 150
wood, adhesion, mechanism, 157-159
    adhesives, 121, 277, 285-286
    bonding, 157-159
    cellulose, formaldehyde on, 61
    cement, 109
    cost particleboard, 167
    extractives, effect, 111
    formaldehyde absoprtion, 59-60
    glue interface, 157
    impregnation, 177
    joints, 111
    lamination, 161-162
    molding, 180-181
    moldings, 212
    particle, thickness, 179
    preservatives, 170
    products, uses, 275
    reactions, with formaldehyde, 59-61, 68
    resin aging, 231-234
    resin interaction, 166
    species, plywood, 221-222
    swelling, 159
wool, excess formaldehyde, 60-61
    shrinkage, 202
workmanship, 256
wound powder, 260
wound powder, foam, 197

Xenotest, weathering, 287
xylidine, 6

yeast, addition, UFR, 110
Yugoslavia, formaldehyde regulations, 256

Zeppelin dirigibles, 19, 196
zinc extenders, 121
zinc fluoride, 105
zinc phthalate, 105
zwitterions, 88